Spectral Analysis of Differential Operators

Interplay Between Spectral and Oscillatory Properties

World Scientific Monograph Series in Mathematics

Published

Vol. 1: Almgren's Big Regularity Paper : Q-Valued Functions Minimizing Dirichlet's Integral and the Regularity of Area-Minimizing Rectifiable Currents up to Codimension 2
Eds. V. Scheffer and J. E. Taylor

Vol. 2: Dynamics and Mission Design Near Libration Points
Vol. I Fundamentals: The Case of Collinear Libration Points
by G. Gómez, J. Llibre, R. Martinez and C. Simó

Vol. 3: Dynamics and Mission Design Near Libration Points
Vol. II Fundamentals: The Case of Triangular Libration Points
by G. Gómez, J. Llibre, R. Martinez and C. Simó

Vol. 4: Dynamics and Mission Design Near Libration Points
Vol. III Advanced Methods for Collinear Points
by G. Gómez, À. Jorba, J. Masdemont and C. Simó

Vol. 5: Dynamics and Mission Design Near Libration Points
Vol. IV Advanced Methods for Triangular Points
by G. Gómez, À. Jorba, J. Masdemont and C. Simó

Vol. 6: Hamiltonian Systems and Celestial Mechanics
Eds. J. Delgado, E. A. Lacomba, E. Pérez-Chavela and J. Llibre

Vol. 7: Spectral Analysis of Differential Operators: Interplay Between Spectral and Oscillatory Properties
by F. S. Rofe-Beketov and A. M. Kholkin
Translated by O. Milatovic

Vol. 8: Twenty Years of Bialowieza: A Mathematical Anthology
Aspects of Differential Geometry Methods in Physics
Eds. S. Twareque Ali, G. G. Emch, A. Odzijewicz, M. Schlichenmaier and S. L. Woronowicz

World Scientific Monograph Series
in Mathematics – Vol. 7

Spectral Analysis of Differential Operators

Interplay Between Spectral and Oscillatory Properties

Fedor S. Rofe-Beketov

Mathematical Division, B.Verkin Institute for Low Temperature Physics of National Academy of Sciences of Ukraine, Kharkov, Ukraine

Aleksandr M. Kholkin

Pryazovskyi State Technical University, Mariupol, Ukraine

Translated by **Ognjen Milatovic**

With Foreword by **Vladimir A. Marchenko**

World Scientific

NEW JERSEY • LONDON • SINGAPORE • BEIJING • SHANGHAI • HONG KONG • TAIPEI • CHENNAI

Published by

World Scientific Publishing Co. Pte. Ltd.

5 Toh Tuck Link, Singapore 596224

USA office: 27 Warren Street, Suite 401-402, Hackensack, NJ 07601

UK office: 57 Shelton Street, Covent Garden, London WC2H 9HE

British Library Cataloguing-in-Publication Data
A catalogue record for this book is available from the British Library.

World Scientific Monograph Series in Mathematics — Vol. 7
SPECTRAL ANALYSIS OF DIFFERENTIAL OPERATORS
Interplay Between Spectral and Oscillatory Properties

ISBN 981-256-276-1

Printed in Singapore by B & JO Enterprise

Foreword

The monograph by F. S. Rofe-Beketov and A. M. Khol'kin is primarily devoted to investigation of the interplay between spectral and oscillatory properties of infinite and, in particular, finite systems of second-order as well as arbitrary-order differential equations. Such systems can be presented in the form of a differential equation with matrix-valued or operator-valued (bounded) coefficients. There is a broad literature on the topic of finite-dimensional generalizations of classical Sturm oscillation theorems. This monograph, however, is the first book on the oscillatory theory of infinite self-adjoint differential systems. The basic mathematical tool in this book is the theory of operators in Hilbert space.

In addition to the main problems, the authors consider a number of related questions that not only play a supporting role in explaining the main theme of the book but are also of independent interest. We will mention a few: construction of a fundamental system of solutions of an operator differential equation with a boundary condition at infinity, investigation of the dependence of the discrete and essential spectrum of an operator boundary-value problem on variations of a (finite or semi-infinite) interval, and a description of self-adjoint extensions for infinite systems of differential equations on a finite or infinite interval using the method of binary relations. The authors also consider applications of the oscillatory theory to the investigation of eigenvalues in the gaps of the essential spectrum of systems of differential equations (including perturbed Hill and one-dimensional Schrödinger operators with almost periodic coefficients).

The book is based on authors' results but it also encompasses many well-known theorems, including classical ones, such as the Morse Index Theorem. This, together with end-of-chapter bibliographical comments, helps expand the circle of possible readers.

The book is intended for graduate students, research mathematicians working in functional analysis and differential equations as well as for physicists interested in spectral theory of differential operators.

V. A. Marchenko

Member of the National Academy of Sciences of Ukraine,

Member of the Russian Academy of Sciences,

Honorary Doctor of the University of Paris 7–Denis Diderot,

Member of the Royal Norwegian Society of Sciences and Letters.

Contents

Preface

The present book focuses on the application of operator theory to oscillatory theory of differential equations and infinite systems of differential equations (the latter can be considered as one differential equation with operator-valued coefficients). Operator theory plays an exceptionally important role in modern mathematics and physics, especially in boundary value problems, quantum mechanics and theory of oscillations. Likewise, spectral analysis of differential operators is one of the most important areas of modern mathematical physics. Schrödinger operator alone (often called Sturm–Liouville operator in one-dimensional case) is the central topic of a series of well-known monographs[1].

Oscillatory theory of differential equations grew out of remarkable papers (called *Mémoires* at that time) [879], [880] and [881] (1829–1836) by Sturm and a short but significant paper [635] (1837) by Liouville and Sturm on second order linear ordinary differential equations with an eigenvalue parameter. With each passing year, the theory of Sturm–Liouville equations and their boundary value problems attracts even greater attention of mathematicians, physicists and engineers across the globe, owing to existing and newly emerging applications of Sturm–Liouville theory to diverse

[1]We will mention some monographs devoted to multi-dimensional Schrödinger operator: F. A. Berezin and M. A. Shubin [27]; K. Chadan and P. Sabatier [40]; H. Cycon, H. R. Froese, W. Kirch and B. Simon [50]; M. S. P. Eastham and H. Kalf [62]; K. Jörgens and J. Weidmann [85]; P. D. Lax and R. S. Phillips [108]; S. P. Merkur'ev and L. D. Faddeev [135]; Kh. Kh. Murtazin and V. A. Sadovnichiy [142]; R. G. Newton [146], [147]; A. G. Ramm [163]; M. Reed and B. Simon [168]; E. W. Schmid and H. Ziegelmann [177]; E. Ch. Titchmarsh [191, Vol. 2]; and J. Weidmann [197]. We will also mention some monographs devoted to Sturm–Liouville operator: Z. S. Agranovich and V. A. Marchenko [1], B. M. Levitan [112], B. M. Levitan and I. S. Sargasyan [113], V. A. Marchenko [121], and E. Ch. Titchmarsh [191, Vol. 1]. (Here we restrict ourselves to mathematical literature.)

areas of science, technology and mathematics, such as the theory of non-linear evolution equations; quantum mechanics, including direct and inverse scattering problems (as already noted, Schrödinger equation coincides in many important cases with Sturm–Liouville equation); theory of oscillations; mathematical physics (classical and modern); theory of periodic and almost periodic systems and theory of stochastic (disordered) systems, including crystal lattices with impurities and quasi-crystals; geometry—in conjunction with the study of vector fields along a geodesic; symplectic geometry and topology. Alongside its numerous applications, the theory of Sturm–Liouville problems itself provides a beautiful arena for application of new mathematical theories and methods. At the same time, the theory of Sturm–Liouville problems stimulates the emergence and development of new mathematical methods (including the development of modern numerical methods and their computer applications [246]) and serves as the touchstone for their comparison and refinement.

In September 2003, on the occasion of the 200-th anniversary of the birth of Jack Charles François Sturm (1803–1855), an international colloquium with participation by numerous renowned mathematicians was organized in Geneva, Switzerland. Central theme of the colloquium was the development of Sturm–Liouville theory from its origin to the present. Highlights of the Sturm Bicentennial Colloquium include B. Simon's lecture on Sturm Oscillation and Comparison Theorems and some applications (see [862]) and W. N. Everitt's lecture [411] devoted to important landmarks in the development of Sturm–Liouville theory. For invited papers by participants of the Sturm Bicentennial Colloquium, see the book [8]. Additionally, W. N. Everitt created a catalogue of Sturm–Liouville differential equations; see [410].

It is well-known (see, for example, [165, Chapter VIII, Sec. 11]) that the investigation of different self-adjoint extensions of symmetric operators and the estimation of the location and multiplicity of their point spectra are among fundamental mathematical problems arising in any quantum-mechanical model. It turns out that the use of the interplay between oscillatory and spectral properties of differential operators, including the operators generated by finite or infinite systems of differential equations (in other words, differential equations with matrix or operator coefficients), provides effective means for determining the location and multiplicity of point spectra. The main part of the present book investigates the interplay between spectral and oscillatory properties for operator differential equations.

Sturm's oscillation theory and its various generalizations for ordinary differential equations and their finite systems, in connection with spectral theory, were considered in a series of well-known monographs[2].

Generalizations of alternation and comparison theorems to the case of finite systems are due to M. Morse [141], G. D. Birkhoff and M. R. Hestens [274], W. T. Reid [757], R. Bott [297] and H. M. Edwards [392]. An interesting topological treatment of Sturm's theorems and their connection with symplectic geometry is considered by V. I. Arnold [238]; however, the aim was not the novelty of mathematical results, for, as noted in [238], in an area as classical as Sturm's theory, it is difficult to keep an eye on all predecessors. In the present book, we compare the topological approach and operator-theoretical approach to Sturm's theorems. Various comparison theorems in Riemannian geometry are discussed in [76, Sec. 6].

In the present book, Sturm's oscillation theorem is transferred, in the corresponding form, to differential equations of arbitrary even order with operator-valued coefficients on finite and infinite intervals[3]. An important

---

[2]See A. G. Aslanyan and V. B. Lidskiy [12]; F. Atkinson [13]; R. Bellman [21]; E. A. Coddington and N. Levinson [46]; N. Dunford and J. T. Schwartz [58]; F. R. Gantmakher and M. G. Krein [69]; I. M. Glazman [70]; A. L. Gol'denveizer, V. B. Lidskiy and P. E. Tovstik [72]; P. Hartman [78]; I. T. Kiguradze and T. A. Chanturiya [89]; A. G. Kostyuchenko and I. S. Sargasyan [94]; K. Kreith [98]; B. M. Levitan and I. S. Sargasyan [113]; L. A. Pastur and A. L. Figotin [154]; G. Sansone [176]; V. N. Shevelo [178]; C. A. Swanson [187]; V. A. Yakubovich and V. M. Starzhinskiy [199] and [200]; J. Weidmann [196]. These monographs contain extensive bibliography on Sturm's oscillation theory and its connection to spectral theory. Note also that a generalization of Sturm's oscillation theorem to finite second-order systems was suggested by G. A. Bliss and I. J. Schoenberg [285]. However, this generalization characterizes oscillatory properties of those systems differently from the present book and connects the oscillatory properties to a sequence of numbers constructed in a special way. Except for the scalar case and related special cases, this sequence of numbers does not coincide with the spectrum.

[3]For scalar semi-bounded second-order problems, generalizations of the classical Sturm's Oscillation Theorem to the case of an infinite interval under various conditions are contained in the paper by H. Weyl [916] and in monographs by N. Dunford and J. T. Schwartz [58] and B. M. Levitan and I. S. Sargasyan [113]. For scalar differential equations of arbitrary even order, the interplay between oscillatory and spectral properties was investigated by M. G. Krein [585] and E. Heinz [506] (see also the monograph by I. M. Glazman [70]). The article [506] of E. Heinz, as noted in its preface by Van Der Waerden, is founded on the legacy of F. Rellich and represents the conclusion of investigations by F. Rellich [761]. For systems of fourth-order differential equations arising in the study of free oscillations of thin elastic shells, a generalization of Sturm's oscillation theorem is contained in the monographs by A. G. Aslanyan and V. B. Lidskiy [12] and A. L. Gol'denveizer, V. B. Lidskiy and P. E. Tovstik [72], and for finite fourth-order systems in the case of simple spectrum, in the article by A. Khalanai and Sh. Shandor [539] which also contains a generalization of results by G. A. Bliss and I. J. Schoenberg. The questions about oscillatory behavior of linear canonical systems were investigated in the

special case of this result is Morse Index Theorem [136], [159] and [160][4]. For differential equations of even order, we give a generalization of comparison and alternation theorems, as well as factorization theorems, generalizing to the considered case a theorem of Frobenius and M. G. Krein–Heinz–Rellich theorem. We note that "scalar" formulations and proofs of those theorems, based on the notion of determinant and the compactness of a finite-dimensional sphere, are not applicable in infinite-dimensional case. Our considerations encompass a broad class of canonical systems. As an application of factorization theorems, we obtain a generalization of Etgen–Pawlowski criterion [78] (second-order case) to equations of arbitrary even order with operator-valued coefficients.

We also establish an analogue of the oscillation theorem for discrete levels in the gap of continuous spectrum for equations of arbitrary even or odd order.

When applied to finite systems or scalar problems, the results of this book for infinite systems turn out to be at least as precise as the known results for the corresponding special cases. Furthermore, some results turn out to be new even in scalar and other special cases.

In conjunction with the investigation of oscillatory problems for infinite systems of differential equations, it was necessary to introduce a construction of the fundamental system of solutions subject to a self-adjoint boundary condition at a finite point or at infinity. In particular, we show that such fundamental solution defines the evolution of Lagrangian plane; more precisely, the evolution of its infinite-dimensional Hermitian analogue in a "doubled" (or, according to the order of the equation, the orthogonal sum of $2n$ identical Hilbert spaces) (separable) Hilbert space. The constructed fundamental solution is convenient in describing self-adjoint extensions of differential operators; in particular, in specifying boundary conditions at infinity. (For second-order scalar differential equation, the solution subject to a self-adjoint boundary condition at a singular end is constructed in [58].)

---

well-known works by V. B. Lidskiy [631] and V. A. Yakubovich [922] (the last two works consider the real case; certain results were transferred to the complex case by V. I. Khrabustovskiy [546]), in the work of V. I. Arnold [238], and, for second order differential equations with operator coefficients, in the works of G. J. Etgen and J. F. Pawlowski [403] and G. J. Etgen and R. T. Lewis [402]. Several criteria for finiteness of the negative spectrum, and, hence, for non-oscillatory behavior of a two-term differential equation of even order for $\lambda = 0$ with compact-operator coefficients were obtained by M. G. Gasymov, V. V. Zhikov and B. M. Levitan [444] and D. R. Yafaev [920].

[4]In [160] M. M. Postnikov noted that it is possible to generalize Morse Index Theorem to the general case of finite Sturm–Liouville systems.

In the present book, for differential equations of arbitrary order with operator coefficients, we use various methods for constructing the fundamental solution of a boundary-value problem and studying its properties.

The proof of oscillation theorems required a clarification of the dependence of eigenvalues and the greatest lower bound of the essential spectrum on the variable end-point of a finite or semi-infinite interval for semi-bounded differential operator of arbitrary even order with operator coefficients. The corresponding results of this book generalize a theorem of Courant and a theorem of M. G. Krein and affirmatively answer the question about strict monotonicity of eigenvalues under the extension of an unbounded domain, asked in [70, Chapter 4, Sec. 46].

In the absolutely indeterminate case, we describe all self-adjoint extensions of a symmetric differential operator of arbitrary (even as well as odd) order with operator coefficients on an infinite interval (real line or half-line). Additionally, we give a parametrization of the characteristic operator function and establish an explicit formula for finding this function. We also construct a solution of the Cauchy problem with the initial data at $(-\infty)$, and, thus, a fundamental solution[5].

Chapter 6 of the present book is devoted to investigation of perturbations of the spectrum and to a study of certain properties of solutions of periodic and almost periodic equations and finite systems. The methods of investigation here are connected to oscillation theory, and Chapter 6 can be considered as an application of those methods. Chapter 6 also contains detailed bibliographical comments.

The main part of the book uses the apparatus of the theory of Hermitian (in other words, self-adjoint) binary relations. For readers' convenience, the book is supplied by an Appendix containing the basics of the theory as well

[5]In the scalar case, a different form of the description of self-adjoint extensions of a second-order operator on an infinite interval was obtained by M. G. Krein [588] and C. Fulton [438]. For a scalar fourth-order equation and a two-term differential expression of an arbitrary even order, the same question was investigated by G. A. Mirzoev [680] and [681] simultaneously and independently of [541]–[543]. A description of all spectral functions of a scalar second-order differential operator was given, with the help of characteristic matrices, in the work of A. V. Shtraus [850] and V. M. Bruk [315], [316]. M. L. Gorbachuk [486] obtained a description of self-adjoint extensions of Sturm–Liouville operators with an operator potential in the absolutely indeterminate case. In the case when the values of deficiency indices coincide neither with the minimal possible ones (limit point case) nor with the maximal possible ones (the quasi-regular case), description of self-adjoint extensions of differential operators was investigated in the works by B. P. Allakhverdiev [221], I. M. Guseinov and R. T. Pashaev [503], F. G. Maksudov and B. P. Allakhverdiev [645], M. M. Malamud and V. I. Mogilevskiy [650], and V. I. Mogilevskiy [691]–[693].

as examples of the use of Hermitian relations. A Hermitian binary relation in a Hilbert space H is equivalent to specifying a Lagrangian plane or hypermaximal neutral lienal with respect to a skew-Hermitian inner product in a "doubled Hilbert space" $H \oplus H$. At the end of the Appendix, the reader can find some bibliographical comments.

Each chapter, and, thus the main part of the book, is based on the authors' results. In end-of-chapter bibliographical comments, we aimed not only to elucidate the history of a considered problem, but also to give a brief review of closely related areas, and, in some cases, to cite some important results of other authors or to review, for readers' convenience, some special facts used in the main text of the book.

In the English translation of the book, Bibliographical Comments are significantly expanded. Bibliographical Comments now contain separate subsections[6] discussing questions, such as a comparison of symplectic interpretation of Sturm's theorems and their operator-theoretical proofs; Marchenko's theorems on characteristic properties of Weyl solutions; a review of the history of Hilbert's 21-st problem and Bolibrukh's counterexample; Everitt–Zettl problem and Brusentsev's example; the results of Everitt, Kogan–Rofe-Beketov, Gilbert and Kalyabin in connection with the problem of possible values of deficiency indices of symmetric differential operators with complex-valued coefficients; Marchenko–Ostrovskiy and Pastur–Tkachenko theorems on the spectra of self-adjoint operators with periodic and limit-periodic coefficients; Batchenko–Gesztesy and Rofe-Beketov theorems on the spectra of non-self-adjoint differential operators (finite-zone type or with periodic coefficients); the results of V. I. Khrabustovskiy, who, at the request of the authors of this book, kindly wrote "The Inverse Sturm–Liouville Problem for the Spectral Matrix on the Whole Axis" (Sec. 6.6.8) and "Characteristic Operator and Boundary Value Problems with Separated Boundary Conditions" (Sec. A.4.3) containing his results; M. L. Gorbachuk theory of generalized boundary values; theory of abstract boundary conditions, connected with the names of J. W. Calkin, A. N. Kochubei and V. M. Bruk; Simon's and Rofe-Beketov's results on the strong resolvent convergence.

In comparison with the Russian edition of the book, the Bibliography now contains more than twice as many entries. The main text of the book

[6]The subsections are enumerated as follows: 4.5.1 denotes Subsection 1 of Section 5 of Chapter 4. The formulas are enumerated as follows: (3.5) denotes the formula 5 of Chapter 3. Theorems, Lemmas, Propositions, Corollaries, Remarks and Examples have separate enumerations containing two numbers, as the enumeration of formulas.

is also expanded, especially Appendix and Chapter 1, where new results are included.

The English translation of our book contains formulations of several open problems that can be used by young mathematicians as research topics.

Thus, the present translation can be considered in fact as a new book, with the title completed by the words "and Related Topics."

Acknowledgments

We are deeply grateful to Professor V. A. Marchenko for valuable remarks and his benevolent attitude with respect to our work. We are also deeply grateful to Professor M. A. Shubin for his advice and great help in our work on the English edition of the book. We thank very much the translator of the book Professor O. Milatovic for his diligent work on the translation and useful remarks. We offer sincere thanks to our friends and colleagues for personal conversations and correspondence by mail or e-mail, which stimulated our work, and for their valuable advice and remarks. We thank all those who sent us their reprints, preprints, and sometimes even books, or communicated to us important scientific information. In this regard, with great appreciation, we are pleased to mention Professors
W. O. Amrein, E. D. Belokolos, D. P. Belozorov, Ch. Bennewitz,
Yu. M. Berezanskiy, M. Sh. Birman, V. M. Bruk, A. G. Brusentsev,
V. P. Burskiy, I. D. Chueshov, L. N. Davydov, V. G. Drinfeld,
V. K. Dubovoy, Yu. M. Dyukarev, M. S. P. Eastham, I. E. Egorova,
W. N. Everitt, M. G. Gasymov, F. Gesztesy, M. L. Gorbachuk,
V. I. Gorbachuk, V. D. Gordevskiy, V. M. Kadets, H. Kalf,
V. I. Khrabustovskiy, E. Ya. Khruslov, A. N. Kochubei,
N. D. Kopachevskiy, A. G. Kostyuchenko, V. P. Kotlyarov,
M. A. Kudryavtsev, A. V. Kuzhel', B. M. Levitan, V. B. Lidskiy,
V. E. Lyantse, M. M. Malamud, V. O. Mishchenko, B. S. Mityagin,
A. D. Myshkis, V. Yu. Novokshenov, I. V. Ostrovskiy, I. E. Ovcharenko,
L. A. Pastur, M. A. Perelmuter, A. Ya. Povzner, A. G. Ramm,
A. G. Rutkas, A. L. Sakhnovich, K. M. Schmidt, Yu. A. Semenov,
B. Simon, O. G. Storozh, V. A. Tkachenko, L. A. Vlasenko,
J. Weidmann, V. A. Yurko, L. B. Zelenko, and V. A. Zolotarev.

With profound gratitude we remember all of our Teachers.

F. S. Rofe-Beketov and A. M. Khol'kin

Introduction

The boundary-value problem

$$-y'' + q(x)y = \lambda y \tag{0.1}$$
$$y'(a)\cos\alpha - y(a)\sin\alpha = 0 \tag{0.2}$$
$$y'(b)\cos\beta + y(b)\sin\beta = 0 \tag{0.3}$$

is known in the literature as the Sturm–Liouville problem. (More generally, one can consider the Sturm–Liouville problem of the form (1.1)–(1.3) below.)

The boundary-value problem (0.1)–(0.3) is said to be regular if the interval $[a, b]$ is finite and the function $q(x)$ is integrable on $[a, b]$. If the interval $[a, b]$ is infinite or $q(x)$ is not integrable on $[a, b]$, the problem (0.1)–(0.3) is said to be singular.

For the regular Sturm–Liouville problem, classical Sturm Oscillation, Comparison and Alternation Theorems have been well known for a long time.

Theorem 0.1 **(Oscillation Theorem)** *There exists an unbounded increasing sequence of eigenvalues $\lambda_1 < \lambda_2 < \cdots < \lambda_n < \cdots$ of the boundary-value problem (0.1)–(0.3). The eigenfunction corresponding to the eigenvalue λ_n has exactly $n-1$ zeros in the interval (a, b).*

Theorem 0.2 **(Alternation Theorem)** *The zeros of the n-th and $(n+1)$-st eigenfunctions of the boundary-value problem (0.1)–(0.3) alternate, i.e. between any two consecutive zeros of the n-th eigenfunction, there is a zero of the $(n+1)$-st eigenfunction.*

Theorem 0.2 is a particular case of the following fundamental Sturm Alternation Theorem.

Theorem 0.3 *Assume that the following two equations are given:*

$$-u'' + q_1(x)u = 0 \tag{0.4}$$

and

$$-v'' + q_2(x)v = 0. \tag{0.5}$$

If $q_1(x) > q_2(x)$ for all $x \in [a, b]$, then between every two zeros of any non-trivial solution of (0.4), there is at least one zero of every solution of (0.5).

Theorem 0.4 **(Comparison Theorem)** *Assume that $u(x)$ is a solution of (0.4) satisfying the initial conditions*

$$u(a) = \cos\alpha, \qquad u'(a) = \sin\alpha,$$

and $v(x)$ is a solution of (0.5) with the same initial conditions. Additionally, assume that $q_1(x) > q_2(x)$ for all $x \in [a, b]$. Assume that $u(x)$ has n zeros in the interval $(a, b]$. Then $v(x)$ has at least n zeros in $(a, b]$, and the k-th zero of $v(x)$ is less than the k-th zero of $u(x)$.

Analogous results hold for semi-bounded problems on an infinite interval; see [58] and [113].

Below we formulate Morse Index Theorem mentioned in the Preface. This theorem pertains to Jacobi equation for a vector field along a geodesic γ on a smooth Riemannian manifold. The corresponding Jacobi equation reduces to the equation (0.1) with $\lambda = 0$ and $(m \times m)$-matrix real symmetric coefficient $q(x)$ expressed, in a special way, through the Riemann curvature tensor and tangent vector to the geodesic—"velocity vector" (see [76, Sec. 4.2][7] and [136, Sec. 14]). The boundary conditions (0.2) and (0.3) satisfied by the vector solutions of the corresponding equation (0.1) now have the simplest possible form: $y(0) = y(1) = 0$.

Morse Index Theorem ([76, Sec. 4.6] or [136, Sec. 15])
Index of the Hessian for the action function

$$E_{**} : T\Omega_\gamma \times T\Omega_\gamma \to \mathbb{R}$$

(i.e. maximal dimension of subspaces $T\Omega_\gamma$ on which the form E_{**} is negative definite) is finite and equal to the number of points of the geodesic $\gamma(t)$,

[7] In [76] the Jacobi equation is reduced to a somewhat more general system. However, the corresponding system is equivalent to the above described equation through a unitary transformation in the L^2-space of vector-valued functions.

$0 < t < 1$, *such that* $\gamma(t)$ *is conjugate to* $\gamma(0)$ *along* γ, *where every conjugate point is counted according to its multiplicity. (In Morse terminology, this multiplicity is also called the index of the point* t *with respect to the endpoint* $t = 0$ *and the equation (0.1). The sum of indices of the point* t *along a given geodesic segment is called "Morse index of the given geodesic segment.")*

There are many works devoted to further development of Morse theory. Here we note the works [311], [386], [392], [504], [509], [707], [711], [726], [840], [866], [884], [893], and the recent review by S. P. Novikov [710].

Morse index is a special case of Maslov index [128]–[131]; see also [10], [57], [137], [237] and [238]. Maslov index plays an important role in multidimensional global asymptotic methods of "quasi-classical" type (for instance, multidimensional generalization of the WKB method), in so-called "quantization conditions," *etc.*, which are outside the scope of the present book.

In conclusion, we note that many linear partial differential equations can be successfully investigated when considered as Sturm–Liouville type equations in one unknown but with an unbounded operator coefficient. This approach requires the use of various methods, such as the theory of rigged spaces due to I. M. Gel'fand–A. G. Kostyuchenko [448] and Yu. M. Berezanskiy [24], [263], the theory of L. Schwartz–S. L. Sobolev distributions, and the theory of hyperfunctions and ultradistributions (see, for example, H. Komatsu [567], [568]). Starting with M. L. Gorbachuk [487], this approach was considered and further developed in the monographs [73], [74], [119], the review paper [496] and other works. For another approach to reduction of partial differential equations to differential equations with operator-valued coefficients, see, for example, the books [95] and [105].

The question about the extension of oscillatory theory to equations with unbounded operator-valued coefficients can still be considered as a problem open for investigations.

Chapter 1

Relation Between Spectral and Oscillatory Properties for the Matrix Sturm–Liouville Problem

1.1 Problem on a Finite Interval

Consider the following boundary-value problem for a vector-valued function with values in a finite-dimensional or infinite-dimensional separable Hilbert space H:

$$l[y] := -\left(P(x)y'\right)' + Q(x)y = \lambda W(x)y \tag{1.1}$$

$$\cos A \cdot P(0)y'(0) - \sin A \cdot y(0) = 0 \tag{1.2}$$

$$\cos B \cdot P(b)y'(b) + \sin B \cdot y(b) = 0. \tag{1.3}$$

Here $P(x) > 0$, $W(x) > 0$, $Q(x)$, A, B, $P^{-1}(x)$, $P'(x)$, $W^{-1}(x)$ are bounded (for all $x \in [0,b]$) self-adjoint operators in H; the dependence on $x \in [0,b]$ is uniformly continuous; and

$$-\frac{\pi}{2} << A, B \leq \frac{\pi}{2} I. \tag{1.4}$$

(The point $-\frac{\pi}{2}$ belongs to the resolvent sets of A and B. In the case $\dim H < \infty$, this restriction is removed. Here I denotes the identity operator in H.)

Differential expression

$$l_W[y] := W^{-1}(x) l[y] \tag{1.5}$$

with boundary conditions (1.2), (1.3) gives rise to a self-adjoint operator L (see Appendix) in the Hilbert space

$$H(0,b) = L_2\{H; (0,b); W(x)d(x)\} \tag{1.6}$$

with the inner product

$$\langle u, v\rangle = \int_0^b (W(x)u(x), v(x))_H \, dx.$$

Under given conditions (including (1.4)), the operator L is semi-bounded below (see [766, pp. 337–338] and [491]). If $\cos A = 0$ and $\cos B = 0$, the boundary conditions (1.2), (1.3) become

$$y(0) = 0, \tag{1.7}$$

$$y(b) = 0. \tag{1.8}$$

By L° we denote the operator corresponding to the boundary value problem (1.1), (1.2), (1.8). In the case $\dim H = m < \infty$, the problem (1.1)–(1.3) and the corresponding self-adjoint operator L have purely discrete spectrum

$$\lambda_1 \leq \lambda_2 \leq \lambda_3 \leq \cdots$$

However, this is not the case when $\dim H = \infty$ (see [766, p. 340][1] and [73, Chapter 3, Theorem 4.4]). Let

$$\lambda_e = \inf \sigma_e(L), \quad \lambda_e^\circ = \inf \sigma_e(L^\circ) \tag{1.9}$$

be the greatest lower bounds of the essential spectrum σ_e of operators L and L° (the eigenvalues of infinite multiplicity belong to σ_e; for $\dim H < \infty$, we have $\lambda_e = +\infty$).

If λ is an eigenvalue of the problem (1.1)–(1.3), we denote its multiplicity by $æ(\lambda)$. If λ is not an eigenvalue of (1.1)–(1.3), we set $æ(\lambda) = 0$. By $N(\lambda)$ we denote the number of eigenvalues $\lambda_k < \lambda$ of the problem (1.1)–(1.3) counting multiplicities. The corresponding quantities $N(\lambda)$, $æ(\lambda)$, λ_n for the operator L° will be denoted by $N^\circ(\lambda)$, $æ^\circ(\lambda)$, λ_n° respectively.

Let $Y(x, \lambda)$ be the fundamental matrix or operator solution to (1.1) with the initial conditions

$$Y(0, \lambda) = \cos A, \quad P(0)Y'(0, \lambda) = \sin A. \tag{1.10}$$

[1] From the example given there, we see that the spectrum of the problem (1.1)–(1.3) can even be absolutely continuous; see Sec. 1.3.1 in Bibliographical Comments to Chapter 1. It is interesting to note that among non-self-adjoint extensions of the minimal operator corresponding to (1.5), there are operators with purely discrete spectrum (see [650] and [693]) in the case $\dim H = \infty$.

Set

$$\text{nul}\, Y(x,\lambda) = \dim\ker Y(x,\lambda), \quad \text{def}\, Y(x,\lambda) = \dim\text{coker}\, Y(x,\lambda).$$

Theorem 1.1 *For $\lambda \le \lambda_e^\circ \le \infty$, we have*

$$\sum_{x\in(0,b)} \textit{nul}\, Y(x,\lambda) = N^\circ(\lambda)(\le\infty), \tag{1.11}$$

$$(\forall\lambda\in\mathbb{R}) \quad \textit{nul}\, Y(x,\lambda) = \textit{def}\, Y(x,\lambda) \tag{1.12}$$

(the summation is taken over all $x \in (0,b)$ such that nul $Y(x,\lambda) \ne 0$).

Theorem 1.2 *For the problem (1.1)–(1.3), the following estimates hold. If $\dim H < \infty$, we have*

$$N(\lambda) - \min\{\textit{rank}\ \cos B, \dim H - æ(\lambda)\} \le N^\circ(\lambda) \le N(\lambda). \tag{1.13}$$

Assume that $\dim H = \infty$ and $\lambda \le \lambda_e$.

(1) If rank $\cos B < \infty$, then $\lambda_e = \lambda_e^\circ$ and

$$N(\lambda) - \textit{rank}\ \cos B \le N^\circ(\lambda) \le N(\lambda).$$

(2) If rank $\cos B = \infty$, then $\lambda_e \le \lambda_e^\circ$ and

$$0 \le N^\circ(\lambda) \le N(\lambda).$$

The equalities may take place in the leftmost inequalities of this Theorem if the left hand sides of these inequalities are non-negative.

Corollary 1.1

(1) In the scalar case ($\dim H = 1$), for the problem (1.1)–(1.3), by (1.11) and (1.13) we have

$$\sum_{x\in(0,b)} \textit{nul}\, Y(x,\lambda_n) = n-1, \tag{1.14}$$

and this is equivalent to the classical Sturm Oscillation Theorem (see Theorem 0.1 in the Introduction).

(2) If $\dim H < \infty$ and $\lambda_n > \lambda_{n-1}$ and $æ(\lambda_n) = \dim H$, the formula (1.14) holds.

(3) If $\dim H \le \infty$ and $æ^\circ(\lambda_n^\circ) = 1$, then (1.14) holds for the problem (1.1), (1.2), (1.8) (since, in this case, $N^\circ(\lambda_n^\circ) = n-1$).

(4) From (1.1) and (1.11) it is easy to see that the number $N^\circ(0)$ of negative eigenvalues of the problem (1.1), (1.2), (1.8) does not depend on $W(x)$. By Proposition 1.1 below, the same is true for the number $N(0)$ of negative eigenvalues of the problem (1.1)–(1.3).

Note that if $\dim H > 1$ and if the general boundary condition (1.3) holds, then the equation (1.14) may not hold as shown in the following example:

Example 1.1 For the diagonal system ($k = 1, 2$)

$$-y_k'' = \lambda y_k, \quad y_k'(0) = 0, \quad y_k'(1) + (-1)^k y_k(1) = 0,$$

we have $Y(x, \lambda) = I_2 \cos(x\sqrt{\lambda})$. The eigenvalues are simple and are determined from the equation $\lambda(\tan\sqrt{\lambda})^2 = 1$, $(\lambda_1 < 0)$. Here

$$\sum_{x\in(0,1)} \text{nul}\, Y(x, \lambda_{2n}) = 2n - 2,$$

contrary to (1.14).

Proposition 1.1 *Denote*

$$Y[x, \lambda; B] := \cos B \cdot P(x)Y'(x, \lambda) + \sin B \cdot Y(x, \lambda). \tag{1.15}$$

Assume that $\lambda \le \lambda_e (\le \infty)$ *and*

$$\cos B \cdot \{-Q(x) + \lambda W(x)\} \cdot \cos B + \sin B \cdot (P(x))^{-1} \cdot \sin B >> 0, \qquad 0 \le x \le b.$$

Assume that, in addition to the conditions (1.4), the operator

$$h_a : D(h_a) \to \overline{D(h_a)}$$

is semi-bounded below:

$$h_a := -(\cos A \cdot \sin B + \sin A \cdot \cos B)^{-1} (\cos A \cdot \cos B - \sin A \cdot \sin B) \ge c \cdot I_{\overline{D(h_a)}}$$

with some $c \in \mathbb{R}$.

Then

$$\sum_{x\in(0,b)} \textit{nul}\, Y[x, \lambda; B] = N(\lambda)(\le\infty) \tag{1.16}$$

and

$$\textit{nul}\, Y[x, \lambda; B] = \textit{def}\, Y[x, \lambda; B].$$

(For the proof of Proposition 1.1, see the Bibliographical Comments to Chapter 4.)

Remark 1.1 Let $\dim H < \infty$. Consider the problem (1.1), (1.7). In this case, in the terminology of Morse, nul $Y(x, \lambda)$ coincides with the index of the point x with respect to the end-point $x = 0$ and the equation (1.1) for the given λ. On the other hand, the number $N^\circ(\lambda)$ of eigenvalues of the problem (1.1), (1.7), (1.8) less than the given λ is the maximal dimension of the subspace of $H(0, b)$ on which the corresponding Dirichlet integral

$$D_\lambda[u] = \int_0^b \{(P(x)u', u') + ((Q(x) - \lambda W(x))u, u)\}\, dx$$

is negative (the Morse quadratic form (i.e. Hessian of the action function (see [136, Sec. 13]) or index form of the geodesic (see [76, Sec. 4.5])) can be reduced to $D_\lambda[u]$). Thus, as noted in the Preface, Theorem 1.1 contains Morse Index Theorem; see the Introduction.

For the proof of Theorem 1.1 we need some preliminary lemmas and Theorem 1.3.

Lemma 1.1 *For all x and $\lambda \in \mathbb{R}$, the operators*

$$V_\pm(x, \lambda) = P(x)Y'(x, \lambda) \pm iY(x, \lambda) \tag{1.17}$$

have bounded inverses defined on all of H.

The operator

$$\theta(x, \lambda) = V_+(x, \lambda)V_-^{-1}(x, \lambda) \tag{1.18}$$

is unitary.

Proof. Let $V_-(\xi, \lambda)l = 0$ for some $x = \xi$, $\lambda \in \mathbb{R}$, $l \in H$. We will show that $l = 0$, i.e. the operator $V_-(\xi, \lambda)$ is invertible. Set $y(x) = Y(x, \lambda)l$. Clearly,

$$P(\xi)y'(\xi) - iy(\xi) = 0. \tag{1.19}$$

By (1.1) we have

$$\text{Im} \int_0^\xi (l[y], y)\, dx = \text{Im}\ \lambda \int_0^\xi (W(x)y, y)\, dx = 0.$$

On the other hand, integrating by parts once and using (1.2) and (1.19), we have

$$\text{Im} \int_0^\xi (l[y], y)\, dx = -\text{Im}\ (P(x)y', y)|_0^\xi = -(y(\xi), y(\xi))$$

since $\operatorname{Im}(P(0)y'(0), y(0)) = 0$ (this holds because from (1.2) and Lemma A.2, we have

$$P(0)y'(0) = \sin A \cdot h, \quad y(0) = \cos A \cdot h \tag{1.20}$$

for some $h \in H$.)

Thus $y(\xi) = y'(\xi) = 0$, i.e. $y(x) \equiv 0$ and, hence, $l = 0$.

We now prove that the range of $V_-(\xi, \lambda)$ is dense. By $Y_1(x, \lambda)$ we denote the solution of (1.1) with the following conditions:

$$Y_1(\xi, \lambda) = I, \quad P(\xi)Y_1'(\xi, \lambda) = -iI.$$

Note that

$$\begin{aligned}&\cos A \cdot P(0)Y_1'(0, \lambda) - \sin A \cdot Y_1(0, \lambda)\\&= W\{Y^*, Y_1\}|_{x=0} = W\{Y^*, Y_1\}|_{x=\xi} = -V_-^*(\xi, \lambda),\end{aligned}$$

where $W\{Y^*, Y_1\} = Y^*(x, \lambda)P(x)Y_1'(x, \lambda) - (Y')^*(x, \lambda)P(x)Y_1(x, \lambda)$ does not depend on x. Thus if $V_-^*(\xi, \lambda)l = 0$, then $y(x) = Y_1(x, \lambda)l$ satisfies the boundary conditions (1.2) and $P(\xi)y'(\xi) + iy(\xi) = 0$, implying, as above, that $l = 0$. The density of the range of $V_-(\xi, \lambda)$ follows since any vector l orthogonal to the range of $V_-(\xi, \lambda)$ is annihilated by $V_-^*(\xi, \lambda)$. In the same way, we can prove that $\operatorname{nul} V_+(\xi, \lambda) = \operatorname{def} V_+(\xi, \lambda) = 0$.

Note that

$$\begin{aligned}&V_-^*(x, \lambda)V_-(x, \lambda) = V_+^*(x, \lambda)V_+(x, \lambda)\\&= (Y')^*(x, \lambda)P^2(x)Y'(x, \lambda) + Y^*(x, \lambda)Y(x, \lambda) \equiv M(x, \lambda).\end{aligned} \tag{1.21}$$

We will show that the operator $M(x, \lambda) = M^*(x, \lambda) >> 0$, i.e. $M(x, \lambda)$ has a bounded inverse in H. This, together with what was proven above, shows that zero is not in the essential spectrum of $V_\pm(x, \lambda)$, and, hence, $V_\pm^{-1}(x, \lambda)$ exist and are bounded in H.

Assume the opposite, i.e. assume that there exists a sequence $l_n \in H$, $|l_n| = 1$, such that $M(\xi, \lambda)l_n \to 0$ as $n \to \infty$. Then $y_n(\xi) \to 0$, $y_n'(\xi) \to 0$, where $y_n(x) = Y(x, \lambda)l_n$. But by Lemma A.10, there exists a homeomorphism between the Cauchy data $y \oplus y' \in H^2$ at both endpoints of $[0, \xi]$ and the corresponding solutions of (1.1) on this interval, and, hence, between $y_n(0) \oplus y_n'(0) \in H^2$ and $y_n(\xi) \oplus y_n'(\xi) \in H^2$. Therefore, $y_n(0) \to 0$ and $y_n'(0) \to 0$, i.e. $M(0, \lambda)l_n \to 0$. This is impossible since $M(0, \lambda) = I$.

By (1.18) and (1.21) it follows that

$$\theta^*(x,\lambda) = V_-^{*-1}(x,\lambda)V_+^*(x,\lambda) = V_-(x,\lambda)V_+^{-1}(x,\lambda) = \theta^{-1}(x,\lambda).$$

This shows that $\theta(x,\lambda)$ is unitary, and the Lemma is proven. □

Lemma 1.2 *1a) For fixed x, as λ increases, every isolated eigenvalue $\omega(x,\lambda)$ with finite multiplicity of the unitary operator $\theta(x,\lambda)$ in (1.18) moves continuously and strictly monotonely*[2] *along the unit circle in the counterclockwise direction (possibly splitting into several eigenvalues moving in the same direction as the original eigenvalue, with the sum of their multiplicities equal to the multiplicity of the original eigenvalue).*

1b) Assume that $+1$ does not belong to the essential spectrum $\sigma_e(\theta(x,\lambda))$. Then for fixed $\lambda \in \mathbb{R}$, as x increases, the eigenvalue $\omega(x,\lambda)$ can pass through the point $+1$ only in the counterclockwise direction, i.e. for $\omega(x,\lambda) = 1$, arg $\omega(x,\lambda)$ is continuous and strictly increasing with respect to x. (In this case, as in 1a), $\omega(x,\lambda)$ can possibly split into several eigenvalues moving in the same direction as the original eigenvalue, with the sum of their multiplicities equal to the multiplicity of the original eigenvalue).

2) The endpoints $\omega_e(x,\lambda)$ of gaps in $\sigma_e(\theta(x,\lambda))$ move along the unit circle in the same way as in 1a) and 1b), i.e. for fixed x, and as λ increases, the motion of $\omega_e(x,\lambda)$ along the unit circle is continuous and strictly monotone; for fixed $\lambda \in \mathbb{R}$, and as x increases, $\omega_e(x,\lambda)$ can pass through the point $+1$ on the unit circle only in the counterclockwise direction.

Proof. The proof follows from the equations

$$\frac{\partial}{\partial\lambda}\theta(x,\lambda) = i\theta(x,\lambda)\Omega_+(x,\lambda), \tag{1.22}$$

$$\theta'(x,\lambda) = i\theta(x,\lambda)\Omega(x,\lambda), \tag{1.23}$$

where

$$\Omega_+(x,\lambda) = 2V_-^{*-1}(x,\lambda)\left[\int_0^x Y^*(t,\lambda)W(t)Y(t,\lambda)\,dt\right]V_-^{-1}(x,\lambda) >> 0, \tag{1.24}$$

[2] i.e. if $\lambda_1 < \lambda_2$, then arc $\omega(x,\lambda_1) <$ arc $\omega(x,\lambda_2)$, where arc $\omega(x,\lambda)$ denotes the length of the arc on the unit circle corresponding to $\omega(x,\lambda)$.

$$\Omega(x,\lambda) = 2V_-^{*-1}(x,\lambda)\left[(Y')^*P(x)Y' + Y^*(\lambda W(x) - Q(x))Y\right]V_-^{-1}(x,\lambda). \tag{1.25}$$

The assertion 1a) is simpler than 1b) since $\Omega_+(x,\lambda)$ is positive. We first consider 1b). Note that

$$\theta(x,\lambda) - I = 2iY(x,\lambda)V_-^{-1}(x,\lambda), \tag{1.26}$$

and, hence, for all $h \in H$, $h \neq 0$ such that $\theta(x,\lambda)h = h$, we have

$$Y(x,\lambda)V_-^{-1}(x,\lambda)h = 0. \tag{1.27}$$

Hence, for sufficiently small $\delta > 0$, we have

$$(\Omega(x,\lambda)h,h) \geq \delta|h|^2 > 0. \tag{1.28}$$

This can be shown as follows. From (1.27) and (1.25) we have

$$(\Omega(x,\lambda)h,h) = 2(M_1(x,\lambda)V_-^{-1}(x,\lambda)h, V_-^{-1}(x,\lambda)h),$$

where

$$M_1(x,\lambda) = Y'^*(x,\lambda)P(x)Y'(x,\lambda) + Y^*(x,\lambda)Y(x,\lambda) >> 0. \tag{1.29}$$

Here, the positivity of $M_1(x,\lambda)$ can be proven as in (1.21).

Let $e^{i\alpha} \notin \sigma(\theta(x,\lambda))$. Then

$$A(x,\lambda) = i(e^{i\alpha}I + \theta(x,\lambda))(e^{i\alpha}I - \theta(x,\lambda))^{-1} \tag{1.30}$$

is a bounded self-adjoint operator. Moreover, to isolated eigenvalues $\omega(x,\lambda)$ of $\theta(x,\lambda)$ correspond isolated eigenvalues

$$\gamma(x,\lambda) = i(e^{i\alpha} + \omega(x,\lambda))(e^{i\alpha} - \omega(x,\lambda))^{-1}$$

of $A(x,\lambda)$, and, as $\omega(x,\lambda)$ moves counterclockwise along the unit circle, $\gamma(x,\lambda) \in \mathbb{R}$ is increasing. We have

$$A'(x,\lambda) = 2i\left(\{e^{i\alpha}I - \theta\}^*\right)^{-1}\Omega(x,\lambda)\{e^{i\alpha}I - \theta(x,\lambda)\}^{-1}. \tag{1.31}$$

When $\omega(x,\lambda) = 1$, for all eigenvectors $h = \theta(x,\lambda)h \neq 0$, we have

$$(e^{i\alpha}I - \theta(x,\lambda))^{-1}h = (e^{i\alpha} - 1)^{-1}h.$$

Thus, using (1.31) and (1.28), we have

$$(A'(x,\lambda)h,h) = 2\left|e^{i\alpha} - 1\right|^{-2}(\Omega(x,\lambda)h,h) \geq \mu|h|^2 > 0, \tag{1.32}$$

for some $\mu > 0$.

To prove 1b), it remains to invoke Lemma 1.3 (1) below: by (1.32), as x increases, the eigenvalues $\gamma(x,\lambda)$, when passing through $i(e^{i\alpha}+1)(e^{i\alpha}-1)^{-1} = \cot\frac{\alpha}{2}$, also increase; therefore, the corresponding $\omega(x,\lambda)$ pass through the point $+1$ on the unit circle in the counterclockwise direction. Similarly, to prove 1a), we use $\frac{\partial}{\partial\lambda}A(x,\lambda) >> 0$ together with Lemma 1.3 (1).

We now prove 2). Note that if, instead of (1.27), we require

$$\left|Y(x,\lambda)V_-^{-1}(x,\lambda)h\right| \leq \epsilon|h| \tag{1.33}$$

for sufficiently small ϵ, then, using (1.25) and (1.29), we see that (1.28) still holds. Thus if $E_\theta(-\epsilon,\epsilon)$ is the spectral projection of the operator $\theta(x,\lambda)$ corresponding to the arc $e^{i\phi}$, $-\epsilon < \phi < \epsilon$, then by (1.26) the inequalities (1.33) and (1.28) hold for all

$$0 \neq h \in E_\theta(-\epsilon,\epsilon)H. \tag{1.34}$$

Let $\gamma_e(A) = \inf\sigma_e(A(x,\lambda))$ be the image of one of the endpoints $\omega_e(x,\lambda)$ of the spectral gap in $\sigma_e(\theta(x,\lambda))$ containing $e^{i\alpha}$. Then

$$\gamma_e(A) = i(e^{i\alpha}+\omega_e(\theta))(e^{i\alpha}-\omega_e(\theta))^{-1},$$

and, as $\omega_e(\theta)$ moves counterclockwise along the unit circle, $\gamma_e(A) \in \mathbb{R}$ increases.

For $\omega_e(\theta)(x,\lambda) = 1$, for sufficiently small $\epsilon' > 0$, and for all

$$0 \neq h \in E_A(\gamma_e-\epsilon',\gamma_e+\epsilon')H \subseteq E_\theta(-\epsilon,\epsilon)H,$$

we have

$$(A'(x,\lambda)h,h) \geq \mu|h|^2 > 0,$$

for sufficiently small $\mu > 0$. This follows from (1.28), which holds under the condition (1.34), and from $\{e^{i\alpha}I-\theta(x,\lambda)\}^{-1}h = (e^{i\alpha}-1)^{-1}h + o_\epsilon(1)h$, where $o_\epsilon(1)$ is an operator quantity and $\|o_\epsilon(1)\| \to 0$ as $\epsilon \to 0$. Similarly, $\frac{\partial}{\partial\lambda}A(x,\lambda) >> 0$ for all x, $\lambda \in \mathbb{R}$.

With the observations made above, the assertion 2) immediately follows from Lemma 1.3 (2) below. □

Lemma 1.3 *Let J be an interval in $\mathbb{R}$. Assume that for all $t \in J$, $A(t)$ is a bounded self-adjoint operator and $t \mapsto A(t)$ is continuously differentiable on J.*

1) Then every isolated eigenvalue $\gamma(t)$ of $A(t)$ with finite multiplicity is continuously increasing in $\mathbb{R}$, possibly splitting into several increasing eigenvalues with the sum of their multiplicities equal to the multiplicity of the original eigenvalue, at every point $t = t_0 \in J$ satisfying the following property: for every eigenvector $h \neq 0$ of $A(t_0)$ corresponding to the eigenvalue $\gamma(t_0)$, the following holds:

$$A(t_0)h = \gamma(t_0)h \Rightarrow (A'(t_0)h, h) > 0. \tag{1.35}$$

2) Also, the endpoints $\gamma_e(t)$ of spectral gaps in $\sigma_e(A(t))$, including $\inf \sigma_e(A(t))$ and $\sup \sigma_e(A(t))$, are continuously increasing at every point $t = t_0 \in J$ satisfying the following property: there exist $\beta > 0$, $\mu > 0$ such that

$$h \in E_0(\lambda_e(t_0) - \beta, \lambda_e(t_0) + \beta)H \Rightarrow (A'(t_0)h, h) \geq \mu |h|^2, \tag{1.36}$$

where $E_0(\lambda)$ is the resolution of identity of $A(t_0)$.

Proof. 1) We can apply mini-max theory [58, Chapter XIII, Sec. 9D] to discrete eigenvalues $\gamma(t)$ lying below $\inf \sigma_e(A(t))$ and above $\sup \sigma_e(A(t))$, and for those eigenvalues, the proof proceeds similarly as in the finite-dimensional case [13, Theorem V.3.2]. Now suppose that the eigenvalues $\gamma(t)$ lie in a bounded gap of $\sigma_e(A(t))$. Consider the operator $B(t) = (\alpha I - A(t))^{-1}$, where $\alpha \notin \sigma(A(t))$ and α belongs to the same gap of $\sigma_e(A(t))$ as $\gamma(t)$. Then $B(t_0)h = (\alpha - \gamma(t_0))^{-1}h$, and by (1.35) and since the eigenspace corresponding to $\gamma(t_0)$ is finite-dimensional, we have

$$\begin{aligned}(B'(t_0)h, h) &= ((\alpha I - A(t_0))^{-1} A'(t_0)(\alpha I - A(t_0))^{-1}h, h) \\ &= |\alpha - \gamma(t_0)|^{-2}(A'(t_0)h, h) \geq \mu_0 |h|^2 > 0,\end{aligned} \tag{1.37}$$

where $\mu_0 > 0$. To the eigenvalues $\gamma(t)$ from the given spectral gap correspond the eigenvalues $(\alpha - \gamma(t))^{-1}$ of $B(t)$ lying either below $\inf \sigma_e(B(t))$ or above $\sup \sigma_e(B(t))$. Using (1.37) and the observation made at the start of the proof, we see that the assertion 1) of the Lemma holds for the eigenvalues $(\alpha - \gamma(t))^{-1}$, so for $t = t_0$, the eigenvalues $(\alpha - \gamma(t))^{-1}$ are increasing. But $\gamma(t)(\neq \alpha)$ and $(\alpha - \gamma(t))^{-1}$ move in the same direction; hence, $\gamma(t)$ is increasing for $t = t_0$. This proves the assertion 1).

We now prove the assertion 2). Set $\gamma_e(t) = \inf \sigma_e(A(t))$,

$$E_1 = E_0((-\infty, \gamma_e(t_0) - \beta]), \quad E_2 = E_0(\gamma_e(t_0) - \beta, \gamma_e(t_0) + \beta),$$

$$E_3 = E_0([\gamma_e(t_0) + \beta, \infty)).$$

Let $h \perp E_1H$, i.e. $h = E_2h + E_3h$. Then from (1.36), for sufficiently small $(t - t_0) > 0$, we have

$$\begin{aligned}(A(t)h,h) &= (A(t_0)h,h) + (t-t_0)(A'(t_0)h,h) + \mathrm{o}(t-t_0)|h|^2 \\ &\geq (\gamma_e(t_0) - \beta)|h|^2 + \beta|E_3h|^2 + (t-t_0)\{\mu|E_2h|^2 \\ &\quad - 2\|A'(t_0)\| \cdot |E_2h| \cdot |E_3h| - \|A'(t_0)\| \cdot |E_3h|^2\} + \mathrm{o}(t-t_0)|h|^2 \\ &\geq \left\{\gamma_e(t_0) - \beta + \frac{1}{2}(t-t_0)\mu + \mathrm{o}(t-t_0)\right\}|h|^2 \\ &\geq \left\{\gamma_e(t_0) - \beta + \frac{1}{3}(t-t_0)\mu\right\}|h|^2.\end{aligned}$$

Taking $0 < \beta < \frac{1}{6}(t-t_0)\mu$, we obtain

$$(A(t)h,h) > \left\{\gamma_e(t_0) + \frac{1}{6}(t-t_0)\mu\right\}|h|^2,$$

for $h \perp E_1H$.

But $\dim E_1H < \infty$ for all $\beta > 0$; hence,

$$\gamma_e(t) \geq \gamma_e(t_0) + \frac{1}{6}(t-t_0)\mu > \gamma_e(t_0).$$

However, if $(t_0 - t) > 0$ is sufficiently small, then

$$(A(t)h,h) \leq \left\{\gamma_e(t_0) - \frac{1}{6}(t-t_0)\mu\right\}|h|^2,$$

for $h \in (E_1 + E_2)H$. Since $\dim(E_1 + E_2)H = \infty$ for all $\beta > 0$, in this case we have $\gamma_e(t) < \gamma_e(t_0)$. This concludes the proof of assertion 2) for $\inf \sigma_e(A(t))$. The proof of assertion 2) for $\sup \sigma_e(A(t))$ proceeds similarly. The problem of the motion of an endpoint of a finite gap in $\sigma_e(A(t))$ can be reduced to the above considered problems by passing to the operator $B(t) = (\alpha I - A(t))^{-1}$, where $\alpha \notin \sigma(A(t_0))$ and α lies in the given spectral gap whose endpoints, after passing from $A(t)$ to $B(t)$, become $\inf \sigma_e(B(t))$ and $\sup \sigma_e(B(t))$ (cf. the proof of the assertion 1)). This concludes the proof of the Lemma. □

Theorem 1.3 *(A generalization of Courant's Theorem). Assume that the operator L_b^0 is determined by the boundary value problem (1.1), (1.2), (1.8) on the interval $[0,b]$. Then*

1) As b decreases, every isolated eigenvalue $\lambda^0(b)$ of L_b^0 with finite multiplicity strictly increases (possibly splitting into several strictly increasing *eigenvalues with the sum of their multiplicities equal to the multiplicity of the original eigenvalue).*

2) As b decreases, the endpoints of gaps in $\sigma_e(L_b^0)$ (including $\lambda_e^0(b) = \inf \sigma_e(L_b^0)$ under the condition (1.4) for A in (1.2)) (continuously) strictly increase.

3) Under the condition (1.4), as $b \searrow 0$, the greatest lower bound of the operator L_b^0 approaches $+\infty$ in a strictly monotone and continuous way.

Remark 1.2 The possibility $0 \in \sigma_e(Y(x, \lambda'))$ for fixed λ' and for all x such that $0 < a_1 \le x \le a_2$ is not excluded. We can construct an example in the form of a diagonal system for which λ' is an eigenvalue of the r-th equation on the interval $[0, x_r]$, where x_r ranges over all rational points of $[a_1, a_2]$.

Proof. 1), 2). Clearly, $æ^0(\lambda) = \text{nul}\ Y(b, \lambda)$. It is known (see [766], [486]) that[3]

$$\lambda \in \sigma_e(L_b^0) \Leftrightarrow 0 \in \sigma_e(Y(b, \lambda)).$$

Using (1.26) and Lemma 1.1, it follows that $æ_{\theta(b,\lambda)}(1) = æ^0(\lambda)$, where $æ_\theta(\omega)$ is the multiplicity of ω as an eigenvalue of θ. Besides, $\lambda \in \sigma_e(L_b^0) \Leftrightarrow 1 \in \sigma_e(\theta(b, \lambda))$, and if λ is an endpoint of a gap in $\sigma_e(L_b^0)$, then by Lemma 1.2 (2), the point $+1$ is the endpoint $\omega_e(b, \lambda)$ of a gap in $\sigma_e(\theta(b, \lambda))$. Thus by Lemma 1.2, the equation $\omega(\xi, \lambda) = 1$ in λ, under the condition $\lambda(b) = \lambda'$, $\lambda' \notin \sigma_e(L_b^0)$, defines $æ^0(\lambda')$ continuous and strictly decreasing functions $\lambda(\xi)$ in a neighborhood of $\xi = b$, where $\lambda(\xi)$ are eigenvalues of L_ξ^0, and if λ' is an endpoint of a gap in $\sigma_e(L_b^0)$, the equation $\omega_e(\xi, \lambda) = 1$, $\lambda(b) = \lambda'$ determines in a neighborhood of b a strictly decreasing function $\lambda(\xi)$ such that $\lambda(\xi)$ is an endpoint of a continuously changing gap in $\sigma_e(L_\xi^0)$.

3) Note that

$$\inf \sigma(L_b^0) = \inf_{\substack{y \in D(L_b^0) \\ y \neq 0}} \frac{(P(0)y'(0), y(0)) + \int_0^b [(Py', y') + Q(y, y)]\, dx}{\int_0^b (Wy, y)\, dx}.$$

By (1.4), we have $A = A_+ \oplus A_-$, where $0 \le A_+ \le \frac{\pi}{2} I$, $-\frac{\pi}{2} I << A_- < 0$, and, hence, $\|\tan A_-\| < \infty$, $(\sin A_+ \cos A_+) \ge 0$. Thus from (1.2) and (1.20) it follows that $P(0)y'(0) = \sin A_+ h + \tan A_- y(0)$ and

$$\begin{aligned}(P(0)y'(0), y(0)) &= (\sin A_+ h, \cos A_+ h) + (\tan A_- y(0), y(0)) \\ &\ge -C|y(0)|^2.\end{aligned}$$

[3]This fact is a special case of Lemma 2.4 below.

For $0 \le x \le b$ we have from (1.8):

$$|y(x)|^2 \le \left\{ \int_0^b |y'|\, dx \right\}^2 \le b \int_0^b |y'|^2\, dx.$$

Besides,

$$(P(x)y', y') \ge \alpha |y'|^2, \quad (W(x)y, y) \le \beta |y|^2, \quad |(Q(x), y, y)| \le \beta |y|^2,$$

for some $\alpha > 0$ and $\beta > 0$.

Thus

$$\inf \sigma(L_b^0) \ge (\alpha - bc - b^2\beta)\beta^{-1}b^{-2} \to +\infty, \quad \text{as } b \searrow 0.$$

This concludes the proof of the Theorem. □

Proof of Theorem 1.1. By L_ξ^0 we denote the operator corresponding to the boundary problem (1.1), (1.2), $y(\xi) = 0$ on the interval $0 \le x \le \xi (\le b)$. Let

$$\lambda_1^0(\xi) \le \lambda_2^0(\xi) \le \cdots \le \lambda_n^0(\xi) \le \cdots (< \lambda_e^0(\xi) \le \infty)$$

denote the discrete spectrum of L_ξ^0 in the interval $(-\infty, \lambda_e^0(\xi))$ (in this indexing multiplicities are counted).

Let $\lambda_e^0(b) < \infty$, i.e. $\dim H = \infty$. By Theorem 1.3, through every point $(b, \lambda_j^0(b))$ in the (x, λ)-plane, passes a continuous, strictly monotone curve $\lambda = \lambda_j^0(x)$. This curve can be continued to its intersection with the line $\lambda = \lambda_e^0(b)$ for $0 < x < b$. The half-strip $\lambda \le \lambda_e^0(b)$, $0 < x \le b$, does not contain any of the points from $\bigcup_{\xi \in (0,b]} \sigma_e \left(L_\xi^0 \right)$ except $(b, \lambda_e^0(b))$. The curves $\lambda = \lambda_j^0(x)$ may completely or partially coincide in accordance with the multiplicities of eigenvalues represented by these curves. The total multiplicity of the curves emanating from the point (b, μ) is equal to the multiplicity $æ^0(\mu)$ of the eigenvalue μ of L_b^0. If $\lambda_e^0(b) = \infty$ ($\dim H < \infty$), the curves $\lambda = \lambda_j^0(x)$ can be continued to $x = +0$ and, as $x \searrow 0$, these curves tend to $+\infty$.

By $x_j(\lambda)$ we denote the inverses of $\lambda = \lambda_j^0(x)$ for $\lambda_j^0(b) \le \lambda \le \lambda_e^0(b)$. Let $\lambda_j^0(b) < \lambda' \le \lambda_e^0(b)$. Then the curve $\lambda = \lambda_j^0(x)$ intersects the line $\lambda = \lambda'$ at precisely one point $x_j(\lambda') \in (0, b)$. The number λ' is an eigenvalue of the operator $L^0_{x_j(\lambda')}$. On one hand, the multiplicity of λ' is equal to $\text{nul}\, Y(x_j(\lambda'), \lambda')$. On the other hand, the multiplicity of λ' is equal to the number of curves $\lambda = \lambda_k(x)$ having a common point $(x_j(\lambda'), \lambda')$. If for all k, $x_k(\lambda') \ne x \in (0, b)$, then $\text{nul}\, Y(x, \lambda') = 0$. Thus for $\lambda' \le \lambda_e^0(b)$, the

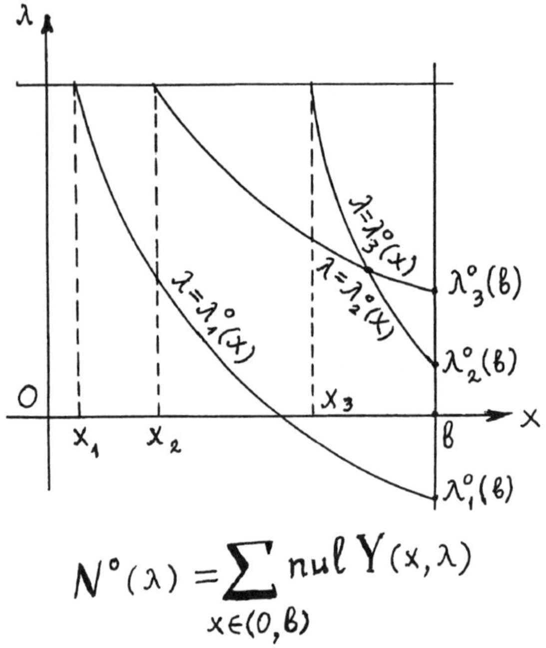

Fig. 1.1

expression $\sum\limits_{x\in(0,b)}$ nul $Y(x,\lambda')$ is equal to the number of curves $\lambda = \lambda_j^0(x)$ (counting multiplicities), intersecting the line $\lambda = \lambda'$ for $0 < x < b$. In other words, $\sum\limits_{x\in(0,b)}$ nul $Y(x,\lambda')$ is equal to the number $N_0(\lambda')$ of eigenvalues $\lambda_j^0(b) < \lambda'$ (if $\lambda_j^0(b) \geq \lambda'$, the curve $\lambda = \lambda_j^0(x)$, $0 < x < b$, does not intersect the line $\lambda = \lambda'$). This proves (1.11).

Since $\theta(x,\lambda)$ is a unitary operator and $V_-(x,\lambda)$ and $V_-^{-1}(x,\lambda)$ are bounded operators in H (see Lemma 1.1), the equality (1.12) follows directly from (1.26). This concludes the proof of Theorem 1.1. □

Lemma 1.4 *Let A_2 and A_1 be self-adjoint extensions in H of a semi-bounded symmetric operator A with finite deficiency indices (p,p).*

Then for all $\lambda \leq \lambda_e = \inf \sigma_e(A)$, we have

$$N_1(\lambda) - p \leq N_2(\lambda). \tag{1.38}$$

If, in addition to the above hypotheses, $æ_A(\lambda) = 0$ *and the operator* A *is closed, then*

$$N_1(\lambda) + æ_1(\lambda) - p \leq N_2(\lambda). \tag{1.39}$$

(In the formulation of Lemma 1.4, notational indices 2, 1 and A correspond to operators A_2, A_1 and A respectively. For $\lambda = \lambda_e$ both sides of (1.38) and (1.39) can be infinite (but only simultaneously); see [70, Sec. 1, Theorem 14] or [586, I]).

Proof. We shall prove (1.39) for $\lambda = \lambda_e$. Set

$$M_1(\lambda) = E_1(\lambda + 0)H, \quad M_A(\lambda) = M_1(\lambda) \cap \mathrm{D}(A). \tag{1.40}$$

Clearly,

$$\dim M_1(\lambda) = N_1(\lambda) + æ_1(\lambda) \leq \dim M_A(\lambda) + p \tag{1.41}$$

and

$$A|_{M_A(\lambda)} = A_2|_{M_A(\lambda)} = A_1|_{M_A(\lambda)} \leq \lambda I. \tag{1.42}$$

We now prove that the inequality (1.42) is strict. If $h \in M_A(\lambda)$ and $(A_1h, h) = \lambda(h, h)$, then from (1.40) it follows that

$$h \in (E_1(\lambda + 0) - E_1(\lambda - 0))H \qquad \text{and} \qquad A_1h = \lambda h.$$

Then $Ah = A_1h = \lambda h$, and since $æ_A(\lambda) = 0$, we have $h = 0$. Therefore,

$$(A_2h, h) < \lambda(h, h), \quad \text{for all } 0 \neq h \in M_A(\lambda),$$

and, hence,

$$\dim M_A(\lambda) \leq \dim E_2(\lambda - 0)H = N_2(\lambda).$$

This together with (1.41) gives (1.39). The inequality (1.38) is simpler and can be proven using similar arguments. This concludes the proof of the Lemma. □

Proof of Theorem 1.2. By variational principles, the eigenvalues $\lambda_j < \lambda_e$ and the number λ_e corresponding to the problem (1.1)–(1.3) do not exceed the eigenvalues λ_j^0 of (1.1), (1.2), (1.8) and λ_e^0 respectively. Therefore, $N_0(\lambda) \leq N(\lambda)$ for $\lambda \leq \lambda_e \leq \lambda_e^0$.

Note that if rank $\cos B < \infty$, then the operators L and L^0 are finite dimensional self-adjoint extensions of their maximal common part Λ. The operator Λ is defined on vector-valued functions $y \in D(L^0)$ satisfying the boundary conditions (1.2), (1.8) and $\cos B \cdot y'(b) = 0$. Deficiency indices of Λ are (rank $\cos B$, rank $\cos B$). Thus $\lambda_e = \lambda_e^0$, and by Lemma 1.4 the estimate (1) from Theorem 1.2 and a part of (1.13) hold. (Here, we take $A_2 = L^0$, $A_1 = L$ and $A = \Lambda$.)

Let Λ' denote the minimal operator with respect to the endpoint $x = b$ defined by (1.5) on the set of vector-valued functions satisfying the conditions (1.2) and $y(b) = y'(b) = 0$. The deficiency indices of the operator Λ' are $(\dim H, \dim H)$, and Λ' does not have any eigenvalues. If $\dim H < \infty$ the operators L and L^0 can be viewed as finite-dimensional extensions of Λ'. Using Lemma 1.4 and the estimate (1) from Theorem 1.2, we get (1.13). Theorem 1.2 is proven. □

1.2 Problem on the Half-Line

By L' we will denote the minimal with respect to $x = \infty$ operator in $H(0, \infty)$ from (1.6) defined by the expression $l_W[y]$ from (1.5), (1.1) and the boundary condition (1.2) at $x = 0$. We assume that L' is semi-bounded below and denote by L_F the Friedrichs extension of L'. We also use the notation $\lambda_e^F = \inf \sigma_e(L_F) \leq \infty$, and we denote the eigenvalues of L_F by

$$\lambda_1^F \leq \lambda_2^F \leq \cdots (< \lambda_e^F).$$

Consider the problem (1.1), (1.2), (1.8) on the interval $[0, b]$, and denote the eigenvalues of this problem by $\lambda_{n,b}^0$:

$$\lambda_{1,b}^0 \leq \lambda_{2,b}^0 \leq \cdots (< \lambda_{e,b}^0).$$

Lemma 1.5 *As $b \to \infty$, we have*

$$\lambda_{n,b}^0 \searrow \lambda_n^F \quad (n = 1, 2, \ldots) \tag{1.43}$$

Remark 1.3 The choice of the boundary condition (1.8) for problems on the intervals $[0, b]$, where $b \to \infty$, is important for the convergence of the corresponding eigenvalues $\lambda_{n,b}$ to λ_n^F with the same number n to take place, even in the case when the semi-bounded minimal (closed) operator L' is self-adjoint, and, therefore, possesses the unique spectral function. For another choice of the sequence of boundary conditions of the type (1.8) at the points $x = b$ as $b \to \infty$ there is a possibility of occurrence of "superfluous" limiting values μ_n of the eigenvalues $\lim \lambda_{n,b} = \mu_n$, which are not the points of the spectrum of the operator L' or L^F. For example, let $\dim H = 1$ and

$L'y = L^F y = -y''$, $y(0) = 0$, $0 < x < \infty$. On the interval $[0, b]$, consider the boundary problem

$$-y'' = \lambda y, \qquad y(0) = 0, \quad y'(b) - (\coth b) y(b) = 0.$$

This problem has a negative eigenvalue $\lambda_{1,b} = -1$ at all $b > 0$, although $\sigma(L') = \sigma_e(L') = [0, \infty)$ and the discrete spectrum of L' or L^F is empty. Evidently, the jump at the point $\lambda = -1$ of the scalar spectral function $\rho_b(\lambda)$ of the considered problem on $[0, b]$ tends to zero when $b \to \infty$ if the eigenfunctions of this problem are normalized by the condition $y'(0) = 1$. The spectral projector $E_b(\{-1\})$ of the same problem tends to 0 strongly.

For the same equation on $[0, b]$, we can consider the boundary value problem $y(0) = 0$, $y'(b) - y(b) = 0$, where the boundary conditions do not depend on b. Then $\lambda_{1,b} \to -1$, and the jump of the spectral function $\rho_b(\lambda)$ at the point $\lambda_{1,b}$ tends to zero when $b \to \infty$, as before.

Proof of Lemma 1.5.[4] From mini-max theory of eigenvalues (see [58, pp. 710–711]) it follows that for any self-adjoint extension L of L' in $H(0, \infty)$, we have

$$\lambda^0_{n,b} > \lambda_n(L), \quad \lambda^0_{e,b} > \lambda_e(L), \quad \lambda^0_{n,b_1} > \lambda^0_{n,b_2} \quad \text{for all } 0 < b_1 < b_2. \tag{1.44}$$

(see also Theorem 1.3).

Therefore,

$$\lim_{b \to \infty} \lambda^0_{n,b}$$

exists, and we denote this limit by λ^0_n.

Taking the operator L_F as L, we see that the following inequality holds:

$$\lambda^0_n \geq \lambda^F_n. \tag{1.45}$$

To get the equality in (1.45), we will prove that $\lambda^F_n \geq \lambda^0_n$. To this end, consider the non-decreasing matrix-valued or operator-valued spectral distribution function $\rho_b(\lambda)$ of the problem (1.1), (1.2), (1.8). The function $\rho_b(\lambda)$ is constructed similarly as in [766, pp. 336–341] or in [486] in the case $P(x) = W(x) \equiv I$. There it was shown how the well-known method for constructing the spectral function of a scalar-valued Sturm–Liouville problem on the half-line is transferred to the case of an equation with an operator-valued potential by considering the corresponding problems on finite intervals and passing to the limit (see [113]). Repeating, with minor

[4] A different proof of an analogous result in a more general case is given in Theorem 3.1.

modifications, the construction from [766, Chapter III, Sec. 7.2], we see that there exists a sequence $b_j \to \infty$ such that $\rho_{b_j}(\lambda) \to \rho(\lambda)$. Here, $\rho(\lambda)$ is a non-decreasing operator-valued function generating the Parseval's equality for vector-valued functions from $H(0,\infty)$ and their Y-transformations with the weight $W(x)$, where $Y(x,\lambda)$ is the solution of the problem (1.1), (1.10).

From the construction we see that for $\lambda < \lambda_e^F$, $\rho(\lambda)$ is a step function whose jumps are finite-rank operators, and the jumps are possible only at λ_n^0 $(n = 1, 2, \ldots)$. Denote by $\mu_1 \leq \mu_2 \leq \cdots$ the jump points (counting multiplicities). Clearly,

$$\lambda_n^0 \leq \mu_n \quad (n = 1, 2, \ldots). \tag{1.46}$$

(Strict inequality can occur if the jump size at any of the points λ_k^0 in (1.45) turns out to be zero in the limit, or if a reduction in the rank of the limiting jump at λ_k^0 as $b \to \infty$ occurs).

For any finite interval Δ on the λ-axis, the kernel

$$G_\Delta = \int_\Delta Y(x,\lambda)\rho(d\lambda)Y^*(t,\lambda)W(t)$$

defines a bounded non-negative self-adjoint integral operator G_Δ in $H(0,\infty)$. The operator $G(\lambda) = G_{(-\infty,\lambda]}$ is a generalized (in the sense of M. A. Naimark) spectral function of the symmetric operator L'; see [6] and [145]. For $\lambda < \lambda_e^F$, $G(\lambda)$ is a step function with the same jump-points μ_n (and with the same multiplicity) as $\rho(\lambda)$:

$$\operatorname{rank} \rho(\mu_n - 0, \mu_n + 0) = \operatorname{rank} G(\mu_n - 0, \mu_n + 0).$$

Using Lemma 1.6 below, we have $\mu_n \leq \lambda_n^F$. This together with (1.45) and (1.46) shows that

$$\mu_n = \lambda_n^0 = \lambda_n^F, \tag{1.47}$$

and the Lemma is proven. □

Lemma 1.6 *Assume that A is a densely-defined, semi-bounded below and symmetric operator in a Hilbert space H. Let A_F be the Friedrichs extension of A, and let $\lambda_j(A_F) < \lambda_e(A_F)$ $((j = 1, 2, \ldots))$ be as at the beginning of this section. Let $G(\lambda)$ be the generalized (in the sense of Naimark [145]) spectral function of A, and let*

$$\mu_1(G) \leq \mu_2(G) \leq \cdots < \mu_e(G)$$

be the jump-points of $G(\lambda)$ (the ordering takes into account the multiplicities, i.e. ranks). Here, $\mu_e(G)$ denotes the greatest lower bound of the essential spectrum corresponding to the spectral function $G(\lambda)$ (the essential spectrum is understood as the set of non-isolated points of increase of $G(\lambda)$ and jumps with infinite multiplicities). Then

$$\mu_j(G) \le \lambda_j(A_F) \quad (j = 1, 2, \ldots), \quad \mu_e(G) \le \lambda_e(A_F). \tag{1.48}$$

Proof. In the case $\mu_1(G) = -\infty$, the statement of the Lemma is trivial. If $G(\lambda)$ is the resolution of identity corresponding to a self-adjoint extension of the operator A without exiting H, then the corresponding statement of the Lemma is contained in [58, Chapter 13, Sec. 9.D3]. Finally, assume that $G(\lambda)$ is determined by a semi-bounded self-adjoint extension B^+ of A (with exiting H) in $H^+ \supset H$. Denote by $P^+ : H^+ \to H^+$ the orthogonal projection onto H: $PH^+ = H$; $\mathrm{D}(B^+) \cap H \supseteq \mathrm{D}(A)$. By mini-max principle [58], we have

$$\lambda_n(B^+) = \sup_{H^+_{n-1}} \inf_{\substack{x \perp H^+_{n-1} \\ 0 \ne x \in \mathrm{D}(B^+)}} (B^+x, x)/(x, x),$$

and

$$\lambda_n(A_F) = \sup_{H_{n-1}} \inf_{\substack{x \perp H_{n-1} \\ 0 \ne x \in \mathrm{D}(A)}} (Ax, x)/(x, x),$$

where $H_{n-1} \subset H$, $H^+_{n-1} \subset H^+$ are all possible subspaces of dimension $n-1$. To any H^+_{n-1} corresponds $H_j = P^+H^+_{n-1} \subset H$, where $j = \dim H_j \le n-1$, and

$$\mathrm{D}(B^+) \cap (H^+ \ominus H^+_{n-1}) \supseteq \mathrm{D}(A) \cap (H \ominus H_j).$$

Therefore

$$\lambda_n(B^+) \le \lambda_{j+1}(A_F) \le \lambda_n(A_F). \tag{1.49}$$

The operator B^+ (which defines G_λ) can be chosen so that it does not have reducing subspaces orthogonal to H (see [6]). Thus the function $G(\lambda)$ has jumps at the same points $\lambda_n(B^+)$ (and with the same multiplicities) as the resolution of identity $E^+(\lambda)$ of B^+:

$$\operatorname{rank} E^+(\lambda_n - 0, \lambda_n + 0) = \operatorname{rank} G(\lambda_n - 0, \lambda_n + 0)$$

because $G(\lambda) = P^+E^+(\lambda)$. Therefore, $\mu_j(G) = \lambda_j(B^+)$, and, hence, from (1.49) we obtain (1.48). The Lemma is proven. □

Theorem 1.4 *Let L' denote the minimal with respect to $x = \infty$ operator in $H(0, \infty) = L_2\{H; (0, \infty); W(x)dx\}$ defined by the expression $W^{-1}(x)l[y]$ (where $l[y]$ is as in (1.1)) and the boundary condition (1.2). Assume that L' is semi-bounded below. Let $N_F(\lambda)$ be the number of eigenvalues $\lambda_n^F < \lambda$ corresponding to the Friedrichs extension L_F of L', and let $N_0(\lambda)$ be the number of jumps $\lambda_n^0 < \lambda$ (counting multiplicities) of the spectral operator-valued function $\rho(\lambda)$ obtained from spectral functions of the problems of type (1.1), (1.2), $y(b_j) = 0$ by passing to the limit as $b = b_j \to \infty$.*

Then for $\lambda \leq \lambda_e^F$, we have

$$\sum_{x \in (0,\infty)} nul\, Y(x, \lambda) = N_F(\lambda) = N_0(\lambda), \tag{1.50}$$

where, as before, $Y(x, \lambda)$ is the fundamental solution to the equation (1.1) with the initial data (1.10).

Remark 1.4 It can be shown that the spectral operator-valued function $\rho(\lambda)$, constructed as above, corresponds to the Friedrichs extension of L'. This is particularly easy to see in the case of a scalar problem: in the limit-point case, the extension is unique, while in the limit-circle case at infinity, the spectrum is discrete, and the equality of eigenvalues $\lambda_n^F = \lambda_n^0$ implies the equality of eigenfunctions, and, hence, boundary conditions at $x = \infty$. In the general case, i.e. when $\dim H = \infty$, one needs to use a generalization of Theorem VIII.1.15 from [87] to sequences of Hermitian relations or of non-densely defined operators (see [782, Theorem 6][5]). Another option is to use an analogous generalization of Theorem VIII.24 (b) from [165] on strong convergence of resolutions of identity for convergent (in the strong resolvent sense) sequences of densely defined operators.

Note also that the spectral function $\rho(\lambda)$ is unique, independently of its construction, for the equation

$$l_1[y] = -y'' + Q(x)y = \lambda y, \quad 0 \leq x < \infty, \tag{1.51}$$

with the operator coefficient $Q^*(x) = Q(x)$ and the boundary condition (1.2) if the corresponding minimal with respect to $x = \infty$ operator L' is semi-bounded below, because, in this case, L' is essentially self-adjoint. (This generalization of Povzner–Wienholtz Theorem [752] (see also [918], [70]) for an operator equation on the real line was proven by Vainerman and Gorbachuk in [896].)

[5]This Theorem is quoted in Bibliographical Comments to Appendix; see Theorem A.15 and Corollary A.6.

On the other hand, the scalar Sturm–Liouville operator with rapidly increasing leading coefficient

$$-(x^4y')' - 2x^2y = \lambda y$$

is semi-bounded below; however, this operator is in the limit-circle case for $x = \infty$ (see [327]).

A sufficient condition for the essential self-adjointness of a positive elliptic second-order operator with variable leading coefficients is given by a theorem of Berezanskiy [25], [268] (see also [220], [787] and [793]). This theorem, when applied to equations with operator-valued coefficients on the half-line or the whole real line, can give different conditions for the uniqueness of the spectral operator-valued function $\rho(\lambda)$. See, for example, S. Clark–F. Gesztesy theorem [342] for the operator of the type (1.5), (1.1) with the matrix-valued coefficients on the whole axis or half-axis. (This theorem is quoted in bibliographical comments for this chapter.)

Proof of Theorem 1.4. Let $N_F(\lambda) = n$. Then $\lambda_n^F < \lambda$ and $(\lambda_n^F, \lambda) \cap \sigma(L_F) = \emptyset$. By Lemma 1.5, we have $\lambda_{n,b}^0 \searrow \lambda_n^F$ as $b \to \infty$. Hence, for sufficiently large values of b, we have $\lambda_{n,b}^0 < \lambda$, $(\lambda_{n,b}^0, \lambda) \cap \sigma(L_B^0) = \emptyset$, and, thus, by Theorem 1.1, for all such b, we have

$$\sum_{x\in(0,b)} \text{nul}\, Y(x,\lambda) = N_b^0(\lambda) = n = N_F(\lambda).$$

Letting $b \to \infty$ and using (1.47) and (1.45), we obtain (1.50). If $N_F(\lambda_e^F) = \infty$, then for all $n > 0$, there exists $b_n > 0$ such that $\lambda_{n,b}^0 < \lambda_e^F$ for $b > b_n$, i.e. $N_b^0(\lambda_e^F) \geq n \to \infty$ as $b \to \infty$, and we again get (1.50). This concludes the proof of the Theorem. □

Theorem 1.5 *In the hypotheses of Theorem 1.4, let Λ denote a self-adjoint extension of L' in $H(0,\infty)$ such that on the left of a fixed $\mu < \lambda_e^F$,*

$$\lambda_n(\Lambda) = \lambda_n^F (< \mu) \quad (n = 1, 2, \ldots).$$

Let L be an arbitrary semi-bounded self-adjoint extension of L' in $H(0,\infty)$, let

$$p = \min_{\Lambda} Def \left\{ L|_{D(L)\cap D(\Lambda)} \right\},$$

where $Def\,\{T\}$ denotes the deficiency number[6] *of an operator T, and let $N_L(\lambda)$ be the function counting the eigenvalues $< \lambda$ of the operator L.*

[6]For a semi-bounded operator, the two deficiency numbers are equal.

Then for all $\lambda \leq \mu$, the following holds:

$$N_L(\lambda) - p \leq \sum_{x\in(0,\infty)} nul\, Y(x,\lambda) = N_F(\lambda) = N_\Lambda(\lambda) \leq N_L(\lambda). \tag{1.52}$$

If Def $\{L'\} = d$ and λ is not an eigenvalue of the (closed) operator L', then for all $\lambda \leq \mu$, we have

$$N_L(\lambda) - \min\{p, d - æ_L(\lambda)\} \leq \sum_{x\in(0,\infty)} nul\, Y(x,\lambda) = N_F(\lambda) \leq N_L(\lambda). \tag{1.53}$$

Corollary 1.2 *If* $\dim H = 1$, *then (in accordance with [58], [113]), we have*

$$\sum_{x\in(0,\infty)} nul\, Y(x, \lambda_n^L) = n - 1.$$

The proof of Theorem 1.5 follows directly from Theorem 1.4 and Lemma 1.4. The proof of Corollary 1.2 follows from (1.53) because in the limit-circle case, the operator L' does not have any eigenvalues and $d = æ(\lambda_n^L) = 1$. In the limit-point case, we have $p = 0$ and we can use (1.52).

Remark 1.5 In the case $\dim H < \infty$ it is known (see [70, p. 12]) that in order for the system to be non-oscillatory for the given λ, it is necessary and sufficient that $N(\lambda) < \infty$. On the other hand, the necessary and sufficient conditions for the system to be non-oscillatory requiring the finite number of zeros of $\det Y(x,\lambda)$ for $0 < x < \infty$ and $\dim H < \infty$ are proven in the real case in [922, p. 25] and in the complex case in [546, p. 129]. This result is partially contained in [78, p. 457]. Similarly, if $\dim H = \infty$, we have the following Corollary which follows from Theorems 1.1, 1.3, and 1.4:

Corollary 1.3 *For all $x \in (0,\infty)$ and $\lambda \notin \sigma_e(L_x^0)$, the operator $Y(x,\lambda)$ is Fredholm (with zero index). In this case, $N_b^0(\lambda_{e,b}^0) < \infty$ if and only if the number of points $x \in (0,b)$ at which nul $Y(x,\lambda_{e,b}^0) \neq 0$ is finite ($b \leq \infty$).*

The following Corollary obviously follows from Theorem 1.4 and Lemma 1.4:

Corollary 1.4 *Assume the hypotheses of Theorem 1.4. Assume also that rank* $\cos A$ *in (1.2) is finite and the restriction L'' of the operator L' by the condition $y(0) = 0$ is non-negative (the last condition is satisfied, if, for* *example, $Q(x)$ is a non-negative operator for all x). Then*

$$\sum_{x\in(0,\infty)} nul\, Y(x,0) \leq rank\ \cos A. \tag{1.54}$$

Indeed, Friedrichs extensions L_1 and L_0 of the operators L' and L'' respectively are at the same time finite dimensional extensions of $\Lambda = L_1 \cap L_0$, and the deficiency indices of Λ are (rank $\cos A$, rank $\cos A$). Thus by Lemma 1.4, using (1.38), we have

$$N_1(0) - \text{rank } \cos A \leq N_0(0) = 0,$$

where the last equation holds since L_0 is non-negative. Now (1.54) follows using the second equality in (1.50) with $\lambda = 0$.

Remark 1.6 From Corollary 1.4 (see (1.54)), for $\dim H < \infty$ we get a Hermitian generalization and improvement of the non-oscillatory theorem [238] (see Theorem 1.9 in the bibliographical comments to this chapter) since rank $\cos A \leq \dim H$ and since, in the terminology of [238], "the verticality moments of the Lagrangian plane", where the evolution of the plane from the initial position (1.2) is given by (1.1) with $\lambda = 0$, coincide with the points x where nul $Y(x,0) > 0$ and their multiplicities coincide with nul $Y(x,0)$. Note that if the equation (1.1) is given on the whole x-axis, and the solution $Y(x,0)$ is defined by Cauchy data (1.10) for $x = 0$, then the formula (1.54), with $A(a)$ instead of A, holds for any interval (a,∞) if the boundary condition (1.2) at the point a is given by the operator $A(a) = A^*(a)$, where $A(a)$ is chosen so that this boundary condition is satisfied by the solution $Y(x,0)$ (this is always possible[7]). Therefore, for any Lagrangian plane whose evolution is given by (1.1), the number of "verticality moments" on the real line counting multiplicities does not exceed $\dim H$, that is

$$\sum_{x\in(-\infty,\infty)} Y(x,0) \leq \dim H. \tag{1.55}$$

Several criteria for the finiteness of the number of negative eigenvalues (counting multiplicities) $N_\infty(0)$ or the discreteness of the negative part of the spectrum (which is equivalent to the equation $\lambda_{e,\infty} = 0$) for the equation (1.51) on the half-line with compact $Q(x)$ are obtained in [444] and [920].

[7]See Definition 2.2, Proposition 2.1 below and Appendix.

We will state without proof two theorems of M. G. Gasymov, V. V. Zhikov and B. M. Levitan; see [444].

Theorem 1.6 **(Gasymov–Zhikov–Levitan [444])** *Assume that for almost all x the operator-valued function $Q(x)$ and the operator A are compact in H. Assume also that $Q(x)$ is locally strongly Bochner integrable*[8] *and*

$$\int_0^\infty \|Q(x)\|^2\, dx < \infty. \tag{1.56}$$

Then the negative spectrum of the operator (1.51) with the boundary condition

$$y'(0) - Ay(0) = 0 \tag{1.57}$$

is discrete.

Theorem 1.7 **(Gasymov–Zhikov–Levitan [444])** *Assume that the local properties of $Q(x)$ are the same as in the hypotheses of the previous theorem, and instead of (1.56) the following condition holds:*

$$\int_0^\infty (1+x)\|Q(x)\|\, dx < \infty. \tag{1.58}$$

Then the operator (1.51) with the boundary condition (1.7): $y(0) = 0$ has a finite negative spectrum.

1.3 Bibliographical Comments

In the case of ordinary differential equations and their finite systems, Sturm-type theorems and their generalizations concerning spectral theory and theory of extensions, are considered in a series of well-known monographs on spectral theory of differential operators, such as [12], [13], [58], [70], [72], [94] and [113]. Each of these references also contains an extensive bibliography. Some other related bibliographical comments are also contained in the Preface of this book.

Basic results of this chapter belong to the authors of [794].

Oscillatory theorem for matrix Sturm–Liouville problems on the real line and for symmetric first-order systems (in particular, Dirac systems) was established in [540]. Proposition 1.1 was recently proven in [792].

[8] For the notion of Bochner integrability see the Bibliographical Comments for this chapter.

Lemma 1.1, the assertions 1a) and 1b) of Lemma 1.2 and 1) of Lemma 1.3 for $\dim H < \infty$ are well-known; see [632] (real case), [13], [407], and [902]. The assertion 2) of Lemma 1.2 and the assertion 2) of Lemma 1.3 do not have a finite-dimensional analogue.

The assertion 1) of Lemma 1.3 can be generalized although we do not need a more general version in this book. If $\dim H < \infty$, this generalization has the following form:

Proposition 1.2 **([238, Sec. 4, Lemma 2])** *If the derivative of the quadratic form with respect to the parameter is non-negative and if for the critical value of the parameter, the restriction of the derivative of the quadratic form on its kernel has rank k, then, when passing the critical value, the number of positive squares (the number of quadratic terms with positive coefficients when the quadratic form is reduced to the sum of quadratic terms) increases by at least k.*

Here, the critical value of the parameter is the value at which the quadratic form is singular. We consider the case when the parameter increases continuously and, after values that are less than the critical one, takes values greater than the critical value. Under this process, the parameter takes the critical value once (and crosses it).

It may happen that the increasing parameter has several critical values. For example, let the quadratic form have the matrix

$$\operatorname{diag}\,(a_1 + t, a_2 + t, \ldots, a_n + t),$$

where t is a parameter and $a_1 < a_2 < \cdots < a_n$.

Here, the critical values of the parameter t are

$$-a_n < -a_{n-1} < \cdots < -a_1,$$

and, as t increases from $-\infty$ to $+\infty$, it takes once each of the critical values.

In reference to Lemma 2 from [238, Sec. 4], we now give a simple example of the case $\dim\ker M(t_0) = m > 1$ and $\operatorname{rank}\,\{M'(t_0)|_{\ker M(t_0)}\} = k < m < n$, where $M(t)$ denotes the matrix of the quadratic form depending on the parameter t. Let

$$M(t) = \operatorname{diag}\,(a_1{+}t, a_2{+}t, \ldots, a_k{+}t, a_{k+1}{+}t^3, \ldots, a_m{+}t^3, a_{m+1}{+}t, \ldots, a_n{+}t)$$

and let

$$a_1 = a_2 = \cdots a_k = \cdots = a_m = 0 < a_{m+1} < \cdots < a_n.$$

Then we have simple critical values of the parameter t

$$-a_n < -a_{n-1} < \cdots < -a_{m+1}$$

and multiple critical value $t_0 = 0$ of t such that

$$M(0) = \text{diag }(0, \ldots, 0, a_{m+1}, \ldots, a_n),$$

with $\dim \ker M(0) = m > 1$.

The matrix of the orthogonal projection P_0 onto $\ker M(0)$ is

$$P_0 = \text{diag }(1, \ldots, 1, 0, \ldots, 0), \quad \text{rank } P_0 = m,$$

$$M'(t) = \text{diag }(1, \ldots, 1, 3t^2, \ldots, 3t^2, 1, \ldots, 1),$$

$$M'(0) = \text{diag }(1, \ldots, 1, 0, \ldots, 0, 1, \ldots, 1),$$

where $\dim \ker M'(0) = m - k$ and $\text{rank } M'(0) = k + n - m$. The range of $M'(0)$ is not contained in $\ker M(0)$ but it has a non-trivial intersection with $\ker M(0)$, and

$$\text{rank } \{M'(0)|_{\ker M(0)}\} = k.$$

The matrix $M'(t)$ is non-negative for all t, and, in this example, the number of positive squares of $M(t)$ is n at $t > 0$ and $n - m$ at $-a_{m+1} < t < 0$. Therefore, the number of positive squares of $M(t)$ increases by $m > k = \text{rank } \{M'(0)|_{\ker M(0)}\}$ when the parameter t crosses the critical value $t_0 = 0$ (in accordance with Lemma 2 from [238, Sec. 4]).

With regard to Lemma 1.4, a related result is contained in [365].

Theorem 1.8 (about non-oscillation [238, Sec. 1]) has the following formulation there:

If the potential energy is non-positive, then the number of verticality moments does not exceed the number of degrees of freedom n.

(This statement has to do with the evolution of Lagrangian planes under the action of systems of the form (1.1) when $\dim H = n < \infty$; see Remark 1.6.)

Note that if for two different problems of the form (1.1)–(1.3) the spectra are discrete ($\dim H < \infty$) and the corresponding functions $N_1(\lambda)$ and $N_2(\lambda)$ coincide—$N_1(\lambda) = N_2(\lambda)$, $\lambda \in \mathbb{R}$, then the eigenvalues of these problems coincide (and so do their multiplicities). In such case, the corresponding problems are said to be isospectral. Under additional restrictions, properties of such pairs of matrix isospectral Sturm–Liouville problems were considered, for example, in [523] (see also references there).

Here we quote the S. Clark–F. Gesztesy theorem mentioned in Remark 1.4.

Theorem 1.9 (**[342]**) *Let $\Delta = \mathbb{R}^+$ or $\mathbb{R}$ and suppose that P, Q, W are $(m \times m)$-matrix-valued functions, (Lebesgue) measurable on Δ, and $P > 0$, $W > 0$, $Q = Q^*$ self-adjoint, P^{-1}, Q, $W \in L_{1,loc}(\Delta; dx)$ (element by element), $P \in AC_{loc}(\Delta)$ (locally absolutely continuous). If the minimal operator $L_{\min,\Delta}$ generated by the expression (1.5)*

$$l_W = W^{-1}(-(d/dx)P(d/dx) + Q) \quad \text{in } L_2(H; \Delta; W(x)dx)$$

is semi-bounded from below and

$$\operatorname{ess} \sup_{r/2<|x|<r} \|W^{-1/2}PW^{-1/2}\| = O(r^2), \qquad r \to \infty,$$

then l_W (1.5) is in limit-point cases at $\pm\infty$ respectively.

Note that very detailed study of essential self-adjointness of general Schrödinger operators of type $D^*D + V$, where D is a first-order elliptic differential operator acting in the space of sections of a Hermitian vector bundle over a manifold, is due to M. Braverman, O. Milatovic and M. Shubin [309]. Note also the papers by Yu. B. Orochko [721], G. M. Zhislin [940], E. F. Zhizhenkova and G. M. Zhislin [941], and A. G. Brusentsev's recent results [325] and [326] on the essential self-adjointness of semi-bounded Schrödinger operators of a general form.

In the case $\dim H < \infty$, Theorem 1.7 is a well-known result proven in [1] using the decomposition method of [70]. Using the asymptotics of the form $Cx + D$ for matrix solutions of the equation (1.51) under the condition (1.58), when $\lambda = 0$ and $x \to \infty$, as obtained in [1], Theorem 1.7 in the case $\dim H < \infty$ can be derived immediately from Theorem 1.4.

The papers [432]–[436] by E. V. Foliadova are devoted to the oscillatory theory of problems with a spectral parameter in the boundary condition.

1.3.1 *An Example of the Operator of (1.1)–(1.3) Type with the Purely Absolutely Continuous Spectrum*

Consider the operator L corresponding to the following system in $H(0,\pi)$:

$$-y'' = \lambda y, \qquad y'(0) = 0, \quad y'(\pi) + By(\pi) = 0, \tag{1.59}$$

where B is a bounded self-adjoint operator on H.

Then the fundamental solution is $Y(x,\lambda) = I \cdot \cos(x\lambda^{1/2})$, and the spectrum of L coincides with the set of those values of λ at which the operator

$$\Omega(\lambda) := -I \cdot \lambda^{1/2} \sin(\pi\lambda^{1/2}) + B\cos(\pi\lambda^{1/2}) \tag{1.60}$$

does not have a bounded inverse operator $\Omega(\lambda)^{-1}$ defined on the whole H; see Lemma 2.4 below. Thus, if a point μ belongs to the spectrum of B, then to the spectrum of L belong all values of λ satisfying the equation

$$\lambda^{1/2} \tan(\pi\lambda^{1/2}) = \mu, \tag{1.61}$$

and, at those points λ, the spectrum of L has the same type as the spectrum of B at the corresponding points μ (continuous, discrete, *etc.*). Those values of λ exhaust the whole spectrum of L when μ runs through the whole spectrum of B.

We can easily obtain the spectral operator-valued distribution function $\rho(\lambda)$ (that gives rise to Parseval equality), for example, by passing to the limit from spectral functions of finite systems of the same type; see [766]. Moreover, if $E(t)$ is the spectral resolution of the identity corresponding to the operator B, i.e. if

$$B = \int_{-\infty}^{\infty} t\, dE(t), \tag{1.62}$$

then

$$\rho(\lambda) = \frac{2}{\pi} \int_{\infty}^{'\lambda} \left(1 + \frac{\sin(2\pi t^{1/2})}{2\pi t^{1/2}}\right)^{-1} d_t\, E(t^{1/2} \tan(\pi t^{1/2})), \tag{1.63}$$

where the sign "prime" on top of the integral indicates that it is necessary to reject all negative jumps (sudden changes) that the operator-valued measure undergoes under the integral at the points $t = (n+(1/2))^2$, where $n = 0,1,2,\ldots$. The domains of integrations are actually finite in both integrals because the operator B is assumed to be bounded. From the last formula for $\rho(\lambda)$ we see that if the spectrum of B is purely absolutely continuous, then so is the spectrum of L.

We also consider the opposite case, i.e. when the operator B has a total system of eigenvectors in H (i.e. the linear span of the corresponding eigenvectors is dense in H). Then the corresponding operator L in $H(0,\pi)$ has the same property. But the eigenvalues $\{\mu_n\}_{n=1}^{\infty}$ of the operator B must have at least one limit point in the closed interval $|\mu| \leq \|B\|$, and, hence, even in this case, the spectrum of L has a non-empty set of the points of the essential spectrum, and this set is unbounded from above.

Analogous reasoning can be applied to the problem (1.59) in the case of an unbounded self-adjoint operator B. Thus, if, for example, B has absolutely continuous spectrum that fills the whole real axis, then, by formulas (1.61)–(1.63), the operator L (1.59) also has absolutely continuous spectrum that fills the whole real axis.

If the operator B has purely discrete spectrum bounded neither from below nor from above, then the operator L (1.59) has a total system of eigenvectors in $H(0,\pi)$. The set of eigenvalues of L is not bounded from below and does not have limit points on the negative half-axis. The set of eigenvalues of L (1.59) is not bounded from above, but there is an infinite set of limit points of eigenvalues of L. Those limit points $\sigma_e(n)$ form the essential spectrum of L (1.59):

$$\sigma_e(L) = \{\sigma_e(n)\}_{n=0}^{\infty}, \quad \text{where} \quad \sigma_e(n) = (n+(1/2))^2.$$

1.3.2 *Oscillatory Theory of Partial Differential Equations*

In the present book, we do not consider the oscillatory theory of partial differential equations in detail. This theory is much more complicated and less developed than oscillatory theory for ordinary differential equations. For example (see the monograph R. Courant–D. Hilbert [48, Vol.I, Chapter 6, Sec. 6]), even for a simple boundary problem such as Dirichlet homogeneous problem for the Laplace operator in the square domain

$$\begin{aligned} -\Delta u &= \lambda u, \qquad 0\le x\le\pi, \qquad 0\le y\le\pi, \\ u(0,y) &= u(x,0) = 0, \qquad u(\pi,y) = u(x,\pi) = 0, \end{aligned}$$

there are arbitrarily large values of n to which correspond eigenfunctions $u_n(x,y)$ such that their nodal lines ($u_n(x,y)=0$) divide the main domain of the problem (in this example—the square) only in two parts. In the considered case, such eigenvalues are of the form $\lambda = 4m^2+1$, and the corresponding eigenfunctions are of the form

$$\sin(2mx)\cdot\sin y + \nu\sin(2my)\cdot\sin x.$$

The eigenfunctions have only one nodal line if $\nu=\nu(m)$ is a positive constant sufficiently close to 1. This nodal line can be obtained by a continuous deformation from the system of nodal lines corresponding to the case $\nu=1$ and forming a rectangular lattice.

A generalization of the Sturm oscillatory theory for ordinary differential equations to the case of solutions of second order elliptic partial differential equations is contained, among other results, in the books [28], [70, Chapter IV], [98], [182], and [187].

Here, we would like to quote another simple generalization of the Sturm Oscillation Theorem ([782, Theorem 2] or [783, Theorem 9]). Let $\Omega \subseteq \mathbb{R}^n$ be a finite or infinite region with piecewise smooth boundary $\Gamma = \partial\Omega$, let $\widetilde{\Omega}$ be $\Omega \cup \Gamma$ with cuts along the portions of Γ that are interior for $\Omega \cup \Gamma$, and let

$$l[u] = \lambda W(x)u, \qquad \det W(x) \neq 0, \qquad W = W^* >> 0, \qquad x \in \Omega, \quad (1.64)$$

be a strongly elliptic system of arbitrary order on $\widetilde{\Omega}$ generating on $C^\infty_{0,\Gamma}(\widetilde{\Omega})$ (i.e. on vector-valued functions in $C^\infty_0(\widetilde{\Omega})$ satisfying a local-type boundary condition on Γ) a symmetric semi-bounded below operator L with $L \geq \gamma$ in $L_2(\widetilde{\Omega}; W\,dx)$. Let L^F be the Friedrichs extension of L and let

$$L_\xi = L|_{C^\infty_{0,\Gamma}(\widetilde{\Omega}) \cap \{u:\, \text{supp}\, u \in \widetilde{\Omega}_\xi\}},$$

where $\Omega_\xi \subset \Omega$ is a family of domains not necessarily monotone with respect to ξ, $0 < \xi < \infty$.

If at $\lambda < \beta$ the spectrum of L_F is discrete, so are the spectra of operators L^F_ξ. (As always, eigenvalues are counted according to their multiplicity.)

Theorem 1.10 **([782], [783])** *Under the conditions above , let $\lambda_j(L^F_\xi)$ be*

(1) continuous in ξ, $0 < \xi < \infty$,
(2) strictly decreasing with respect to ξ, and
(3) $\lambda_1(L^F_\xi) \to +\infty$ as $\xi \downarrow 0$.

Let $N(\lambda) := \dim E_{\lambda-0}(L^F)$, where $E_\lambda(A)$ denotes the resolution of identity of some operator $A = A^*$ (here, $A = L^F$), and let $\kappa_\xi(\lambda) \geq 0$ be the multiplicity of λ as an eigenvalue of the operator L^F_ξ. Then

$$N(\lambda) = \sum_{\xi \in (0,\infty)} \kappa_\xi(\lambda), \qquad \lambda < \beta. \quad (1.65)$$

Here, we make additional assumptions about properties of the family Ω_ξ and about the set of functions belonging to the domain of the considered symmetric differential operator:

(i) If Ω is an infinite domain, then

$$(\Omega \cap K_N(0)) \subset \Omega_\xi, \quad \xi > \xi(N), \qquad K_N(x) := \{y : |x - y| < N\}, \quad (1.66)$$

and $D(L) = C^{\infty}_{0,\Gamma}(\widetilde{\Omega})$, *i.e. every* $u \in D(L)$ *is identically equal to zero in a neighborhood of infinity, but it is not necessarily equal to zero on* Γ.

(ii) If Ω *is a bounded region, then we consider two subcases:*

(ii a) $D(L) = C^{\infty}_{0}(\Omega)$ *and, instead of (1.66), take*

$$\Omega = \bigcup_{k=0}^{\infty} \bigcap_{\xi=k}^{\infty} \Omega_{\xi},$$

(ii b) $D(L) = C^{\infty}_{0,\Gamma}(\widetilde{\Omega}\backslash(\cup\{x_j\}))$, $x_j \in \widetilde{\Omega}$, $j = 1, \ldots, M$, *and, instead of (1.66), take*

$(\Omega\backslash \cup K_{\epsilon}(x_j)) \subset \Omega_{\xi}$, *for* $\xi > \xi(\epsilon)$ *and for all positive sufficiently small* ϵ.

For the scalar Sturm–Liouville problem, Theorem 1.10 is equivalent to the Sturm Oscillation Theorem, and in the matrix case—to its generalization which contains the Morse Index Theorem (see Remark 1.1 or [794]). As shown in [48] and [642], the condition (2) of the Theorem ($\lambda_j(L^F_{\xi})$ is strictly decreasing with respect to ξ) is guaranteed by uniqueness of the solution of the Cauchy problem for (1.64). The conditions that guarantee uniqueness of the solution of the Cauchy problem for (1.64) belong to Holmgren, Carleman (see [36] and [337]), Landis [613], M. M. Lavrent'ev [617] (see also [139], [156] and [701]). In the case of a homogeneous Dirichlet problem, the condition (3) of the Theorem is guaranteed by the condition due to Kondrat'ev (see [569] and [570]) and Landis (see [614]): there exists $\epsilon > 0$ such that

$$\operatorname{diam}^{(\epsilon)}_{(\mathrm{int})} \Omega_{\xi} \to 0, \qquad \text{as } \xi \downarrow 0,$$

or, more explicitly,

$$\sup_{x \in \mathbb{R}^n} \inf\{r : \operatorname{mes}(K_r(x)\backslash \Omega_{\xi}) > \epsilon r^n\} \to 0 \text{ if } \xi \to 0.$$

For the notation $K_r(x)$, see (1.66). Here, our notation $\operatorname{diam}^{(\epsilon)}_{(\mathrm{int})} \Omega_{\xi}$ denotes the interior diameter of the domain Ω_{ξ} "up to measure $\epsilon > 0$." (The interior diameter of a domain is the supremum of diameters of the balls that can be inscribed in this domain.)

For the regular problem on Ω_{ξ}, the condition (3) of the Theorem is also guaranteed by a simpler Vishik condition (see [900, Theorem 5, Sec. 5]):

$$\operatorname{mes}(\Omega_{\xi}) \to 0, \qquad \text{as } \xi \downarrow 0.$$

With regard to condition (1), there is a generalization [782] of R. Courant's Theorem [48]; see the Bibliographical Comments to Appendix of the present book—in particular, Theorem A.18. With the help of embedding theorems (see [184] and [148]), this generalization enables one to establish the continuity of eigenvalues for substantially more general variations of the region (without the condition of proximity of boundaries of perturbed and unperturbed regions) and for a broader class of equations. (See Example A.18 and Open Question A.1.) In some sense, Theorem 1.10 is analogous to S. Smale's Theorem [866], but these theorems do not contain each other.

Remark 1.7 **([782])** The positivity of the Schrödinger operator

$$Lu := -\Delta u + q(x)u$$

in $L_2(\Omega)$ at $u|_\Gamma = 0$ does not guarantee its semi-boundedness at $\frac{\partial u}{\partial \nu}|_\Gamma = 0$. (Here ν is the normal to Γ.)

For example, let $x = (x_1, x_2)$, $q(x) = -|x|^2 = -(x_1^2 + x_2^2)$, and let

$$\Omega = \left\{(x_1, x_2) \in \mathbb{R}^2\colon -\infty < x_1 < \infty,\, 0 < x_2 < (1 + x_1^2)^{-1}\right\}.$$

In both cases the operator L is essentially self-adjoint (see [774]), and, under Dirichlet conditions, L is positive, because for all $u \in C^\infty_{0,\Gamma}(\Omega)$ we have

$$\begin{aligned}(Lu, u)_{L^2(\Omega)} &= \int_{-\infty}^{\infty} dx_1 \int_0^{(1+x_1^2)^{-1}} (|\nabla u|^2 - |x|^2|u|^2)\, dx_2 \\ &\geq (\pi^2 - 1)\|u\|^2_{L_2(\Omega)} > 0.\end{aligned}$$

Here, to estimate the interior integral, we used the following classical variational inequality (see [15] and [77]) for functions $u(t)$ of $C^1[0, b]$, where $u(0) = u(b) = 0$:

$$\int_0^b |u'(t)|^2\, dt \geq \left(\frac{\pi}{b}\right)^2 \int_0^b |u(t)|^2\, dt.$$

However, under Neumann conditions, for the sequence of functions $u_n(x) \in C^\infty_{0,\Gamma}(\Omega)$, $|\nabla u_n(x)| < \text{const}$,

$$u_n(x_1, x_2) = \begin{cases} 0, & x_1 \notin [n, 2n], \\ 1, & n + 1 \leq x_1 \leq 2n - 1, \\ 0 \leq u_n \leq 1, & x_1 \in (n, n+1) \cup (2n - 1, 2n), \end{cases}$$

we have

$$\frac{(Lu_n, u_n)_{L^2(\Omega)}}{\|u_n\|^2_{L^2(\Omega)}} \to -\infty, \qquad \text{as } n \to \infty.$$

1.3.3 *Bochner Integral*

For more on the notion of Bochner integral of vector-valued functions with values in a Banach space X and, in particular, on Bochner integral of operator-valued functions with values in the space $B(X)$ of bounded linear operators in X (such functions can be considered as vector-valued functions with values in $B(X)$), see, for example, [26], [51], [73], [74], [81, Chapter III, Sec. 1] and [201]. This integral is defined as a limit of integrals of vector-valued functions whose range consists of at most countable set of elements. A function $x(t)$ with values in a separable Hilbert space $X = H$ is locally integrable if and only if the function $(x(t), f)$ is measurable for all vectors $f \in H$ and the norm $\|x(t)\|$ is locally integrable.

For an operator-valued function $Y(t)$, the notion of Bochner integrability in uniform operator topology (when $Y(t)$ is considered as a function in $B(X)$ with the norm $\|Y(t)\|$) differs from the notion of Bochner integrability in strong operator topology (when the integral $\int_a^b Y(t)\,dt$ is understood as an operator $V \in B(X)$ such that

$$Vf := \int_a^b Y(t)f\,dt$$

for all $f \in X$; see, for example, [81, Sec. 3.8]).

Since the Banach space $B(X)$ is not separable (even when $X = H$ is a separable Hilbert space), the operator-valued function $Y(t)$ with values in $B(H)$ may fail to be locally Bochner integrable (in uniform operator topology) even if the norm $\|Y(t)\|$ is locally integrable and the vector function $Y(t)f$ is locally Bochner integrable for all $f \in H$ (and, hence, $Y(t)$ is strongly Bochner integrable [9]).

However, it is important to keep in mind that the set of compact operators $S_\infty(H)$ in a separable Hilbert space H is also separable, and, in this case, the situation is simpler[10].

[9] An example of operator Hill's equation, showing how stable boundedness of its solutions depends on integrability type of the operator-valued coefficient $P(t)$ of the considered equation, is given in Bibliographical Comments for Chapter 6.

[10] However, the set $S_\infty(B)$ can be non-separable even if B is separable. Professor

In connection with the spectral theory of operator-valued measures in a Hilbert space and their applications in functional analysis, see also the recent paper [649].

V. M. Kadets kindly communicated to the authors a simple example. Let B be l^1—the space of all summable sequences. (It is well known that B is separable.) Then the set of compact one-dimensional operators $\{F\} \subset S_\infty(l^1)$ of the form $Fh = (f(h), 0, 0, \ldots)$, where $f \sim F$ is a (bounded linear) functional on l^1 in one-to-one correspondence with the operator F, is not separable, because the space l^∞ consisting of all bounded sequences (the dual of l^1) is not separable.

Chapter 2

Fundamental System of Solutions for an Operator Differential Equation with a Singular Boundary Condition

2.1 Separated Self-Adjoint Boundary Conditions

In this chapter we will construct a fundamental solution of a certain differential equation satisfying a self-adjoint boundary condition at one of the end-points of (a, b), $-\infty \leq a \leq \infty$, and we will prove the consistency of this fundamental solution.

Let H be a separable Hilbert space with the inner product $(\cdot, \cdot)$ and the norm $|\cdot|$, $\dim H \leq \infty$. We denote by

$$H(a,b) = L_2\{H; (a,b); W(x)\,dx\}$$

the Hilbert space of vector-valued functions $y(x)$ with values in H with the inner product

$$\langle y, z\rangle_{(a,b)} = \int_a^b (W(x)y(x), z(x))\,dx, \quad -\infty \leq a < b \leq \infty$$

and the corresponding norm $\|\cdot\|_{(a,b)}$, where $W(x) = W^*(x) >> 0$ is a bounded operator in H with the positive greatest lower bound for all $x \in (a, b)$ (here the dependence on x is continuous in the strong sense). When there is no ambiguity, we will omit the interval (a, b) in the definition of the inner product and the norm in $H(a, b)$.

In what follows, $B(H)$ denotes the set of all bounded linear operators on the space H.

Consider the self-adjoint differential equation of order $r \geq 1$ with operator-valued coefficients in $B(H)$:

$$l[y] = \sum_{k=0}^{r} i^k l_k[y] = \lambda W(x)y, \tag{2.1}$$

where

$$l_{2j} = D^j p_j(x) D^j, \quad p_j^*(x) = p_j(x),$$

$$l_{2j-1} = 1/2 D^{j-1}\{Dq_j(x) + q_j^*(x)D\}D^{j-1}, \quad D = d/dx.$$

Here, the operator-valued coefficients $p_j(x)$, $q_j(x)$, together with their derivatives of order up to and including j, uniformly continuously depend on x, and the coefficient with the highest order derivative ($p_n(x)$ for $r = 2n$, and $\mathrm{Re}\, q_n(x) = \frac{1}{2}(q_n(x) + q_n^*(x))$ for $r = 2n - 1$) in the equation (2.1) has a bounded inverse in H for all $x \in (a, b)$, including a and b if they are finite.

By L we denote the minimal differential operator in $H(a, b)$ determined by the differential expression

$$l_W[y] = W^{-1}(x) l[y]. \tag{2.2}$$

A linear condition of the form

$$U_a[y] = 0, \tag{2.3}$$

where U_a is a linear map from $H(a, c)$, $a < c \le b$, into a Hilbert space $\mathbf{H}_a$, will be called a boundary condition at the point a if every two vector-valued functions which coincide in a neighborhood of a satisfy or do not satisfy this condition simultaneously.

By L_ξ we will denote the closure of the operator in $H(a, \xi)$, $a < \xi \le b$, determined by the differential expression (2.2) and the condition (2.3) on the smooth functions vanishing in a neighborhood of the point ξ. We will call L_ξ the minimal operator with respect to the end-point ξ corresponding to (2.2) and (2.3).

We will say that the boundary condition $U_a[y] = 0$ is self-adjoint at the point a if the operator L_b is symmetric and the functions $y \in \mathrm{D}(L_b^*)$ satisfy this boundary condition. Then L_ξ is symmetric for all $\xi \in (a, b)$.

Similarly, we define the self-adjoint condition

$$U_b[y] = 0, \tag{2.4}$$

at point b.

The general form of self-adjoint conditions for equations of arbitrary order with bounded operator-valued coefficients is derived in the Appendix. On an infinite interval, in the absolutely indeterminate case, this was done in Chapter 5. If $\dim H < \infty$, then absolutely indeterminate case and quasi-regular case are the same.

An operator L in $H(a,b)$ has a self-adjoint extension determined by separated boundary conditions if and only if there exist self-adjoint boundary conditions for this operator at points a and b. The restriction of the operator L^* to a manifold satisfying those conditions is a self-adjoint extension L. In the case of a second order differential equation (or a system of second order differential equations), separated boundary conditions and the Sturm-type boundary conditions are the same.

A semi-bounded differential operator in $H(a,b)$ always has self-adjoint extensions with separated boundary conditions; for example, the Friedrichs extension.

On a bounded interval, without semi-boundedness assumption on the operator, the operators of even order always have separated self-adjoint boundary conditions (see Appendix). However, on the half-line $(0,\infty)$, it may happen that an operator of even order does not have self-adjoint separated boundary conditions when $\dim H \leq \infty$. As an example, we may consider a diagonal infinite system of second order equations in scalar-valued functions, where the system of the first two equations determines a symmetric operator with deficiency indices $(2,3)$ or $(3,2)$ (see [529]), and the remaining equations form a diagonal system where all scalar operators have deficiency indices $(1,1)$. Then the operator L has deficiency indices (∞,∞) and it has self-adjoint extensions. However, the extension L_1 of L determined by the self-adjoint boundary condition at $x=0$ has deficiency indices $(0,1)$ or $(1,0)$ and, therefore, L_1 does not have self-adjoint extensions. Consequently, L does not have self-adjoint extensions with the separated boundary conditions (at $x=0$ and $x=\infty$).

Consider the case of an operator of order $r=2n-1$ on a bounded interval. If $\dim H<\infty$ and $r\geq 1$, or $\dim H=\infty$ and $r=1$, separated boundary conditions exist if and only if (see Theorem A.9)

$$\dim H_x^+ = \dim H_x^- , \tag{2.5}$$

where H_x^+ and H_x^- are reducing subspaces of the operator Re $q_n(x)$, such that Re $q_n(x)$ and $-$Re $q_n(x)$ are positive definite on H_x^+ and H_x^- respectively. If $\dim H=\infty$ and $r\geq 3$, then separated self-adjoint boundary conditions always exist.

In what follows, we assume that for the minimal operator L in $H(\alpha,\beta)$ there exist self-adjoint separated boundary conditions on any interval $(\alpha,\beta)\subseteq(a,b)$ (and, hence, $r\cdot\dim H=\infty$ or even).

For equations of arbitrary order with bounded coefficients on (a,b), self-adjoint separated boundary conditions have the following form (see (A.66) and Theorem A.9):

$$U_a[y] \equiv \cos A \cdot y^{\vee}(a) - \sin A \cdot y^{\wedge}(a) = 0, \tag{2.6}$$

$$U_b[y] \equiv \cos B \cdot y^{\vee}(b) + \sin B \cdot y^{\wedge}(b) = 0, \tag{2.7}$$

$-\frac{\pi}{2}I < A, B \le \frac{\pi}{2}I$ are self-adjoint operators in $\widetilde{\mathbf{H}}$ ($-\pi/2$ is not an eigenvalue of A, B, and I is the identity operator in $\widetilde{\mathbf{H}}$). Here $\widetilde{\mathbf{H}}$, $y^{\wedge}(x)$, and $y^{\vee}(x)$ are defined as follows: for $r = 2n$,

$$\widetilde{\mathbf{H}} = H^n = H \oplus H \oplus \ldots \oplus H,$$

$$\begin{aligned} y^{\wedge}(x) &= \text{col}\ \{y(x), y'(x), \ldots, y^{(n-1)}(x)\}, \\ y^{\vee}(x) &= \text{col}\ \{y^{[2n-1]}(x), y^{[2n-2]}(x), \ldots, y^{[n]}(x)\}, \end{aligned} \tag{2.8}$$

($y^{[2n-j]}$ is a quasi-derivative in (2.11)).

If $r = 2n - 1$ and if (2.5) holds, we define

$$\widetilde{\mathbf{H}} = H^+ \oplus H^{n-1},$$

$$\begin{aligned} y^{\wedge}(x) = \text{col}\ \{&(V^+(x)q_+^{1/2}(x) + V^-(x)q_-^{1/2}(x))y^{(n-1)}(x), \\ &y(x), y'(x), \ldots, y^{(n-2)}(x)\} \in \widetilde{\mathbf{H}}, \end{aligned} \tag{2.9}$$

$$\begin{aligned} y^{\vee}(x) = \text{col}\ \{&i(V^-(x)q_-^{1/2}(x) - V^+(x)q_+^{1/2}(x))y^{(n-1)}(x), \\ &y^{[2n-2]}(x), y^{[2n-3]}(x), \ldots, y^{[n]}(x)\} \in \widetilde{\mathbf{H}}, \end{aligned}$$

where

$$q_+(x) = \frac{1}{2}(\text{Re}\ q_n(x))P_x^+, \quad q_-(x) = -\frac{1}{2}(\text{Re}\ q_n(x))P_x^-,$$

P_x^+ and P_x^- are the orthogonal projections on H_x^+ and H_x^-. Here H^+ is an arbitrary Hilbert space such that $\dim H^+ = \dim H_x^+$ for all $x \in (a,b)$, and $V^{\pm}(x)$ are isometries mapping $H_x^{\pm}$ onto H^+.

If $\dim H = \infty$ and $r = 2n - 1 \ge 3$, we define

$$\widetilde{\mathbf{H}} = H^+ \oplus H^{n-1},$$

$$y^{\wedge}(x) = \text{col}\,\{(V_1^+(x)y^+(x) + V_1^-(x)y^-(x)), y(x), \dots, y^{(n-3)}(x)\}, \quad (2.10)$$

$$y^{\vee}(x) = \text{col}\,\{i(V_1^-(x)y^-(x) - V_1^+(x)y^+(x)), y^{[2n-2]}(x), \dots, y^{[n+1]}(x)\},$$

where

$$y^+(x) = \{q_+^{1/2}(x)y^{(n-1)}, 1/2(y^{[n]}(x) - iy^{(n-2)}(x))\} \in H_x^+ \oplus H,$$
$$y^-(x) = \{q_-^{1/2}(x)y^{(n-1)}, 1/2(y^{[n]}(x) + iy^{(n-2)}(x))\} \in H_x^- \oplus H.$$

Here H^+ is an arbitrary separable Hilbert space, $V_1^{\pm}(x)$ are isometries mapping $H_x^{\pm} \oplus H$ onto $H^+ \oplus H$. (If, in particular, (2.5) holds, then, instead of (2.10), we can use (2.9)).

In (2.8), (2.9), (2.10), $y^{[k]}$ are quasi-derivatives corresponding to the differential expression $l[y]$ in (2.1) of order $r = 2n$ or $r = 2n - 1$:

$$\begin{aligned} y^{[k]} &= y^{(k)}, \ k = 0, 1, \dots, [r/2] - 1, \\ y^{[r-n]} &= p_n y^{(n)} - (i/2) q_n y^{(n-1)} \quad (p_n(x) \equiv 0 \text{ for } r = 2n - 1), \\ y^{[r-k]} &= -d/dx\, y^{[r-k-1]} + p_k y^{(k)} + (i/2)[q_{k+1}^* y^{(k+1)} - q_k y^{(k-1)}], \\ & k = 0, 1, \dots, n - 1 \quad (q_0 \equiv 0). \end{aligned} \quad (2.11)$$

For the problem on unbounded interval (a, ∞), $a \geq -\infty$, in the absolutely indeterminate case, a description of self-adjoint separated boundary conditions is given in Chapter 5.

It can be checked directly (see Appendix) that for any two functions $y, z \in \mathrm{D}(L^*)$ and any interval $[\alpha, \beta] \subseteq (a, b)$, Lagrange's identity can be written as follows:

$$\langle l_W[y], z\rangle_{(\alpha,\beta)} - \langle y, l_W[z]\rangle_{(\alpha,\beta)} = [(y^{\wedge}, z^{\vee}) - (y^{\vee}, z^{\wedge})]\,\big|_{\alpha}^{\beta}. \quad (2.12)$$

2.2 A Construction for the Fundamental Solution of the Boundary-Value Problem

By L_ξ^A we denote the operator in $H(a, \xi)$, $\xi \in (a, b)$, determined by (2.2), (2.3) and the boundary condition

$$y^{\vee}(\xi) - A \cdot y^{\wedge}(\xi) = 0, \quad (2.13)$$

where A is a bounded linear operator in $\widetilde{\mathbf{H}}$ satisfying the following condition:

$$|\mathrm{Im}(Ah, h)| \geq \epsilon \cdot |h|^2, \quad \exists \epsilon > 0, \forall h \in \widetilde{\mathbf{H}}. \quad (2.14)$$

The space $\widetilde{\mathbf{H}}$ is defined as in (2.8), (2.9) and (2.10) by the formula

$$\widetilde{\mathbf{H}} = \begin{cases} H^n & \text{if } r = 2n \\ H^+ \oplus H^{n-1} & \text{if } r = 2n-1. \end{cases} \tag{2.15}$$

By $\sigma_d(T)$ and $\sigma_e(T)$ we denote the discrete and the essential spectrum of T respectively.

Lemma 2.1 *On the real axis, the operator L_ξ^A does not have any eigenvalues and it does not have residual spectrum. Moreover, the following holds*

$$\mathbb{R} \cap \sigma_e(L_\xi^A) = \sigma_e(L_\xi), \tag{2.16}$$

and $\sigma_e(L_\xi)$ does not depend on ξ. (If $a > -\infty$ and the operator is regular, then $\sigma_e(L_\xi) = \emptyset$.)

Proof. Since all functions $y \in \mathrm{D}(L_\xi^A)$ satisfy the self-adjoint boundary condition at a, then using Lagrange's identity (2.12) and the boundary condition (2.13), we get

$$\begin{aligned} &\langle L_\xi^A y, y\rangle_{(a,\xi)} - \langle y, L_\xi^A y\rangle_{(a,\xi)} = (y^\wedge(\xi), y^\vee(\xi)) \\ &-(y^\vee(\xi), y^\wedge(\xi)) = (y^\wedge(\xi), Ay^\wedge(\xi)) - (Ay^\wedge(\xi), y^\wedge(\xi)) \\ &= ((A^* - A)y^\wedge(\xi), y^\wedge(\xi)) = -2i(\text{Im } Ay^\wedge(\xi), y^\wedge(\xi)), \end{aligned} \tag{2.17}$$

where

$$\text{Im } A = \frac{1}{2i}(A - A^*).$$

If $L_\xi^A y = \lambda y$, by (2.17) we get $\text{Im } \lambda = 0$, and by (2.13), (2.14), we get $y^\wedge(\xi) = 0$; hence, $y^\vee(\xi) = 0$, and therefore, $y(x) \equiv 0$.

The operator $(L_\xi^A)^*$ is the extension of L_ξ determined by the boundary condition $y^\vee(\xi) - A^* y^\wedge(\xi) = 0$. As in the case of L_ξ^A, the operator $(L_\xi^A)^*$ does not have real eigenvalues, and, hence, L_ξ^A does not have residual spectrum in $\mathbb{R}$.

Since $\sigma_e(L_\xi) \subseteq \sigma_e(L_\xi^A)$, in order to prove (2.16), it is enough to show that $\mathbb{R} \cap \sigma_e(L_\xi^A) \subseteq \sigma_e(L_\xi)$. Let $\lambda \in \sigma_e(L_\xi^A)$, $\text{Im } \lambda = 0$. Then there exists an orthonormal sequence $y_k \in \mathrm{D}(L_\xi^A)$ such that, as $k \to \infty$,

$$\|L_\xi^A y_k - \lambda y_k\| \to 0.$$

From (2.17) we get $(\text{Im } Ay_k^\wedge(\xi), y_k^\wedge(\xi)) \to 0$. Now by (2.13), (2.14), we have $y_k^\wedge(\xi) \to 0$ and $y_k^\vee(\xi) \to 0$, as $k \to \infty$.

Let $0 \le \phi_j \in C_0^\infty$, $j = 0, 1, \dots, r-1$ be smooth compactly supported scalar-valued functions such that

$$\phi_j^{(k)}(\xi) = \delta_{jk}, \quad k = 0, 1, \dots, r-1, \tag{2.18}$$

and

$$\phi_j(x) = 0 \quad \text{for } x \le \xi - \alpha, \ \alpha > 0.$$

Set

$$\gamma_{k,\xi}(x) = \sum_{j=0}^{r-1} y_k^{(j)}(\xi)\phi_j(x). \tag{2.19}$$

Then $\gamma_{k,\xi}^\wedge(\xi) = y_k^\wedge(\xi)$, $\gamma_{k,\xi}^\vee(\xi) = y_k^\vee(\xi)$ and $z_k(x) := y_k(x) - \gamma_{k,\xi}(x) \in \mathrm{D}(L_\xi)$. As $k \to \infty$, we have

$$\|y_k - z_k\| = \|\gamma_{k,\xi}\| \le c\,(|y_k^\vee(\xi)| + |y_k^\wedge(\xi)|) \to 0.$$

Besides, $\langle z_k, z_j \rangle \to 0$ for $j \ne k$, as $j, k \to \infty$, i.e. the sequence $\{z_k\}$ is almost orthonormal in $H(a, \xi)$ and, hence, non-compact. Since

$$\|l_W[y_k] - l_W[z_k]\| = \|l_W[\gamma_{k,\xi}]\| \le c\,(|y_k^\wedge(\xi)| + |y_k^\vee(\xi)|) \to 0,$$

it follows that $\|L_\xi z_k - \lambda z_k\| \to 0$ as $k \to \infty$. Hence $\lambda \in \sigma_e(L_\xi)$, and (2.16) is proven.

Let us show that for $a > -\infty$ the essential spectrum of L_ξ is empty. Assume the opposite. Then for some λ, $\mathrm{Im}\,\lambda = 0$, there exists an orthonormal sequence $y_k \in \mathrm{D}(L_\xi)$ such that $\|L_\xi y_k - \lambda y_k\| \to 0$ as $k \to \infty$. Set $L_\xi y_k - \lambda y_k = f_k$. The vector function $y_k(x, \lambda)$ is a solution of the problem

$$l_W[y] - \lambda y = f_k(x), \quad y^\wedge(\xi) = y^\vee(\xi) = 0, \tag{2.20}$$

where zero Cauchy data $y^\wedge(\xi)$ and $y^\vee(\xi)$ at the point ξ correspond to the minimality condition for the operator L_ξ with respect to the point ξ.

Let $C(x, t; \lambda)$ be the Cauchy function[1] corresponding to the expression $l_W - \lambda$. Then for all $x \in (a, \xi]$ we have

$$y_k(x) = \int_\xi^x C(x, t; \lambda) f_k(t)\, dt, \tag{2.21}$$

and since $\|f_k\| \to 0$ as $k \to \infty$, then the following holds uniformly in x:

$$|y_k^\wedge(x)| + |y_k^\vee(x)| \to 0 \quad \forall x \in (a, \xi].$$

[1]For the definition of the Cauchy function and its main properties, see the end of Bibliographical Comments to Chapter 2.

Now if $a > -\infty$, this contradicts the equality $\|y_k\|_{(a,\xi)} = 1$. Thus, in this case, $\sigma_e(L_\xi) = \emptyset$. In the case of a semi-bounded second-order operator, this assertion follows from Theorem 1.3.

Now suppose that $a = -\infty$. We will show that $\lambda \in \sigma_e(L_{\xi_1})$ for all $\xi_1 < \xi$. This implies that the essential spectrum of L_ξ does not depend on ξ since $\sigma_e(L_{\xi_1}) \subseteq \sigma_e(L_\xi)$.

Setting $z_k(x) = y_k(x) - \gamma_{k,\xi_1}(x)$, where $\gamma_{k,\xi_1}(x)$ is defined as in (2.18), (2.19) with $\xi = \xi_1$, we get $z_k \in \mathrm{D}(L_{\xi_1})$ and

$$\|z_k - y_k\|_{(-\infty,\xi_1)} \leq c \sum_{k=0}^{r-1} \left| y_k^{(j)}(\xi_1) \right| \to 0$$

as $k \to \infty$. Since

$$\|y_k\|^2_{(-\infty,\xi_1)} = \|y_k\|^2_{(-\infty,\xi)} - \|y_k\|^2_{(\xi_1,\xi)} \to 1,$$

it follows that $\|z_k\|^2_{(-\infty,\xi_1)} \to 1$. Besides, $\langle z_k, z_j \rangle_{(-\infty,\xi_1)} \to 0$, $j \neq k$, $j, k \to \infty$, i.e. the sequence $\{z_k\}$ is almost orthonormal in $H(-\infty, \xi_1)$. Since as $k \to \infty$,

$$\|L_{\xi_1} z_k - \lambda z_k\|_{(-\infty,\xi_1)} \leq \|L_\xi y_k - \lambda y_k\|_{(-\infty,\xi)} + C \sum_{k=0}^{r-1} \left| y_k^{(j)}(\xi_1) \right| \to 0,$$

it follows that $\lambda \in \sigma_e(L_{\xi_1})$, and the Lemma is proven. □

Lemma 2.2 *Let $\|\cdot\|_{G_r}$ be the graph norm given by*

$$\|y\|^2_{G_r} := \|y\|^2 + \|g[y]\|^2 = \|y\|^2_{(\alpha,\beta)} + \|f\|^2_{(\alpha,\beta)},$$

where $y(\cdot) \in H(\alpha,\beta)$, $-\infty < \alpha < \beta < \infty$ are solutions of the non-homogeneous linear differential equation of order r

$$g[y] := y^{(r)} + g_1(x)y^{(r-1)} + \cdots + g_r(x)y = f(x) \in H(\alpha,\beta)$$

with arbitrary operator-valued coefficients $g_k(x)$ depending continuously on $x \in [\alpha,\beta]$ and with the Cauchy data at the point $\xi \in [\alpha,\beta]$

$$\tilde{y}(\xi) = col\,\{y(\xi), y'(\xi), \ldots, y^{(r-1)}(\xi)\} \in H^r. \tag{2.22}$$

Then there exists a one-one bicontinuous correspondence (i.e. a homeomorphism) between the solutions y (considered in the graph norm $\|\cdot\|_{G_r}$) and the Cauchy data (2.22).

Proof. For homogeneous equations with operator-valued coefficients, the proof is given in the Appendix (see Lemma A.10). For non-homogeneous equations with scalar-valued coefficients, the proof is given in [58, Lemma XIII.2.16]. In our case, the proof of the Lemma is similar. □

Definition 2.1 A solution $Y(x,\lambda) \in B(\mathbf{H}, H)$ of the equation (2.1), where $\mathbf{H}$ is any Hilbert space[2], is said to be a fundamental solution of the problem (2.1), (2.3) if the following conditions are satisfied:

1) for all $h \in \mathbf{H}$ and $\xi \in (a,b)$, we have $y = Y(x,\lambda)h \in H(a,\xi)$, and y satisfies the boundary condition (2.3);

2) every solution $y \in H(a,\xi)$ of the problem (2.1), (2.3) can be written as

$$y(x,\lambda) = Y(x,\lambda)h; \tag{2.23}$$

3) for some (and, hence, for every) x, the self-adjoint operator

$$M_Y(x,\lambda) := \sum_{k=0}^{r-1} Y^{(k)^*}(x,\lambda) \cdot Y^{(k)}(x,\lambda) >> 0, \tag{2.24}$$

i.e. $M_Y(x,\lambda)$ is positive and has a bounded inverse in $\mathbf{H}$.

The conditions 2), 3) of Definition 2.1 mean that the system of solutions of the problem (2.1), (2.3) determined by the fundamental solution $Y(x,\lambda)$ is complete and linearly independent.

If $a > -\infty$ then the fundamental solution $Y(x,\lambda)$ of (2.1), (2.6) can be constructed as a solution of (2.1) with the operator-valued Cauchy data

$$Y^\wedge(a,\lambda) = \cos A \cdot K, \quad Y^\vee(a,\lambda) = \sin A \cdot K,$$

where, according to (2.8)–(2.11), the operators $Y^\wedge(a,\lambda)$ and $Y^\vee(a,\lambda)$ are defined by the formulas

$$Y^\wedge h = (Yh)^\wedge, \quad Y^\vee h = (Yh)^\vee, \quad \forall h \in \mathbf{H}, \tag{2.25}$$

and $K \in B(\mathbf{H}, \widetilde{\mathbf{H}})$, $K^{-1} \in B(\mathbf{H}, \widetilde{\mathbf{H}})$.

Since the initial conditions do not depend on λ, it follows that $Y(x,\lambda)$ is an entire function of λ.

If $\mathbf{H} = \widetilde{\mathbf{H}}$, where $\widetilde{\mathbf{H}}$ is as in (2.15), then we can take $K = I_{\mathbf{H}}$ (here $I_{\mathbf{H}}$ is the identity operator in $\mathbf{H}$).

If $a = -\infty$ or a is a finite singular point, then the construction of the fundamental solution of (2.1), (2.3) cannot be reduced to Cauchy problem.

[2]It is often convenient to set $\mathbf{H} = \widetilde{\mathbf{H}}$; see (2.15).

However, in several cases, the fundamental solution can be constructed explicitly; see [541], [543], [634], [783] and [808].

In the following theorem, we will prove the existence of the fundamental solution for the problem (2.1), (2.3) for $a \geq -\infty$.

Theorem 2.1 *Assume that* $\dim \mathbf{H} = \frac{r}{2} \dim H$ *and* $\lambda \in \mathbb{C} \backslash \sigma_e(L_\xi)$[3].

Then there exists a fundamental solution $Y(x,\lambda) \in B(\mathbf{H}, H)$ *of the problem (2.1), (2.3). The fundamental solution* $Y(x,\lambda)$ *can be constructed as an analytic function in* λ *in a neighborhood* Λ_ρ *of the set* $\Lambda := \mathbb{R} \backslash \sigma_e(L_\xi)$ *in the complex plane.*

Proof. For $A = iI_{\widetilde{\mathbf{H}}}$, the operator L_ξ^A will be denoted by L_ξ^i. By Lemma 2.1 the set Λ, and, hence, some neighborhood Λ_ρ of Λ in the complex plane, belongs to the resolvent set of L_ξ^i. This neighborhood can be chosen to be symmetric with respect to the real axis. We define the operator $T_\lambda \colon H(a,\xi) \to \widetilde{\mathbf{H}}$ by the formula

$$T_\lambda f = (R_\lambda f)^\wedge(\xi), \quad \lambda \in \Lambda_\rho,$$

where R_λ is the resolvent of L_ξ^i. The operator T_λ is bounded.

Indeed, if $f_k \to 0$ in $H(a,\xi)$, then $R_\lambda f_k = g_k \to 0$ in $H(a,\xi)$ and $L_\xi^i g_k = \lambda g_k + f_k \to 0$ in $H(a,\xi)$, as $k \to \infty$. Thus by Lemma 2.2 it follows that

$$g_k^{(j)}(\xi) \to 0 \text{ in } H, \quad j = 0, 1, \ldots, r-1,$$

i.e. $g_k^\wedge(\xi) \to 0$, $g_k^\vee(\xi) \to 0$ and $T_\lambda f_k = g_k^\wedge(\xi) \to 0$ in the norm of $\widetilde{\mathbf{H}}$. Therefore, the operator T_λ is closed, and since it is defined on the whole $H(a,\xi)$, then by the Closed Graph Theorem it follows that T_λ is bounded. Therefore the functional

$$\phi_\lambda(f) = ((R_\lambda f)^\wedge(\xi), h), \quad \lambda \in \Lambda_\rho,\ h \in \widetilde{\mathbf{H}}$$

is bounded in $H(a,\xi)$ (cf. [58, pp. 638–640]).

By Riesz Representation Theorem it follows that

$$((R_\lambda f)^\wedge(\xi), h) = \int_a^\xi (W(x) f(x), g_h(x, \bar{\lambda}))\, dx, \tag{2.26}$$

where $g_h(\cdot, \bar{\lambda}) \in H(a,\xi)$ is uniquely defined. This enables us to define an operator $G_\lambda = G(\cdot, \lambda) \colon \widetilde{\mathbf{H}} \to H(a,\xi)$, where $\lambda \in \Lambda_\rho$, by the formula

[3] By Lemma 2.1, $\sigma_e(L_\xi)$ does not depend on ξ, and for $a > -\infty$, $\sigma_e(L_\xi)$ is empty.

$G(x,\lambda)h = g_h(x,\lambda)$. By (2.26) it follows that

$$((R_\lambda f)^\wedge(\xi), h) = \int_a^\xi (W(x)f(x), G(x,\bar{\lambda})h)\,dx, \tag{2.27}$$

Since the resolvent R_λ is analytic in $\lambda \in \Lambda_\rho$, then $G(x,\lambda)$ is weakly analytic with respect to λ, and, hence, (see [87]) $G(x,\lambda)$ is analytic in the uniform operator topology for $\lambda \in \Lambda_\rho$. If $f \in \mathrm{D}(L_\xi)$, then for all $h \in H$ and $\lambda \in \Lambda_\rho$, from (2.27) we have

$$\begin{aligned}&\langle (L_\xi - \lambda I)f, G_{\bar{\lambda}}h\rangle = \langle (L_\xi^i - \lambda I)f, G_{\bar{\lambda}}h\rangle \\ &= ((R_\lambda(L_\xi^i - \lambda I)f)^\wedge(\xi), h) = (f^\wedge(\xi), h) = 0.\end{aligned}$$

Therefore, $G_{\bar{\lambda}}h \in \mathrm{D}((L_\xi)^*)$ and $((L_\xi)^* - \bar{\lambda}I)G(x,\bar{\lambda})h = 0$ for all $\lambda \in \Lambda_\rho$. Therefore, for all $\lambda \in \Lambda_\rho$ and $h \in \widetilde{\mathbf{H}}$, the vector-valued function $G(\cdot,\lambda)h \in H(a,\xi)$ is a solution of (2.1) on the interval (a,ξ), satisfies (2.3) at the point a and analytically depends on λ.

We will show that the operators

$$G^\wedge(x,\lambda),\ G^\vee(x,\lambda)\colon \widetilde{\mathbf{H}} \to \widetilde{\mathbf{H}} \tag{2.28}$$

are bounded for all $x \le \xi$, $\lambda \in \Lambda_\rho$.

Since $G^\wedge(x,\lambda)$ and $G^\vee(x,\lambda)$ are defined on all $\widetilde{\mathbf{H}}$, by the Closed Graph Theorem it suffices to show that they are closed. Let

$$h_k \to 0 \text{ and } G(\cdot,\bar{\lambda})h_k \to g(\cdot,\bar{\lambda}) \text{ in } H(a,\xi), \quad \text{as } k \to \infty.$$

Then for all $f \in H(a,\xi)$ we have

$$0 = \lim_{k\to\infty} ((R_\lambda f)^\wedge(\xi), h_k) = \lim_{k\to\infty} \langle f, G_{\bar{\lambda}}h_k\rangle = \langle f, g\rangle,$$

and, hence, $g(\cdot,\bar{\lambda}) = 0$. Since the vector-valued function $G(x,\bar{\lambda})h$ is a solution of (2.1), it follows that

$$l_W[G_{\bar{\lambda}}h_k] \to 0 \text{ in } H(a,\xi).$$

Thus by Lemma 2.2, for all $x \le \xi$, $\lambda \in \Lambda_\rho$, we have

$$G^\wedge(x,\bar{\lambda})h_k \to 0 \text{ and } G^\vee(x,\bar{\lambda})h_k \to 0 \text{ in } \widetilde{\mathbf{H}}.$$

Thus the operators (2.28) are closed and bounded.

The operator solution $Y_\xi(x,\lambda) \in B(\widetilde{\mathbf{H}}, H)$ of the equation (2.1) on the interval (a,b) with the Cauchy data

$$Y_\xi^\wedge(\xi,\lambda) = G^\wedge(\xi,\lambda), \quad Y_\xi^\vee(\xi,\lambda) = G^\vee(\xi,\lambda) \tag{2.29}$$

satisfies the condition 1) in Definition 2.1.

Let us check the condition 2). Let $y(\cdot,\lambda) \in H(a,\xi)$ be a solution of (2.1) satisfying the boundary condition (2.3). Let $\lambda \in \Lambda_\rho$ and set

$$g(x,\bar{\lambda}) := R_{\bar{\lambda}} f, \quad f \in H(a,\xi).$$

Using (2.12) and noting that $g \in \mathrm{D}(L_\xi^i)$ satisfies the self-adjoint boundary condition (2.3) at the point $a \geq -\infty$ and that $l[g] - \bar{\lambda} W g = Wf$, we have

$$\int_a^\xi (W(x)f(x), y(x,\lambda))\,dx = \int_a^\xi (l[g] - \bar{\lambda} W g, y)\,dx$$
$$= (g^\wedge(\xi,\bar{\lambda}), y^\vee(\xi,\lambda)) - (g^\vee(\xi,\bar{\lambda}), y^\wedge(\xi,\lambda))$$
$$+ \int_a^\xi (g, l[y] - \lambda W y)\,dx = (g^\wedge(\xi,\bar{\lambda}), y^\vee(\xi,\lambda))$$
$$-(g^\vee(\xi,\bar{\lambda}), y^\wedge(\xi,\lambda)) = ((R_{\bar{\lambda}} f)^\wedge(\xi), y^\vee(\xi,\lambda))$$
$$-((R_{\bar{\lambda}} f)^\vee(\xi), y^\wedge(\xi,\lambda)).$$

Since $R_{\bar{\lambda}} f \in \mathrm{D}(L_\xi^i)$, it follows that $(R_{\bar{\lambda}} f)^\vee(\xi) = i(R_{\bar{\lambda}} f)^\wedge(\xi)$ and

$$\int_a^\xi (W(x)f(x), y(x,\lambda))\,dx = ((R_{\bar{\lambda}} f)^\wedge(\xi), y^\vee(\xi,\lambda) + iy^\wedge(\xi,\lambda)). \tag{2.30}$$

By (2.27), with $h = y^\vee(\xi,\lambda) + iy^\wedge(\xi,\lambda)$, we have

$$\int_a^\xi (W(x)f(x), y(x,\lambda))\,dx = \int_a^\xi (W(x)f(x), Y_\xi(x,\lambda)h)\,dx,$$

and (2.23) follows since $f \in H(a,\xi)$ is arbitrary.

We will now check the condition 3) of Definition 2.1. If $M_{Y_\xi}(x_0,\lambda)h = 0$ for some $x_0 \in (a,b)$ and $h \in \widetilde{\mathbf{H}}$, then $y(x,\lambda) := Y_\xi(x,\lambda)h$ satisfies (2.1), $y^{(k)}(x_0,\lambda) = 0$, $k = 0,1,\ldots,r-1$, and, hence, $y(x,\lambda) = Y_\xi(x,\lambda)h \equiv 0$. Then by (2.27) we have

$$0 = \langle f, Y_\xi h\rangle = ((R_{\bar{\lambda}} f)^\wedge(\xi), h)$$

and, hence, $(g^\wedge(\xi), h) = 0$ for all $g = R_{\bar{\lambda}} f \in \mathrm{D}(L_\xi^i)$. Since $g^\wedge(\xi)$ ranges over all of $\widetilde{\mathbf{H}}$, it follows that $h = 0$. Thus $0 \notin \sigma_d(M_{Y_\xi})$. If there exists a sequence $h_k \in \widetilde{\mathbf{H}}$ such that

$$M_{Y_\xi}(x_0,\lambda)h_k \to 0,$$

then $y_k^\wedge(x_0) \to 0$, $y_k^\vee(x_0) \to 0$, where

$$y_k(x) = Y_\xi(x,\lambda)h_k \in H(a,\xi).$$

But there exists a homeomorphism between the Cauchy data $y^\wedge \oplus y^\vee \in \widetilde{\mathbf{H}}^2$ at any of the end-points of $[x_0, \xi]$ (or $[\xi, x_0]$) and the corresponding solutions of the equation (2.1) on this interval. Hence, there exists a homeomorphism between $y_k^\wedge(x_0) \oplus y_k^\vee(x_0) \in \widetilde{\mathbf{H}}^2$ and $y_k^\wedge(\xi) \oplus y_k^\vee(\xi) \in \widetilde{\mathbf{H}}^2$ (see Lemma 2.2). Therefore, $y_k^\wedge(\xi) \to 0$ and $y_k^\vee(\xi) \to 0$ as $k \to \infty$. Setting

$$M_\xi(x,\lambda) = Y_\xi^{\vee^*}(x,\bar{\lambda}) \cdot Y_\xi^\vee(x,\lambda) + Y_\xi^{\wedge^*}(x,\bar{\lambda}) \cdot Y_\xi^\wedge(x,\lambda), \tag{2.31}$$

where $\lambda \in \Lambda_\rho$, we have $M_\xi(\xi,\lambda)h_k \to 0$. We will show that $M_\xi(\xi,\lambda) = I_{\widetilde{\mathbf{H}}}$. This implies that $h_k \to 0$, and, hence, $0 \notin \sigma_e(M_{Y_\xi}(x_0,\lambda))$ for any $x_0 \in (a,b)$, $\lambda \in \Lambda_\rho$. This shows that the condition 3) holds and that $Y_\xi(x,\lambda)$ is a fundamental solution of (2.1), (2.3).

Take

$$y(x,\lambda) = Y_\xi(x,\lambda)h, \tag{2.32}$$

where $\lambda \in \Lambda_\rho$.

Then (2.30) holds. On the other hand, by (2.27) and (2.29) we have

$$\begin{aligned}\int_a^\xi (W(x)f(x), y(x,\lambda))\,dx &= \int_a^\xi (W(x)f(x), Y_\xi(x,\lambda)h)\,dx \\ &= ((R_{\bar{\lambda}}f)^\wedge(\xi), h).\end{aligned} \tag{2.33}$$

Comparing (2.30) and (2.33) for $\lambda \in \Lambda_\rho$, we have

$$((R_\lambda f)^\wedge(\xi), h) = ((R_\lambda f)^\wedge(\xi), y^\vee(\xi,\bar{\lambda}) + iy^\wedge(\xi,\bar{\lambda})). \tag{2.34}$$

Since $(R_\lambda f)^\wedge(\xi)$ ranges over all of $\widetilde{\mathbf{H}}$, it follows that $y^\vee(\xi,\lambda) + iy^\wedge(\xi,\lambda) = h$. Hence from (2.32), for $x = \xi$ and for all $h \in \widetilde{\mathbf{H}}$, $\lambda \in \Lambda_\rho$, we get

$$[Y_\xi^\vee(x,\lambda) + iY_\xi^\wedge(x,\lambda)]h = h,$$

i.e.

$$Y_\xi^\vee(x,\lambda) + iY_\xi^\wedge(x,\lambda) = I_{\widetilde{\mathbf{H}}}. \tag{2.35}$$

For vector-valued functions $y(x,\lambda) = Y_\xi(x,\lambda)h$ and $z(x,\bar{\lambda}) = Y_\xi(x,\bar{\lambda})g$, where h, $g \in \widetilde{\mathbf{H}}$ and $\lambda \in \Lambda_\rho$, we have for all $x \in (a,b)$:

$$\begin{aligned}0 &= \int_a^x [(l[y], z) - (y, l[z])]\,dx \\ &= (y^\wedge(x,\lambda), z^\vee(x,\bar{\lambda})) - (y^\vee(x,\lambda), z^\wedge(x,\bar{\lambda}))\end{aligned}$$

$$= (Y_\xi^\wedge(x,\lambda)h, Y_\xi^\vee(x,\bar\lambda)g) - (Y_\xi^\vee(x,\lambda)h, Y_\xi^\wedge(x,\bar\lambda)g)$$
$$= ((Y_\xi^{\vee^*}(x,\bar\lambda)\cdot Y_\xi^\wedge(x,\lambda) - Y_\xi^{\wedge^*}(x,\bar\lambda)\cdot Y_\xi^\vee(x,\lambda))h, g).$$

Since $h,\ g \in \widetilde{\mathbf{H}}$ are arbitrary, then for all $x \in (a,b)$ and $\lambda \in \Lambda_\rho$ we have

$$W_{x,\lambda}\{Y_\xi^*, Y_\xi\} = 0,$$

where

$$W_{x,\lambda}\{Y^*, Z\} := Y^{\vee^*}(x,\bar\lambda)Z^\wedge(x,\lambda) - Y^{\wedge^*}(x,\bar\lambda)Z^\vee(x,\lambda), \tag{2.36}$$

with $Y(x,\lambda),\ Z(x,\lambda) \in B(\mathbf{H}, H)$.

Thus by (2.35) we have

$$M_\xi(\xi,\lambda) = (Y_\xi^{\vee^*}(\xi,\bar\lambda) - iY_\xi^{\wedge^*}(\xi,\bar\lambda))(Y_\xi^\vee(\xi,\lambda) + iY_\xi^\wedge(\xi,\lambda)) = I_{\widetilde{\mathbf{H}}}, \tag{2.37}$$

and Theorem 2.1 is proven. □

Corollary 2.1 *The fundamental solution $Y(x,\lambda)$ is regular in λ at the points belonging to the discrete spectrum $\sigma_d(\widetilde{L_\xi})$ of any self-adjoint extension $\widetilde{L_\xi}$ of L_ξ and at the points of $\sigma_e(\widetilde{L_\xi})\backslash\sigma_e(L_\xi)$.*

2.3 Self-Consistency of the Fundamental Solution of the Self-Adjoint Problem and the Evolution of the Corresponding Hermitian Relation (or Lagrangian Plane in $H \oplus H$)

Definition 2.2 A solution $Y(x,\lambda) \in B(\mathbf{H}, H)$ of the equation (2.1) for $\lambda \in \mathbb{R}$ is said to be self-consistent if for some $\xi \in (a,b)$ there exists a self-adjoint boundary condition

$$U_{\xi,\lambda}[y] := \cos A_{\xi,\lambda} y^\vee(\xi) - \sin A_{\xi,\lambda} y^\wedge(\xi) = 0 \tag{2.38}$$

which is satisfied for $x = \xi$ by all functions of the form $y(x,\lambda) = Y(x,\lambda)h$ (here $A_{\xi,\lambda}$ is a self-adjoint operator in $\widetilde{\mathbf{H}}$).

If (2.38) holds for some $\xi \in (a,b)$, then it holds for all $\xi \in (a,b)$ since (2.38) is equivalent to a Hermitian relation (see Appendix, Definition A.1) and since the following holds for solutions of (2.1) for $\lambda \in \mathbb{R}$ due to Lagrange's identity:

$$\frac{d}{dx}[(y^\vee, z^\wedge) - (y^\wedge, z^\vee)] = 0.$$

Proposition 2.1 *If $Y(x,\lambda)$ is a fundamental solution of the problem (2.1), (2.3), then (2.38) is equivalent to the following boundary condition for vector-valued functions:*

$$Y^{\wedge^*}(\xi,\lambda) \cdot y^\vee(\xi) - Y^{\vee^*}(\xi,\lambda) \cdot y^\wedge(\xi) = 0. \tag{2.39}$$

The condition (2.39) is generated at every point $\xi \in (a,b)$ by the operator-valued coefficients $Y^{\wedge^*}(\xi,\lambda)$, $Y^{\vee^*}(\xi,\lambda)$ continuously depending on ξ. Therefore, the Hermitian relation (2.38) evolves continuously with respect to ξ. In other words, the Lagrangian subspace of $\dim H$ (or the complex "Lagrangian plane") representing the graph in $H \oplus H$ of the Hermitian relation (2.38) evolves continuously (in the sense of the aperture between subspaces[4]).

The proof follows immediately from Lemma 2.3 below, Theorem 2.2 below and Theorem A.2 from the Appendix.

Remark 2.1 It can be shown that Definition 2.2 of the self-consistent (sometimes called adjoint or self-adjoint) operator solution is equivalent to the definitions used in [78] in the case of a canonical system ($\dim H = 2m < \infty$) and in the case of a scalar equation of even order (for the latter case, see also [70]).

Lemma 2.3 *Let $Y(x,\lambda) \in B(\widetilde{\mathbf{H}}, H)$ be a fundamental solution of the problem (2.1), (2.3) for some $\lambda \in \Lambda = \mathbb{R}\backslash\sigma_e(L_\xi)$. Then for all $x \in (a,b)$, the operators*

$$V_\pm(x,\lambda) = Y^\vee(x,\lambda) \pm iY^\wedge(x,\lambda) \colon \mathbf{H} \to \widetilde{\mathbf{H}} \tag{2.40}$$

have bounded inverses defined on all of $\widetilde{\mathbf{H}}$, and the operator

$$\theta(x,\lambda) = V_+(x,\lambda) \cdot V_-^{-1}(x,\lambda) \tag{2.41}$$

is unitary in $\widetilde{\mathbf{H}}$. If $Y(x,\lambda)$ is defined on some segment $\Delta \subset \Lambda$, then $\theta(x,\lambda)$ admits analytic continuation in λ from Δ onto $\Lambda_\rho \supset \Lambda$.

[4]See [6, Sec. 39].

Proof. We will first prove the Lemma for the fundamental solution $Y_\xi(x,\lambda)$ constructed in Theorem 2.1. Let

$$V_\pm^\xi(x,\lambda) = Y_\xi^\vee(x,\lambda) \pm iY_\xi^\wedge(x,\lambda) \colon \widetilde{\mathbf{H}} \to \widetilde{\mathbf{H}}.$$

Let $V_-^\xi(x_0,\lambda)h = 0$ for some $x_0 \in (a,b)$, $\lambda \in \Lambda$, $h \in \widetilde{\mathbf{H}}$. Then the solution $y(x,\lambda) = Y_\xi(x,\lambda)h$ of (2.1) satisfies the boundary condition (2.3) and

$$y^\vee(x_0,\lambda) - iy^\wedge(x_0,\lambda) = 0, \tag{2.42}$$

i.e. $y(x,\lambda)$ is an eigenfunction of the problem (2.1), (2.3), (2.42). This contradicts Lemma 2.1, and, hence, for all $x \in (a,b)$, $\lambda \in \Lambda$, we have nul $V_-^\xi(x,\lambda) = 0$. Analogously, we obtain nul $V_+^\xi(x,\lambda) = 0$.

By (2.36) and (2.31) the following holds for $Y_\xi(x,\lambda)$, Im $\lambda = 0$:

$$(V_-^\xi)^*(x,\lambda) \cdot V_-^\xi(x,\lambda) = (V_+^\xi)^*(x,\lambda) \cdot V_+^\xi(x,\lambda) = M_\xi(x,\lambda). \tag{2.43}$$

As in the case of $M_Y(x,\lambda)$, for all $x \in (a,b)$ and $\lambda \in \Lambda$, we have

$$M_\xi(x,\lambda) >> 0. \tag{2.44}$$

Thus the operators $V_\pm^\xi(x,\lambda)$ do not have the essential spectrum in a neighborhood of zero.

We will prove that nul $(V_-^\xi)^*(t,\lambda) = 0$ for all $t \in (a,b)$.

To this end, we will construct the fundamental solution $Y_{1,t}(x,\lambda)$ of the problem (2.1), (2.3) with the help of the resolvent of the operator (2.2), (2.3) and $y^\vee(t) + iy^\wedge(t) = 0$. Unlike in (2.35), we have

$$Y_{1,t}^\vee(t,\lambda) - iY_{1,t}^\wedge(t,\lambda) = I_{\widetilde{\mathbf{H}}}. \tag{2.45}$$

The system of solutions to (2.1), (2.3) defined by the operator $Y_{1,t}(x,\lambda)$ is complete.

Let $h \in \widetilde{\mathbf{H}}$ be arbitrary, and consider the function

$$y(x,\lambda) = Y_\xi(x,\lambda)h. \tag{2.46}$$

Then there exists $h_1 \in \widetilde{\mathbf{H}}$ such that

$$y(x,\lambda) = Y_{1,t}(x,\lambda)h_1. \tag{2.47}$$

Conversely, for all $h_1 \in \widetilde{\mathbf{H}}$, if $y(x)$ is defined by the formula (2.47), there exists $h \in \widetilde{\mathbf{H}}$ satisfying (2.46).

Thus we have

$$Y_\xi(x,\lambda)h = Y_{1,t}(x,\lambda)h_1, \tag{2.48}$$

and any of the vectors h, h_1 can be chosen arbitrarily (then the other vector is uniquely determined from the chosen one).

From (2.48) and (2.35) we have

$$h = (Y_{1,t}^{\vee}(\xi, \lambda) + iY_{1,t}^{\wedge}(\xi, \lambda))h_1,$$

and from (2.48) and (2.45), we similarly get

$$h_1 = (Y_\xi^{\vee}(t, \lambda) - iY_\xi^{\wedge}(t, \lambda))h.$$

Since h_1 ranges over all of $\widetilde{\mathbf{H}}$, we have nul $(V_-^\xi)^*(t, \lambda) = 0$.

The equality nul $(V_+^\xi)^*(t, \lambda) = 0$ is proven similarly starting from

$$Y_\xi(x, \lambda)h = Y_{1,t}(x, \lambda)h_1,$$

where h and h_1 range over all of $\widetilde{\mathbf{H}}$, and using (2.35).

Now let $Y(x, \lambda)$ be an arbitrary fundamental solution of the problem (2.1), (2.3). Then for all $x \in (a, b)$, similarly as in (2.48), we have

$$Y_\xi(x, \lambda)h = Y(x, \lambda)g,$$

where the vectors h, g range over all of $\widetilde{\mathbf{H}}$ and $\mathbf{H}$ respectively, and one is uniquely determined from the other. Let $K_\xi(\lambda) \colon \widetilde{\mathbf{H}} \to \mathbf{H}$ be the operator defined by the formula $K_\xi(\lambda)h = g$. Then for all $x \in (a, b)$ we have

$$Y_\xi(x, \lambda) = Y(x, \lambda)K_\xi(\lambda). \tag{2.49}$$

We will show that the operator $K_\xi(\lambda)$ has a bounded inverse on all of $\mathbf{H}$. Indeed, if for some $h \in \widetilde{\mathbf{H}}$, we have $K_\xi(\lambda)h = 0$, then $Y_\xi(x, \lambda)h \equiv 0$, and, hence, $M_{Y_\xi}(x, \lambda)h \equiv 0$, i.e. $h = 0$ by (2.37). If $K_\xi(\lambda)h_j \to 0$, $h_j \in \widetilde{\mathbf{H}}$, then

$$(M_{Y_\xi}(x, \lambda)h_j, h_j) = (K_\xi^*(\lambda)M_Y(x, \lambda)K_\xi(\lambda)h_j, h_j) \to 0.$$

As $j \to \infty$, by (2.37) we have $h_j \to 0$. Therefore, $K_\xi(\lambda)$ has a bounded inverse. Since we can interchange $Y(x, \lambda)$ and $Y_\xi(x, \lambda)$, it follows that the inverse is defined on all $\mathbf{H}$.

From what we have proven for $V_\pm^\xi$, using (2.49) it follows that $V_\pm(x, \lambda)$ in (2.40) have bounded inverses in all of $\widetilde{\mathbf{H}}$.

By (2.41) and (2.43) it follows that the operator

$$\theta(x, \lambda) = V_+(x, \lambda) \cdot V_-^{-1}(x, \lambda) = V_+^\xi(x, \lambda) \cdot (V_-^\xi)^{-1}(x, \lambda),$$

satisfies the following equality for all $x \in (a,b)$, $\lambda \in \Lambda$:

$$\begin{aligned}\theta^*(x,\lambda) &= (V_-^{\xi^*})^{-1}(x,\lambda) \cdot V_+^{\xi^*}(x,\lambda) \\ &= V_-^{\xi}(x,\lambda) \cdot (V_+^{\xi})^{-1}(x,\lambda) = \theta^{-1}(x,\lambda).\end{aligned}$$

Thus the operator $\theta(x,\lambda)$ is unitary, and it admits analytic continuation in λ from $\Delta \subset \Lambda$. The Lemma is proven. □

Theorem 2.2 *For all $\lambda \in \mathbb{R}\backslash\sigma_e(L_\xi)$, every fundamental solution $Y(x,\lambda)$ of (1.1), (1.3) is self-consistent. Besides,*

$$nul\, Y^\wedge(x,\lambda) = nul\, Y^{\wedge^*}(x,\lambda), \tag{2.50}$$

and for all $\lambda \in \mathbb{R}\backslash\sigma_e(L_x^0)$, the operator $Y^\wedge(x,\lambda)$ is Fredholm with zero index. (Here L_x^0 denotes the extension of the operator L_x in $H(a,x)$ defined by the condition $y^\wedge(x) = 0$, i.e. the Friedrichs extension for a semi-bounded operator L_x).

Proof. Given $\xi \in (a,b)$ and $\lambda \in \mathbb{R}\backslash\sigma_e(L_\xi)$, define a binary relation θ in $\widetilde{\mathbf{H}}$ by the following condition:

$$u\theta u' \Leftrightarrow \{\exists h \in \widetilde{\mathbf{H}} \colon u = Y^\wedge(\xi,\lambda)h,\ u' = Y^\vee(\xi,\lambda)h\}. \tag{2.51}$$

We will show that the relation θ is Hermitian (see Definition A.1).

We need to check the following properties:

1) if $u\theta u'$ and $v\theta v'$, then

$$(u',v) - (u,v') = 0. \tag{2.52}$$

2) if there exist $v,\ v' \in \widetilde{\mathbf{H}}$ such that (2.52) holds for all $u,\ u' \in \widetilde{\mathbf{H}}$ satisfying $u\theta u'$, then $v\theta v'$.

Property 1) immediately follows from (2.51), (2.49) and (2.36).

We now check property 2). For all $h \in \mathbf{H}$, the following holds:

$$\begin{aligned}0 = (u'v) - (u,v') &= (Y^\vee h, v) - (Y^\wedge h, v') \\ &= (h, Y^{\vee^*}(\xi,\lambda)v - Y^{\wedge^*}(\xi,\lambda)v').\end{aligned}$$

Thus we have

$$Y^{\vee^*}(\xi,\lambda)v - Y^{\wedge^*}(\xi,\lambda)v' = 0. \tag{2.53}$$

Setting

$$v' + iv = V_+(\xi,\lambda)h = (Y^\vee(\xi,\lambda) + iY^\wedge(\xi,\lambda))h, \tag{2.54}$$

and using (2.40) and (2.36), we obtain

$$\begin{aligned}0 &= Y^{\vee^*}(\xi,\lambda)v - Y^{\wedge^*}(\xi,\lambda)v' \\ &= Y^{\vee^*}(\xi,\lambda)v - Y^{\wedge^*}(\xi,\lambda)\cdot[V_+(\xi,\lambda)h - iv] \\ &= V_-^*(\xi,\lambda)v - Y^{\wedge^*}(\xi,\lambda)V_+(\xi,\lambda)h,\end{aligned}$$

i.e.

$$\begin{aligned}V_-^*(\xi,\lambda)v &= Y^{\wedge^*}(\xi,\lambda)\cdot Y^{\vee}(\xi,\lambda)h + iY^{\wedge^*}(\xi,\lambda)\cdot Y^{\wedge}(\xi,\lambda)h \\ &= Y^{\vee^*}(\xi,\lambda)\cdot Y^{\wedge}(\xi,\lambda)h + iY^{\wedge^*}(\xi,\lambda)\cdot Y^{\wedge}(\xi,\lambda)h \\ &= V_-^*(\xi,\lambda)\cdot Y^{\wedge}(\xi,\lambda)h,\end{aligned}$$

and this shows that $v = Y^{\wedge}(\xi,\lambda)h$. By (2.54), $v' = Y^{\vee}(\xi,\lambda)h$. Thus $v\theta v'$ and θ is Hermitian. By Theorem A.1 on the general form of Hermitian relations, there exists a self-adjoint operator $A_{\xi,\lambda}$ in $\widetilde{\mathbf{H}}$ such that all solutions $y(x,\lambda) = Y(x,\lambda)h$, where $h \in \widetilde{\mathbf{H}}$, satisfy the self-adjoint boundary condition (2.38). The operator $A_{\xi,\lambda}$, under the condition

$$-\frac{\pi}{2}I_{\widetilde{\mathbf{H}}} < A_{\xi,\lambda} \leq \frac{\pi}{2}I_{\widetilde{\mathbf{H}}},$$

is uniquely defined by the relation θ in (2.51), and the self-consistency of $Y(x,\lambda)$ is proven.

Since

$$Y^{\wedge}(x,\lambda) = \frac{1}{2i}(\theta(x,\lambda) - I)\cdot V_-(x,\lambda), \tag{2.55}$$

and since $\theta(x,\lambda)$ is unitary and $V_-(x,\lambda)$ and $V_-^{-1}(x,\lambda)$ are bounded, we obtain (2.50).

Using Lemma 2.4 below with $B = \frac{\pi}{2}I_{\widetilde{\mathbf{H}}}$, it follows that for all $\lambda \in \mathbb{R}\backslash\sigma_e(L_x^0)$, the operator $Y^{\wedge}(x,\lambda)$ is Fredholm. It remains to prove the following Lemma:

Lemma 2.4 *Let $\widetilde{L}$ be the self-adjoint extension of L_ξ determined by the boundary condition*

$$\cos B\cdot y^{\vee}(\xi) + \sin B\cdot y^{\wedge}(\xi) = 0, \quad B = B^*,$$

and let $Y(x,\lambda) \in B(\widetilde{\mathbf{H}}, H)$ be the fundamental solution of the problem (2.1), (2.3). Then the operator

$$\Omega(\lambda) = \cos B\cdot Y^{\vee}(\xi,\lambda) + \sin B\cdot Y^{\wedge}(\xi,\lambda)$$

does not have a bounded inverse on all of $\widetilde{\mathbf{H}}$ *for* $\lambda \notin \sigma_e(L_\xi)$ *if and only if* $\lambda \in \sigma(\widetilde{L})$. *Let* σ_p *denote the point spectrum. The following also hold:* $0 \in \sigma_p(\Omega(\lambda)) \Leftrightarrow \lambda \in \sigma_p(\widetilde{L})$; $0 \in \sigma_e(\Omega(\lambda)) \Leftrightarrow \lambda \in \sigma_e(\widetilde{L})$; $nul\, \Omega(\lambda) = nul\, \Omega^*(\lambda) = nul\, (\widetilde{L} - \lambda I)$.

Proof. (Necessity) If there exists $0 \neq h \in \widetilde{\mathbf{H}}$ such that $\Omega(\lambda)h = 0$, then λ is an eigenvalue of $\widetilde{L}$ and $y(x,\lambda) = Y(x,\lambda)h$ is the corresponding eigenfunction. If $0 \in \sigma_e(\Omega(\lambda))$, then there exists a non-compact sequence $\{h_k\}$, $|h_k| = 1$ such that $\Omega(\lambda)h_k \to 0$ as $k \to \infty$. Vector-valued functions $y_k(x) = Y(x,\lambda)h_k$ satisfy the boundary condition at the point a, but generally do not satisfy the boundary condition at ξ. Set $f_k = \Omega(\lambda)h_k$ and construct a compactly supported vector-valued function $\gamma_k(x)$ such that $\gamma_k^\wedge(\xi) = \sin B \cdot f_k$ and $\gamma_k^\vee(\xi) = \cos B \cdot f_k$ as follows:

$$\gamma_k(x) = \sum_{j=0}^{r-1} \phi_j(x)\{0;\ldots;0;I;0;\ldots;0\}(\widetilde{P})^{-1}(\xi)\begin{pmatrix}\sin B\\ \cos B\end{pmatrix} f_k,$$

where I is in the $j+1$-st place, $j = 0,1,\ldots,r-1$; $\phi_j(x)$ are defined according to (2.18), and $\widetilde{P}(\xi)$ is chosen so that for all smooth vector-valued functions $\gamma(x) \in H$ we have

$$\widetilde{P}(\xi)\tilde{\gamma}(\xi) = \gamma^\Delta(\xi) := \begin{pmatrix}\gamma^\wedge(\xi)\\ \gamma^\vee(\xi)\end{pmatrix}, \tag{2.56}$$

where $\tilde{\gamma}(x) = \mathrm{col}\,\{\gamma(x), \gamma'(x), \ldots, \gamma^{(r-1)}(x)\}$.

For (2.56) to hold in the case $r = 2n$, we need to use (2.8)–(2.11) and set

$$\widetilde{P}(x) = \begin{pmatrix} I_n & 0_n \\ * & P_{22}\end{pmatrix}, \quad P_{22}(x) = \begin{pmatrix} & * & & (-1)^{n-1}p_n(x) \\ & & \cdot^{\cdot^{\cdot}} & \\ & -p_n(x) & & \\ p_n(x) & & & 0 \end{pmatrix}$$

(all elements of the matrix P_{22} under the diagonal are equal to 0; I_n and 0_n respectively denote the identity and the zero operator in H^n).

In the case $r = 2n-1$, under the condition (2.5), we set

$$\widetilde{P}(x) = \left(\begin{array}{c|c|c} & V^+(x)q_+^{1/2}(x) + V^-(x)q_-^{1/2}(x) & \\ \hline I_{n-1} & & \\ \hline & -i[V^+(x)q_+^{1/2}(x) - V^-(x)q_-^{1/2}(x)] & \\ \hline * & * & Q(x) \end{array}\right)$$

$$Q(x) = \begin{pmatrix} * & & & (-1)^{(n-2)} i\mathrm{Re}\ q_n(x) \\ & & \cdot^{\cdot^{\cdot}} & \\ & -i\mathrm{Re}\ q_n(x) & & \\ i\mathrm{Re}\ q_n(x) & & & \mathbf{0} \end{pmatrix}$$

(all unwritten elements of $\widetilde{P}(x)$ are equal to zero, and all elements of $Q(x)$ under the diagonal are equal to zero).

For $r = 2n - 1$ and $\dim H = \infty$, the operator $\widetilde{P}(x)$ has an analogous structure and has a bounded inverse on the whole H^r for all x. Then we get

$$\tilde{\gamma}_k(\xi) = \widetilde{P}^{-1}(\xi) \begin{pmatrix} \sin B \\ \cos B \end{pmatrix} f_k, \tag{2.57}$$

and comparing (2.56) and (2.57), we obtain

$$\gamma_k{}^{\wedge}(\xi) = \sin B \cdot f_k, \quad \gamma_k{}^{\vee}(\xi) = \cos B \cdot f_k.$$

Thus the vector-valued function $z_k(x) = y_k(x) - \gamma_k(x)$ satisfies the following boundary condition at ξ:

$$\begin{aligned} &\cos B \cdot z_k{}^{\vee}(\xi) + \sin B \cdot z_k{}^{\wedge}(\xi) \\ &= [\cos B \cdot y_k{}^{\vee}(\xi) + \sin B \cdot y_k{}^{\wedge}(\xi)] - [\cos B \cdot \gamma_k{}^{\vee}(\xi) + \sin B \cdot \gamma_k{}^{\wedge}(\xi)] \\ &= \Omega(\lambda) h_k - f_k = 0. \end{aligned}$$

Furthermore, as $k \to \infty$, we have

$$\begin{aligned} &\|z_k - y_k\| = \|\gamma_k\| \le c \cdot |f_k| \to 0 \\ &\|\widetilde{L} z_k - \lambda z_k\| \le c \cdot |f_k| \to 0. \end{aligned} \tag{2.58}$$

As $k \to \infty$, we have $\|y_k\| \not\to 0$; otherwise, since the initial data continuously depend on the solution of the equation (see Lemma 2.2), we would have $y_k^{(j)}(x) \to 0$, $j = 0, 1, \ldots, r-1$ for all $x \in (a, \xi)$, and this contradicts (2.24). Therefore, $\|z_k\| \not\to 0$ as $k \to \infty$, and, hence, $\lambda \in \sigma(\widetilde{L})$. We will now show that $\lambda \in \sigma_e(\widetilde{L})$. Since $Y(\cdot, \lambda)\colon H \to H(a, \xi)$ is a closed operator defined on all of H, by the Closed Graph Theorem it follows that $Y(\cdot, \lambda)$ is bounded, i.e. for all $\lambda \in \mathbb{R} \backslash \sigma_e(L_\xi)$ we have

$$\|Y(\cdot, \lambda) h\|_{(a,\xi)} \le C(\lambda) \cdot |h|_H. \tag{2.59}$$

Therefore, the sequences $\{y_k\}$ and $\{z_k\}$ are bounded since $|h_k| = 1$. We will now show that these sequences are non-compact. This and (2.58) shows that $\lambda \in \sigma_e(\widetilde{L})$.

Since $\{h_k\}$ is a non-compact sequence, by the Hausdorff compactness criterion, there exists $\epsilon > 0$ such that $\{h_k\}$ contains a subsequence with the following property: the distance between any two elements of this subsequence is at least ϵ. We will denote the chosen subsequence also by $\{h_k\}$, so $\{h_k\}$ has no convergent subsequence. If the sequence $\{y_k(\cdot,\lambda)\}$ happened to be compact, then it would contain a convergent subsequence. Suppose that the whole sequence $\{y_k\}$ converges. By Lemma 2 from [70, pp. 40–41], from any bounded non-compact sequence $\{h_k\}$, one can choose a subsequence $\{h_{n_k}\}$ such that the sequence of differences $\{h_{n_{k+1}} - h_{n_k}\}$ is also non-compact. But sequence $\{y_{n_{k+1}} - y_{n_k}\}$ converges to 0, and this is impossible due to non-compactness of $\{h_{n_{k+1}} - h_{n_k}\}$ and the property (2.24) of the fundamental solution. Non-compactness of $\{z_k\}$ follows from non-compactness of $\{y_k\}$.

Finally, if the range of $\Omega(\lambda)$ is not dense in $\widetilde{\mathbf{H}}$, then the corresponding λ is again an eigenvalue of $\widetilde{L}$. Indeed, similarly as in (2.55), we have

$$Y^{\vee}(x,\lambda) = \frac{1}{2}(\theta(x,\lambda) + I)\cdot V_-(x,\lambda);$$

hence,

$$\Omega(\lambda) = \frac{1}{2}[e^{-iB}\theta(x,\lambda) + e^{iB}]V_-(\xi,\lambda).$$

Therefore,

$$\begin{aligned}
\Omega^*(\lambda) &= \frac{1}{2}V_-^*(\xi,\lambda)[\theta^*(\xi,\lambda)e^{iB} + e^{-iB}] \\
&= \frac{1}{2}V_-^*(\xi,\lambda)\theta^*(\xi,\lambda)e^{iB}[e^{iB} + e^{-iB}\theta]e^{-iB} \\
&= \frac{1}{2}V_-^*(\xi,\lambda)\theta^*(\xi,\lambda)e^{iB}[e^{iB}V_-(\xi,\lambda) + e^{-iB}V_+(\xi,\lambda)]V_-^{-1}(\xi,\lambda)e^{-iB} \\
&= V_-^*(\xi,\lambda)\theta^*(\xi,\lambda)e^{iB}[\cos B\cdot Y^{\vee}(\xi,\lambda) + \sin B\cdot Y^{\wedge}(\xi,\lambda)] \\
&\cdot V_-^{-1}(\xi,\lambda)e^{-iB} = V_-^*(\xi,\lambda)\theta^*(\xi,\lambda)e^{iB}\Omega(\lambda)\cdot V_-^{-1}(\xi,\lambda)e^{-iB}.
\end{aligned}$$

Now if $\Omega^*(\lambda)h = 0$, then $\Omega(\lambda)\cdot V_-^{-1}(\xi,\lambda)e^{-iB}h = 0$, and, thus, the vector-valued function $y(x,\lambda) = Y(x,\lambda)\cdot V_-^{-1}(\xi,\lambda)e^{-iB}h$ is an eigenfunction of $\widetilde{L}$.

(Sufficiency) If λ is an eigenvalue of $\widetilde{L}$, then 0 is an eigenvalue of $\Omega(\lambda)$. Let $\lambda \in \sigma_e(\widetilde{L})\backslash\sigma_e(L_\xi)$. Then there exists a non-compact normalized sequence $z_k \in \mathrm{D}(\widetilde{L})$ such that $\widetilde{L}z_k - \lambda z_k := f_k \to 0$ as $k \to \infty$. For arbitrary $h \in \mathbf{H}$, the vector-valued function $y(x,\lambda) = Y(x,\lambda)h$ is a solution of (2.1) and satisfies the boundary condition at the point a. On one hand, by (2.59) we have

$$|\langle L_\xi^* z_k, y\rangle - \langle z_k, L_\xi^* y\rangle| = |\langle \lambda z_k + f_k, y\rangle - \langle z_k, \lambda y\rangle|$$
$$= |\langle f_k, y\rangle| \le C(\lambda)|h| \cdot \|f_k\|,$$

and, on the other hand,

$$\langle L_\xi^* z_k, y\rangle - \langle z_k, L_\xi^* y\rangle = (z_k^\wedge(\xi), y^\vee(\xi,\lambda)) - (z_k^\vee(\xi), y^\wedge(\xi,\lambda))$$
$$= (z_k^\wedge(\xi), Y^\vee(\xi,\lambda)h) - (z_k^\vee(\xi), Y^\wedge(\xi,\lambda)h)$$
$$= (Y^{\vee^*}(\xi,\lambda)z_k^\wedge(\xi) - Y^{\wedge^*}(\xi,\lambda)z_k^\vee(\xi), h).$$

Since h is arbitrary, we have, as $k \to \infty$,

$$\mathbf{H} \ni g_k := Y^{\vee^*}(\xi,\lambda)z_k^\wedge(\xi) - Y^{\wedge^*}(\xi,\lambda)z_k^\vee(\xi) \to 0$$

because $|g_k| \le C(\lambda) \cdot \|f_k\| \to 0$.

By Lemma 2.3 there exists $e_k \in \mathbf{H}$ such that

$$z_k^\vee(\xi) + iz_k^\wedge(\xi) = (Y^\vee(\xi,\lambda) + iY^\wedge(\xi,\lambda))e_k. \tag{2.60}$$

Proceeding similarly as when we used (2.54) to obtain a representation for v and v', from (2.60) we have

$$z_k^\wedge(\xi) = Y^\wedge(\xi,\lambda)e_k + V_-^{*-1}(\xi,\lambda)g_k,$$
$$z_k^\vee(\xi) = Y^\vee(\xi,\lambda)e_k - V_-^{*-1}(\xi,\lambda)g_k. \tag{2.61}$$

Note that as $k \to \infty$, we have $e_k \not\to 0$; otherwise, we have $z_k^\wedge(\xi) \to 0$, $z_k^\vee(\xi) \to 0$ and $\lambda \in \sigma_e(L_\xi)$ (see the proof of Lemma 2.1), and this contradicts the hypothesis.

Since $z_k(x)$ satisfies the boundary condition at ξ, we have, as $k \to \infty$,

$$\Omega(\lambda)e_k = \cos B(Y^\vee(\xi,\lambda)e_k - z_k^\vee(\xi)) + \sin B(Y^\wedge(\xi,\lambda)e_k - z_k^\wedge(\xi))$$
$$= i(\cos B + i\sin B)V_-^{*-1}g_k \to 0,$$

and, hence, $0 \in \sigma(\Omega(\lambda))$.

We will show that the sequence $\{e_k\}$ is bounded and non-compact, and, hence, $0 \in \sigma_e(\Omega(\lambda))$. Boundedness of $\{e_k\}$ follows from boundedness of the left hand side of (2.60), which in turn follows from continuous dependence of the Cauchy data at $x = \xi$ on the solutions (considered in the graph norm) $z_k(t)$ of a linear differential equation ($z_k(t)$ are normalized); see

Lemma 2.2. Non-compactness of $\{e_k\}$ follows from non-compactness of $\{z_k(\cdot)\}$ and Lemma 2 from [70, pp. 40–41] which enables us to chose a subsequence $\{z_{n_k}(\cdot)\}$ with the following property: $\{u_k(\cdot) := z_{n_{k+1}}(\cdot) - z_{n_k}(\cdot)\}$ is non-compact and, at the same time, the sequence $e_{n_{k+1}} - e_{n_k} \to 0$ if $\{e_k\}$ is compact. But then from (2.61) we would have $u_k^{\wedge}(\xi) \to 0$ and $u_k^{\vee}(\xi) \to 0$, and, hence, $\lambda \in \sigma_e(L_\xi)$, contradicting the hypothesis. This concludes the proof of Lemma 2.4 and Theorem 2.2 □

2.4 A Different Construction for the Fundamental Solution of the Boundary-Value Problem

The idea of the proof of Theorem 2.1 for a scalar second-order equation is contained in [58, pp. 638–640]. In this section, we will use another method to construct the fundamental solution $Y(x, \lambda)$ of (2.1), (2.3) which turns out to be analytic in λ in a (generally different) neighborhood $\Lambda_\rho^{'} \subset \mathbb{C}$ of the set $\Lambda = \mathbb{R} \backslash \sigma(L_\xi)$. This allows the analytic continuation of both fundamental solutions to the union of neighborhoods Λ_ρ and $\Lambda_\rho^{'}$.

By R_λ^0 we will denote the resolvent of the self-adjoint operator L_ξ^0 representing the extension in $H(a,\xi)$ of the symmetric operator L_ξ with the boundary condition $y^{\wedge}(\xi) = 0$. For all $\lambda \in \mathbb{C}\backslash\sigma(L_\xi^0)$, the resolvent R_λ^0 analytically depends on λ. We define an operator $T_\lambda^0 \colon H(a,\xi) \to \widetilde{\mathbf{H}}$ by the formula

$$T_\lambda^0 f = (R_\lambda^0 f)^{\vee}(\xi), \quad \lambda \in \mathbb{C}\backslash\sigma(L_\xi^0).$$

As in Sec. 2.2, the operator T_λ^0 is bounded, and the following functional is also bounded:

$$\phi_\lambda^0(f) := ((R_\lambda^0 f)^{\vee}(\xi), h), \quad \lambda \in \mathbb{C}\backslash\sigma(L_\xi^0),\ h \in \widetilde{\mathbf{H}}.$$

Hence by the Riesz Representation Theorem, we have

$$((R_\lambda^0 f)^{\vee}(\xi), h) = \langle f(\cdot), g_h(\cdot, \bar{\lambda})\rangle_{(a,\xi)}, \tag{2.62}$$

where $g_h(\cdot, \bar{\lambda}) \in H(a,\xi)$ is uniquely defined. This allows us to define the operator $G_\lambda^0 = G^0(\cdot, \lambda) \colon \widetilde{\mathbf{H}} \to H(a,\xi)$, where $\lambda \in \mathbb{C}\backslash\sigma(L_\xi^0)$, by the formula $G_\lambda^0 h = g_h(x, \lambda)$. This operator analytically depends on λ in the uniform operator topology.

Similarly as in the proof of Theorem 2.1, we can show that the operator-valued solution $Y_\xi(x,\lambda) \in B(\widetilde{\mathbf{H}}, H)$ of (2.1) on the interval (a,b) with the

Cauchy data

$$Y_\xi^\wedge(\xi,\lambda) = (G^0)^\wedge(\xi,\lambda), \quad Y_\xi^\vee(\xi,\lambda) = (G^0)^\vee(\xi,\lambda)$$

satisfies the conditions 1)–3) of Definition 2.1, i.e. $Y_\xi(x,\lambda)$ is a fundamental solution of the problem (2.1), (2.3). In the proof, we use (cf. (2.35))

$$Y_\xi^\wedge(\xi,\lambda) = -I, \quad \forall\lambda \in \mathbb{C}\backslash\sigma(L_\xi^0).$$

Indeed, let $y(x,\lambda) = Y_\xi(\xi,\lambda)h$ for an arbitrary $h \in \widetilde{\mathbf{H}}$ with $\lambda \in \mathbb{C}\backslash\sigma(L_\xi^0)$. Then by (2.62), for all $f \in H(a,\xi)$, we have

$$\langle f(\cdot), y(\cdot,\lambda)\rangle_\xi = \langle f(\cdot), Y_\xi(\cdot,\lambda)h\rangle_\xi = ((R_{\bar\lambda}^0 f)^\vee(\xi), h). \tag{2.63}$$

For arbitrary $f \in H(a,\xi)$, set $R_{\bar\lambda}^0 f = g(x,\bar\lambda)$. Noting that $g \in \mathrm{D}(L_\xi^0)$ and y satisfy the boundary conditions (2.3) at the point a and that $l[g] - \bar\lambda W g = Wf$, we have

$$\begin{aligned}
\langle f(\cdot), y(\cdot,\lambda)\rangle_\xi &= \int_a^\xi (l[g] - \bar\lambda W g, y)\,dx \\
&= (g^\wedge(\xi,\bar\lambda), y^\vee(\xi,\lambda)) - (g^\vee(\xi,\bar\lambda), y^\wedge(\xi,\lambda)) + \int_a^\xi (g, l[y] - \lambda W y)\,dx \\
&= (g^\wedge(\xi,\bar\lambda), y^\vee(\xi,\lambda)) - (g^\vee(\xi,\bar\lambda), y^\wedge(\xi,\lambda)) \\
&= ((R_{\bar\lambda}^0 f)^\wedge(\xi,\lambda), y^\vee(\xi,\lambda)) - ((R_{\bar\lambda}^0 f)^\vee(\xi,\lambda), y^\wedge(\xi,\lambda)) \\
&= -((R_{\bar\lambda}^0 f)^\vee(\xi,\lambda), y^\wedge(\xi,\lambda)).
\end{aligned} \tag{2.64}$$

Comparing (2.62) and (2.64) we get

$$((R_{\bar\lambda}^0 f)^\vee(\xi), h) = -((R_{\bar\lambda}^0 f)^\vee(\xi,\lambda), y^\wedge(\xi,\lambda)).$$

Thus

$$y^\wedge(\xi,\lambda) = -h,\ Y_\xi^\wedge(\xi,\lambda)h = -h,\ Y_\xi^\wedge(\xi,\lambda) = -I,\ \forall\lambda \in \mathbb{C}\backslash\sigma(L_\xi^0).$$

The solution $Y_\xi(\xi,\lambda)$ is self-consistent,

$$\mathrm{nul}\ Y_\xi^\wedge(\xi,\lambda) = \mathrm{nul}\ Y_\xi^{\wedge^*}(\xi,\lambda),$$

and for all $\lambda \in \mathbb{R}\backslash(\sigma(L_\xi^0) \cup \sigma(L_x^0))$, the operator $Y_\xi^\wedge(\xi,\lambda)$ is Fredholm. Unlike the fundamental solution constructed in Sec. 2.2, which is analytic in λ in a neighborhood $\Lambda_\rho \subset \mathbb{C}$ of the set $\Lambda = \mathbb{R}\backslash\sigma(L_\xi)$, the above obtained solution is regular with respect to λ in the upper and the lower half plane and it has poles at the points λ_k^0 of the discrete spectrum of L_ξ^0.

The operator

$$V_-^\xi(\xi,\lambda) = Y_\xi^\vee(\xi,\lambda) - iY_\xi^\wedge(\xi,\lambda) = Y_\xi^\vee(\xi,\lambda) + iI$$

has a bounded inverse defined on all of $\widetilde{\mathbf{H}}$ for all $\lambda \in \mathbb{R}\backslash\sigma(L_\xi^0)$. The points $\lambda = \lambda_k^0$ are the poles of the analytic operator-valued function $V_-^\xi(\xi,\lambda)$. It can be shown that the fundamental solution

$$Y^0(x,\lambda) = Y_\xi(x,\lambda)\cdot(V_-^\xi)^{-1}(\xi,\lambda)$$

coincides for all $\lambda \in \mathbb{R}\backslash\sigma_e(L_\xi)$ with the fundamental solution $Y(x,\lambda)$ which is analytic in λ for all $\lambda \in \Lambda_\rho$ and was constructed in Sec. 2.2. This allows us to analytically continue in λ both of these fundamental solutions to $\mathbb{C}\backslash(\sigma(L_\xi)\cup\sigma(V_-^\xi(\xi,\lambda)))$ (recall that $\sigma(L_\xi)$ is independent of ξ).

Consider the following operator in $\widetilde{\mathbf{H}}$:

$$\Gamma(\lambda) = \int_a^\xi Y_\xi^*(x,\bar\lambda)W(x)Y_\xi(x,\lambda)\,dx,\ \lambda \in \mathbb{C}\backslash\sigma(L_\xi^0).$$

The operator $\Gamma(\lambda)$ analytically depends on λ. We will prove that for all $\lambda \in \mathbb{R}\backslash\sigma(L_\xi^0)$, the operator $\Gamma(\lambda)$ has a bounded inverse. Indeed, suppose that there exists $\lambda \in \mathbb{R}\backslash\sigma(L_\xi^0)$ and $h \in \widetilde{\mathbf{H}}$ such that $\Gamma(\lambda)h = 0$. Then from the formula

$$(\Gamma(\lambda)h,h) = \int_a^\xi (W(x)G^0(x,\lambda)h, G^0(x,\lambda)h)\,dx \tag{2.65}$$

we have $G^0(x,\lambda)h \equiv 0$, and, hence, $h = 0$.

If $\Gamma(\lambda)h_k \to 0$, where $k \to \infty$ and $|h_k| = 1$, then from (2.65) it follows that $\|G^0(\cdot,\lambda)h_k\| \to 0$ as $k \to \infty$, and this implies $(G^0)^\wedge(\xi,\lambda)h_k \to 0$, $(G^0)^\vee(\xi,\lambda)h_k \to 0$, which contradicts the equality $Y_\xi^\wedge(\xi,\lambda) = -I$.

Consider the Laurent's expansion of the resolvent R_λ^0 and the solution $Y_\xi(x,\lambda)$ in a neighborhood of an eigenvalue λ_k of the operator L_ξ^0:

$$R_\lambda^0 = \frac{P_{-1}}{\lambda-\lambda_k} + \sum_{m=0}^\infty P_m(\lambda-\lambda_k)^m, \tag{2.66}$$

$$Y_\xi(\xi,\lambda) = \frac{g_{-1}(x,\lambda_k)}{\lambda-\lambda_k} + \sum_{m=0}^\infty g_m(x,\lambda_k)(\lambda-\lambda_k)^m. \tag{2.67}$$

Lemma 2.5 *For all $m = 0, 1, 2, \ldots$, the following holds:*

$$\int_a^\xi g_{-1}^*(x, \lambda_k) W(x) g_m(x, \lambda_k)\, dx$$
$$= \int_a^\xi g_m^*(x, \lambda_k) W(x) g_{-1}(x, \lambda_k)\, dx = 0. \tag{2.68}$$

Proof. Substituting (2.66) and (2.67) into (2.62) and equating the coefficients with the same powers of $(\lambda - \lambda_k)$, we get

$$(P_m f)^\vee(\xi) = \int_a^\xi g_m^*(x, \lambda_k) W(x) f(x)\, dx, \tag{2.69}$$

where $m = -1, 0, 1, 2, \ldots$

It is known that $(-P_{-1})$ is the projection onto the eigenspace H_k corresponding to the eigenvalue λ_k. If

$$\phi_1(x), \phi_2(x), \ldots, \phi_{æ}(x) \tag{2.70}$$

is an orthonormal basis of H_k, then $(-P_{-1})$ is an integral operator in $H(a, \xi)$ with the kernel $K(x,t) = \sum_{j=1}^{æ} \phi_j(x) \phi_j^*(t)$:

$$(-P_{-1} f)(x) = \int_a^\xi K(x,t) W(t) f(t)\, dt.$$

Then from (2.69) we get (for $m = -1$)

$$g_{-1}(x, \lambda_k) = \sum_{j=1}^{æ} \phi_j(x) \phi_j^{\vee *}(\xi). \tag{2.71}$$

It is known that the following equations hold for the resolvent expansion of a closed symmetric operator (see, for example, [201]):

$$P_{-1} \cdot P_m = P_m \cdot P_{-1} = 0, \quad m = 0, 1, 2 \ldots$$

and

$$P_m = {P_0}^{m+1}, \quad m = 1, 2, \ldots, \quad {P_{-1}}^2 = -P_{-1}.$$

Therefore, for any function $f(x) \in H(a, \xi)$, we have

$$\int_a^\xi {g_{-1}}^*(x, \lambda_k) W(x) P_m f(x)\, dx = (P_{-1} P_m f)^\vee(\xi) = 0, \tag{2.72}$$

$$\int_a^\xi g_m{}^*(x,\lambda_k)W(x)P_{-1}f(x)\,dx = (P_m P_{-1}f)^\vee(\xi) = 0, \tag{2.73}$$

$$\int_a^\xi g_{-1}{}^*(x,\lambda_k)W(x)P_{-1}f(x)\,dx = (P_{-1}P_{-1}f)^\vee(\xi)$$
$$= -(P_{-1}f)^\vee(\xi) = -\int_a^\xi g_{-1}{}^*(x,\lambda_k)W(x)f(x)\,dx, \tag{2.74}$$

$$\int_a^\xi g_m{}^*(x,\lambda_k)W(x)P_0f(x)\,dx = (P_m P_0 f)^\vee(\xi) = (P_{m+1}f)^\vee(\xi)$$
$$= \int_a^\xi g_{m+1}{}^*(x,\lambda_k)W(x)f(x)\,dx. \tag{2.75}$$

From (2.75) we have

$$\int_a^\xi g_1{}^*(x,\lambda_k)W(x)f(x)\,dx = \int_a^\xi g_0{}^*(x,\lambda_k)W(x)P_0f(x)\,dx.$$

Take $f = P_0h$. Applying (2.75) twice, we get

$$\int_a^\xi g_2{}^*(x,\lambda_k)W(x)h(x)\,dx = \int_a^\xi g_1{}^*(x,\lambda_k)W(x)P_0h(x)\,dx$$
$$= \int_a^\xi g_0{}^*(x,\lambda_k)W(x)(P_0)^2h(x)\,dx.$$

Thus, for any function $f(x) \in H(a,\xi)$ we obtain

$$\int_a^\xi g_m{}^*(x,\lambda_k)W(x)f(x)\,dx = \int_a^\xi g_0{}^*(x,\lambda_k)W(x)(P_0)^m f(x)\,dx.$$

Consider the operator $T\colon H(a,\xi) \to \widetilde{\mathbf{H}}$ defined by the formula

$$Tf = \int_a^\xi g_{-1}{}^*(x,\lambda_k)W(x)f(x)\,dx.$$

Then for all $h \in \widetilde{\mathbf{H}}$, we have

$$(Tf,h) = \left(\int_a^\xi g_{-1}{}^*(x,\lambda_k)W(x)f(x)\,dx, h\right)$$

$$= \int_a^\xi (W(x)f(x), g_{-1}(x,\lambda_k)h)\,dx = \langle f, g_{-1}h\rangle.$$

We define the adjoint operator $T^*\colon \widetilde{\mathbf{H}} \to H(a,\xi)$ by the formula

$$T^*h = g_{-1}(x,\lambda_k)h. \tag{2.76}$$

We define the operator $T_1\colon H(a,\xi) \to \widetilde{\mathbf{H}}$ by the formula

$$T_1 f = \int_a^\xi g_{-1}{}^*(x,\lambda_k)W(x)P_{-1}f(x)\,dx.$$

As above, we get

$$T_1^*h = P_{-1}{}^*[g_{-1}{}^*(x,\lambda_k)h] = P_{-1}[g_{-1}(x,\lambda_k)h]. \tag{2.77}$$

By (2.74) we have $T_1 = -T$, and, hence, from (2.76) we get

$$P_{-1}[g_{-1}(x,\lambda_k)h] = -g_{-1}(x,\lambda_k)h.$$

Similarly, it can be shown that for all $h \in \widetilde{\mathbf{H}}$, the following holds:

$$g_m(x,\lambda_k)h = P_0{}^m(g_0(x,\lambda_k)h). \tag{2.78}$$

Using (2.77) and (2.78), for all $h \in \widetilde{\mathbf{H}}$ we have

$$\int_a^\xi g_{-1}{}^*(x,\lambda_k)W(x)g_m(x,\lambda_k)h\,dx$$
$$= \int_a^\xi g_m{}^*(x,\lambda_k)W(x)g_{-1}(x,\lambda_k)h\,dx = 0,$$

where $m = 0, 1, 2, \ldots$. This equation is equivalent to (2.68), and the Lemma is proven. □

By Lemma 2.5, for all $\lambda \in \mathbb{C}\backslash\sigma(L_\xi^0)$, we get

$$\Gamma(\lambda) = \frac{1}{(\lambda-\lambda_k)^2}[A + (\lambda-\lambda_k)^2 B(\lambda)],$$

where

$$A = \int_a^\xi g_{-1}{}^*(x,\lambda_k)W(x)g_{-1}(x,\lambda_k)\,dx,$$

$$B(\lambda) = \sum_{m,n=0}^{\infty} \int_a^\xi g_m{}^*(x,\lambda_k)W(x)g_n(x,\lambda_k)\,dx(\lambda-\lambda_k)^{m+n}. \tag{2.79}$$

Lemma 2.6 *1) Let $\phi_1(x), \ldots, \phi_{æ}(x)$ be an orthonormal basis of the eigenspace $H_k = (-P_{-1})H(a,\xi)$. Let $\mathbf{H}_k$ denote the subspace of $\widetilde{\mathbf{H}}$ spanned by $\phi_1^{\vee}(\xi), \ldots, \phi_{æ}^{\vee}(\xi)$, and let A_k denote the restriction of A to $\mathbf{H}_k$. Then the operator A_k has a bounded inverse.*

2) The operator $B(\lambda)$ in $\widetilde{\mathbf{H}}$ has a bounded inverse for all λ in some neighborhood of λ_k.

Proof. 1) By (2.71) we have

$$\begin{aligned} A &= \int_a^{\xi} g_{-1}{}^*(x,\lambda_k) W(x) g_{-1}(x,\lambda_k)\, dx \\ &= \sum_{i,j=1}^{æ} \phi_j^{\vee}(\xi) \int_a^{\xi} \phi_j{}^*(x,\lambda_k) W(x) \phi_i(x,\lambda_k)\, dx\, \phi_i^{\vee^*}(\xi) \\ &= \sum_{j=1}^{æ} \phi_j^{\vee}(\xi) \phi_j^{\vee^*}(\xi). \end{aligned}$$

Vectors $\phi_j^{\vee}(\xi) \in \widetilde{\mathbf{H}}$, $j = 1, 2, \ldots, æ$, and, hence, $A \neq 0$ on the subspace $\mathbf{H}_k \subset \widetilde{\mathbf{H}}$.

2) Suppose that for some $\lambda' \neq \lambda_k$, $\operatorname{Im} \lambda' = 0$ and $h \in \widetilde{\mathbf{H}}$, we have $B(\lambda')h = 0$. Since

$$(B(\lambda')h, h) = \int_a^{\xi} \left| W^{1/2}(x) G_0(x,\lambda',\lambda_k) h \right|^2 dx, \tag{2.80}$$

where

$$G_0(x,\lambda,\lambda_k) = \sum_{m=0}^{\infty} g_m(x,\lambda_k)(\lambda - \lambda_k)^m,$$

it follows that $G_0(x,\lambda',\lambda_k)h = 0$, and hence

$$G^0(x,\lambda')h = \frac{g_{-1}(x,\lambda_k)h}{\lambda' - \lambda_k}.$$

Since $g_{-1}(x,\lambda_k)h$ is a solution of $l[y] = \lambda W(x) y$ for $\lambda = \lambda_k$, and since $G^0(x,\lambda')h$ is a solution of the same equation with $\lambda = \lambda'$ and $\lambda' \neq \lambda_k$, it follows that $G^0(x,\lambda')h \equiv 0$. Hence $h = 0$. Let $\lambda' = \lambda_k$. If $B(\lambda_k)h = \int_a^{\xi} g_0{}^*(x,\lambda_k) W(x) g_0(x,\lambda_k) h\, dx = 0$, then $g_0(x,\lambda_k)h \equiv 0$, and by (2.78) we get $g_m(x,\lambda_k)h \equiv 0$. Therefore, $B(\lambda)h = 0$ for all λ from some neighborhood of λ_k. But this contradicts what we proved above for $\operatorname{Im} \lambda = 0$. Suppose that, as $j \to \infty$, $B(\lambda)h_j \to 0$, where $|h_j| = 1$. Then from (2.79) we have $\|G_0(\cdot,\lambda,\lambda_k)h_j\| \to 0$. Denoting $P_{-1}^{\perp}(\lambda) = \sum_{m=0}^{\infty} P_m(\lambda-\lambda_k)^m$, for all $f \in H(a,\xi)$ we have, as $j \to \infty$,

$$\begin{aligned}
\left|((P_{-1}^{\perp}f)^{\vee}(\xi),h_j)\right| &= \left|\sum_{m=0}^{\infty}((P_m f)^{\vee}(\xi),h_j)(\lambda-\lambda_k)^m\right| \\
&= \left|\sum_{m=0}^{\infty}\left(\int_a^{\xi} g_m{}^*(x,\lambda_k)W(x)f(x)\,dx, h_j\right)(\lambda-\lambda_k)^m\right| \\
&= \left|\left(\int_a^{\xi}\sum_{m=0}^{\infty} g_m{}^*(x,\lambda_k)(\lambda-\lambda_k)^m W(x)f(x)\,dx, h_j\right)\right| \qquad (2.81)\\
&= \left|\left(\int_a^{\xi} G_0{}^*(x,\lambda,\lambda_k)W(x)f(x)\,dx, h_j\right)\right| \\
&= \left|\int_a^{\xi}(W(x)f(x),G_0(x,\lambda,\lambda_k)h_j)\,dx\right| \le \|f(\cdot)\|\cdot\|G_0(\cdot,\lambda,\lambda_k)h_j\| \to 0.
\end{aligned}$$

For all $h_j \in \widetilde{\mathbf{H}}$, using Lemma 2.7 below, choose a sequence of functions $f_{j,m} \in H(a,\xi)$. Then by (2.81) with $f = f_{j,m}$, for all m, we get

$$\left|((P_{-1}^{\perp}f_{j,m})^{\vee}(\xi),h_j)\right| \le c|h_j|\cdot\|G_0(\cdot,\lambda,\lambda_k)h_j\| \to 0, \text{ as } j\to\infty. \qquad (2.82)$$

However, by Lemma 2.7, for sufficiently large m and for all j, we have

$$\begin{aligned}
\left|((P_{-1}^{\perp}f_{j,m})^{\vee}(\xi),h_j)\right| &= \left|((P_{-1}^{\perp}f_{j,m})^{\vee}(\xi)-h_j,h_j)+1\right| \\
&\ge 1-\left|(P_{-1}^{\perp}f_{j,m})^{\vee}(\xi)-h_j\right| > \frac{1}{2}. \qquad (2.83)
\end{aligned}$$

The conditions (2.82) and (2.83) contradict each other.

Lemma 2.7 *For any vector $h \in \widetilde{\mathbf{H}}$, there exists a sequence of functions $f_m \in H(a,\xi)$ such that for all m, the following hold: $\|f_m\| \le C\cdot|h|$ and $|(P_{-1}^{\perp}f_m)^{\vee}(\xi)-h| \to 0$ as $m\to\infty$, uniformly in h.*

Proof. Set $H_k^{\perp} = \mathrm{D}(L_\xi^0)\ominus H_k$. Since $P_{-1}^{\perp}P_{-1} = P_{-1}P_{-1}^{\perp} = 0$, and

$$R_\lambda^0 H(a,\xi) = \left(\frac{P_{-1}}{\lambda-\lambda_k}+P_{-1}^{\perp}\right)H(a,\xi),$$

it follows that $P_{-1}^{\perp}H(a,\xi) \subseteq H_k^{\perp}$. However, since for all $f \in H_k^{\perp}$,

$$P_{-1}L_\xi^0 f = \left(R_\lambda^0 - \frac{P_{-1}}{\lambda-\lambda_k}\right)L_\xi^0 f = f - \frac{1}{\lambda-\lambda_k}P_{-1}L_\xi^0 f = f,$$

it follows that $P_{-1}^{\perp} H(a,\xi) = H_k^{\perp}$.

We will adjust the eigenfunctions $\phi_j(x)$ at ξ so that

$$\phi_{j,m}^{\wedge}(\xi) = \phi_{j,m}^{\vee}(\xi) = 0, \quad \|\phi_j(\cdot) - \phi_{j,m}(\cdot)\| \to 0, \tag{2.84}$$

$\|L_\xi^0 \phi_j - L_\xi^0 \phi_{j,m}\| \to 0$, as $m \to \infty$. The functions $\phi_{j,m}$ can be chosen so that for $i \neq j$ we have

$$\langle \phi_j, \phi_{i,m} \rangle = 0. \tag{2.85}$$

(This is possible by Lemma 3 in Sec. 6 of [70]). Note that from (2.85) it follows that

$$\langle \phi_j, \phi_{j,m} \rangle \to 1, \quad \text{as } m \to \infty. \tag{2.86}$$

Using the vector $h \in \widetilde{\mathbf{H}}$, we construct the function $\psi(x) \in C_0^\infty$ such that $\psi^\wedge(\xi) = 0$, $\psi^\vee(\xi) = h$. Then $P_{-1}\psi(x) = \sum_{j=1}^{æ} c_j \cdot \phi_j(x)$, where $c_j = \langle \psi, \phi_j \rangle$, and the following inequalities hold:

$$|c_j| \leq \|\psi\| = \left(\int_a^\xi \left| W^{1/2}(x)\psi(x) \right|^2 dx \right)^{\frac{1}{2}} \leq C \cdot |h|. \tag{2.87}$$

Set $y_m(x) = \psi(x) - \sum_{j=1}^{æ} c_j \cdot \phi_{j,m}(x)$. Then $y_m \in \mathrm{D}(L_\xi^0)$, $y_m^\wedge(\xi) = 0$, $y_m^\vee(\xi) = h$ and $y_m(x) = y_m^\perp(x) + \sum_{j=1}^{æ} \langle y, \phi_j \rangle \phi_j(x)$, where $y_m^\perp(x)$ is the projection of $y_m(x)$ onto $\mathbf{H}_k^\perp$. By (2.85) and (2.86) it follows that

$$\langle y_m, \phi_j \rangle = \langle \psi - \sum_{i=1}^{æ} c_i \phi_{i,m}, \phi_j \rangle = \langle \psi, \phi_j \rangle - \sum_{i=1}^{æ} c_i \langle \phi_{i,m}, \phi_j \rangle = c_j \cdot \epsilon_{j,m},$$

where $\epsilon_{j,m} \to 0$ as $m \to \infty$.

Therefore,

$$y_m(x) = y_m^\perp(x) + \sum_{j=1}^{æ} \epsilon_{j,m} c_j \cdot \phi_j(x).$$

Since $y_m(x) \in \mathrm{D}(L_\xi^0)$, there exists a function $f_m(x) \in H(a, \xi)$ such that $R_\lambda^0 f_m = y_m$, and the following holds:

$$(P_{-1}^\perp f_m)^\wedge(\xi) = 0, \quad \left|(P_{-1}^\perp f_m)^\vee(\xi) - h\right| \leq \sum_{j=1}^{æ} |\epsilon_{j,m} c_j \cdot \phi_j^\vee(\xi)| \to 0,$$

as $m \to \infty$, uniformly in h. Besides, by (2.87), for all m we have

$$\|f_m\| - \|(L_\xi^0 - \lambda I)y_m\| \leq \|(L_\xi^0 - \lambda I)\psi\| + \sum_{j=1}^{\infty} |c_j| \cdot \|(L_\xi^0 - \lambda I)\phi_{j,m}\| \leq c \cdot |h|,$$

and Lemma 2.7 and Lemma 2.6 are proven. □

By Lemma 2.6 the operator A has æ positive eigenvalues. Thus, there exists a sufficiently small neighborhood Λ_k of λ_k such that for all $\lambda \in \Lambda_k$, the operator $\widetilde{\Gamma}(\lambda) = A + (\lambda - \lambda_k)^2 B(\lambda)$ has æ eigenvalues which are greater than some $\beta > 0$. Let $E(t, \lambda)$ be the resolution of identity corresponding to the operator $\widetilde{\Gamma}(\lambda)$. By $P_1(\lambda) = E(\beta, \lambda)$, $P_2(\lambda) = I - E(\beta, \lambda)$ we denote the projections onto the invariant subspaces $\mathbf{H}_1(\lambda) = P_1(\lambda)\widetilde{\mathbf{H}}$, $\mathbf{H}_2(\lambda) = P_2(\lambda)\widetilde{\mathbf{H}}$ ($\dim \mathbf{H}_2(\lambda) =$ æ for all $\lambda \in \Lambda_k$).

As shown in [201], the operators $P_1(\lambda)$ and $P_2(\lambda)$ analytically depend on $\lambda \in \Lambda_k$. Since the projection operators $P_1(\lambda)$ and $P_2(\lambda)$ commute with $\widetilde{\Gamma}(\lambda)$, we have

$$\widetilde{\Gamma}(\lambda) = P_1(\lambda)\widetilde{\Gamma}(\lambda)P_1(\lambda) + P_2(\lambda)\widetilde{\Gamma}(\lambda)P_2(\lambda).$$

By $\Omega(\lambda, \lambda_k)$ we denote the evolution operator (see [51]) which keeps track of the rotation of invariant subspaces $\mathbf{H}_1(\lambda)$, $\mathbf{H}_2(\lambda)$ of $\widetilde{\Gamma}(\lambda)$ as λ varies in the neighborhood Λ_k. The operator $\Omega(\lambda, \lambda_k)$ is unitary, differentiable in λ and satisfies the rotation property, i.e.

$$P_j(\lambda) = \Omega(\lambda, \lambda_k) P_j(\lambda_k) \Omega^*(\lambda, \lambda_k).$$

Therefore, $\widetilde{\Gamma}(\lambda) = \Omega(\lambda, \lambda_k)[\Gamma_1(\lambda) \oplus \Gamma_2(\lambda)]\Omega^*(\lambda, \lambda_k)$, where $\Gamma_j(\lambda)$, $j = 1, 2$ are self-adjoint operators in $\mathbf{H}_j(\lambda_k)$:

$$\Gamma_j(\lambda) = P_j(\lambda_k)\Omega^*(\lambda, \lambda_k)\widetilde{\Gamma}(\lambda)\Omega(\lambda, \lambda_k)P_j(\lambda_k).$$

Then

$$\Gamma^{1/2}(\lambda) = (1/(\lambda - \lambda_k))\Omega(\lambda, \lambda_k)[\Gamma_1^{1/2}(\lambda) \oplus \Gamma_2^{1/2}(\lambda)]\Omega^*(\lambda, \lambda_k).$$

Since

$$\Gamma_1^{1/2}(\lambda) = (\lambda - \lambda_k)P_1(\lambda_k)B(\lambda_k)P_1(\lambda_k) + (\lambda - \lambda_k)^2 \cdot \widetilde{\Gamma}_1(\lambda),$$

and

$$\Gamma_2^{1/2}(\lambda) = A^{1/2} + (\lambda - \lambda_k) \cdot \widetilde{\Gamma}_2(\lambda),$$

where $\Gamma_1(\lambda)$ and $\Gamma_2(\lambda)$ are differentiable operator-valued functions, then for all $\lambda \neq \lambda_k$, in some neighborhood of λ_k, there exits the inverse operator $\Gamma^{-1/2}(\lambda)$. Set

$$Y_\Gamma(x, \lambda) = Y_\xi(x, \lambda) \cdot \Gamma^{-1/2}(\lambda).$$

The operator-valued function $Y_\xi(x, \lambda)$ is analytic for all $\lambda \in \mathbb{C}\backslash\sigma(L_\xi^0)$ and has poles at the points $\lambda = \lambda_k$. Thus the operator-valued function $Y_\Gamma(x, \lambda)$ has either a pole or a removable singularity at $\lambda = \lambda_k$. If $\lambda = \lambda_k$ were a pole, then as $\lambda \to \lambda_k$ in any direction, the limit would be infinite. However, for all $\lambda \in \mathbb{R}\backslash\sigma(L_\xi^0)$ and $h \in \widetilde{\mathbf{H}}$, we have

$$\|Y_\Gamma(\cdot, \lambda)h\|^2 = \left(\left(\int_a^\xi {Y_\Gamma}^*(x, \lambda)W(x)Y_\Gamma(x, \lambda)\, dx\right) h, h\right) = |h|^2.$$

Thus the operator $Y_\Gamma(x, \lambda) \colon \widetilde{\mathbf{H}} \to H$ is uniformly bounded in λ. Moreover, the operator $Y_\Gamma(x, \lambda) \colon \widetilde{\mathbf{H}} \to H(a, \xi)$ is bounded. Therefore, the singularity at $\lambda = \lambda_k$ is removable.

Defining

$$Y_\Gamma(x, \lambda_k) = \lim_{\lambda \to \lambda_k} Y_\Gamma(x, \lambda),$$

we get a non-singular (in the sense of (2.24)) operator-valued solution $Y_\Gamma(x, \lambda)$ depending analytically on λ in some neighborhood of $\Lambda_\rho' \subset \mathbb{C}$ of the set $\mathbb{R}\backslash\sigma_e(L_\xi^0)$.

Theorem 2.3 *The problem (2.1), (2.3) has a fundamental solution $Y_\Gamma(x, \lambda) \in B(\widetilde{\mathbf{H}}, H)$ which is analytic with respect to λ in some neighborhood $\Lambda_\rho' \subset \mathbb{C}$ of the set $\mathbb{R}\backslash\sigma_e(L_\xi^0)$.*

(Note that here $\dim \widetilde{\mathbf{H}} = (r/2) \dim H$, by virtue of (2.8)–(2.10), and, by assumption, the right hand side of this equality must be either an integer or infinity.)

2.5 Bibliographical Comments

Section 2.1 is based on results from [772] and [773] (see the Appendix). The description of extensions of symmetric operators in terms of abstract boundary conditions is contained in [58], [318] and [556]. See also [73], [74]; the review [496] containing an extensive bibliography; and more recent works [360], [361], [646] and [647]. In particular, a description of the Friedrichs and Krein extensions in terms of abstract boundary conditions and abstract Weyl function is obtained in [360] and [361]. (See also [874], [875] and Sec. A.4.1 of the present book.)

In the case of a scalar second-order equation, a construction of the solution satisfying the boundary condition at the singular end-point is given in [58]. An explicit construction of the finite for $x = 0$ and entire with respect to λ solution to a scalar Schrödinger equation with highly singular repulsive potential at $x = 0$ is given in [634] and [808] (see also [305], [806] and [807]). In the case of an operator potential, an analogous construction is given in [783]. In the absolutely indeterminate case, a construction of the fundamental solution for an equation of arbitrary order with operator-valued coefficients on the real line is given in [541] and [543]. The existence of a fundamental solution with boundary condition at infinity for differential equations of arbitrary order with operator-valued coefficients was first proclaimed and then proven in [795] and [796]. These papers contain the proof of self-consistency of the corresponding fundamental solution. Another method for the construction of a fundamental solution for a differential-operator equation is given in [544]. The fundamental system of solutions of a differential-operator equation without any boundary conditions (this case is simpler than the one considered in the present book) was considered in [666]. In the reference [666], several properties of the above mentioned fundamental system of solutions, specific to the case $\dim H = \infty$, were investigated. Lemma 2.4 is contained in [796], and for a second-order problem on a bounded interval, this lemma was proven in [486] and [766].

Fundamental solutions of a broad class of first order systems with polynomial dependence on the spectral parameter and rational matrix potentials are considered in [830].

* * *

Here we recall the main properties of the Cauchy function.

The Cauchy function $C(x, t; \lambda)$ gives the solution of the Cauchy problem (2.20) for the equation with an arbitrary right hand side and zero initial data at the prescribed point ξ by the formula (2.21). The Cauchy function satisfies the following equation with respect to x (at every t):

$$(l_W - \lambda)[C(x,t;\lambda)] = 0,$$

and the following initial conditions at $x = t$:

$$\left(\frac{d^j}{dx^j}C(x,t;\lambda)\right)|_{x=t} = 0, \qquad j = 0,1,\ldots,r-2;$$

$$\left((W(x))^{-1}s_r(x)\left(\frac{d^{(r-1)}}{dx^{(r-1)}}C(x,t;\lambda)\right)\right)|_{x=t} = 1.$$

Here, $s_r(x)$ denotes the leading coefficient of the differential expression $l[y]$ (2.1). For example, if $r = 2n$, then $s_r(x) = (-1)^n p_n(x)$.

Chapter 3

Dependence of the Spectrum of Operator Boundary Problems on Variations of a Finite or Semi-Infinite Interval

3.1 Dependence of Eigenvalues and the Greatest Lower Bound of the Spectrum for a Semi-Bounded Below Differential Operator on Variations of the Interval

Throughout this chapter we assume that the minimal with respect to the end-point $b \leq \infty$ operator L_b in $H(a, b)$ corresponding to the differential expression $l_W[y]$ in (2.2), (2.1) and the self-adjoint boundary condition (2.3), where $a \geq -\infty$, is semi-bounded below. Note that semi-bounded below differential operators have even order, and their leading coefficient has constant sign.

By $\widetilde{L}$ we denote the self-adjoint extension of L_b corresponding to the boundary condition (2.4) (for $b < \infty$ this condition can be represented as in (2.7)).

In the case of a bounded interval (a, b), if the operator $p_n(x) >> 0$ for all $x \in [a, b]$, and if in the boundary conditions (2.6), (2.7),

$$-\frac{\pi}{2}I << A, B \leq \frac{\pi}{2}I \tag{3.1}$$

(i.e. the point $-\frac{\pi}{2}$ belongs to the resolvent set of operators A, B; for $\dim H < \infty$ this restriction is removed), then the operator L is semi-bounded below.[1]

In the case $\dim H < \infty$, the problem (2.1), (2.6), (2.7) on a bounded

[1] For second-order operators on a bounded interval, the lower estimate is given in the proof of part 3) of the Theorem 1.3; see also [766, pp. 337–338]. In the general case, see [491].

interval and the corresponding operator $\widetilde{L}$ have purely discrete spectrum

$$\lambda_1 \le \lambda_2 \le \lambda_3 \le \cdots$$

However, if $\dim H = \infty$, $a > -\infty$, $b < \infty$, the spectrum is not purely discrete (see [766, p. 340] and [73, Chapter III, Theorem 4.4]). Let $\lambda_e = \inf \sigma_e(\widetilde{L})$ be the greatest lower bound of the essential spectrum of $\widetilde{L}$ (eigenvalues of infinite multiplicity are included in the essential spectrum).

By L_ξ^0 we denote the operator in $H(a,\xi)$ corresponding to the problem (2.1), (2.3) and the condition $y^\wedge(\xi) = 0$.

Theorem 3.1 *The eigenvalues $\lambda_j(\xi)$, $j = 1, 2, \ldots$ of L_ξ^0, where*

$$\lambda_1(\xi) \le \lambda_2(\xi) \le \lambda_3(\xi) \le \cdots < \sigma_e(L_\xi^0),$$

continuously and strictly monotonically decrease with respect to ξ (possibly splitting or merging while conserving the total multiplicity), and as $\xi \nearrow b \le \infty$,

$$\lambda_j(\xi) \searrow \lambda_j(L_b^F)\,(< \inf \sigma_e(L_b^F)), \tag{3.2}$$

where L_b^F denotes the Friedrichs extension of L_b (if $\xi < \infty$, $L_\xi^0 = L_\xi^F$).

Proof. By variational principles it follows that the eigenvalues $\lambda_j(\xi)$ ($< \lambda_e(\xi)$) do not increase with respect to $\xi \in (a,b)$. We will show their strict monotonicity. Assume the opposite, i.e. assume that there exists a ξ-interval $\Delta \subset (a,b)$ on which $\lambda_j(\xi) = \lambda' = \text{const}$. Choose any sequence $\xi_k \nearrow \xi_0$, ξ_k, $\xi_0 \in \Delta$, and the corresponding eigenfunctions $y_k \in H(a,\xi_k)$, $y_k^\wedge(\xi_k) = 0$. Clearly, $y_k(x) = Y(x,\lambda')h_k$, where $Y(x,\lambda) \in B(H^n, H)$, is a fundamental self-consistent solution of the problem (2.1), (2.3). Let

$$z_k(x) = \begin{cases} y_k(x) & x \le \xi_k, \\ 0 & x \ge \xi_k. \end{cases}$$

The set $\{z_k\}_{k=1}^\infty$ is linearly independent by the choice of supports of z_k ($k = 1, 2, \ldots$). The functions $z_k(x)$ and, hence, any linear combination $z(x) = \sum_{k=1}^m \alpha_k z_k(x)$, $m < \infty$, belong to the domain $\mathrm{D}(F)$ of the quadratic form

$$F(z) = \int_a^{a_1} (l[z], z)\,dx + (z^\vee(a_1), z^\wedge(a_1)) + \sum_{k=0}^n \int_{a_1}^b (p_k z^{(k)}, z^{(k)})\,dx$$

$$+\operatorname{Im}\sum_{k=1}^{n}\int_{a_1}^{b}(q_k z^{(k-1)}, z^{(k)})\,dx, \tag{3.3}$$

where a_1 is an (arbitrarily) fixed number in the interval (a, ξ_1). (If $a > -\infty$, we take $a_1 = a$). On the other hand,

$$\begin{aligned}F(z) &= \int_a^{a_1}(l[z], z)\,dx + \int_{a_1}^{\xi_1}(l[z], z)\,dx + \sum_{k=1}^{m-1}\int_{\xi_k}^{\xi_{k+1}}(l[z], z)\,dx\\ &+\sum_{k=1}^{m}\alpha_k(y_k{}^{\vee}(\xi_k), z_k{}^{\wedge}(\xi_k)) = \lambda'\|z\|^2_{(a,\xi_0)}\\ &+\sum_{k,j=1}^{m}\alpha_k\bar{\alpha}_j(y_k{}^{\vee}(\xi_k), z_j{}^{\wedge}(\xi_k)).\end{aligned} \tag{3.4}$$

Note that $z_j^{\wedge}(\xi_k) = 0$ for $j \le k$, and for $j > k$, we have $z_j^{\wedge}(\xi_k) = y_j^{\wedge}(\xi_k)$. Since the solution $Y(x,\lambda)$ is self-consistent, i.e. $Y^{\wedge *}(x,\lambda)Y^{\vee}(x,\lambda) - Y^{\vee *}(x,\lambda)Y^{\wedge}(x,\lambda) \equiv 0$, it follows that $(y_k^{\vee}(x), y_j^{\wedge}(x)) - (y_k^{\wedge}(x), y_j^{\vee}(x)) = 0$. Hence, for $x = \xi_k$ we have $(y_k^{\vee}(\xi_k), y_j^{\wedge}(\xi_k)) = 0$ because $y_k^{\wedge}(\xi_k) = 0$. Thus for all j, k, we have $(y_k^{\vee}(\xi_k), z_j^{\wedge}(\xi_k)) = 0$. Therefore, on the infinite-dimensional linear manifold spanned by the vector-valued functions $z_1(x)$, $z_2(x)$,..., $z_m(x)$,..., we have

$$F(z) = \lambda'\|z\|^2_{(a,\xi_0)},$$

contradicting the condition $\lambda' < \inf\sigma_e(L^0_{\xi_0})$ and the semi-boundedness of $L^0_{\xi_0}$. Strict monotonicity is proven[2]. Continuity of $\lambda_j(\xi)$ follows from [782, Corollary 2 (see Proposition 3.1 in the bibliographical comments for Chapter 3)], where the family A^F_{ξ} is replaced by the above considered operators L^0_{ξ}.

The property (3.2) follows from [782, Lemma 2][3] which generalizes the result of Theorem VIII.3.15 in [87] about the stability of the discrete (for $\lambda < \beta$) part of the spectrum[4] of the strong resolvent limit[5] of a convergent from above sequence of self-adjoint uniformly semi-bounded below operators, for the case when the terms of the sequence can be non-densely defined operators which are self-adjoint on the closures of their initial domains. (In

[2]If $\dim H < \infty$, this property can be deduced from the following assertion from [546]: on any ξ-interval, we have $\det Y^{\wedge}(\xi,\lambda) \not\equiv 0$.

[3]See Proposition 3.2 and Definition 3.1 in the Bibliographical Comments at the end of Chapter 3.

[4]In our case, $\beta = \inf\sigma_e(L^F_b)$.

[5]In our case, L^F_b.

our case, those operators are just L^0_ξ with dense domains in $H(a,\xi)$. However, $H(a,\xi)$ is not dense in $H(a,b)$ for $\xi < b$, when it is considered as subspace of $H(a,b)$ consisting of vector-functions that are identically equal to zero at $\xi < x \le b$.) The Theorem is proven. □

Theorem 3.2 *The greatest lower bound* $\inf \sigma(L^0_\xi)$ *of the spectrum and the greatest lower bound* $\inf \sigma_e(L^0_\xi)$ *of the essential spectrum have the same limit as* $\xi \searrow a$*:*

$$\inf \sigma(L^0_\xi) \nearrow \sup_{t\in(a,b)} \{\inf \sigma_e(L^0_t)\} \le \infty. \tag{3.5}$$

If $a > -\infty$*, this limit is equal to infinity.*

Proof. The greatest lower bound $\inf \sigma(L^0_\xi) > -\infty$ and is non-increasing with respect to $\xi \in (a,b)$. Therefore, there exists $\lim_{\xi \searrow a} \{\inf \sigma(L^0_\xi)\} = \mu \le \infty$. Clearly, $\mu \le \sup_{\xi\in(a,b)} \{\inf \sigma_e(L^0_\xi)\}$. If $\mu < \sup_{\xi\in(a,b)} \{\inf \sigma_e(L^0_\xi)\}$, then there exists $\xi_0 \in (a,b)$ such that

$$\mu < \inf \sigma_e(L^0_{\xi_0}). \tag{3.6}$$

But then for all $\xi \le \xi_0$, we have

$$\inf \sigma(L^0_\xi) \le \mu < \inf \sigma_e(L^0_{\xi_0}) \le \inf \sigma_e(L^0_\xi),$$

and, hence, as $\xi \searrow a$, we get $\inf \sigma(L^0_\xi) = \lambda_1(\xi) \nearrow \mu \le \infty$, where $\lambda_1(\xi) < \sigma_e(L^0_\xi)$ is the smallest eigenvalue (with finite multiplicity) of the operator L^0_ξ. Consider the sequence $\{\mu - \delta_k\}$, where $\delta_k \searrow 0$ as $k \to \infty$. For every $\mu - \delta_k$, there exists a point ξ_k such that $\lambda_1(\xi_k) = \mu - \delta_k$ and a vector $h_k \in H^n$, $h_k \ne 0$, such that $Y^\wedge(\xi_k, \mu - \delta_k)h_k = 0$. The vector-valued function $y_k(x, \mu - \delta_k) = Y(x, \mu - \delta_k)h_k$ is an eigenfunction of the operator $L^0_{\xi_k}$. For arbitrary $m < \infty$, let

$$z_k(x) \;=\; \begin{cases} y_k(x, \mu - \delta_k) & x \le \xi_k, \\ 0 & x \ge \xi_k, \end{cases}$$

and let $z(x) = \sum_{k=1}^{m} \alpha_k z_k(x)$. Similarly to (3.4), we have

$$F(z) = \int_a^{a_1} (l[z], z)\, dx + \int_{a_1}^{\xi_1} (l[z], z)\, dx + \sum_{k=2}^{m} \int_{\xi_k}^{\xi_{k-1}} (l[z], z)\, dx$$

$$+\sum_{k=1}^{m}\alpha_k(y_k{}^{\vee}(\xi_k,\mu-\delta_k),z_k{}^{\wedge}(\xi_k))=\mu\|z\|^2_{(a,\xi_0)}-\sum_{k,j=1}^{m}\alpha_k\bar{\alpha}_j\delta_k\langle z_k,z_j\rangle_{(a,\xi_0)}$$

$$+\sum_{k,j=1,j<k}^{m}\alpha_k\bar{\alpha}_j(y_k{}^{\vee}(\xi_k,\mu-\delta_k),y_j{}^{\wedge}(\xi_k,\mu-\delta_j)). \tag{3.7}$$

For $j<k$, on one hand, we have

$$\langle l_W[y_k],y_j\rangle_{(a,\xi_k)}-\langle y_k,l_W[y_j]\rangle_{(a,\xi_k)}=\langle(\mu-\delta_k)y_k,y_j\rangle_{(a,\xi_k)}$$
$$-\langle y_k,(\mu-\delta_j)y_j\rangle_{(a,\xi_k)}=(\delta_j-\delta_k)\langle y_k,y_j\rangle_{(a,\xi_k)}. \tag{3.8}$$

On the other hand, we have

$$\langle l_W[y_k],y_j\rangle_{(a,\xi_k)}-\langle y_k,l_W[y_j]\rangle_{(a,\xi_k)}$$
$$=(y_k{}^{\wedge}(\xi_k,\mu-\delta_k),y_j{}^{\vee}(\xi_k,\mu-\delta_j))-(y_k{}^{\vee}(\xi_k,\mu-\delta_k),y_j{}^{\wedge}(\xi_k,\mu-\delta_j))$$
$$=-(y_k{}^{\vee}(\xi_k,\mu-\delta_k),y_j{}^{\wedge}(\xi_k,\mu-\delta_j)). \tag{3.9}$$

Comparing (3.7)–(3.9), we obtain

$$F(z)=\mu\|z\|^2_{(a,\xi_0)}-\sum_{k,j=1,j>k}^{m}\alpha_k\bar{\alpha}_j\delta_k\langle z_k,z_j\rangle-\sum_{k,j=1,j<k}^{m}\alpha_k\bar{\alpha}_j\delta_j\langle z_k,z_j\rangle$$
$$=\mu\sum_{k,j=1}^{m}\alpha_k\bar{\alpha}_j\langle z_k,z_j\rangle-\sum_{k,j=1}^{m}a_{kj}\alpha_k\bar{\alpha}_j\langle z_k,z_j\rangle, \tag{3.10}$$

where $a_{kj}=a_{jk}=\delta_l$, $l=\min\{j,k\}$.

For all $\epsilon>0$, we can choose δ_1, δ_2, $\delta_3,\ldots$, δ_m so small that on the m-dimensional linear manifold spanned by $z_1(x)$, $z_2(x)$, $z_3(x),\ldots$, $z_m(x)$, we have

$$F(z)\le(\mu+\epsilon)\|z\|^2_{(a,\xi_0)}.$$

Since $\epsilon>0$ is arbitrary, it follows that the operator $L^0_{\xi_0}$ has at least m eigenvalues $\lambda_j\le\mu$, $j=1,2,\ldots,m$. Since m is also arbitrary, we have a contradiction with (3.6), and (3.5) is proven. For $a>-\infty$, we have the equality in the right hand side of (3.5). This can be shown similarly to the part 3) of Theorem 1.3: there exists $C>0$ such that, as $\xi\searrow a$, we have

$$\inf\sigma(L^0_\xi)=\inf_{0\ne y\in \mathrm{D}(L^0_\xi)}[F(y)\|y\|^{-2}]\ge C(\xi-a)^{-2n},$$

because for all $y \in D(L_\xi^0)$ and $-\infty < a \le x \le \xi$,

$$\left|y^{(k)}(x)\right|^2 \le (\xi - a)^{2(n-k)-1} \int_a^\xi \left|y^{(n)}(x)\right|^2 dx, \; k = 0, 1, \ldots, n-1,$$

and $-\frac{\pi}{2} I_n << A \le \frac{\pi}{2} I_n$ in (3.1). The Theorem is proven. □

3.2 Continuity and Monotonicity of the Greatest Lower Bound of the Essential Spectrum for Semi-Bounded Below Differential Operators

Theorem 3.3 *The greatest lower bound of the essential spectrum, $\lambda_e(\xi) := \inf \sigma_e(L_\xi^0)$, is continuous in ξ and is non-increasing. For $a > -\infty$, $\lambda_e(\xi)$ is strictly decreasing in ξ.*

Proof. We will first prove that $\lambda_e(\xi)$ is continuous from the left. Let $\lambda' = \inf \sigma_e(L_{\xi_0}^0)$ for some $\xi_0 \in (a, b)$. Then there exists an orthonormal sequence $y_k \in D(L_{\xi_0}^0)$ such that $\|(L_{\xi_0}^0)y_k - \lambda' y_k\| \to 0$ as $k \to \infty$. For sufficiently small $\delta > 0$, we will construct a sequence $z_k \in D(L_{\xi_0 - \delta}^0)$ close to y_k in the norm of the quadratic form of the operator $L_{\xi_0}^0$. To this end, set $(L_{\xi_0}^0)y_k - \lambda' y_k = f_k$. Replacing, if necessary, the sequence $\{y_k\}$ by a subsequence, for every $\epsilon' > 0$, we may assume

$$\sum_{k=1}^{\infty} \|f_k\| < \epsilon'. \tag{3.11}$$

Take $\alpha > 0$ so small that the greatest lower bound of the operator $L_{\xi_0 - \alpha, \xi_0}^0$ in $H(\xi_0 - \alpha, \xi_0)$ corresponding to $l_W[y]$, with the conditions

$$y^\wedge(\xi_0 - \alpha) = y^\wedge(\xi_0) = 0, \tag{3.12}$$

is greater than λ'.

Let $Y_\delta(x) \in B(H^n, H)$, $0 \le \delta < \alpha$, be the operator-valued solution of $l_W[y] = \lambda' y$ with the initial data: $Y_\delta^\wedge(\xi_0 - \delta) = 0$, $Y_\delta^\vee(\xi_0 - \delta) = I_n$. The operator $Y_\delta^\wedge(x)$ has a bounded inverse in H^n for all $x \in [\xi_0 - \alpha, \xi_0] \backslash \{\xi_0 - \delta\}$.

Let $u_{k,\delta}(x)$ be the solution of the problem

$$l_W[u] - \lambda' u = f_k, \quad u^\wedge(\xi_0 - \alpha) = y_k^\wedge(\xi_0 - \alpha), \quad u^\wedge(\xi_0 - \delta) = 0,$$

and let $C(x,t)$ be the Cauchy function of this equation. Then

$$u_{k,\delta}(x) = \int_{\xi_0-\delta}^{x} C(x,t) f_k(t)\, dt + Y_\delta(x)(Y^\wedge{}_\delta)^{-1}(\xi_0-\alpha)[y_k^\wedge(\xi_0-\alpha) - \int_{\xi_0-\delta}^{\xi_0-\alpha} C_x{}^\wedge(\xi_0-\alpha,t) f_k(t)\, dt]. \tag{3.13}$$

For $\delta = 0$, we have $u_{k,0}(x) = y_k(x)$ for all $x \in (\xi_0-\alpha, \xi_0-\delta)$. In the case $\delta \to 0$, we have $|y_k(x) - u_{k,\delta}(x)| \to 0$ for all $x \in (\xi_0-\alpha, \xi_0-\delta)$, and, hence,

$$\|y_k - u_{k,\delta}\|_{(\xi_0-\alpha,\xi_0-\delta)} \to 0. \tag{3.14}$$

Let

$$z_k(x) = \begin{cases} y_k(x) & a \le x \le \xi_0-\alpha, \\ u_{k,\delta}(x) & \xi_0-\alpha \le x \le \xi_0-\delta, \\ 0 & \xi_0-\delta \le x \le b, \end{cases} \tag{3.15}$$

and let

$$z(x) = \sum_{k=1}^{m} \alpha_k z_k(x), \quad \sum_{k=1}^{m} |\alpha_k|^2 = 1, \quad m < \infty.$$

As $\delta \to 0$, we have

$$\|z_k - y_k\|_{(\alpha,\xi_0)} \to 0. \tag{3.16}$$

Similarly as in (3.4), we have

$$\begin{aligned} F(z) &= \int_a^{\xi_0-\alpha} (l[y], y)\, dx + \int_{\xi_0-\alpha}^{\xi_0-\delta} (l[u_\delta], yu_\delta)\, dx \\ &+(y^\vee(\xi_0-\alpha) - u_\delta^\vee(\xi_0-\alpha), y^\wedge(\xi_0-\alpha)) = \lambda' \int_a^{\xi_0-\delta} (W(x)z, z)\, dx \\ &+ \int_a^{\xi_0-\alpha} (W(x)f, y)\, dx + \int_{\xi_0-\alpha}^{\xi_0-\delta} (W(x)f, u_\delta)\, dx \\ &+(y^\vee(\xi_0-\alpha) - u_\delta^\vee(\xi_0-\alpha), y^\wedge(\xi_0-\alpha)), \end{aligned} \tag{3.17}$$

where

$$y(x) = \sum_{k=1}^{m} \alpha_k y_k, \quad u_\delta(x) = \sum_{k=1}^{m} \alpha_k u_{k,\delta}(x), \quad f(x) = \sum_{k=1}^{m} \alpha_k f_k(x).$$

Since $\|y\|_{(a,\xi_0)} = 1$, then by (3.11), (3.14) and (3.15), we have

$$|\langle f, y\rangle_{(a,\xi_0-\alpha)}| \le \epsilon', \quad |\langle f, u_\delta\rangle_{(\xi_0-\alpha,\xi_0-\delta)}| \le \epsilon'. \tag{3.18}$$

By Lemma 2.2, Cauchy data on the interval (a_1, ξ_0), where $a < a_1 < \xi_0 - \alpha$, continuously depend in the graph norm on solutions of a linear equation. Hence $|y^{(n)}(t)| < C(\xi_0)$ for all $t \in (a_1, \xi_0)$. Since $y^{(k)}(\xi_0) = 0$, $k = 0, 1, \dots, n-1$, we have $|y^{(n-k)}(t)| \le C(\xi_0 - t)^k$. For $\xi_0 - \alpha \le t \le \xi_0$, we have $|y^{(n-k)}(t)| \le C\alpha^k$, and, hence, for $\alpha < 1$, we get $|y^{\wedge}(t)| < C_1\alpha$, where $C_1 > 0$ is independent of α. By (3.13) we have $|y^{\vee}(\xi_0-\alpha) - u_\delta^{\vee}(\xi_0-\alpha)| \to 0$ as $\delta \to 0$, and, hence,

$$|(y^{\vee}(\xi_0-\alpha) - u_\delta^{\vee}(\xi_0-\alpha), y^{\wedge}(\xi_0-\alpha))| \le C_1\alpha|y^{\vee}(\xi_0-\alpha) - u_\delta^{\vee}(\xi_0-\alpha)| \to 0.$$

Now from (3.17) and (3.18) we have

$$F(z) \le \lambda'\|z\|^2_{(a,\xi_0-\delta)} + 2\epsilon' + \gamma(\delta),$$

where $\gamma(\delta) \to 0$ as $\delta \to 0$.

We will prove that for all $0 < \delta < \alpha$, we have

$$\|z\|_{(a,\xi_0-\delta)} \ge C(\xi_0), \tag{3.19}$$

where $C(\xi_0) > 0$ does not depend on z and δ. On one hand, (3.19) implies that the linear manifold M spanned by $\{z_k\}_{k=1}^{\infty}$ is infinite-dimensional since for every non-trivial linear combination we have $\|\sum \alpha_k z_k\|_{(a,\xi_0-\delta)} \ne 0$. On the other hand, (3.19) implies that for every $\epsilon > 0$, there exist ϵ' and δ_1 such that for all $0 < \delta < \delta_1 < \alpha$ and $z(x)$ in (3.15), we have

$$F(z) \le (\lambda' + \epsilon)\|z\|^2_{(a,\xi_0-\delta)}. \tag{3.20}$$

Since the inequality is homogeneous, it holds on the whole linear manifold M, and, hence, $\lambda_e(\xi_0 - \delta) \le \lambda' + \epsilon$ for all $0 < \delta < \delta_1$. This shows that $\lambda_e(\xi)$ is continuous from the left at any point $\xi_0 \in (a, b)$.

To prove (3.19), we split the set of functions $\{y(x)\}$ into two disjoint subsets

$$M_1 = \left\{y\colon \|y\|_{(a,a_1)} \le \frac{\sqrt{2}}{2}\right\}, \quad M_2 = \left\{y\colon \|y\|_{(a,a_1)} > \frac{\sqrt{2}}{2}\right\}.$$

If $y \in M_2$, then by (3.15) we have

$$\|z\|_{(a,\xi_0-\delta)} \ge \frac{\sqrt{2}}{2}. \tag{3.21}$$

Let $y \in M_1$. Then we have $\|y\|_{(a_1,\xi_0)} \geq \sqrt{2}/2$. Since $y(x)$ is a solution of the equation $l_W[y] - \lambda y = f$, whose right hand side is sufficiently small in $H(a, \xi)$, then by continuous dependence of solutions on initial data and on the right hand side of the equation, we get

$$\frac{\sqrt{2}}{2} \leq \|y\|_{(a_1,\xi_0)} \leq C_1(\xi_0)\{|y^\wedge(a_1)| + |y^\vee(a_1)| + \|f\|_{(a_1,\xi_0)}\}, \tag{3.22}$$

where $C_1(\xi_0) > 0$. From (3.11), with $\epsilon' < \frac{\sqrt{2}}{4}C_1^{-1}(\xi_0)$, we have

$$|y^\wedge(a_1)| + |y^\vee(a_1)| \geq \frac{\sqrt{2}}{4}C_1^{-1}(\xi_0). \tag{3.23}$$

Conversely, by continuous dependence of initial data on a solution considered in the graph norm, for sufficiently small ϵ' in (3.11), we have

$$|y^\wedge(a_1)| + |y^\vee(a_1)| \leq C_2(\xi_0 - \delta)\|y\|_{(a_1,\xi_0-\delta)}, \tag{3.24}$$

where $C_2(\xi_0 - \delta) > 0$. Comparing (3.23) and (3.24), we get $\|y\|_{(a_1,\xi_0-\delta)} \geq C_3(\xi_0, \delta)$, where $C_3(\xi_0, \delta) = \frac{\sqrt{2}}{4}C_1^{-1}(\xi_0)C_2^{-1}(\xi_0 - \delta)$ is uniformly bounded for small $\delta > 0$ and bounded away from 0. Therefore, $\|y\|_{(a,\xi_0-\delta)} \geq C_3(\xi_0)$ for all $0 < \delta < \delta_1$. By (3.16) and (3.21) we have (3.19) for all $z(x)$ in (3.15), where $C(\xi_0) = \min\{\sqrt{2}/4; C_3(\xi_0)\}$. This proves that the infimum $\lambda_e(\xi)$ of the essential spectrum is continuous from the left. We will now show that $\lim_{\xi \to \xi_0+0} \lambda_e(\xi) = \lambda_e(\xi_0)$. This will prove the continuous dependence of $\lambda_e(\xi)$ on ξ. Since the infimum of the essential spectrum is non-increasing in ξ, we have $\lambda_e(\xi_0 + 0) := \lambda_+ \leq \lambda_e(\xi_0)$. Suppose that $\lambda_+ < \lambda_e(\xi_0)$. Then by Theorem 2.2 the operator $Y^\wedge(\xi_0, \lambda_+)$ is Fredholm, where $Y(x, \lambda)$ is the fundamental solution of the problem (2.1), (2.2). By Theorem IV.5.17 from [87] (the stability of Fredholm property for operators), it follows that the operator $Y^\wedge(x, \lambda)$ is Fredholm in a neighborhood of (ξ_0, λ_+) in the (x, λ)-plane. On the other hand, in any neighborhood of (ξ_0, λ_+), there exists a point (ξ_1, λ_1), $\xi_1 > \xi_0$, $\lambda_1 \leq \lambda_+ < \inf \sigma(L_\xi)$ such that $\lambda_1 \in \sigma_e(L^0_{\xi_1})$. Therefore, $0 \in \sigma_e(Y^\wedge(\xi_1, \lambda_1))$, and, hence, $Y^\wedge(\xi_1, \lambda_1)$ is not Fredholm. We reached a contradiction, so $\lambda_+ = \lambda_e(\xi_0)$, and the continuity in ξ of $\lambda_e(\xi)$ is proven.

We will now prove that for all $a > -\infty$, $\lambda_e(\xi)$ is strictly monotone[6]. Let $u_k(x, \lambda)$ be a solution of the following Cauchy problem:

$$l_W[u] - \lambda u = f_k; \quad u^\wedge(a, \lambda) = y_k^\wedge(a),\ u^\vee(a, \lambda) = y_k^\vee(a). \tag{3.25}$$

[6]For $a = -\infty$, the infimum of the essential spectrum of an operator may be independent of ξ; for example, if $\dim H < \infty$.

Then

$$u_k(x,\lambda') = y_k(x), \quad u_k^\wedge(\xi_0,\lambda') = 0. \tag{3.26}$$

For sufficiently small δ, the greatest lower bound of the operator $L^0_{\xi_0,\xi_0+\delta}$ in (3.12) is greater than λ'. We choose such δ and we fix it. Then for all $\lambda \leq \lambda'$, there exists a solution $v_k(x,\lambda)$ of the boundary problem

$$l_W[v] = \lambda v, \quad v^\wedge(\xi_0,\lambda) = u_k^\wedge(\xi_0,\lambda'-\epsilon),\, v^\wedge(\xi_0+\delta,\lambda) = 0, \tag{3.27}$$

where $\epsilon > 0$. Let

$$z_k(x) = \begin{cases} u_k(x,\lambda'-\epsilon) & a \leq x \leq \xi_0, \\ v_k(x,\lambda'-\epsilon) & \xi_0 \leq x \leq \xi_0+\delta, \\ 0 & \xi_0+\delta \leq x \leq b. \end{cases} \tag{3.28}$$

Similarly as in (3.17), we have

$$\begin{aligned} F(z) = &\int_a^{\xi_0} (l[u],u)\,dx + \int_{\xi_0}^{\xi_0+\delta} (l[v],v)\,dx \\ &+ (u^\vee(\xi_0,\lambda'-\epsilon) - v^\vee(\xi_0,\lambda'-\epsilon), u^\wedge(\xi_0,\lambda'-\epsilon)). \end{aligned}$$

Since $F(z)$ is real (as the closure of a real quadratic form on smooth functions), it follows that $(u^\vee(\xi_0,\lambda'-\epsilon) - v^\vee(\xi_0,\lambda'-\epsilon), u^\wedge(\xi_0,\lambda'-\epsilon))$ is also real. By (3.25) and (3.26) we have

$$\begin{aligned} F(z) = &(\lambda'-\epsilon)\|z\|^2_{(a,\xi_0+\delta)} + \langle f,u\rangle_{(a,\xi_0)} \\ &+ (u^\vee(\xi_0,\lambda'-\epsilon) - v^\vee(\xi_0,\lambda'-\epsilon), u^\wedge(\xi_0,\lambda'-\epsilon)), \end{aligned} \tag{3.29}$$

and, hence, $\langle f,u\rangle_{(a,\xi_0)}$ is also real. Since

$$\|u(\cdot,\lambda')\|_{(a,\xi_0)} = \left\|\sum_{k=1}^m \alpha_k y_k\right\|_{(a,\xi_0)} = \left(\sum_{k=1}^m |\alpha_k|^2\right)^{1/2} = 1, \tag{3.30}$$

then using (3.11) we have

$$\langle f,u\rangle_{(a,\xi_0)} \leq \epsilon'. \tag{3.31}$$

Since $\phi(x) = u(x,\lambda') - u(x,\lambda'-\epsilon)$ is a solution of the Cauchy problem $l_W[\phi] - (\lambda'-\epsilon)\phi = \epsilon u(x,\lambda')$, $\phi^\wedge(a) = \phi^\vee(a) = 0$, it follows that

$$\phi(x) = \epsilon \int_a^x C(x,t)u(t,\lambda')\,dt,$$

where $C(x,t)$ is the Cauchy function of this problem. Now since $\|u(\cdot,\lambda')\|_{(a,\xi_0)} = 1$, we have $|\phi^\wedge(\xi_0)| \le \alpha\epsilon$, where $\alpha = \text{const} > 0$. By (3.26) it follows that $u^\wedge(\xi_0,\lambda') = 0$; hence,

$$|v^\wedge(\xi_0,\lambda'-\epsilon)| = |u^\wedge(\xi_0,\lambda'-\epsilon)| = |\phi^\wedge(\xi_0)| \le \alpha\epsilon.$$

So by Lemma 3.1 below, we get $|v^\vee(\xi_0,\lambda'-\epsilon)| \le \alpha\beta\epsilon$. Therefore,

$$|(v^\vee(\xi_0,\lambda'-\epsilon), u^\wedge(\xi_0,\lambda'-\epsilon))| \le \alpha^2\beta\epsilon^2, \tag{3.32}$$

where α, β are independent of ϵ, u, v.

Note that by (3.25), where $y_k \in \mathrm{D}(L^0_{\xi_0})$, it follows that $u(x,\lambda)$ satisfies the boundary condition (2.6). Then by Lemma A.2 we have $u^\wedge(a,\lambda) = \cos A\cdot h$ and $u^\vee(a,\lambda) = \sin A \cdot h$, for some $h \in H^n$. Since $2|h| \ge |u^\wedge(a,\lambda)| + |u^\vee(a,\lambda)|$, then similarly as in (3.23), we get $8|h| \ge \sqrt{2}C_1^{-1}(\xi_0)$, and, hence, $|h| \ge C_1 = \text{const} > 0$ for all $a < \xi_0 < C < b$. Thus by (3.26) and by Lemma 3.2 below, for sufficiently small $\epsilon > 0$, we have

$$\mathrm{Re}\,(u^\vee(\xi_0,\lambda'-\epsilon), u^\wedge(\xi_0,\lambda'-\epsilon)) < -C_2\epsilon,$$

where $C_2 > 0$ is independent of ϵ and u. This and (3.32) give

$$(u^\vee(\xi_0,\lambda'-\epsilon) - v^\vee(\xi_0,\lambda'-\epsilon), u^\wedge(\xi_0,\lambda'-\epsilon)) < 0.$$

Therefore, from (3.29) and (3.31) we have

$$F(z) \le (\lambda'-\epsilon)\|z\|^2_{(a,\xi_0+\delta)} + \epsilon'.$$

Repeating the proof of (3.19) with the condition $\sum |\alpha_k|^2 = 1$, we obtain

$$\|z(\cdot,\lambda'-\epsilon)\|^2_{(a,\xi_0+\delta)} \ge C(\xi_0), \tag{3.33}$$

where $C(\xi_0) > 0$ is independent of ϵ and z. Choosing $\epsilon > 0$ and $\epsilon' > 0$ so small that the inequality $\epsilon'' := \epsilon - \epsilon' C^{-1}(\xi_0) > 0$ holds, we get

$$F(z) < (\lambda' - \epsilon'')\cdot \|z\|^2_{(a,\xi_0+\delta)}. \tag{3.34}$$

The inequality (3.34) is homogeneous; hence, it also holds on the linear manifold spanned by $\{z_k\}_{k=1}^\infty$ in (3.28), which is infinite-dimensional by (3.33). Therefore, for sufficiently small $\delta > 0$, we have $\lambda_e(\xi_0+\delta) < \lambda_e(\xi_0)$. Since $\xi_0 \in (a,b)$ is arbitrary, and since $\lambda_e(\xi)$ is non-increasing in ξ, it follows that $\lambda_e(\xi)$ is strictly decreasing. Indeed, if $\xi_1 < \xi_2$, we can take $\xi_1 < \xi_0 < \xi_0 + \delta < \xi_2$, and, hence, $\lambda_e(\xi_2) \le \lambda_e(\xi_0+\delta) < \lambda_e(\xi_0) \le \lambda_e(\xi_1)$.

Lemma 3.1 *For all sufficiently small $\delta > 0$ (such that the greatest lower bound of the operator $L^0_{\xi,\xi+\delta}$ in (3.12) is greater than $\lambda' = \lambda_e(\xi_0) \in \mathbb{R}$), there exists $\beta > 0$ such that the solutions of*

$$l_W[y] = \lambda y, \quad y^\wedge(\xi) = g, \quad y^\wedge(\xi+\delta) = 0, \forall g \in H^n \tag{3.35}$$

satisfy the estimate

$$|y^\vee(\xi,\lambda)| \leq \beta \cdot |y^\wedge(\xi,\lambda)|, \quad \textit{for all } \lambda \leq \lambda', \tag{3.36}$$

where β is independent of g.

Proof. To every $g \in H^n$ corresponds a unique solution y_g of (3.35). Define the operator $T: H^n \to H$ by the formula $Tg = y_g^\vee(\xi)$. The operator T is defined on all of H^n. We will prove that T is closed, and, hence, bounded by the Closed Graph Theorem. Let $g_k \to 0$ and $Tg_k \to f$. We will show that $f = 0$. Let $y_k(x)$ be a solution of (3.35) with $g = g_k$, and let $z(x) = y_f(x)$ be a solution of (3.35) with $g = f$. Then, as $k \to \infty$, we have

$$0 = \langle l[y_k], z\rangle - \langle y_k, l[z]\rangle = \{(y_k^\wedge(x), z^\vee(x)) \\ -(y_k^\vee(x), z^\wedge(x))\}|_\xi^{\xi+\delta} = (Tg_k, f) - (g_k, z^\vee(\xi)) \to (f, f).$$

Therefore $f = 0$, and the operator T is closed. The Lemma is proven. □

Lemma 3.2 *Let $u(x,\lambda)$ ($Im\ \lambda = 0$) be a solution of the non-homogeneous equation*

$$l_W[u] = \lambda u + f \tag{3.37}$$

with the initial data $u^\wedge(a,\lambda) = \cos A \cdot h$, $u^\vee(a,\lambda) = \sin A \cdot h$, $h \in H^n$, $-\frac{\pi}{2}I_n < A \leq \frac{\pi}{2}I_n$. Suppose that there exists λ' such that $u^\wedge(\xi,\lambda') = 0$.

Then for all $\|f\| \leq \epsilon$, $|\lambda - \lambda'| < \epsilon$, where $\epsilon > 0$ is a sufficiently small fixed number, we have

$$Re\,\frac{\partial}{\partial\lambda}(u^\vee, u^\wedge)|_{x=\xi} \geq \gamma|h|^2; \left|Im\,\frac{\partial}{\partial\lambda}(u^\vee, u^\wedge)|_{x=\xi}\right| \leq \epsilon_1|h|^2, \tag{3.38}$$

where $\gamma > 0$ is independent of h, $f(x)$, and $\epsilon_1 \to 0$, as $\epsilon \to 0$.

Proof. Suppose, first, that $f = 0$. Let $Y(x,\lambda) \in B(H^n, H)$ be the operator solution to (2.1) with the initial data

$$Y^\wedge(a,\lambda) = \cos A, \quad Y^\vee(a,\lambda) = \sin A. \tag{3.39}$$

It is well-known [562] (see also [78]) that for all real λ, the equation (2.1) can be reduced to a symmetric system of the form

$$J\frac{dz}{dx} = (\lambda A(x) + B(x))z, \tag{3.40}$$

where $z(x,\lambda) = y^{\wedge}(x,\lambda) \oplus y^{\vee}(x,\lambda) \in H^{2n}$,

$$J = \begin{pmatrix} 0 & -I_n \\ I_n & 0 \end{pmatrix}, \; A(x) = \begin{pmatrix} A_{11} & 0 \\ 0 & 0 \end{pmatrix}, \; B(x) = B^*(x) = \begin{pmatrix} B_{11} & B_{12} \\ B_{21} & B_{22} \end{pmatrix},$$

are operators in H^{2n}, where

$$A_{11}(x) = W(x) \oplus \underbrace{0 \oplus \cdots \oplus 0}_{n-1}; \quad B_{jk}\colon H^n \to H^n,\; j,k = 1,2, \tag{3.41}$$

and

$$B_{22}(x) = \underbrace{0 \oplus 0 \oplus \cdots \oplus 0}_{n-1} \oplus p_n^{-1}(x).$$

Differentiating the system (3.40), we get

$$\frac{\partial^2 Y^{\wedge}}{\partial x \partial \lambda} = B_{21}\frac{\partial Y^{\wedge}}{\partial \lambda} + B_{22}\frac{\partial Y^{\vee}}{\partial \lambda}$$

and

$$\frac{\partial^2 Y^{\vee}}{\partial x \partial \lambda} = -A_{11}Y^{\wedge} - (\lambda A_{11} + B_{11})\frac{\partial Y^{\wedge}}{\partial \lambda} - B_{12}\frac{\partial Y^{\vee}}{\partial \lambda}.$$

Hence for all real λ we get (see [13, p. 352])

$$\frac{\partial}{\partial x}\left(Y^{\vee *}\frac{\partial}{\partial \lambda}Y^{\wedge} - Y^{\wedge *}\frac{\partial}{\partial \lambda}Y^{\vee}\right) = Y^{\wedge *}A_{11}Y^{\wedge}. \tag{3.42}$$

Since $Y^{\wedge}(a,\lambda)$ and $Y^{\vee}(a,\lambda)$ do not depend on λ, integrating (3.42) on the interval (a,x), we get

$$\begin{aligned} &Y^{\vee *}\frac{\partial}{\partial \lambda}Y^{\wedge}(x,\lambda) - Y^{\wedge *}\frac{\partial}{\partial \lambda}Y^{\vee}(x,\lambda) \\ &= \int_a^x Y^{\wedge *}(t,\lambda)A_{11}(t)Y^{\wedge}(t,\lambda)\,dt. \end{aligned}$$

If for some $\lambda' \in \mathbb{R}$ and $h \in H^n$, we have $Y^{\wedge}(\xi,\lambda')h = 0$, then

$$\frac{\partial}{\partial \lambda}(Y^{\vee *}Y^{\wedge}h,h)\Big|_{\substack{\lambda=\lambda' \\ x=\xi}} = \left(\left(Y^{\vee *}\frac{\partial}{\partial \lambda}Y^{\wedge} - Y^{\wedge *}\frac{\partial}{\partial \lambda}Y^{\vee}\right)h,h\right)\Big|_{\substack{\lambda=\lambda' \\ x=\xi}}$$

$$= \left(\left(\int_a^\xi Y^{\wedge *}(t,\lambda') A_{11}(t) Y^{\wedge}(t,\lambda')\, dt \right) h, h \right).$$

We will show that the operator

$$K_\xi(\lambda) = \int_a^\xi Y^{\wedge *}(t,\lambda) A_{11}(t) Y^{\wedge}(t,\lambda)\, dt >> 0 \tag{3.43}$$

for any fixed λ, Im $\lambda = 0$ and $\xi \in (a,b)$. This implies that the following inequality holds for the solution $u(x,\lambda') = Y(x,\lambda')h$ of the homogeneous equation (2.1):

$$\frac{\partial}{\partial \lambda}(u^\vee, u^\wedge)\Big|_{\substack{\lambda = \lambda' \\ x = \xi}} \geq \gamma |h|^2,$$

where $\gamma > 0$ is independent of h. By continuity, this inequality holds for all λ sufficiently close to λ', and also for solutions of the non-homogeneous equation (3.37) for small $\|f\|$.

Assume the opposite, i.e. assume that there exist $\lambda = \lambda'$, $\xi = \xi_0$ such that the operator $K_\xi(\lambda)$ does not have a bounded inverse. If $K_{\xi_0}(\lambda')h = 0$, then by continuity of the integrand, we get $A_{11}(t)Y^\wedge(t,\lambda')h = 0$ for all $t \in [a, \xi_0]$. Thus by (3.41) we have $Y(t,\lambda')h = 0$ for all $t \in [a,\xi_0]$, and, hence, $h = 0$. If $K_{\xi_0}(\lambda')h_m \to 0$, as $m \to \infty$, then $Y(t,\lambda')h_m \to 0$ for all $t \in [a,\xi_0]$. Therefore, $\|y_m\|_{(a,\xi_0)} \to 0$, where $y_m(x) = Y(x,\lambda)h_m$ is a solution of (2.1). Since the initial data continuously depend on the solution in the graph norm (see Lemma 2.2 or Lemma A.10), we get $|y^\wedge{}_m(a,\lambda)| + |y^\vee{}_m(a,\lambda)| \to 0$. By (3.39) we have $y^\wedge{}_m(a) = \cos A \cdot h_m$ and $y^\vee{}_m(a) = \sin A \cdot h_m$, so $h_m \to 0$ as $m \to \infty$, and (3.43) is proven. This concludes the proof of Lemma 3.2 and Theorem 3.3. □

3.3 Bibliographical Comments

The results of this chapter are due to the authors; see [795] and [798]. In the case of a Sturm–Liouville problem with operator-valued coefficients on a bounded interval, the same questions were investigated in Chapter 1 (see also [794]) by different methods. Those methods are not directly transferrable to the present case because the equation (2.1) is reduced to a canonical system having a more general form than the canonical system to which a second order equation is reduced (unlike in the case $r = 2$, in the case $n > 1$, $r = 2n$, the corresponding Hamiltonian may fail to be strictly positive on any Lagrangian plane).

Theorem 3.1 generalizes Courant's theorem and, in one-dimensional case, gives an affirmative answer to the question from Glazman's book [70, Chapter 4, No. 46, p. 229] about the strict monotonicity of eigenvalues with respect to extension of the infinite domain (of course, if this domain does not coincide with the whole space). (In the case of a second order equation and $\dim H < \infty$, this answer was obtained in [540].)

In the case of a scalar real operator with $a > -\infty$, $y^\wedge(a) = 0$, the last assertion of Theorem 3.2 becomes one of the theorems of M. G. Krein [585, Theorem 21].

Regarding the reference to [562] in the proof of Lemma 3.2, we also mention the works containing a further development and generalization of [562] in different directions: [317], [472], [475], [619] and [621]. (The references [472] and [475] contain a significant step forward from the results of [562] and [561] in the direction of a solution of the Everitt's problem [409] concerning the deficiency indices of symmetric differential operators with complex coefficients.)

We will formulate a Proposition (see [782, Corollary 2]) used in the proof of Theorem 3.1.

Proposition 3.1 *Let A be a semi-bounded below symmetric operator, $A \geq \gamma I$, and let D_ξ, D_η be linear manifolds such that for all $\xi < \eta$, $D_\xi \subseteq D_\eta \subseteq D(A)$. Then the family $A_\xi^F = (A|D_\xi)^F$ is continuous at $\xi = \xi_0$ (in the sense of strong resolvent convergence) if*

$$\left[\bigcup_{\xi<\xi_0} D_\xi\right]_A^{\sim} = \bigcap_{\xi>\xi_0} [D_\xi]_A^{\sim},$$

where $[D]_A^{\sim}$ denotes the closure of the set D in A-metric, i.e. in the metric of the quadratic form of the operator $A - (\gamma - 1)I$.

If $\beta \in \mathbb{R}$ is such that $\dim E_\beta(A^F) < \infty$, where $E_\lambda(A^F)$ is resolution of the identity of the operator A^F, then the eigenvalues $\lambda_j(A_\xi^F)$ lying below β are continuous in ξ at the point $\xi = \xi_0$.

Note that the Friedrichs extension A_ξ^F of a non-densely defined semi-bounded operator A_ξ is not an operator but rather a Hermitian relation whose resolvent is defined by the formula due to Ando, Nishio and Shtraus (see [225], [855], [856] and also [782]):

$$R_\lambda(A_\xi^F) = R_\lambda^{(\xi)}((P_\xi A_\xi)^{F_\xi})P_\xi, \tag{3.44}$$

where $R_\lambda^{(\xi)}$ is the resolvent in $H_\xi := \overline{D(A_\xi)} = \overline{D_\xi}$, P_ξ is the orthogonal projection on H_ξ, and the symbol F_ξ denotes an ordinary Friedrichs extension in H_ξ (there, the operator $P_\xi A_\xi$ is densely defined).

Corresponding to Lemma 2 in [782], we also formulate a generalization of Theorem VIII.3.15 from [87] used in the proof of Theorem 3.1.

Proposition 3.2 *Let $A_n : D(A_n) \to \overline{D(A_n)} =: H_n \subseteq H$ be a sequence of self-adjoint semi-bounded below operators in H_n. Assume that A_n converges from above in the strong resolvent sense to a self-adjoint semi-bounded below operator A whose lower part of the spectrum for $\lambda < \beta$ is discrete, i.e. consists of isolated eigenvalues with finite multiplicities $\lambda_1(A) \le \lambda_2(A) \le \cdots (< \beta)$, enumerated counting multiplicities. Then these eigenvalues are stable in the sense of Kato (see Definition 3.1 below) and*

$$\lambda_j(A_n) \to \lambda_j(A) + 0, \quad \text{where } \lambda_j(A) < \beta,\ n \to \infty,$$

and, if the sequence A_n is monotone (non-increasing in the sense of quadratic forms), the convergence of $\lambda_j(A_n)$ is also monotone:

$$\lambda_j(A_n) \searrow \lambda_j(A) < \beta, \quad \text{as } n \to \infty.$$

(The above formulated Proposition can be extended to sequences of Hermitian relations.)

Definition 3.1 Relative to sequences of non-densely defined in H self-adjoint operators A_n in H_n converging in the strong resolvent sense to a self-adjoint operator A in $\overline{D(A)}$, the stability in the sense of Kato is defined as follows: an isolated eigenvalue λ' of the operator A is said to be stable (with respect to the given limiting process or perturbation) if some punctured neighborhood $\Delta\backslash\{\lambda'\}$ of λ' belongs to the domain of strong convergence of $\{R_\lambda(A_n)P_n\}$, where P_n is the orthogonal projection onto H_n, $R_\lambda(A_n)$ is the resolvent in H_n, and for spectral projections corresponding to Δ, the following uniform convergence relation holds (P is the orthogonal projection onto $\overline{D(A)}$):

$$\|E_\Delta(A_n)P_n - E_{\{\lambda'\}}(A)P\| \to 0 \quad \text{as } n \to \infty,$$

and, consequently, for sufficiently large n,

$$\dim E_\Delta(E_n) = \dim E_{\{\lambda'\}}(A).$$

For the notion of resolvent convergence (which is also called "generalized convergence" in [87]) see also [165], and for its generalizations to non-densely defined sequences of operators and sequences of relations, see [782], [784], [860] and [861].

More complete exposition of the results and questions touched upon in these Bibliographical Comments are quoted, with proofs, in the section "Strong Resolvent Convergence of Hermitian Relations and Operators With Non-dense Domains and Resolution of the Identity. Stability of Eigenvalues" in Bibliographical Comments to the Appendix.

In conclusion, we would like to recall the following well-known fact. Beside singularities of a differential operator caused by the infinitude of the operator domain, there are also singularities caused by local behavior of coefficients of the operator. In relation to such kind of singularities, we note the recent works [517], [834], [887], [903] and some old papers related to the scattering problem on the potential with a high singularity at the origin: [305], [634], [783] and [806]–[808].

In [834], the authors considered somewhat equivalent methods of determining the Sturm–Liouville operator with a singular potential. One of the methods considered in [834], namely, approximate method, is extended in [903] to a wide class of singular boundary-value problems of arbitrary even order (there, a scalar problem on the finite interval is considered).

Chapter 4

Relation Between Spectral and Oscillatory Properties for Operator Differential Equations of Arbitrary Order

4.1 Oscillatory Theorem for Operator Differential Equations of Even Order

In this section we transfer the classical oscillatory Sturm's theorem to boundary-value problems of arbitrary even order with matrix or operator-valued coefficients on a bounded or unbounded interval.

Suppose that the minimal operator L_b with respect to the endpoint $b \leq \infty$, corresponding to the equation of the form (2.1), is semi-bounded below, $r = 2n$.

It is easy to show that semi-bounded below differential operator with operator-valued coefficients must be of even order $r = 2n$, and its leading coefficient $p_n(x)$, which by the hypothesis has bounded inverse for every $x \in [a, b]$, must be strictly positive for every x, i.e. $p_n(x) >> 0$. At the same time, the requirement that only the essential spectrum of the differential operator of the form (2.1) is semi-bounded below is not sufficient for the leading coefficient $p_n(x)$ of this differential operator to be strictly positive or at least non-negative, as the following Example shows.

Example 4.1 Suppose that a differential operator L with operator-valued coefficients is an orthogonal sum of two operators

$$L = L_{(1)} \oplus L_{(2)},$$

where the operator $L_{(1)}$ acts in the space of square integrable with weight $W_{(1)}$ vector-valued functions with values in a separable Hilbert space H_1:

$$H_1(a, b) = L_2\{H_1; (a, b); W_{(1)}(x)\, dx\},$$

and $L_{(1)}$ is semi-bounded below (hence, its essential spectrum is semi-bounded below). Here, $\dim H_1 \le \infty$.

Let $L_{(2)}$ be an operator acting in the space

$$H_2(a,b) = L_2\{H_2; (a,b); W_{(2)}(x)\,dx\},$$

where $\dim H_2 < \infty$, and $L_{(2)}$ has strictly negative leading coefficient and purely discrete spectrum (which is, of course, not semi-bounded below).

In this case, the operator $L = L_{(1)} \oplus L_{(2)}$ acting in the space

$$H(a,b) = H_1(a,b) \oplus H_2(a,b) = L_2\{H_1 \oplus H_2; (a,b); (W_{(1)} \oplus W_{(2)})\,dx\}$$

has the same essential spectrum as the operator $L_{(1)}$:

$$\sigma_e(L) = \sigma_e(L_{(1)}) > -\infty,$$

i.e. the essential spectrum of L is semi-bounded below, while the leading coefficient of L is not a positive operator–function. For every $x \in (a,b)$ the leading coefficient represents an orthogonal sum of a positive operator from $B(H_1)$ and a negative operator from $B(H_2)$. (We recall that $B(H)$ denotes the set of all bounded linear operators in the space H.)

As noted in Chapter 1, for $b = \infty$, even in the scalar case, it may happen that semi-bounded below operator L_b has deficiency indices different from zero; for example, the (scalar) Sturm–Liouville operator

$$-(x^4y')' - 2x^2y = \lambda y$$

is semi-bounded below. However, this operator is in the limit circle case for $b = \infty$.

In the sequel, by Def$\{T\}$ we will denote the deficiency number[1] of an operator T.

We denote by $N(\lambda)$ the number of eigenvalues $\lambda_k < \lambda$ of the operator $\widetilde{L}$ corresponding to the problem (2.1), (2.3), (2.4)[2] counting multiplicities. We denote the multiplicity of λ_k by æ(λ_k). If λ is not an eigenvalue of

[1]Every symmetric operator T has certain deficiency indices (m,n), where m and n are the deficiency numbers of the operator T under consideration. Here $0 \le m, n \le \infty$. If $m = n = 0$, then T is essentially self-adjoint and the closure of T is a self-adjoint operator. The condition $m = n \le \infty$ is necessary and sufficient for a symmetric operator T to have self-adjoint extensions. Under this condition, we may write Def$\{T\} = m (= n)$. The equality $m = n$ is guaranteed when T is semi-bounded or if there exists at least one gap in the essential spectrum $\sigma_e(T)$ of T. In other words, $m = n$ is guaranteed if not whole real axis belongs to $\sigma_e(T)$.

[2]In the case of regular end-points a or b, the self-adjoint boundary conditions at those points are of the form (2.6), (2.7).

the given problem, then we set $æ(\lambda) = 0$. The quantities $N(\lambda)$, $æ(\lambda)$, λ_k corresponding to the Friedrichs extension L_b^F of L_b will be denoted by $N^F(\lambda)$, $æ^F(\lambda)$, λ_k^F respectively (for $b < \infty$, we have $L_b^F = L_b^0$, where L_b^0 is the operator corresponding to the problem (2.1), (2.3), $y^\wedge(b) = 0$, so, parallel with the superscript F, we will also use the superscript 0).

We also recall that $Y(x, \lambda)$ is the fundamental solution of the problem (2.1), (2.3), and $Y^\wedge$ is defined in accordance with (2.8) and (2.25).

Theorem 4.1 *For all $\lambda < \lambda_e(\widetilde{L}) := \inf \sigma_e(\widetilde{L})$, we have*

$$N(\lambda) - p \leq \sum_{x \in (a,b)} nul\, Y^\wedge(x, \lambda) = N^F(\lambda) \leq N(\lambda), \tag{4.1}$$

where $p = Def\,\{\widetilde{L}|_{D(L_b^F) \cap D(\widetilde{L})}\}$. If λ is not an eigenvalue of L_b, then for all $\lambda < \lambda_e(\widetilde{L})$, we have

$$N(\lambda) - \min\{p, def\, L_b - æ(\lambda)\} \leq \sum_{x \in (a,b)} nul\, Y^\wedge(x, \lambda)$$
$$= N^F(\lambda) \leq N(\lambda) \tag{4.2}$$

(in the case of a regular end-point $b < \infty$, $p = rank\, \cos B \leq \infty$, where B is as in (2.7), $def\, L_b = n \cdot \dim H$). If the condition

$$\lambda_e(L_\xi^F) > \lambda_e(L_b^F), \quad for\ \xi \in (a, b), \tag{4.3}$$

holds; in particular, if the endpoint $a > -\infty$ is regular, then (4.1) holds also for $\lambda = \lambda_e(\widetilde{L})$, $p < \infty$. (The sum is taken over the values of $x \in (a, b)$ such that $nul\, Y^\wedge(x, \lambda) \neq 0$. The lower estimate in (4.1), (4.2) can be attained if it is non-negative.)

Remark 4.1 If $a > -\infty$ and if for some $\lambda \in \mathbb{R}$, there exists a fundamental solution $Y^\wedge(x, \lambda)$ of the problem (2.1), (2.6) such that the operator $Y^\wedge(x, \lambda)$ is Fredholm for all $x \in (a, b)$, then $\lambda \leq \lambda_e(L_b^F)$.

Lemma 4.1 [3] *Assume that A_2 and A_0 are self-adjoint extensions of a semi-bounded operator A with finite deficiency indices (p, p) in a Hilbert space H. Then*

$$N_1(\lambda) - p \leq N_2(\lambda) \tag{4.4}$$

[3] Lemma 4.1 is the same as Lemma 1.4, and the proof is given there. We repeat the formulation for reader's convenience. A closely related result was established in [365].

for all $\lambda \le \lambda_e = \inf \sigma_e(A)$. *If, in addition,* $æ_A(\lambda) = 0$ *and the operator* A *is closed, then*

$$N_1(\lambda) + æ_1(\lambda) - p \le N_2(\lambda) \tag{4.5}$$

(indices 2, 1, A *correspond to the operators* A_2, A_1 *and* A *respectively). For* $\lambda = \lambda_e$, *both sides of the inequalities (4.4), (4.5) may be infinite (but only at the same time; see [70, Sec. 1, Theorem 14] or [586, I]).*

Proof. We will first prove the Theorem in the case $b < \infty$ and $\cos B = 0$ in (2.7). Let L^0_ξ be the operator corresponding to (2.1), (2.3), $y^\wedge(\xi) = 0$ on the interval (a, ξ), $\xi \le b$. Let

$$\lambda^0_1(\xi) \le \lambda^0_2(\xi) \le \cdots \le \lambda^0_n(\xi) \le \cdots (< \lambda^0_e(\xi) \le \infty)$$

be the discrete spectrum of L^0_ξ in the interval $(-\infty, \lambda^0_e(\xi))$, enumerated so that the multiplicities are counted. Let $\lambda^0_e(b) < \infty$, i.e. $\dim H = \infty$ or $a = -\infty$. By Theorem 3.1, through every point $(b, \lambda^0_j(b))$ in the (x, λ)-plane, passes a continuous strictly monotone curve $\lambda = \lambda^0_j(x)$. If the condition (4.3) holds, in particular, if $a > -\infty$, the curve $\lambda = \lambda^0_j(x)$ can be continued until its intersection with the line $\lambda = \lambda^0_e(b)$, $a < x < b$. The half-strip $\lambda < \lambda^0_e(b)$, $a < x \le b$ does not contain any points of the set $\bigcup_{\xi \in (a,b]} \sigma_e(L^0_\xi)$. If (4.3) holds, in particular, if $a > -\infty$, by strict monotonicity of the infimum of the essential spectrum in this case, we have $\lambda^0_e(\xi) > \lambda^0_e(b)$ for all $\xi < b$. Therefore, the half-strip $\lambda \le \lambda^0_e(b)$, $a < x \le b$ does not contain any points of $\bigcup_{\xi \in (a,b]} \sigma_e(L^0_\xi)$ except $(b, \lambda^0_e(b))$. The curves $\lambda = \lambda^0_j(x)$ may completely or partially coincide in accordance with the multiplicities of eigenvalues represented by those curves. The total multiplicity of all curves emanating from (b, μ) is equal to the multiplicity $æ^0(\mu)$ of number μ viewed as an eigenvalue of the operator L^0_b. If $\lambda^0_e(b) = +\infty$, from (3.5) it follows that the curves $\lambda = \lambda^0_j(x)$ can be continued to $x = a+0$ and, as $x \searrow a$, $\lambda^0_j(x)$ tend to $+\infty$ ($a \ge -\infty$).

By $x_j(\lambda)$ we denote the inverse of $\lambda = \lambda^0_j(x)$, where $\lambda^0_j(b) \le \lambda < \lambda^0_e(b)$ ($\le \lambda^0_e(b)$ if (4.3) holds; in particular, if $a > -\infty$). Let $\lambda^0_j(b) < \lambda' < \lambda^0_e(b)$ ($\le \lambda^0_e(b)$ if (4.3) holds; in particular, if $a > -\infty$). Then the curve $\lambda = \lambda^0_j(x)$ intersects the line $\lambda = \lambda'$ at exactly one point $x_j(\lambda') \in (a, b)$. The number λ' is an eigenvalue of the operator $L_{x_j(\lambda')}$. The multiplicity of λ' is equal, on one hand, to the value of nul $Y^\wedge(x_j(\lambda'), \lambda')$, and, on the other hand, to the number of curves $\lambda = \lambda_j(x)$ containing a common point $(x_j(\lambda'), \lambda')$. If $(a, b) \ni x \ne x_j(\lambda')$ for all j, then nul $Y^\wedge(x, \lambda') = 0$.

Therefore, $\sum_{x\in(a,b)}$ nul $Y^\wedge(x,\lambda')$, where $\lambda' < \lambda_e^0(b)$ ($\leq \lambda_e^0(b)$ if (4.3) holds; in particular, if $a > -\infty$), is equal to the number of curves $\lambda = \lambda_j(x)$, $a < x < b$, counting multiplicities, intersecting the line $\lambda = \lambda'$, i.e. to the number $N^0(\lambda')$ of eigenvalues $\lambda_j^0(b) < \lambda'$ (when $\lambda_j^0(b) \geq \lambda'$, the curve $\lambda = \lambda_j^0(x)$, $a < x < b$, does not intersect the line $\lambda = \lambda'$). Therefore, we have shown that if $b < \infty$ and $\lambda < \lambda_e^0(b)$ ($\leq \lambda_e^0(b)$ if (4.3) holds; in particular, if $a > -\infty$), then

$$\sum_{x\in(a,b)} \text{nul } Y^\wedge(x,\lambda) = N^0(\lambda).$$

By variational principles, the eigenvalues $\lambda_j < \lambda_e$ and the number λ_e for the problem (2.1), (2.3), (2.7) do not exceed the corresponding eigenvalues λ_j^0 and λ_e^0 of (2.1), (2.3), $y^\wedge(b) = 0$. Therefore, $N^0(\lambda) \leq N(\lambda)$ for all $\lambda \leq \lambda_e \leq \lambda_e^0$. If $p = \infty$, we have

$$0 \leq \sum_{x\in(a,b)} \text{nul } Y^\wedge(x,\lambda) \leq N(\lambda).$$

If $p < \infty$, then the operators L_b^0 and $\widetilde{L}$ are finite-dimensional self-adjoint extensions of their maximal common part $\Lambda = \widetilde{L}|_{\mathrm{D}(L_b^0)\bigcap \mathrm{D}(\widetilde{L})}$. Therefore, $\lambda_e = \lambda_e^0$, and by Lemma 4.1, the inequality (4.1) and a part of (4.2) hold. Note that Λ is an operator defined on vector-valued functions $y \in \mathrm{D}(L_b^0)$ satisfying the boundary conditions (2.3), $y^\wedge(b) = 0$, and, besides, $\cos B \cdot y^\vee(b) = 0$. Thus, deficiency indices of Λ are (rank $\cos B$, rank $\cos B$).

Minimal operator L_b with respect to the end-point $x = b < \infty$ given by (2.2) on the vector-valued functions satisfying (2.3) and $y^\wedge(b) = y^\vee(b) = 0$ has deficiency indices $(n \cdot \dim H, n \cdot \dim H)$. Moreover, L_b does not have any eigenvalues. If $\dim H < \infty$, the operators $\widetilde{L}$ and L_b^0 can be considered as finite-dimensional extensions of L_b. Using Lemma 4.1 again, together with (4.1), we get (4.2).

To prove the Theorem in the case $b = \infty$, we pass to the limit in the case of finite b.

Let $N^F(\lambda) = k$. Then $\lambda_k^F < \lambda$ and $(\lambda_k^F, \lambda)\bigcap \sigma(L_b^F) = \emptyset$. By Theorem 3.1 we have $\lambda_k^0(\xi) \searrow \lambda_k^F$, as $\xi \to \infty$. Thus, for sufficiently large $\xi > \xi_0$, we have $\lambda_k^0(\xi) < \lambda$, $(\lambda_k^0(\xi), \lambda)\bigcap \sigma(L_\xi^0) = \emptyset$. Therefore, for all $\xi > \xi_0$, the following holds:

$$\sum_{x\in(a,\xi)} \text{nul } Y^\wedge(x,\lambda) = N_\xi^0(\lambda) = k = N^F(\lambda). \tag{4.6}$$

Using (3.2), as $\xi \to \infty$, we get

$$\sum_{x\in(a,\infty)} \text{nul}\, Y^\wedge(x,\lambda) = N^F(\lambda). \tag{4.7}$$

By Lemma 4.1 we get (4.1) and (4.2), and the Theorem is proven. □

We will now prove the assertion of Remark 4.1. If $\lambda > \lambda_e(L_b^F)$, then for $a > -\infty$, by Theorems 3.2 and 3.3, there exists $\xi \in (a,b)$ such that $\lambda = \inf \sigma_e(L_\xi^0)$. In this case, as noted above, $0 \in \sigma_e(Y^\wedge(\xi,\lambda))$, and, hence, the operator $Y^\wedge(\xi,\lambda)$ is not Fredholm. □

Corollary 4.1 *If $\lambda_k > \lambda_{k-1}$ for the operator L_b^F, then for this value of k, the following holds:*

$$\sum_{x\in(a,b)} nul\, Y^\wedge(x,\lambda_k) = k-1. \tag{4.8}$$

If $æ(\lambda_k) = n\cdot \dim H < \infty$ and $\lambda_k(\widetilde{L}) > \lambda_{k-1}(\widetilde{L})$, $b<\infty$, then (4.8) also holds for $\widetilde{L}$.

In particular, for a scalar second order equation, $Y^\wedge(x,\lambda) = Y(x,\lambda)$, and (4.8) extends the classical Sturm Oscillation Theorem to the case allowing non-real coefficients of the first derivative term (and hence allowing non-real solution $Y(x,\lambda)$), and also to the case of an unbounded interval because, for $r=2$, (4.8) also holds for $\widetilde{L}$ when $-\infty \le a < b \le \infty$.

If $b<\infty$, the Corollary is obvious. If $b=\infty$, $a>-\infty$, $r=2$, the Corollary was proven in Chapter 1 (see Corollary 1.2).

The equality (4.8) means that for the eigenvalues λ_k of the problem (2.1), $y^\wedge(a) = y^\wedge(b) = 0$ $(-\infty < a < b < \infty)$ such that

$$\text{nul}\, Y^\wedge(x,\lambda_k) \le 1 \quad \text{for all } x \in (a,b), \tag{4.9}$$

b is the k-th adjoint to a point (relative to the equation (2.1) with $\lambda = \lambda_k$). However, without (4.9), this is, generally, not true[4]. An **example** of a boundary problem $y^{IV} + 10y'' + 9y = \lambda y$, $y^\wedge(0) = y^\wedge(\pi) = 0$, where $\lambda = 0$ is an eigenvalue of multiplicity two, is given in [12, p. 111]. In the reference [13], one can find a sufficient condition for the simplicity of eigenvalues of a scalar fourth order equation.

[4]In this regard, some inaccuracies sometimes occur in literature. For example, part B of Theorem 3 in [539] requires some correction (other results in [539] are not directly related to spectrum).

Note that if $n \cdot \dim H > 1$ and if the general boundary condition (2.7) holds, the equality (4.8) may not hold without additional assumptions (see Example 1.1 in Chapter 1).

Remark 4.2 Morse Index Theorem (see Remark 1.1 and Introduction) is a special case of Theorem 4.1 when $r = 2$, $\dim H < \infty$, $-\infty < a < b < \infty$, $\cos A = \cos B = 0$.

4.2 Comparison and Alternation Theorems

Let $r = 2n$. To the differential expression $l[y]$ we associate the quadratic form $l_\Delta[y, y]$—Dirichlet's integral over the interval Δ:

$$l_\Delta[y,y] = \int_\Delta \left\{ \sum_{j=0}^{n} (p_j(x)y^{(j)}, y^{(j)}) \right.$$

$$\left. + \frac{i}{2} \sum_{j=0}^{n-1} \left[(q_{j+1}^*(x)y^{(j+1)}, y^{(j)}) - (q_{j+1}(x)y^{(j)}, y^{(j+1)}) \right] \right\} dx.$$

Theorem 4.2 *Let $-\infty < a < b < \infty$, and let $Y_1(x)$, $Y_2(x)$ be fundamental solutions of two problems of the type (2.1), (2.6): $l^{(k)}[Y_k] = 0$, $k = 1, 2$ respectively, with $A = A_k$ as in (2.6) respectively. Suppose that the corresponding operators $L_b^{(k)0}$ with additional boundary condition $y^\wedge(b) = 0$ are semi-bounded below, and*

$$\inf(\sigma_e(L_b^{(2)0})) > 0, \quad W_k(x) \equiv I, \quad r = 2n,$$

$$l_{(a,b)}^{(1)}[y, y] \le l_{(a,b)}^{(2)}[y, y], \tag{4.10}$$

for all vector-valued functions vanishing in a neighborhood of b.

Suppose that

$$rank \left\{ Y_2^{\vee *} Y_1^\wedge - Y_2^{\wedge *} Y_1^\vee \right\}_{x=a} = m < \infty \tag{4.11}$$

or, equivalently,

$$rank \left\{ \sin A_2 \cdot \cos A_1 - \cos A_2 \cdot \sin A_1 \right\} = m < \infty. \tag{4.12}$$

Then for all $\beta \in (a, b]$, we have

$$\sum_{x \in (a,\beta]} nul\, Y_1^\wedge(x) \ge \sum_{x \in (a,\beta]} nul\, Y_2^\wedge(x) - m. \tag{4.13}$$

The condition (4.10) holds if, in particular, for $j = 0, 1, \ldots, n$,

$$p_j^{(1)}(x) \le p_j^{(2)}(x), \quad q_j^{(1)}(x) = q_j^{(2)}(x), \quad x \in (a,b). \tag{4.14}$$

If $l^{(1)} = l^{(2)}$, *then the sum in (4.13) can be taken over* $x \in [\alpha, \beta] \subset (a,b]$. *If* $\dim H < \infty$ *and (4.14) holds and* m *is replaced by* $n \cdot \dim H$, *then the sum in (4.13) can also be taken over* $x \in [\alpha, \beta]$:

$$\sum_{x\in[\alpha,\beta]} nul\ Y_1^{\wedge}(x) \ge \sum_{x\in[\alpha,\beta]} nul\ Y_2^{\wedge}(x) - n \cdot \dim H. \tag{4.15}$$

Corollary 4.2 *If* $l^{(1)} = l^{(2)}$, *then for all* $[\alpha,\beta] \subset (a,b)$, *we have*

$$\left| \sum_{x\in[\alpha,\beta]} nul\ Y_1^{\wedge}(x) - \sum_{x\in[\alpha,\beta]} nul\ Y_2^{\wedge}(x) \right| \le m, \tag{4.16}$$

where m *is defined by (4.11) and (4.12).*

If $\sum_{x\in[\alpha,\beta]} nul\ Y_1^{\wedge}(x) \ge m+1$, *then the interval* $[\alpha,\beta]$ *contains at least one point at which* $nul\ Y_2^{\wedge}(x) \ge 1$.

In particular, if $\dim H < \infty$, we have $m \le n \cdot \dim H$. Therefore, Theorem 4.2 implies the comparison and alternation theorems of Heinz–Rellich (see [506] and [70, no. 44]) as well as a Comparison Theorem closely related to the formulation of V. I. Arnold [238]. Corollary 4.2 contains a more precise and more general version of the Theorem about Zeros [238] . In addition, Arnold's form of the Alternation Theorem [238] can be obtained from Theorem 4.1 (see the bibliographical comments at the end of this chapter).

Proof. By $M_\beta^{(2)0}$ we denote the self-adjoint operator in $H(\alpha,\beta)$, $\beta \in (a,b]$, corresponding to the differential expression $l^{(2)}[y]$ and self-adjoint boundary conditions $y^{\wedge}(\beta) = 0$ and (2.6) with $A = A_1$. By (2.39), this boundary condition is equivalent to

$$Y_1^{\wedge^*} \cdot y^{\vee}(a) - Y_1^{\vee^*} \cdot y^{\wedge}(a) = 0. \tag{4.17}$$

Note that if $y^{\wedge}(b) = 0$, then $\langle l[y[, y\rangle_{(a,b)} = (y^{\vee}, y^{\wedge})|_a + l_{(a,b)}[y,y]$, and that the quantities $y^{\wedge(1)}$, $y^{\wedge(2)}$ constructed from $l^{(1)}$ and $l^{(2)}$ coincide. Therefore, if both pairs $y^{\vee(1)}$, $y^{\wedge(1)}$ and $y^{\vee(2)}$, $y^{\wedge(2)}$ satisfy (2.6) with the same A, then $(y^{\vee(1)}, y^{\wedge(1)}) = (y^{\vee(2)}, y^{\wedge(2)})$. Thus from (4.10) we have $L_\beta^{(1)0} \le M_\beta^{(2)0}$, and, hence,

$$N_1^F(0) \geq N_M^F(0),\ N_1^F(0) + æ_1^F(0) \geq N_M^F(0) + æ_M^F(0). \tag{4.18}$$

(The quantities $N^F(\lambda)$, $æ^F(\lambda)$ corresponding to the operators $L_\beta^{(1)0}$, $L_\beta^{(2)0}$, $M_\beta^{(2)0}$ are equipped with the indices 1, 2, M respectively). The operators $L_\beta^{(2)0}$, $M_\beta^{(2)0}$ are self-adjoint extensions of their maximal common part Λ defined on vector-valued functions $y \in \mathrm{D}(M_\beta^{(2)0})$ satisfying the condition

$$Y_2^{\wedge^*} \cdot y^\vee(a) - Y_2^{\vee^*} \cdot y^\wedge(a) = 0. \tag{4.19}$$

Lemma 4.2 *Deficiency indices of Λ are (m, m), where*[5]

$$m = rnk\ W_a\{Y_2^*, Y_1\} = rnk\ \{\sin A_2 \cdot \cos A_1 - \cos A_2 \cdot \sin A_1\},$$

$$W_x\{Y_2^*, Y_1\} \equiv \left(Y_2^{\vee^*} Y_1^\wedge - Y_2^{\wedge^*} Y_1^\vee\right)(x).$$

Proof of Lemma 4.2. If $g_1, g_2, \ldots, g_m$ is a basis of the range of W_a, and $h_1, h_2, \ldots h_m \in H$ are such that $W_a h_k = g_k$, $k = 1, 2, \ldots m$, then the vector-valued function $y_k(x) = \phi(x) Y_1(x) h_k$, where $0 \leq \phi(x) \in C_0^\infty$ is a scalar-valued function equal to 1 in a neighborhood of the point a and equal to 0 in a neighborhood of β, belong to $\mathrm{D}(M_\beta^{(2)0})$ and are linearly independent modulo $\mathrm{D}(\Lambda)$. Indeed, if $\sum_{k=1}^{m} c_k y_k \in \mathrm{D}(\Lambda)$, then

$$0 = \sum_{k=1}^{m} c_k [Y_2^{\wedge^*} y_k^\vee(a) - Y_2^{\vee^*} y_k^\wedge(a)] = -\sum_{k=1}^{m} c_k W_a h_k = -\sum_{k=1}^{m} c_k g_k.$$

Thus $c_k = 0$, $k = 1, 2, \ldots, m$, and, hence, $\dim \mathrm{D}(M_\beta^{(2)0})/\mathrm{D}(\Lambda) \geq m$.

Let $z_1(x)$, $z_2(x)$, $\ldots$, $z_{m+1}(x)$ be arbitrary vectors from $\mathrm{D}(M_\beta^{(2)0})$. By (4.17) and by Lemma A.2 on the parametric representation of Hermitian binary relations, we have

$$z_k^\wedge(a) = Y_1^\wedge(a) e_k, \quad z_k^\vee(a) = Y_1^\vee(a) e_k, \quad k = 1, 2, \ldots, m+1,$$

for some $e_k \in H$.

[5] Therefore, rnk $W_a\{Y_2^*, Y_1\} =$ rnk $W_a\{Y_1^*, Y_2\}$.

The vectors $W_a e_1, W_a e_2, \ldots, W_a e_{m+1}$ are linearly dependent. Thus there exists a non-trivial linear combination $\sum_{k=1}^{m+1} \alpha_k W_a e_k = 0$. Since

$$Y_2^{\wedge^*} \sum_{k=1}^{m+1} \alpha_k z_k^{\vee}(a) - Y_2^{\vee^*} \sum_{k=1}^{m+1} \alpha_k z_k^{\wedge}(a) = - \sum_{k=1}^{m+1} \alpha_k W_a e_k = 0,$$

we have $\sum_{k=1}^{m+1} \alpha_k z_k(x) \in \mathrm{D}(\Lambda)$.

Thus $\dim \mathrm{D}(M_\beta^{(2)0})/\mathrm{D}(\Lambda) = m$, and, hence, Def $\{\Lambda\} = m$. This concludes the proof of the Lemma.

Since $M_\beta^{(2)0}$ and $L_\beta^{(2)0}$ are finite-dimensional extensions of Λ, it follows that $\sigma_e(M_\beta^{(2)0}) = \sigma_e(L_\beta^{(2)0})$. By hypotheses of the theorem, we have $\inf \sigma_e(L_\beta^{(2)0}) > 0$. Hence, for all $\beta \in (a, b]$ we have $\inf \sigma_e(M_\beta^{(2)0}) > 0$. Using Lemma 4.1, we get $N_M^F(0) + æ_M^F(0) \geq N_2^F(0) + æ_2^F(0) - m$. Comparing with (4.18), we have $N_1^F(0) + æ_1^F(0) \geq N_2^F(0) + æ_2^F(0) - m$, and, hence, from Theorem 4.1, we obtain (4.13) because

$$\sum_{x \in (\alpha, \beta)} \text{nul } Y_j^{\wedge}(x) + æ_j^F(0) = \sum_{x \in (\alpha, \beta]} \text{nul } Y_j^{\wedge}(x), \; j = 1, 2. \tag{4.20}$$

If $l^{(1)} = l^{(2)}$, then we consider the operators $L_\beta^{(1)0}$, $L_\beta^{(2)0}$, $M_\beta^{(2)0}$ in $H(\alpha - \epsilon, \beta)$, $\epsilon > 0$, with self-adjoint boundary conditions of the form (4.17), (4.19) at the point $\alpha - \epsilon$ defined by the same fundamental solutions $Y_1(x)$, $Y_2(x)$ for $x = \alpha - \epsilon \geq a$ instead of $x = a$. Since $Y_1^{\wedge}(x)$, $Y_2^{\wedge}(x)$ are Fredholm operators (see Theorem 2.2) and since $\inf \sigma_e$ of differential operators on (a, b) is positive, we see from Remark 4.1 that for the above considered differential operators on $H(\alpha - \epsilon, \beta)$, $\inf \sigma_e$ is also positive. If $l^{(1)} = l^{(2)}$, then $W_x\{Y_2^*, Y_1\}$ is independent of x, and, hence, rank $W_{\alpha - \epsilon} =$ rank $W_a = m$.

Therefore, using (4.20), we get (4.13), with the sum taken with respect to $x \in (\alpha - \epsilon, \beta]$. Now letting $\epsilon \to 0$, it follows that

$$\sum_{x \in [\alpha, \beta]} \text{nul } Y_1^{\wedge}(x) \geq \sum_{x \in (\alpha, \beta]} \text{nul } Y_2^{\wedge}(x) - m. \tag{4.21}$$

The assertion of the Theorem in the case $\dim H < \infty$ follows from the inequality $m \leq n \cdot \dim H$. This concludes the proof of the Theorem. □

Corollary 4.2 follows from Theorem 4.2 because we can interchange $Y_1^{\wedge}(x)$ and $Y_2^{\wedge}(x)$ in (4.13) when $l^{(1)} = l^{(2)}$.

4.3 Multiplicative Representation of Positive Differential Operators and Its Applications

Frobenius Factorization Theorem [70, Sec. 44, Theorem 10] can be generalized to the case of finite and infinite system of differential equations as follows:

Theorem 4.3 *Let $\omega(x) \in B(H^n, H)$ be a self-consistent solution of (2.1) with $\lambda = 0$ and $r = 2n$ (definiteness of $p_n(x)$ and semi-boundedness of l is not assumed here). Then at every point $x \in (a,b)$ such that $(\omega^\wedge)^{-1}(x) \in B(H^n)$ exists, the differential expression l in (2.1) admits the multiplicative representation*

$$l = \mu^+ p_n(x)\mu \quad or \quad l[y] = \mu^+[p_n(x)\mu[y]], \tag{4.22}$$

where

$$\mu[y] = y^{(n)} - \omega^{(n)}(x)(\omega^\wedge)^{-1}(x)y^\wedge, \tag{4.23}$$

μ^+ is the formal adjoint of the differential expression $\mu[y]$ in $L_2(H;(a,b);dx)$. Here

$$\mu(\omega) = 0, \quad \mu^+\left[\left\{\underbrace{0_H;\ldots;0_H}_{n-1};I_H\right\}(\omega^\wedge)^{*-1}(x)\right] = 0. \tag{4.24}$$

Recall that for $r = 2n$, $y^\wedge(x)$ and $y^\vee(x)$ are defined by the formulas (2.8), where $y^{[2n-j]}$ are quasi-derivatives.

Proof. By (4.23) we clearly have $\mu[\omega] = 0$. Since the differential expression μ in (4.23) can be written as

$$\begin{aligned}
\mu[y] &= y^{(n)} - \{0_H;\ldots;0_H;I_H\}(d\omega^\wedge/dx)(\omega^\wedge)^{-1}y^\wedge \\
&= \{0_H;\ldots;0_H;I_H\}\left\{I_{H^n}D - \left(\frac{d}{dx}\omega^\wedge\right)(\omega^\wedge)^{-1}\right\}\begin{pmatrix} I \\ ID \\ \vdots \\ ID^{n-1}\end{pmatrix} y,
\end{aligned}$$

where $D = \frac{d}{dx}$, it follows the expression μ^+ which is formally adjoint to μ in $L_2(H; (a,b); dx)$ has the following form:

$$\mu^+[y] = -\left\{I; -ID; \dots; (-1)^{n-1}D^{n-1}\right\}\left\{I_{H^n}D + (\omega^\wedge)^{-1^*}\frac{d}{dx}(\omega^\wedge)^*\right\}\begin{pmatrix}0\\0\\\vdots\\I\end{pmatrix}y.$$

Therefore,

$$\begin{aligned}
&\mu^+\left[\{0;0;\dots;0;I\}(\omega^\wedge)^{*-1}\right] = \left\{-I; ID; \dots; (-1)^nD^{n-1}\right\}\\
&\times\left\{I_{H^n}D + (\omega^\wedge)^{*-1}\frac{d}{dx}(\omega^\wedge)^*\right\}\begin{pmatrix}0\\0\\\vdots\\I\end{pmatrix}\cdot(0;0;\dots;0;I)(\omega^\wedge)^{*-1}\\
&= \left\{-I; ID; \dots; (-1)^nD^{n-1}\right\}\left\{I_{H^n}D + (\omega^\wedge)^{*-1}\frac{d}{dx}(\omega^\wedge)^*\right\}\\
&\times\left[I_{H^n} - \begin{pmatrix}I&&&&\\&I&&&\\&&\ddots&&\\&&&I&\\&&&&0\end{pmatrix}\right](\omega^\wedge)^{*-1}\\
&= \left\{-I; ID; \dots; (-1)^nD^{n-1}\right\}\left[\frac{d}{dx}(\omega^\wedge)^{*-1} + (\omega^\wedge)^{*-1}\left(\frac{d}{dx}(\omega^\wedge)^*\right)(\omega^\wedge)^{*-1}\right.\\
&\left.-\begin{pmatrix}D&&&&\\&D&&&\\&&\ddots&&\\&&&D&\\&&&&0\end{pmatrix}(\omega^\wedge)^{*-1} - (\omega^\wedge)^{*-1}\cdot(\omega^\wedge)^*\begin{pmatrix}0&&&&\\I&0&&&\\&I&\ddots&&\\&&\ddots&0&\\&&&I&0\end{pmatrix}(\omega^\wedge)^{*-1}\right]\\
&= \left\{I; -ID; \dots; (-1)^{n-1}D^{n-1}\right\}\begin{pmatrix}D&&&&\\I&D&&&\\&I&\ddots&&\\&&\ddots&D&\\&&&I&0\end{pmatrix}(\omega^\wedge)^{*-1} = 0.
\end{aligned}$$

It can be directly checked that

$$V(x,t) = \omega(x) \cdot (\omega^\wedge)^{-1}(t) \cdot \begin{pmatrix} 0 \\ \vdots \\ 0 \\ I \end{pmatrix} : H \to H$$

is an operator-valued Cauchy function for $\mu[y]$, and so is $-V^*(t,x)$ for the formal adjoint $\mu^+[y]$. In particular, by (4.24), we have $\mu_x^+[V^*(t,x)] = 0$, and

$$\mu_x^+ \left[\int_\xi^x V^*(t,x) l_t[y]\, dt \right] = -l_x[y], \tag{4.25}$$

provided that at the point $\xi \in (a,b)$ the operator $\omega^\wedge(x)$ has a bounded inverse on the whole H^n and x belongs to a neighborhood of ξ in which $\omega^\wedge(x)$ has a bounded inverse on the whole H^n (such a neighborhood of ξ exists since $\omega^\wedge(x)$ is continuous). Integrating by parts on the left hand side of (4.25), we get

$$\begin{aligned} &\int_\xi^x V^*(t,x) l_t[y]\, dt = \int_\xi^x l_t\,[V(t,x)]^*\, y(t)\, dt \\ &+ \left[V^{\vee^*}(t,x) y^\wedge(t) - V^{\wedge^*}(t,x) y^\vee(t) \right] |_{t=\xi}^x \\ &= \{0,\ldots,0,I\} \left\{ (\omega^\wedge)^{*-1}(x) \cdot (\omega^\vee)^*(x) y^\wedge(x) - y^\vee(x) \right\} \\ &-\{0,\ldots,0,I\} (\omega^\wedge)^{*-1}(x) \left\{ (\omega^\vee)^*(\xi) y^\wedge(\xi) - (\omega^\wedge)^*(\xi) y^\vee(\xi) \right\}. \end{aligned} \tag{4.26}$$

Since $\omega(x)$ is a self-consistent solution, we have

$$(\omega^\vee)^*(\omega^\wedge) - (\omega^\wedge)^*(\omega^\vee) = 0,$$

and, hence, $(\omega^\wedge)^{*-1}(\omega^\vee)^* = \omega^\vee(\omega^\wedge)^{-1}$. Using (2.8) and the equality $y^{[n]} = p_n y^{(n)} - \frac{i}{2} q_n y^{(n-1)}$, the first half of the right hand side of (4.26) is equal to

$$\begin{aligned} &\{0,\ldots,0,I\} (\omega^\vee)(x) \cdot (\omega^\wedge)^{-1}(x) \cdot y^\wedge(x) - y^{[n]}(x) \\ &= \omega^{[n]}(x) \cdot (\omega^\wedge)^{-1}(x) \cdot y^\wedge(x) - y^{[n]}(x) \\ &= p_n(x) \left\{ \omega^{(n)}(x) \cdot (\omega^\wedge)^{-1}(x) \cdot y^\wedge(x) - y^{(n)}(x) \right\} \\ &-\frac{i}{2} q_n(x) \left\{ \omega^{(n-1)}(x) \cdot (\omega^\wedge)^{-1}(x) \cdot y^\wedge(x) - y^{(n-1)}(x) \right\} \\ &= -p_n(x)\mu[y] - \frac{i}{2} q_n(x) \left\{ \{0,\ldots,0,I\} y^\wedge(x) - y^{(n-1)}(x) \right\} \\ &= -p_n(x)\mu(y). \end{aligned}$$

Using (4.25), (4.26) and (4.24), we get (4.22). The Theorem is proven. □

As an application of Theorem 4.2, we will establish a sufficient condition for the oscillatory property of $2n$-th order equations.

Definition 4.1 The $2n$-th order equation (2.1) with a positive leading coefficient is said to be oscillatory for $\lambda = 0$ in a neighborhood of $x = \infty$ if it does not have self-consistent solutions $Y(x,0) \in D(H^n, H)$ such that $(Y^\wedge)^{-1}(x,0) \in B(H^n)$ exists for all sufficiently large x.

If $\dim H < \infty$, Definition 4.1 is equivalent to the standard definition from [70].

Let g be a positive linear functional on the C^*-algebra $B(H)$. A large class of such functionals has the form $g(T) = \mathrm{Tr}(AT)$, where $A = A^* \geq 0$ is a trace class operator. If $\dim H < \infty$, then all positive functionals on $B(H)$ are of such form; see [910].

We define $g(D^j p(x) D^k) = D^j g(p(x)) D^k$, where $p(x) \in B(H)$.

Theorem 4.4 *If the scalar equation $g(l)[u] = 0$ is oscillatory, then the $2n$-th order operator equation (2.1) with the positive leading coefficient is also oscillatory for $\lambda = 0$.*

Proof. Let $g(l)[u] = 0$ be oscillatory. Then (see [70]) $\forall N > 0$, there exist $\alpha > 0$, $x_2 > x_1 > N$ such that there exists a non-trivial solution $f : \mathbb{R} \to \mathbb{C}$ of the scalar problem $g(l)[f] = -\alpha f$, $f^\wedge(x_1) = f^\wedge(x_2) = 0$. On the other hand, if the equation $l[u] = 0$ is not oscillatory, then it has a solution $\omega(x)$ satisfying the hypotheses of Theorem 4.3 for all $x > N > 0$, where N is a sufficiently large number. Thus for all $x > N$ the differential expression l has a multiplicative representation (4.22). Integrating by parts and applying the functional g to both sides of the equality, we arrive at a contradiction:

$$0 \leq g\left(\int_{x_1}^{x_2} (\mu[f])^* p_n(x) \mu[f]\, dx\right) = \int_{x_1}^{x_2} \overline{f}(x) g(l) f(x)\, dx$$
$$= -\alpha \int_{x_1}^{x_2} |f|^2\, dx < 0.$$

This concludes the proof of the Theorem. □

The following theorem generalizes a theorem of M. G. Krein, Heinz and Rellich; see [70, no. 44], [506], [585] and [761].

Theorem 4.5 *Let $p_n(x) >> 0$, $r = 2n$. Then the minimal operator L is non-negative if and only if one of the following two conditions holds:*

1) the differential expression l in (2.1) can be represented in the form (4.22), where $\mu[y] = y^{(n)} + s_{n-1}(x)y^{(n-1)} + \cdots + s_0(x)y$, $a < x < b$, $s_k(x) \in C^n(a,b)$, $k = 0, 1, \ldots, n-1$.

2) there exists $\omega(x)$ satisfying the conditions of Theorem 4.3, including the invertibility of $\omega^\wedge(x)$, for all $x \in (a,b)$.

Remark 4.3 If the above considered differential operator is positive definite ($L >> 0$), then it also admits the factorization in the closure of (a,b) and $(\omega^\wedge)^{-1} \in B(H^n)$ on $[a,b]$. If $-\infty < a < b < \infty$, then the converse also holds.

Proof. If any of the conditions 1), 2) holds, then the differential expression l admits the multiplicative representation (4.22), and $l = \mu^+ p_n \mu$ is clearly non-negative (even for $p_n \geq 0$). If $-\infty < a < b < \infty$ and if l admits a multiplicative representation in $[a,b]$ with $(\omega^\wedge)^{-1} \in B(H^n)$ on $[a,b]$, then the minimal operator L is positive definite. Otherwise, the self-adjoint extension L_b^0 determined by the conditions $y^\wedge(b) = 0$ and

$$\omega^{\wedge^*}(a)y^\vee(a) - \omega^{\vee^*}(a)y^\wedge(a) = 0, \tag{4.27}$$

would satisfy the inequality $\inf \sigma(L_b^0) \leq 0$. Since $\inf \sigma(L_\xi^0)$, where L_ξ^0 is determined by the expression l in (2.1) and by the conditions (4.27) and $y^\wedge(\xi) = 0$, continuously and strictly monotonically depends on ξ and since $\inf \sigma(L_\xi^0) \nearrow +\infty$ as $\xi \searrow a$ (see Theorem 3.2; the condition (4.27) is semi-bounded[6]), it would be possible to find a point $\xi \in (a,b)$ such that $0 \in \sigma(L_\xi^0)$. But then by Lemma 2.4 we would have $0 \in \sigma(\omega^\wedge(\xi))$, and this contradicts our assumption.

Now let $L \geq 0$. If a (or b) is finite, then we take $\omega(x)$ to be the fundamental solution of the equation $l[y] = 0$ with the initial data: $\omega^\wedge(a) = 0$, $\omega^\vee(a) = I$ (or, correspondingly, $\omega^\wedge(b) = 0$, $\omega^\vee(b) = I$). If for some $\xi \in (a,b)$ it happened that $0 \in \sigma(\omega^\wedge(\xi))$, then by Lemma 2.4 we would have $0 \in \sigma(L_\xi^0)$, where L_ξ^0 is the self-adjoint operator determined by the problem (2.1), $y^\wedge(\xi) = 0$, $y^\wedge(a) = 0$ ($y^\wedge(b) = 0$). Therefore, since the greatest lower bound of this problem is strictly monotone in ξ (see Theorem 3.3), it would follow that $\inf \sigma(L^F) < 0$, contradicting the non-negativity of L. Thus $(\omega^\wedge)^{-1} \in B(H^n)$ for all $x \in (a,b)$. Now by Theorem 4.3, the representation (4.22) is valid on (a,b), and $\mu[y]$ satisfies the specified properties.

If the operator L is positive definite and if a (or b) is finite, then, choosing at the other end-point the self-adjoint boundary condition $U_b^F[y] = 0$

[6]Since $[\omega^{\wedge^*}(a)]^{-1} \in B(H^n)$.

corresponding to the Friedrichs extension, we have $L_\xi >> 0$, where L_ξ is the minimal operator in the space $H(\xi, b)$ (respectively, in $H(a, \xi)$) with respect to the point $\xi \in [a, b)$ ($\xi \in (a, b]$). Therefore 0 is a point of regular type[7] for L, and there exists a fundamental solution $\omega(x) = Y(x, 0)$ of the problem $l[y] = 0$, $U_b^F[y] = 0$ (or $U_a^F[y] = 0$). Here $\omega^\wedge(x)$ has a bounded inverse on the whole H^n for all $x \in [a, b)$ ($x \in (a, b]$). If a and b are finite, then since the infimum of the spectrum of L^F continuously depends on a and b (see Theorem 3.2), it follows that there exist ϵ, $\epsilon' > 0$ such that the minimal operator in $H(a - \epsilon, b + \epsilon')$ is positive definite. Therefore, there exists a solution $\omega(x)$ satisfying the conditions of Theorem 4.3 on $(a - \epsilon, b + \epsilon') \supset [a, b]$.

Thus, if at least one of the end-points of (a, b) is finite, Theorem 4.5 and Remark 4.3 are proven.

Now let $a = -\infty$ and $b = \infty$. If $0 = \inf \sigma_e(L)$, then the fundamental solution $Y(x, \lambda)$ of the problem (2.1), (2.3) as constructed in Theorems 2.1 and 2.3 might not exist for $\lambda = 0$. On the infimum of the essential spectrum of the minimal operator L, we will construct the self-consistent operator-valued solution $\omega(x) \in B(H^n, H)$ of the equation (2.1) on the real axis satisfying the conditions of Theorem 4.3. This will conclude the proof of Theorem 4.5.

Let $Y_0(x)$, $Y_1(x) \in B(H^n, H)$ be the solutions of (2.1) for $\lambda = 0$ and with the boundary conditions

$$Y_0^\wedge(0) = I_n, \quad Y_0^\vee(0) = 0_n, \quad Y_1^\wedge(0) = 0_n, \quad Y_1^\vee(0) = I_n.$$

Set $\omega_\alpha(x) = Y_0(x) + Y_1(x)C_\alpha$, $0 \neq \alpha \in \mathbb{R}$, where

$$C_\alpha = -(Y_1^\wedge)^{-1}(\alpha) \cdot Y_0^\wedge(\alpha) \in B(H^n) \tag{4.28}$$

is chosen so that $\omega_\alpha^\wedge(\alpha) = 0$. (The existence of $(Y_1^\wedge)^{-1}(\alpha) \in B(H^n)$ is guaranteed for $\alpha \neq 0$ since $L \geq 0$ on the real axis.)

We will show that $\omega_\alpha(x)$ is a self-consistent fundamental solution of

$$l[y] = 0, \quad y^\wedge(\alpha) = 0, \quad \alpha \neq 0, \quad x \in \mathbb{R}. \tag{4.29}$$

First of all, note that $C_\alpha^* = C_\alpha$ since

$$0 = W_a\left\{\omega_a^*, \omega_a\right\} = W_0\left\{\omega_a^*, \omega_a\right\} = C_\alpha^* - C_\alpha.$$

[7] $\lambda \in \mathbb{C}$ is called a point of regular type of a linear operator T if the operator $(T - \lambda I)^{-1}$ exists and is bounded (although T may have the domain which is not dense in H).

Thus, $\omega_\alpha(x)$ satisfies the following self-adjoint boundary condition for $x = 0$:

$$y^\vee(0) - C_\alpha y^\wedge(0) = 0. \tag{4.30}$$

Since $\omega_\alpha^\wedge(0) = I_n$ and $\omega_\alpha^\vee(0) = C_\alpha$, it follows that any solution of $l[y] = 0$ satisfying (4.30) can be represented as $y = \omega_\alpha(x)h$, $h \in H^n$, and

$$M_{\omega_\alpha}(x) \equiv \sum_{k=0}^{2n-1} \omega_\alpha^{(k)^*}(x)\omega_\alpha^{(k)} >> 0, \tag{4.31}$$

because $M_{\omega_\alpha}(0) \geq I_n >> 0$ (see (2.24)). Therefore, $\omega_\alpha(x)$ is a self-consistent fundamental solution of the problem $l[y] = 0$ and (4.30). Thus by Lemma 2.3, there exists $[\omega_\alpha^\vee(x) + i\omega_\alpha^\wedge(x)]^{-1} \in B(H^n)$, and for $x = \alpha$, this implies the existence of $[\omega_\alpha^\vee(\alpha)]^{-1} \in B(H^n)$. Hence, any solution of (4.29) can be represented as $\omega_\alpha(x)h$, $h \in H^n$, and, thus, using (4.31), we see that $\omega_\alpha(x)$ is a self-consistent fundamental solution of (4.29). Therefore, for all $x \neq \alpha$, the operator $\omega_\alpha^\wedge(x)$ has a bounded inverse on all of H^n. This can be seen as follows: if there exists $\xi \neq \alpha$ such that $0 \in \sigma(\omega_\alpha^\wedge(\xi))$, then by Lemma 2.4, we have $0 \in \sigma(L^0_{\alpha,\xi})$, where $L^0_{\alpha,\xi}$ is the self-adjoint operator determined by (2.1) with $y^\wedge(\alpha) = y^\wedge(\xi) = 0$ on the interval $[\alpha, \xi]$ ($[\xi, \alpha]$). But this is impossible since $L \geq 0$ on $(-\infty, \infty)$ and since the infimum of $\sigma(L^0_{\alpha,\xi})$ is strictly decreasing with the expansion of a bounded interval (see Theorem 3.3).

From (4.28) we have

$$\frac{dC_\alpha}{d\alpha} = (Y_1^\wedge)^{-1}\left(\frac{d}{d\alpha}Y_1^\wedge\right)(Y_1^\wedge)^{-1}Y_0^\wedge - (Y_1^\wedge)^{-1}\frac{d}{d\alpha}Y_0^\wedge. \tag{4.32}$$

Using the system notation (3.40) with $\lambda = 0$ for the equation (2.1) with $r = 2n$, from (4.32) we get

$$\begin{aligned}\frac{dC_\alpha}{d\alpha} &= (Y_1^\wedge)^{-1}\left\{(B_{21}Y_1^\wedge + B_{22}Y_1^\vee)(Y_1^\wedge)^{-1}Y_0^\wedge - (B_{21}Y_0^\wedge + B_{22}Y_0^\vee)\right\}\\ &= (Y_1^\wedge)^{-1}B_{22}\left\{Y_1^\vee(Y_1^\wedge)^{-1}Y_0^\wedge - Y_0^\vee\right\}.\end{aligned}$$

Since $0 = W_0\{Y_1^*, Y_1\} = W_\alpha\{Y_1^*, Y_1\}$, it follows that

$$Y_1^\vee(\alpha)(Y_1^\wedge)^{-1}(\alpha) = (Y_1^\wedge)^{*-1}(\alpha)Y_1^{\vee^*}(\alpha).$$

Thus

$$\frac{dC_\alpha}{d\alpha} = (Y_1^\wedge)^{-1}B_{22}\left\{(Y_1^\wedge)^{*-1}(Y_1^\vee)^*Y_0^\wedge - (Y_1^\wedge)^{*-1}(Y_1^\wedge)^*Y_0^\vee\right\}$$

$$= (Y_1^\wedge)^{-1} B_{22} (Y_1^\wedge)^{*-1} W_\alpha \{Y_1^*, Y_0\} = (Y_1^\wedge)^{-1} B_{22} (Y_1^\wedge)^{*-1} W_0 \{Y_1^*, Y_0\}$$
$$= (Y_1^\wedge)^{-1} B_{22} (Y_1^\wedge)^{-1^*} \geq 0 \tag{4.33}$$

(the last inequality holds by (3.41) and since $p_n >> 0$).

We will show that $|C_\alpha| <$ const as $\alpha \to \pm\infty$ (and, generally, for all $\alpha \in \mathbb{R} \setminus (-\epsilon, \epsilon)$, $\epsilon > 0$).

From (4.33) it follows that the upper bound of the operator $C_\alpha = C_\alpha^*$ remains bounded above as $\alpha \to -\infty$. We claim that $\inf \sigma(C_\alpha)$ is bounded below as $\alpha \to -\infty$. Suppose that there exists a sequence $\alpha_k \to -\infty$ such that $\inf \sigma(C_{\alpha_k}) \to -\infty$. Then, for sufficiently large k, the self-adjoint boundary problem

$$l_W[y] = W^{-1} l[y] = \lambda y, \quad y^\vee(0) - C_{\alpha_k} y^\wedge(0) = 0, \quad y^\wedge(1) = 0$$

in $H(0,1) = L_2\{H; (0,1); W(x)dx\}$ has negative greatest lower bound, as seen by considering the corresponding Dirichlet-type quadratic form in $H(0,1)$ with the term $(C_{\alpha_k} y^\wedge(0), y^\wedge(0))$ outside the integral. But then, for such values of k, there exists $x_k \in (0,1)$ such that $0 \in \sigma(\omega_{\alpha_k}^\wedge(x_k))$, contradicting the above proven property $[\omega_\alpha^\wedge(x)]^{-1} \in B(H^n)$ for all $x \neq \alpha$.

Since C_α is bounded and monotone as $\alpha \to -\infty$, there exists a self-adjoint strong limit $C = s - \lim_{\alpha \to -\infty} C_\alpha$.

Let $\omega(x) = Y_0(x) + Y_1(x)C$. Since C_α is non-decreasing for $\alpha \neq 0$, it follows that for all $\xi < 0$, the greatest lower bound of the operator determined by $l_W[y]$ and self-adjoint boundary conditions $y^\wedge(\xi) = 0$, $y^\vee(0) - C_\alpha y^\wedge(0) = 0$ in $H(\xi, 0)$ does not decrease as $\alpha \searrow -\infty$. Therefore, the differential operator determined by (2.1), $y^\wedge(\xi) = 0$, $y^\vee(0) - C y^\wedge(0) = 0$ has a positive greatest lower bound, and, by Theorem 4.1, the operator $\omega^\wedge(x)$ has a bounded inverse on the whole H^n for all $x \leq 0$. On the positive half-line, we will consider the family of operators L_α, where L_α is the Friedrichs extension with respect to $x = \infty$ of the operator L'_α in $H(0,\infty)$ determined by the expression $l_W[y]$ and the boundary condition (4.30) at zero. The operator $L_\alpha \geq 0$ is non-increasing as α decreases by (4.33). Therefore, as $\alpha \to -\infty$, we have $L_{(-\infty)} \geq 0$, and the operator $\omega^\wedge(x)$ has a bounded inverse on H^n for all $x \in \mathbb{R}$. This concludes the proof of the Theorem. □

Corollary 4.3 *Assume that for some non-trivial vector-valued function $h(x) \in C^{2n}(\alpha, \beta)$, $h^{(k)}(\alpha) = h^{(k)}(\beta) = 0$, $k = 0, 1, \ldots, n-1$, $a \leq \alpha < \beta \leq b$, the following holds: $\int_\alpha^\beta (l[h], h)\, dx \leq 0$ (in particular, if α and β are conjugate points). Then for every self-consistent solution ω, $l[\omega] = 0$, there* *exists a point $x \in [\alpha, \beta]$ at which $\omega^\wedge(x)$ does not have a bounded inverse on the whole H^n. If the inequality is strict, then such a point x exists in (α, β).*

Indeed, suppose that there exists a self-consistent solution $\omega(x)$ of the equation $l[\omega] = 0$ such that $\omega^\wedge(x)$ has a bounded inverse on the whole H^n for all $x \in [\alpha, \beta]$. Then by Remark 4.3, the minimal operator in $H(\alpha, \beta)$, and, also, its Friedrichs extension are positive definite, contradicting $h \in \mathrm{D}(L^F)$ and $\langle L^F h, h \rangle \leq 0$. If the inequality is strict, then using similar arguments and Theorem 4.5, we obtain the second assertion of the Corollary.

4.4 Discrete Levels in the Gaps of the Essential Spectrum

In this section, we will require neither the semi-boundedness of the minimal operator $L \subseteq \widetilde{L}$, where $\widetilde{L}$ is a self-adjoint extension of L, nor that the order of the differential equation (2.1) is even (we allow $r = 2n$ and $r = 2n - 1$). We will assume that the coefficients of the differential expression $l_W[y]$ in (2.2) are C^{2r}-differentiable. Then we can consider the expression

$$l_W^2[y] = l_W[l_W[y]] \tag{4.34}$$

as an ordinary differential expression.

By M we will denote the minimal differential operator determined by the differential expression (4.34) in $H(a, b)$. Since the operators M and $(\widetilde{L})^2$ coincide on C_0^∞, it follows that they coincide on the domain of M. Thus $M \subseteq (\widetilde{L})^2$, and, hence, the operator $(\widetilde{L})^2$ is a self-adjoint extension $\widetilde{M}$ of the positive symmetric operator M.

By M_b we denote the restriction of $\widetilde{M}$ by the minimality requirement with respect to the end-point b. In what follows,

$$p = \text{def}\left\{\widetilde{M}|\mathrm{D}(\widetilde{M}) \cap \mathrm{D}(M_b^F)\right\},$$

and $N(\lambda, \mu)$ is the number of eigenvalues $\lambda_k \in (\lambda, \mu)$ of the operator $\widetilde{L}$ counting multiplicities æ(λ_k).

Let $Y(x, \lambda)$ be a fundamental solution of the problem (2.1), (2.3). Let[8]

$$Y(x, \lambda, \mu) = \{Y(x, \lambda); Y(x, \mu)\} : \mathbf{H}^2 \to H \tag{4.35}$$

[8]In (4.36) we could take quasi-derivatives; however, this is not necessary. In this book, the reason why we use quasi-derivatives is not the insufficient smoothness of coefficients, but convenience in writing Lagrange's identity.

$$Y^{\Delta}(x,\lambda,\mu) = \text{col}\ \left\{Y(x,\lambda,\mu); Y'(x,\lambda,\mu); \ldots ; Y^{(r-1)}(x,\lambda,\mu)\right\} : \mathbf{H}^2 \to H^r \tag{4.36}$$

Theorem 4.6 *Assume that (α, β) is a gap in the essential spectrum of the operator $\widetilde{L}$, $\alpha < \lambda < \mu < \beta$. Then*

$$N(\lambda,\mu) - p \leq \sum_{x\in(a,b)} nul\ Y^{\Delta}(x,\lambda,\mu) \leq N(\lambda,\mu), \tag{4.37}$$

and

$$nul\ Y^{\Delta}(x,\lambda,\mu) = nul\ Y^{\Delta^*}(x,\lambda,\mu). \tag{4.38}$$

If $æ_{L_b}(\lambda) = æ_{L_b}(\mu) = 0$, where $L_b \subseteq \widetilde{L}$ is the minimal operator with respect to the end-point b, then, instead of p in (4.37), we can take

$$\min\{p, def\ M_b - æ(\lambda) - æ(\mu)\}.$$

If $a > -\infty$, then the Theorem also holds for $\lambda = \alpha$, $\mu = \beta$.

Remark 4.4 If $b < \infty$, then $0 < p = \frac{r}{2} \dim H$, and def $M_b = r \cdot \dim H$. (From the inequality $p > 0$, we see that for all $b < 0$, the Friedrichs extension M_b^F cannot be obtained as the square of a self-adjoint differential operator $\widetilde{L}$, i.e. the operator $(M_b^F)^{1/2}$ is not differential although $M_b^F \geq 0$.)

Proof. The self-adjoint extension of the minimal (closed) operator M leading to the self-adjoint operator $(\widetilde{L})^2$ is determined by the following boundary conditions:

$$U_a[y] = 0, \quad U_a[l_W[y]] = 0, \tag{4.39}$$

$$U_b[y] = 0, \quad U_b[l_W[y]] = 0, \tag{4.40}$$

$$l_W[y] \in H(a,b). \tag{4.41}$$

The condition (4.41) may completely or partially follow from (4.39), (4.40) and from the requirement $y \in \mathrm{D}(M^*)$. The conditions (4.39), (4.40) together generate a self-adjoint extension of the minimal operator M if and only if for all $y \in \mathrm{D}(M^*)$, (4.39), (4.40) imply that $l_W[y] \in H(a,b)$.

If the deficiency indices of L are finite, then by a theorem about the square of a closed symmetric operator with finite deficiency indices (see [70,

Sec. 1, Theorem 22]), we have def $(L^2) = 2\text{def }(L)$. Since def $M \geq$ def (L^2), we have

$$\text{def } M \geq 2\text{def } L. \tag{4.42}$$

(If def $L = \infty$, then both sides of (4.42) are infinite.) If the inequality is strict, then (4.41) does not completely follow from (4.39), (4.40) and the requirement $y \in \mathrm{D}(M^*)$. The examples in which the strict inequality holds are known already in the scalar case with $r = 2$; see [338]. The conditions (4.39)–(4.41) are a generaliziation of the inequality (4.42) and the conditions when this inequality becomes an equality[9] to the case def $L = \infty$.

The condition (4.41) is linear and separated in the following sense: if $y \in \mathrm{D}(M^*)$ satisfies (4.41), then any function $f \in \mathrm{D}(M^*)$ that coincides with y in a neighborhood of a (or b) and is equal to 0 in a neighborhood of b (or a), respectively, also satisfies (4.41). This implies that any function $\phi \in \mathrm{D}(M^*)$ coinciding with y in neighborhoods of a and b satisfies (4.41). The condition (4.41) can be written in a different way:

$$\int_a \left|W^{1/2} l_W[y]\right|^2 dx < \infty, \tag{4.43}$$

$$\int^b \left|W^{1/2} l_W[y]\right|^2 dx < \infty. \tag{4.44}$$

Without the loss of generality, we will assume that $\lambda = -\mu < 0$. Otherwise, we will consider the differential expression

$$l_c[y] = l_W[y] - cy, \quad c = \frac{\lambda + \mu}{2}.$$

Since $\mu^2 < \inf \sigma_e(M_b) \leq \min\{\alpha^2, \beta^2\}$, by Theorem 2.1, the problem

$$l_W^2[y] = \mu^2 y \tag{4.45}$$

with the conditions (4.39), (4.43) has a fundamental operator-valued solution $Z(x, \mu^2)\colon \mathbf{H}^2 \to H$.

On the other hand, the operator-valued function $Y(x; -\mu; \mu)\colon \mathbf{H}^2 \to H$ in (4.35) is also a solution of (4.45), and for all $h \in \mathbf{H}^2$, the vector-valued

[9] In the case of a scalar equation, the inequality (4.42) and the necessary and sufficient conditions for the equality to hold in (4.42) are given in a remark of A. Zettl on the paper [412]: the equality in (4.42) holds if and only if $y \in \mathrm{D}(M^*)$ implies $l_W[y] \in H(a,b)$. In this book, the sufficient condition is stronger, while the necessary condition is stronger in [412]. In the whole, the necessary and sufficient conditons are equivalent; see also [933].

function

$$y_h(x,\mu) = Y(x;-\mu;\mu)h \tag{4.46}$$

satisfies the conditions (4.39), (4.43). The operator $Y(x;-\mu;\mu) = \{Y(x,-\mu);Y(x,\mu)\}$ is non-singular for all $\mu \neq 0$, i.e. if $Y(x;-\mu;\mu)h \equiv 0$, $h = h_- \oplus h_+ \in \mathbf{H}^2$, then $h = 0$. Indeed, if for some h we have $Y(x;-\mu;\mu)h \equiv 0$, then

$$Y(x,-\mu)h_- + Y(x,\mu)h_+ \equiv 0. \tag{4.47}$$

Then

$$0 = l_w[Y(x,-\mu)h_- + Y(x,\mu)h_+] = -\mu Y(x,-\mu)h_- + \mu Y(x,\mu)h_+. \tag{4.48}$$

Comparing (4.47) and (4.48), for all $\mu \neq 0$, we get $Y(x,-\mu)h_- = Y(x,\mu)h_+ \equiv 0$, and, hence, $h_- = h_+ = 0$.

Since $Z(x,\mu^2)$ is a fundamental solution of (4.45), (4.39), (4.43), it follows that for all $h \in \mathbf{H}^2$ and for every function

$$y_h(x,\mu) = Y(x;-\mu;\mu)h, \tag{4.49}$$

there exists $g \in \mathbf{H}^2$ such that $y_h(x,\mu) = z_g(x,\mu)$, where

$$z_g(x,\mu) = Z(x,\mu^2)g. \tag{4.50}$$

Conversely, we will show that for every $g \in \mathbf{H}^2$, if the function $y(x,\mu) = z_g(x,\mu)$ is defined by the formula (4.50), then there exists $h \in \mathbf{H}^2$ such that (4.49) holds. Then the following equality holds:

$$Y(\cdot;-\mu;\mu)h = Z(\cdot,\mu^2)g, \tag{4.51}$$

where one of the vectors $h,\ g \in \mathbf{H}^2$ can be chosen arbitrarily, while the other is determined from the chosen one.

The vector-valued function

$$u(x,\mu) = z_g(x,\mu) - Y(x,\mu)h_+ - Y(x,-\mu)h_-,\ h_\pm \in \mathbf{H} \tag{4.52}$$

is a solution of the problem (4.45), (4.39), (4.43). We will show that we can choose $h_\pm \in \mathbf{H}$ so that

$$(l_W[u])^\wedge(c) = i\mu u^\vee(c,\mu), \tag{4.53}$$

$$(l_W[u])^\vee(c) = -i\mu u^\wedge(c,\mu), \tag{4.54}$$

where $c \in (a, b)$. Integrating by parts, we get

$$0 = \langle l_W^2[u], u\rangle - \langle u, l_W^2[u]\rangle = ((l_W[u])^\wedge(c), u^\vee(c))$$
$$- ((l_W[u])^\vee(c), u^\wedge(c)) + (u^\wedge(c), (l_W[u])^\vee(c)) - (u^\vee(c), (l_W[u])^\wedge(c))$$
$$= 2i\mu\left[(u^\vee(c), u^\vee(c)) + (u^\wedge(c), u^\wedge(c))\right].$$

If $\mu \neq 0$, then $u^\wedge(c) = u^\vee(c) = 0$, and, hence, by (4.53) and (4.54), we have $(l_W[u])^\wedge(c) = (l_W[u])^\vee(c) = 0$. Therefore, $u(x,\mu) \equiv 0$, and this implies (4.51).

We will now show (4.53) and (4.54). Using

$$l_W[u] = l_W[z_g] - \mu \cdot Y(x,\mu)h_+ + \mu \cdot Y(x,-\mu)h_-$$

and (2.40), the conditions (4.53) and (4.54) become

$$(l_W[z_g])^\wedge(c) - i\mu z_g^\vee(c) = -\mu i V_+(c,\mu)h_+ - \mu i V_-(c,-\mu)h_-$$
$$(l_W[z_g])^\vee(c) + i\mu z_g^\wedge(c) = \mu V_+(c,\mu)h_+ - \mu V_-(c,-\mu)h_-.$$

This system is uniquely solvable with respect to $h_\pm$ since the operators $V_\pm(c, \pm\mu)$ are invertible (see Lemma 2.3). Taking $h_\pm$ in (4.52) to be the solution of this system, it follows that $u(x,\mu)$ satisfies (4.53) and (4.54). Thus, the representation (4.51) is proven.

Since $Z(x,\mu^2)$ is a fundamental solution of the $2r$-th order problem (4.45), (4.39), (4.43), it follows that

$$Z^\wedge(x,\mu^2) = \text{col}\left\{Z(x,\mu^2), Z'(x,\mu^2), \ldots, Z^{(r-1)}(x,\mu^2)\right\} : \mathbf{H}^2 \to H^r.$$

By (4.36) and (4.51) we have

$$Z^\wedge(x,\mu^2)g = Y^\Delta(x,-\mu,\mu)h. \tag{4.55}$$

We will show that for all $\mu \neq 0$, we have

$$\text{nul } Z^\wedge(x,\mu^2) = \text{nul } Y^\Delta(x,-\mu,\mu). \tag{4.56}$$

If for some $x \in (a,b)$ there exist linearly independent vectors h_1, h_2, …, $h_k \in \mathbf{H}^2$ such that $Y^\Delta(x,-\mu,\mu)h_i = 0$, $i = 1, 2, \ldots, k$, then the corresponding vectors g_1, $g_2, \ldots$, $g_k \in \mathbf{H}^2$ defined by the formula (4.51) are also linearly independent. Otherwise, by (4.51), there exists a non-trivial linear combination $c_1h_1 + c_2h_2 + \cdots + c_kh_k$ such that $Y(x,-\mu,\mu)(c_1h_1 + c_2h_2 + \cdots + c_kh_k) \equiv 0$, and since the operator solution $Y(x,-\mu,\mu)$ is non-singular (in the sense of the definition given below the equation (4.46)), we have $c_1h_1 + c_2h_2 + \cdots + c_kh_k = 0$, contrary to the assumption on linear independence of h_1, h_2, ..., h_k. Therefore,

$$\text{nul } Y^{\Delta}(x, -\mu, \mu) \leq \text{nul } Z^{\wedge}(x, \mu^2). \tag{4.57}$$

Analogously, it can be shown that

$$\text{nul } Z^{\wedge}(x, \mu^2) \leq \text{nul } Y^{\Delta}(x, -\mu, \mu). \tag{4.58}$$

Comparing (4.57) and (4.58), we get (4.56).

If for some $x \in (a, b)$ we have $h_0 \in \text{coker } Y^{\Delta}(x, -\mu, \mu)$, then for all $h \in \mathbf{H}^2$ we have $(Y^{\Delta}(x, -\mu, \mu)h, h_0) = 0$. By (4.51), we have $(Z^{\wedge}(x, \mu^2)g, h_0) = 0$ for all $g \in \mathbf{H}^2$, and, hence, $\text{coker } Y^{\Delta}(x, -\mu, \mu) \subseteq \text{coker } Z^{\wedge}(x, \mu^2)$. The opposite inclusion can be proven similarly, and, hence,

$$\text{coker } Y^{\Delta}(x, -\mu, \mu) = \text{coker } Z^{\wedge}(x, \mu^2). \tag{4.59}$$

By Theorem 2.2, the fundamental solution $Z^{\wedge}(x, \mu^2)$ is Fredholm. Comparing (4.56), (4.59) and applying (2.50) to $Z^{\wedge}(x, \mu^2)$, we get (4.38).

Denote by $N_1(\mu^2)$ the number of eigenvalues $\mu_k^2 < \mu^2$ of $\widetilde{M}$ counting multiplicities $æ_1(\mu_k^2)$. Note that

$$N_1(\mu^2) = N(-\mu, \mu) \quad (\mu > 0), \tag{4.60}$$

whereby $æ_1(0) = æ(0)$, and

$$æ(-\mu_k) + æ(\mu_k) = æ_1(\mu_k^2). \tag{4.61}$$

Therefore, by (4.56) and by Theorem 4.1 for the operator $\widetilde{M} = (\widetilde{L})^2$, we have (4.37).

If $b < \infty$, then the self-adjoint extension $\widetilde{L}$ of the minimal operator L_b with respect to the end-point $x = b$ is determined by the boundary condition of the form

$$\cos B \cdot y^{\vee}(b) + \sin B \cdot y^{\wedge}(b) = 0, \tag{4.62}$$

and the self-adjoint extension $\widetilde{M} = (\widetilde{L})^2$ of the operator M_b is determined by the boundary conditions (4.62) and

$$\cos B \cdot (l_W[y])^{\vee}(b) + \sin B \cdot (l_W[y])^{\wedge}(b) = 0. \tag{4.63}$$

The operators $\widetilde{M}$ and M_b^F are the self-adjoint extensions of their maximal common part Λ defined on vector-valued functions $y \in \mathrm{D}(M^*)$ satisfying the conditions (4.39), (4.43), $y^{\wedge}(b) = y^{\vee}(b) = 0$ and (4.63). The deficiency indices of Λ are $(\text{rank }\{\cos B, \sin B\}, \text{rank }\{\cos B, \sin B\})$, i.e. def $\{\widetilde{M}|\mathrm{D}(\widetilde{M}) \cap \mathrm{D}(M_b^F)\} = \frac{r}{2} \dim H$. This concludes the proof of Theorem 4.6 and Remark 4.4. □

4.5 Bibliographical Comments

The results given in this chapter are due to the authors. The corresponding results were first announced in [795], and their detailed exposition with proofs was first given in the preprint [797], whose contents were subsequently published in the volumes [799] and [800]. (In the case of second-order equations, many of the corresponding results were published in [794], and those results comprise Chapter 1 of the present book.)

In the case of a scalar second-order differential equation with singular boundary condition, Theorem 4.1 is essentially contained in [58]. However, our assertion about sufficient conditions for the formula (4.1) with $\lambda = \lambda_e(\widetilde{L})$ is new in the case of a scalar Sturm–Liouville equation with two singular end-points. For scalar differential equations of even order, the interplay between spectral and oscillatory properties was investigated by M. G. Krein [585] and E. Heinz [506] (see also [70]). A generalization of Sturm Oscillation Theorem to the case of finite systems of fourth order differential equations on a bounded interval is contained in the monographs by A. G. Aslanyan and V. B. Lidskiy [12] and A. L. Gol'denveizer, V. B. Lidskiy, and P. E. Tovstik [72]. In the case canonical systems, the property of being oscillatory (in the sense of Definition 4.1) was investigated in the works of V. B. Lidskiy [632], V. A. Yakubovich [922], V. I. Khrabustovskiy [546], and, in the case of second order equations with operator coefficients, in the works of G. Etgen, J. Pawlowski [403] and G. Etgen, R. Lewis [402].

Theorem 4.2 is an infinite dimensional generalization of the Comparison Theorem and the Alternation Theorem of Heinz–Rellich; see [70] and [506]. The Alternation Theorem in this chapter is, in a way, a stronger version of certain generalizations of Sturm's Alternation Theorem in the finite-dimensional case (as explained in the present chapter).

General idea of the proof of Theorem 4.3 agrees with the proof of the Frobenius Factorization Theorem by M. G. Krein [585] (see also [70]), although the latter proof is based on the use of determinants. In this regard, even in the scalar case, our scheme is different from the well-known schemes, including the formulation of Frobenius Theorem, which is replaced by an equivalent one.

Theorem 4.4 gives a sufficient condition for differential equations of even order to be oscillatory. In the case of second order equations with operator-valued coefficients, Theorem 4.4 generalizes a theorem of Etgen and Pawlowski [403] because it allows operators containing terms with the first derivative.

Theorem 4.5 generalizes a theorem of M. G. Krein–Heinz–Rellich; see [70, Sec. 44], [506], [585] and [761].

In the finite dimensional case, for the fourth order equations, Corollary 4.3 implies a result of [609] with weaker requirements than in [609].

We also mention the works of V. N. Everitt and A. Zettl [415], [931], [932] and F. Gesztesy and M. Ünal [464] dealing with the above types of questions, and also a work of Yu. M. Arlinskiy [234] dealing with the factorization and reduction of abstract operators to a divergent form.

In the case $r = 1$, $\dim H < \infty$, $-\infty < x < \infty$, Theorem 4.6 is contained in [542]. Another approach to the investigation of discrete levels in spectral gaps, based on the method of phase functions, is developed for the scalar Schrödinger operator in [786, Theorem 4]. (See also Chapter 6 of the present book). The paper [463] by F. Gesztesy, B. Simon and G. Teschl contains a result for a scalar Sturm–Liouville equation related to Theorem 4.6.

Another approach to the definition of the notion of oscillatory vector-solution of an operator differential equation is offered by Yu. I. Domshlak [382]–[384].

4.5.1 *Symplectic Interpretation of Sturm-Type Theorems and Their Operator-Theoretical Proofs*

We will mention the Theorem about Zeros from [238]. Let H be a real Hilbert space, $\dim H < \infty$, $\lambda = 0$, let the matrices $P(x)$ and $Q(x)$ in (1.1) be real and symmetric, and let $P(x)$ be positive definite for all $x \in [0, b]$. A Lagrangian plane in $H \oplus H$ is said to be vertical if it contains a non-zero vector col $\{y, z\}$ with $y = 0$. Lagrangian plane, whose evolution under stated conditions is determined by (1.1), is defined due to (2.39) by the equation

$$Y^*(x,0)z - Y'^*(x,0)P(x)y = 0, \quad \{y,z\}H \oplus H, \tag{4.64}$$

for some $A = A^*$ in (1.10), while the verticality moments of the given Lagrangian plane are those $x \in (0, b)$ for which $\det Y(x, 0) = 0$. (Clearly, the verticality moments in [238] are considered taking into account their geometric multiplicities.)

Theorem 4.7 (Theorem about Zeros [238]) *On the interval containing* $1 + \dim H$ *verticality moments of one Lagrangian plane, all other Lagrangian planes (corresponding by (4.64) to the solution* $Y_1(x, 0)$ *instead of* $Y(x, 0)$ *and with* A_1 *instead of* A *in (1.10)) become vertical at least once. Moreover, on an arbitrary interval of the* x*-axis, the difference in the number of verticality moments for any two Lagrangian planes evolving according to the same system (of the type (1.1),* $\dim H < \infty$*) does not exceed* $\dim H$*.*

(It is interesting to compare this Theorem with our Corollary 4.2.)

We now consider the Comparison Theorem from [238], where Hamiltonian systems with $2n \times 2n$ matrices $H_1(t)$, $H_2(t)$ are considered. For each t, the matrices $H_1(t)$, $H_2(t)$ are positive definite on the fixed Lagrangian plane α. The corresponding systems have the form

$$J\frac{d}{dt}\begin{pmatrix} u \\ v \end{pmatrix} = H_j(t)\begin{pmatrix} u \\ v \end{pmatrix}, \tag{4.65}$$

where $j = 1, 2$, vectors u, v belong to the same space H with $\dim H = n$, and matrix $J = \begin{pmatrix} 0_n & -I_n \\ I_n & 0_n \end{pmatrix}$, where I_n, 0_n are the identity and zero $n \times n$ matrix respectively. Denote by $N(H_j)$ the number of non-transversality moments with respect to the Lagrangian plane α of the Lagrangian planes evolving under the action of the systems (4.65) with the corresponding $j = 1, 2$.

Theorem 4.8 (Comparison Theorem [238]) *If* $H_1(t) \geq H_2(t)$*, then* $N(H_1) \geq N(H_2) - n$*.*

We will show that a variant of the Comparison Theorem, very close to the above stated theorem, follows, as a special case for $\dim H < \infty$ and $r = 2$, from Theorem 4.2 concerning an equation of even order r in an infinite-dimensional space H. Let the canonical systems (4.65) be given in an infinite-dimensional doubled Hilbert space $H \oplus H$ to which the vectors $\operatorname{col}\{u(t)\, v(t)\}$ belong for all t.

Each Lagrangian plane α in $H \oplus H$ can be defined by the equation

$$\cos A \cdot v - \sin A \cdot u = 0, \tag{4.66}$$

where $A = A^*$ is an arbitrary self-adjoint operator in H. For any $A = A^*$, the equation (4.66) defines some Lagrangian plane in $H \oplus H$ with respect to the skew-Hermitian scalar product $\left[\begin{pmatrix} u \\ v \end{pmatrix}, \begin{pmatrix} s \\ w \end{pmatrix}\right] = (v, s) - (u, w)$; see the Appendix, Sec. A.1.

We make a substitution

$$\begin{pmatrix} y \\ z \end{pmatrix} = \Phi_A \begin{pmatrix} u \\ v \end{pmatrix}, \tag{4.67}$$

where Φ_A is an operator in $H \oplus H$ with the block 2×2 matrix

$$\Phi_A = \begin{pmatrix} \sin A & -\cos A \\ \cos A & \sin A \end{pmatrix},$$

and the operator A is chosen in accordance with (4.66), corresponding to the given Lagrangian plane α on which the matrices $H_j(t)$ are positive by the hypothesis. Then, to non-transversality moments with respect to the plane α of the Lagrangian plane evolving under the action of (4.65), correspond those instants t where nul $Y(t) > 0$ holds for the matrix (operator) solution $\begin{pmatrix} Y \\ Z \end{pmatrix}$ of the new equation defining the evolution of Lagrangian plane. The new equation (4.69) is obtained from (4.65) by the substitution (4.67).

Here nul $Y(t) := \dim\ker Y(t)$, where nul $Y^*(t) =$ nul $Y(t)$,

$$Y^*(t)Z(t) - Z^*(t)Y(t) = 0,$$
$$Y^*(t)Y(t) + Z^*(t)Z(t) > 0,$$

and the evolution of the Lagrangian plane determined by (4.65) and (4.66) is given by the formula

$$Z^*(t)y - Y^*(t)z = 0, \tag{4.68}$$

where the evolving Lagrangian plane for each t consists of all possible vectors $\begin{pmatrix} y \\ z \end{pmatrix}$ satisfying (4.68).

By the substitution (4.67), the equations (4.65) become

$$J\frac{d}{dt}\begin{pmatrix} y \\ z \end{pmatrix} = H_j^A(t)\begin{pmatrix} y \\ z \end{pmatrix}, \tag{4.69}$$

where $H_j^A(t) = \Phi_A H_j(t) \Phi_A^*$, and the Lagrangian plane α is now defined by the equation $y = 0$. Since the forms corresponding to the operators $H_j(t)$ are positive on the plane α, it follows that the operators $H_{22}^{jA}(t)$ are positive in the space H, where

$$H_j^A(t) = \begin{pmatrix} H_{11}^{jA}(t) & H_{12}^{jA}(t) \\ H_{21}^{jA}(t) & H_{22}^{jA}(t) \end{pmatrix}. \tag{4.70}$$

Here $(H_j^A(t))^* = H_j^A(t)$. The equations (4.69), for $j = 1, 2$, are equivalent to Sturm–Liouville systems of the form

$$l^j[Y] = -(P_j Y' + R_j Y)' + R_j^* Y' + Q_j Y = 0, \tag{4.71}$$

where $P_j = P_j^* = (H_{22}^{jA})^{-1} > 0$, $Q_j = Q_j^*$. Conversely, corresponding to (4.71), we obtain the system (4.69), (4.70) by putting $z = P_j y' + R_j y$ and

$$H_j^A(t) = \begin{pmatrix} -Q_j + R_j^* P_j^{-1} R_j & -R_j^* P_j^{-1} \\ -P_j^{-1} R_j & P_j^{-1} \end{pmatrix}. \tag{4.72}$$

The quantities $P_j = (H_{22}^{jA})^{-1}$, Q_j, R_j are in a bijective correspondence with H_{11}^{jA}, $H_{22}^{jA} > 0$, $H_{12}^{jA} = (H_{21}^{jA})^*$, which, in turn, are in a bijective correspondence with the operator blocks of $H_j(t)$ due to the equation $H_j^A(t) = \Phi_A H_j(t) \Phi_A^*$.

Note that by (4.72) the quadratic form of $H_j^A(t)$ can be represented as

$$\left(H_j^A(t) \begin{pmatrix} y \\ z \end{pmatrix}, \begin{pmatrix} y \\ z \end{pmatrix} \right) = (P_j^{-1}(z - R_j y), z - R_j y) - (Q_j y, y). \tag{4.73}$$

On the other hand, the Dirichlet integral for (4.71) is the integral of the quadratic form

$$D_j \left[\begin{pmatrix} y \\ y' \end{pmatrix}, \begin{pmatrix} y \\ y' \end{pmatrix} \right] := (P_j(t)y', y') + (R_j y, y') + (R_j^* y', y) + (Q_j y, y), \tag{4.74}$$

with the 2×2 block matrix

$$D_j(t) = \begin{pmatrix} Q_j & R_j^* \\ R_j & P_j \end{pmatrix}. \tag{4.75}$$

The hypotheses of the Comparison Theorem from [238] require the inequality

$$H_1^A(t) \geq H_2^A(t) \tag{4.76}$$

for matrices (4.72) or for the forms (4.73). If, in addition, we require $R_1(t) = R_2(t)$, then (4.76) is equivalent to

$$D_1(t) \leq D_2(t), \tag{4.77}$$

which is sufficient for the hypothesis (4.10) of Theorem 4.2 to hold (here $r = 2$), and, hence, for its result (4.13). Thus, under the additional assumption $R_1(t) = R_2(t)$, the Comparison Theorem from [238] follows from Theorem 4.2 as a special case for $r = 2$, $\dim H < \infty$, because then $m \leq \dim H$. However, if we do not require $R_1 = R_2$, then the inequalities (4.76) and (4.77) are not equivalent, and the corresponding corollary of Theorem 4.2 neither contains the Comparison Theorem from [238] nor is contained in this theorem. As for the general case when $r > 2$, Theorem 4.2 covers Hamiltonian systems with matrices which may fail to be strictly positive on any Lagrangian plane, unlike the hypotheses of [238].

Proof of Proposition 1.1 ([792]). Let us write the equation (2.1) at $r = 2$ in the form

$$-(P(t)y' + R(t)y)' + R^*(t)y' + Q(t)y = \lambda W(t)y. \tag{4.78}$$

Denote by $Y(t, \lambda)$ the fundamental solution of (4.78) satisfying the following conditions at some $a \in \mathbb{R}$ and $A = A^* \in B(H)$:

$$Y(a, \lambda) = \cos A, \qquad Y^{[1]}(a, \lambda) = \sin A, \tag{4.79}$$

where

$$Y^{[1]}(t, \lambda) := P(t)Y'(t, \lambda) + R(t)Y(t, \lambda)$$

is the quasi-derivative. Then, by Theorem 4.1, for the boundary-value problem for (4.78) with the boundary conditions

$$\cos A \cdot y^{[1]}(a) - \sin A \cdot y(a) = 0, \tag{4.80}$$

$$y(b) = 0, \tag{4.81}$$

where $a < b$, we have the following equality:

$$\sum_{t \in (a,b)} \text{nul } Y(t, \lambda) = N^0(\lambda). \tag{4.82}$$

(If $R(t) \equiv 0$, this gives Theorem 1.1.)

The equation (4.78) is equivalent to the system

$$J \frac{d}{dt} \text{col}(y, y^{[1]}) = H(t, \lambda) \text{col}(y, y^{[1]}), \tag{4.83}$$

where $H(t, \lambda)$ coincides with the matrix (4.72), with the index j omitted and Q replaced by $Q - \lambda W$.

We now consider the boundary-value problem (4.78), (4.80) and

$$\cos B \cdot y^{[1]}(b) + \sin B \cdot y(b) = 0, \tag{4.84}$$

and put

$$\mathrm{col}(z, z^{[1]}) = \Phi_B^* \mathrm{col}(y, y^{[1]}). \tag{4.85}$$

Then to the boundary condition (4.84) corresponds the condition

$$z(b) = 0, \tag{4.86}$$

and to the condition (4.80) corresponds the condition

$$\begin{aligned}&(\cos A \cdot \sin B + \sin A \cdot \cos B) z^{[1]}(a)\\ &+(\cos A \cdot \cos B - \sin A \cdot \sin B) z(a) = 0,\end{aligned} \tag{4.87}$$

which, as easily seen, is generated by a Hermitian relation; see Theorem A.2 and Corollary A.1. Therefore, the condition (4.87) is a self-adjoint boundary condition for the point $t = a$.

By virtue of (4.85), the system (4.83) is equivalent to the system

$$J \frac{d}{dt} \mathrm{col}(z, z^{[1]}) = H(t, \lambda; B) \mathrm{col}(z, z^{[1]}), \tag{4.88}$$

where

$$H(t, \lambda; B) = \Phi_B^* H(t, \lambda) \Phi_B. \tag{4.89}$$

A direct calculation shows that

$$\begin{aligned}&H_{2,2}(t, \lambda; B) = \cos B \cdot \{-Q(t) + \lambda W(t) + R^*(t) P^{-1}(t) R(t)\} \cos B\\ &+ \sin B \cdot P^{-1}(t) R(t) \cos B + \cos B \cdot R^*(t) P^{-1}(t) \sin B\\ &+ \sin B \cdot P^{-1}(t) \sin B.\end{aligned} \tag{4.90}$$

If the condition

$$H_{2,2}(t, \lambda; B) >> 0, \qquad a \leq t \leq b, \tag{4.91}$$

is satisfied, that is, if the transformed Hamiltonian (4.89) is positive on the Lagrangian plane (4.86) or, equivalently, if the Hamiltonian (4.83) is positive on the corresponding Lagrangian plane (4.84), then we can pass from the Hamiltonian system (4.88) to an equivalent second order system of the type (4.78):

$$-(P(t, \lambda, B) z' + R(t, \lambda, B) z)' + R^*(t, \lambda, B) z' + Q(t, \lambda, B) z = 0, \tag{4.92}$$

where

$$P(t, \lambda, B) = (H_{2,2}(t, \lambda, B))^{-1}, \tag{4.93}$$

and, at $\lambda \in \mathbb{R}$ and $B^* = B$,

$$P^*(t, \lambda, B) = P(t, \lambda, B), \qquad Q^*(t, \lambda, B) = Q(t, \lambda, B).$$

Let λ be fixed here and let the right hand side of (4.92) be $\mu z(t, \lambda)$, where $\mu = 0$ is a new spectral parameter. Then, under some additional conditions, we can apply the Oscillatory Theorem (namely, the formula of the type (4.82)) to the boundary value problem (4.92), (4.87), (4.86) if to the Hermitian binary relation (4.87) corresponds an operator h_a semi-bounded below on its domain $\mathrm{D}(h_a)$:

$$\begin{aligned} h_a &:= -(\cos A \cdot \sin B + \sin A \cdot \cos B)^{-1}(\cos A \cdot \cos B - \sin A \cdot \sin B) \\ &\geq c \cdot I_{\overline{\mathrm{D}(h_a)}}, \qquad c \in \mathbb{R}. \end{aligned} \tag{4.94}$$

Note that this condition is always satisfied in the case of finite systems, that is, if $\dim H < \infty$.

Denote by $Z(t, \lambda, B; \mu = 0)$ the fundamental solution of the problem (4.92), (4.87), and by $N_\lambda^0(0)$—the number of negative eigenvalues μ of the problem (4.92), (4.87), (4.86) if the problem has the discrete spectrum at $\mu < 0$.

Under the above conditions, the Oscillatory Theorem 4.1 gives:

$$\sum_{t \in (a,b)} \mathrm{nul}\ Z(t, \lambda, B; 0) = N_\lambda^0(0). \tag{4.95}$$

Denoting

$$Y[t, \lambda, B] := \cos B \cdot Y^{[1]}(t, \lambda) + \sin B \cdot Y(t, \lambda), \tag{4.96}$$

by substitution (4.85) we have

$$\sum_{t \in (a,b)} \mathrm{nul}\ Y[t, \lambda, B] = N(\lambda). \tag{4.97}$$

This concludes the proof of Proposition 1.1 for the equation (4.78). The equation (1.1) is a special case of (4.78) with $R(t) \equiv 0$. □

We now consider the Arnold Alternation Theorem. Using our definitions, this theorem is formulated as follows in [238]:

Theorem 4.9 (on the alternation; V. I. Arnold [238]) *If the Hamiltonian function $H(t)$ is positive definite on Lagrangian planes α and β, then the numbers ν of non-transversality instants with respect to the planes α*

and β of the Lagrangian plane that evolves in correspondence to the equation (4.83) can differ among themselves in an arbitrary interval at most by the number of degrees of freedom:

$$|\nu_\alpha - \nu_\beta| \leq \dim H (< \infty). \tag{4.98}$$

(In [238] a real finite-dimensional space H is considered.)

Corollary 4.4 **([238])** *On an interval containing* $\dim H + 1$ *non-transversality instants with respect to α, there is a non-transversality instant with respect to β.*

We will now prove the Alternation Theorem for the cases $\dim H \leq \infty$ using our considerations. Let at a fixed λ the Lagrangian plane directed by the equation (4.83) be defined by the boundary condition (4.80) with the corresponding $A = A^*$. Let the Lagrangian planes α and β, relative to which the non-transversality instants of the evolving Lagrangian plane are considered, be given by the condition (4.84) with B and B_1 correspondingly.

Suppose that for a given A and, correspondingly, B and B_1, the semi-boundedness condition (4.94) is satisfied (in the case $\dim H < \infty$, this condition is satisfied automatically). Since the Hamiltonian $H(t, \lambda)$ (4.83) is positive on both Lagrangian planes α and β, we have (4.91) and, analogously,

$$H_{2,2}(t, \lambda, B_1) >> 0, \qquad a \leq t \leq b. \tag{4.99}$$

From here we get (4.97) and, correspondingly,

$$\sum_{t \in (a,b)} \text{nul}\, Y[t, \lambda; B_1] = N_1(\lambda), \tag{4.100}$$

where $Y[t, \lambda; B_1]$ is defined by the formula (4.96) with B_1 instead of B, and $N_1(\lambda)$ is the analogue of $N(\lambda)$ for the case of the condition (4.84) with B_1 instead of B. (We remind the reader that we assume that the conditions (1.4) for A, B and B_1 are also satisfied.) Now by Lemma 1.4(=4.1) we have

$$|N(\lambda) - N_1(\lambda)| \leq p := \text{Def}\ \{\Lambda\} \leq \dim H, \tag{4.101}$$

where Λ is a symmetric operator equal to the common part of two self-adjoint operators corresponding to the boundary-value problems (4.78), (4.80), (4.84) with B and, correspondingly, B_1 in (4.84). Hence, by Lemma 4.2 we have

$$\text{Def}\ \{\Lambda\} = \text{rnk}\ \{\sin B_1 \cdot \cos B - \cos B_1 \cdot \sin B\}. \tag{4.102}$$

From here, together with (4.101), (4.100) and (4.97), we obtain the following generalization of the Alternation Theorem 4.9.

Theorem 4.10 **([792])** *Under the above conditions, we have:*

$$\left| \sum_{t\in(a,b)} nul\, Y[t,\lambda;B] - \sum_{t\in(a,b)} nul\, [t,\lambda;B_1] \right| \le Def\,\{\Lambda\}$$
$$= rnk\,\{\sin B_1 \cdot \cos B - \cos B_1 \cdot \sin B\} \le \dim H \le \infty. \qquad (4.103)$$

Corollary 4.5 *On an interval containing* $1 + Def\,\{\Lambda\}$ *points where* $nul\, Y[t,\lambda;B] > 0$*, there is a point where* $nul\, Y[t,\lambda;B_1] > 0$ *(if* $Def\,\{\Lambda\} < \infty$*).*

* * *

Below we give two theorems of D. R. Yafaev [920] guaranteeing discreteness or finiteness of the negative part of the spectrum of $2n$-th order operator with two terms.

Theorem 4.11 **(Yafaev [920])** *Let* $Q(x)$ *be a compact, non-negative operator acting in a separable Hilbert space* H*. Let* $Q(x)$ *be locally Bochner integrable, and let* h *be a positive parameter. Then the operator* L *on the half line*

$$L = (-1)^n (d^{2n}/dx^{2n}) - hQ(x), \quad 0 \le x < \infty,$$

with boundary conditions $u(0) = u'(0) = \cdots = u^{(n-1)}(0) = 0$*, for all* $h > 0$*, is semi-bounded below and has only discrete negative part of the spectrum if and only if the following condition holds:*

$$\lim_{x\to\infty} \int_x^{x+l} Q(t)\,dt = 0.$$

Theorem 4.12 **(Yafaev [920])** *The operator* L *has only finite negative part of the spectrum for all* $h > 0$ *if and only if the following condition holds:*

$$\lim_{x\to\infty} x^{2n-l} \int_x^{\infty} Q(t)\,dt = 0.$$

In the scalar case, Theorems 4.11 and 4.12 turn into a result of M. Sh. Birman [276], and for $\dim H < \infty$, these are the results of V. V. Martynov [662]. In the case $n = 1$ and $\dim H = \infty$, Theorems 4.11 and 4.12 generalize Theorems 1.6 and 1.7 in the present book.

The works by L. B. Zelenko [927] and R. S. Ismagilov [518] are dedicated to conditions for discreteness of the spectrum and investigation of spectral asymptotics for the Sturm–Liouville operator in the case $\dim H < \infty$, but with taking into account the velocity of rotation of eigenvectors of the potential matrix.

Some conditions for pure discreteness and non-semi-boundedness of the spectrum of the scalar two-term operator with oscillating potential were obtained in the works by R. S. Ismagilov [519] (second order) and M. G. Gimadislamov [477] (order $2n$).

Chapter 5

Self-Adjoint Extensions of Systems of Differential Equations of Arbitrary Order on an Infinite Interval in the Absolutely Indefinite Case

In this chapter, with the help of boundary conditions at singular and regular end-points, we describe all self-adjoint extensions of arbitrary-order symmetric differential operators with matrix-valued or operator-valued coefficients on the real line or half-line in the absolutely indeterminate case. We give an explicit formula for finding the characteristic operator-valued function $M(\lambda)$ (see [145]) for boundary-value problems, and we also give a parametrization of $M(\lambda)$.

5.1 A Description of Self-Adjoint Extensions of Differential Operators of Arbitrary Order with Operator-Valued Coefficients on an Infinite Interval

Consider the differential equation (2.1) of order $r \geq 1$ on the interval (a, ∞), $a \geq -\infty$.

Assume that the following conditions are satisfied:

(A) The operator-valued coefficients $p_j(x)$, $q_j(x)$, together with their derivatives of order up to and including j uniformly continuously depend on x; the coefficient with the highest-order derivative ($p_n(x)$ if $r = 2n$ and $\operatorname{Re} q_n(x)$ if $r = 2n-1$) in the equation (2.1) has a bounded inverse on H for all $x \in (a, \infty)$; the weight $W(x) = W^*(x) >> 0$.

(B) The coefficients of the equation satisfy the following property: for every operator-valued solution $Y_i(x, \lambda)$ of the equation (2.1) with the initial data $Y_i^{(k-1)}(0, \lambda) = \delta_{ik} I$, i, $k = 1, 2, \ldots, r$ (I is the identity operator in H), the following integral converges for some $\lambda = \lambda_0$, Im $\lambda_0 > 0$ and $\lambda = \overline{\lambda_0}$:

$$\int_a^\infty |W^{1/2}(x)Y_i(x,\lambda)|^2\, dx < \infty.$$

(For a Sturm–Liouville equation with operator-valued coefficients, this requirement is fulfilled if conditions of V. B Lidskiy [486], [631] hold). Then the above integrals converge for all λ, i.e. , in accordance with the definition from [262] and [486] for the second-order equation, we have the absolutely indeterminate case. Moreover, the deficiency indices of the minimal operator determined by the differential expression $l_W[y]$ (see (2.2)) in $H(a,\infty)$ are equal to $(r \cdot \dim H, r \cdot \dim H)$, $\dim H \leq \infty$.

(C) For the differential operator L in $H(\alpha,\beta)$, there exist self-adjoint separated boundary conditions on any interval $(\alpha,\beta) \subseteq (a,\infty)$. In this case, for a differential expression of odd order, either (2.5) holds for all $x \in (a,\infty)$, or $\dim H = \infty$ and $r \geq 3$ (see Sec. 2.1).

It is well-known (see [562]) that the system (2.1) is equivalent to the first-order system (compare with (3.40))

$$\frac{i}{2}\left\{\frac{d}{dx}(Q_1(x)\tilde{y}) + Q_1^*(x)\frac{d\tilde{y}}{dx}\right\} = \{\lambda A(x) + B_1(x)\}\,\tilde{y}, \tag{5.1}$$

where

$$B_1(x) = B_1^*(x), \quad A(x) = \underbrace{W(x) \oplus 0 \oplus \cdots \oplus 0}_{r}; \tag{5.2}$$

for $r = 2n$,

$$\tilde{y} = y \oplus y' \oplus \cdots \oplus y^{(n-1)} \oplus y^{[2n-1]} \oplus \cdots \oplus y^{[n+1]} \oplus y^{[n]} \tag{5.3}$$

$$Q_1(x) = \begin{pmatrix} 0 & iI_n \\ -iI_n & 0 \end{pmatrix} = \frac{J}{i}, \tag{5.4}$$

where I_n is the identity operator in H^n; and for $r = 2n-1$,

$$\tilde{y} = y \oplus y' \oplus \cdots \oplus y^{(n-2)} \oplus y^{[2n-2]} \oplus \cdots \oplus y^{[n+1]} \oplus y^{[n]} \oplus (-iy^{(n-1)}), \tag{5.5}$$

$$Q_1(x) = \left(\begin{array}{cc|c} 0 & iI_{n-1} & \\ -iI_{n-1} & 0 & \\ \hline & & q_n(x) \end{array}\right) = Q(x) + iK(x)$$

(all unspecified operators are equal to 0). Here

$$Q(x) = \text{Re } Q_1(x) = \left(\begin{array}{cc|c} 0 & iI_{n-1} & \\ -iI_{n-1} & 0 & \\ \hline & & \text{Re } q_n(x) \end{array}\right), \tag{5.6}$$

where Re $T = \frac{1}{2}(T + T^*)$. If $q_n^* \neq q_n$, then $Q_1^* \neq Q_1$. But the system (5.1) is equivalent to the system

$$\frac{i}{2}\left\{\frac{d}{dx}(Q(x)\tilde{y}) + Q(x)\frac{d\tilde{y}}{dx}\right\} = \{\lambda A(x) + B(x)\}\,\tilde{y}, \tag{5.7}$$

where $B(x) = B_1(x) + \frac{1}{2}K'(x)$. (For $r = 2n$, $Q(x) = Q_1(x)$, $B(x) = B_1(x)$). Let $Y(x,\lambda)$, $Z(x,\lambda)$ be arbitrary operator-valued solutions of (2.1). Define the operator $\widetilde{Y}(x,\lambda)$ by the formula $\widetilde{Y}(x,\lambda)h = \widetilde{(Yh)}(x,\lambda)$. It is well-known [562] that

$$(\widetilde{Y})^*(x,\bar{\lambda})Q(x)\widetilde{Z}(x,\lambda) = (\widetilde{Y})^*(0,\bar{\lambda})Q(0)\widetilde{Z}(0,\lambda) \tag{5.8}$$

for all x. This is easily checked by differentiating the left hand side of (5.8) and using (5.7).

Definition 5.1 (Compare with Definition 2.1). An operator-valued function $Y(x,\lambda) \in B(H^r, H)$ satisfying the equation (2.1) is said to be a fundamental solution of (2.1) if

1) the solution system of (2.1) defined by the operator $Y(x,\lambda)$ is complete in the following sense: any vector-valued solution $y(x,\lambda)$ of (2.1) can be represented as

$$y(x,\lambda) = Y(x,\lambda)h, \quad h \in H^r, \tag{5.9}$$

2) for all x, λ, the operator

$$M(x,\lambda) = \sum_{k=0}^{r-1} Y^{(k)*}(x,\lambda)\cdot Y^{(k)}(x,\lambda) >> 0, \tag{5.10}$$

i.e. $M(x,\lambda)$ is positive and has a bounded inverse on the whole H^r.

Let $Y_0(x)\colon H^r \to H$ be the fundamental solution of (2.1) with $\lambda = 0$, satisfying the initial condition

$$\widetilde{Y}_0(0) = \left[Q^2(0)\right]^{-1/4}.$$

Then for all x, we have

$$(\widetilde{Y}_0)^*(x) \cdot Q(x) \cdot \widetilde{Y}_0(x) = \frac{J}{i}, \tag{5.11}$$

where

$$J = i \text{ sign } Q(0) = iQ(0) \left[Q^2(0)\right]^{-1/2}. \tag{5.12}$$

Note that

$$J^* = -J, \quad J^2 = -I_r. \tag{5.13}$$

Let

$$(Sy)(x) = (\widetilde{Y}_0)^{-1}(x) \cdot \tilde{y}(x), \tag{5.14}$$

where $y(x)$ is a vector-valued functions with values in H. Non-singular linear substitution

$$\tilde{y}(x, \lambda) = \widetilde{Y}_0(x) \cdot (Sy)(x, \lambda) \tag{5.15}$$

transforms the system (5.7) (see Lemma I.I in [562]) into the system

$$J(Sy)' = \lambda H(x)(Sy), \tag{5.16}$$

where $H(x) = (\widetilde{Y}_0)^*(x) \cdot A(x) \cdot \widetilde{Y}_0(x)$. By (5.2) we have $H(x) = (Y_0)^*(x) \cdot W(x) \cdot Y_0(x)$. Therefore, using the condition (B), we have

$$\int_a^\infty |H(t)| \, dt < \infty. \tag{5.17}$$

Lemma 5.1 *Assume that the coefficients of the equation (2.1) satisfy the conditions (A), (B), (C). Then the following hold:*

1) for all $y \in D(L^*)$, the following strong limit exists:

$$\lim_{x \to \infty} (Sy)(x) = (Sy)(\infty);$$

2) if $y(x, \lambda)$ is a solution of (2.1), then for all $x \geq 0$ we have

$$\begin{aligned} |(Sy)(x, \lambda)| \cdot \exp(-|\lambda|\sigma(x)) &\leq |(Sy)(\infty, \lambda)| \\ &\leq |(Sy)(x, \lambda)| \cdot \exp(|\lambda|\sigma(x)), \end{aligned} \tag{5.18}$$

$$|(Sy)(x, \lambda) - (Sy)(\infty, \lambda)| \leq |(Sy)(0, \lambda)| \cdot |\lambda|\sigma(x) \exp(|\lambda|\sigma(0)), \tag{5.19}$$

where $\sigma(x) = \int_x^\infty |H(x)| \, dx$;

3) for all $h \in H^r$ there exists a unique solution $y(x, \lambda)$ of the equation (2.1) satisfying the following initial condition at the singular end-point:

$$\lim_{x \to \infty} (Sy)(x, \lambda) = h.$$

Proof. 1) Let $y \in D(L^*)$. Then $l_W[y] - \lambda y = f \in H(a, \infty)$. Consider the equation

$$l[u] - \lambda W u = W f. \tag{5.20}$$

Recall that the homogeneous equation (2.1) is equivalent to the system (5.7). By substitution (5.14), the system (5.7) is transformed into (5.16). Similarly, the non-homogeneous equation (5.20) is reduced to

$$J(Su)' = \lambda H(x)(Su) + F \tag{5.21}$$

or

$$(Su)' = -\lambda J H(x)(Su) - JF, \tag{5.22}$$

where $F(x) = Y_0^*(x)W(x)f(x)$. The solution of this equation is written as

$$(Su)(x, \lambda) = U(x, 0)(Su)(0, \lambda) - \int_0^x U(x, \xi) J F(\xi)\, d\xi, \tag{5.23}$$

where $U(x, \xi)$ is the evolution operator of the equation (5.21) (see [51, p. 148]). By Lemma 2.4 from [51, p. 162], for all $0 \le \xi \le x$ we have

$$\exp\left[-|\lambda| \int_\xi^x |H(t)|\, dt\right] \le |U(x, \xi)| \le \exp\left[|\lambda| \int_\xi^x |H(t)|\, dt\right], \tag{5.24}$$

and by (5.17) we get

$$\exp\left[-|\lambda|\sigma(0)\right] \le |U(x, \xi)| \le \exp\left[|\lambda|\sigma(0)\right]. \tag{5.25}$$

Since $f \in H(0, \infty)$ and $|W^{1/2}(x)Y_0(x)| = |Y_0^*(x)W^{1/2}(x)|$, then using condition (B) we have

$$\begin{aligned}
&\int_0^\infty |F(x)|\, dx = \int_0^\infty |Y_0^*(x)W(x)f(x)|\, dx \\
&\le \int_0^\infty |Y_0^*(x)W^{1/2}(x)| \cdot |W^{1/2}(x)f(x)|\, dx \\
&\le \sqrt{\int_0^\infty |Y_0^*(x)W^{1/2}(x)|^2\, dx} \cdot \sqrt{\int_0^\infty |W^{1/2}(x)f(x)|^2\, dx} < \infty.
\end{aligned}$$

Thus, from (5.23) and (5.25) we have

$$|(Su)(x,\lambda)| \leq \exp\left[|\lambda|\sigma(0)\right] \cdot \left\{|(Su)(0,\lambda)| + \int_0^\infty |F(x)|\,dx\right\} = M < \infty.$$

Therefore, every solution $(Su)(x,\lambda)$ of the equation (5.22) is bounded with respect to $x \in (0,\infty)$. From (5.22) we get

$$|(Su)'| \leq |H(x)| \cdot |\lambda| \cdot M + |F(x)|, \tag{5.26}$$

and, hence, the integral $\int_0^\infty |(Su)'(x,\lambda)|\,dx$ converges, and the following strong limit exists:

$$\lim_{x\to\infty} (Su)(x,\lambda) = (Su)(\infty,\lambda).$$

Indeed, as $x \to \infty$, we have

$$\begin{aligned} |(Sy)(x,\lambda) - (Sy)(\infty,\lambda)| &= \left|\int_x^\infty (Su)'(x,\lambda)\,dx\right| \\ &\leq \int_x^\infty |(Su)'|\,dx \to 0. \end{aligned} \tag{5.27}$$

Since the vector-valued function $(Sy)(x)$ is a solution of (5.22), it follows that for all $y \in \mathrm{D}(L^*)$, the following strong limit exists:

$$\lim_{x\to\infty} (Su)(x) = (Su)(\infty).$$

The assertions 2) and 3) follow from (5.17), (5.23) and (5.25) (see also [51, p. 166] and [78, p. 327]).

Theorem 5.1 *In the absolutely indeterminate case, the following hold:*

1) for all $y,\ z \in D(L^*)$, we have the Green's formula:

$$\langle L^*y, z\rangle - \langle y, L^*z\rangle = (\Gamma_1 y, \Gamma_2 z)_{\widetilde{H}^2} - (\Gamma_2 y, \Gamma_1 z)_{\widetilde{H}^2}. \tag{5.28}$$

Here, for the equation on the half-line,

$$\begin{aligned} \Gamma_1 y &= -P_1 \cdot (Sy)(0) \oplus P_1 \cdot (Sy)(\infty), \\ \Gamma_2 y &= P_2 \cdot (Sy)(0) \oplus P_2 \cdot (Sy)(\infty), \end{aligned} \tag{5.29}$$

and for the equation on the whole real line,

$$\Gamma_1 y = -P_1 \cdot (Sy)(-\infty) \oplus P_1 \cdot (Sy)(\infty), \tag{5.30}$$

$$\Gamma_2 y = P_2 \cdot (Sy)(-\infty) \oplus P_2 \cdot (Sy)(\infty),$$

where $(Sy)(\pm\infty)$, $(Sy)(0)$ *are defined from* $y(x) \in D(L^*)$ *by the formula (5.14). The operators* P_1 *and* P_2 *are defined as follows: for* $r = 2n$, *we set*

$$P_1 = (I_n; 0_n), \quad P_2 = (0_n; I_n) \colon H^r \to \widetilde{\mathbf{H}};$$

for $r = 2n-1$ *and assuming* $\dim H_x^- = \dim H_x^+$ *(see (2.5)), we set*

$$P_1 = \left(\begin{array}{c|c|c} & & \frac{i}{\sqrt{2}}\left[V^+(0)P_0^+ + V^-(0)P_0^-\right] \\ \hline I_{n-1} & 0_{n-1} & \end{array} \right),$$

$$P_2 = \left(\begin{array}{c|c|c} & & \frac{1}{\sqrt{2}}\left[V^+(0)P_0^+ - V^-(0)P_0^-\right] \\ \hline 0_{n-1} & I_{n-1} & \end{array} \right).$$

If $\dim H_x^- \neq \dim H_x^+$ *and* $r = 2n-1 \geq 3$ *and* $\dim H = \infty$, *we can set*

$$P_1 = \left(\begin{array}{c|c|c|c|c} & P_{12}^{(1)} & & P_{14}^{(1)} & P_{15}^{(1)} \\ \hline I_{n-2} & & 0_{n-2} & & \end{array} \right);$$

$$P_2 = \left(\begin{array}{c|c|c|c|c} & P_{12}^{(2)} & & P_{14}^{(2)} & P_{15}^{(2)} \\ \hline 0_{n-2} & & I_{n-2} & & \end{array} \right);$$

$$P_{12}^{(1)} = P_{14}^{(2)} = -\frac{i}{2}\left[V_1^+(0)\begin{pmatrix} 0_+ \\ I_1 \end{pmatrix} - V_1^-(0)\begin{pmatrix} 0_- \\ I_1 \end{pmatrix}\right];$$

$$P_{14}^{(1)} = -P_{12}^{(2)} = \frac{1}{2}\left[V_1^+(0)\begin{pmatrix} 0_+ \\ I_1 \end{pmatrix} + V_1^-(0)\begin{pmatrix} 0_- \\ I_1 \end{pmatrix}\right];$$

$$P_{15}^{(1)} = \frac{i}{\sqrt{2}}\left[V_1^+(0)\begin{pmatrix} P_0^+ \\ 0 \end{pmatrix} + V_1^-(0)\begin{pmatrix} P_0^- \\ 0 \end{pmatrix}\right];$$

$$P_{15}^{(2)} = \frac{1}{\sqrt{2}}\left[V_1^+(0)\begin{pmatrix} P_0^+ \\ 0 \end{pmatrix} - V_1^-(0)\begin{pmatrix} P_0^- \\ 0 \end{pmatrix}\right].$$

Here, in accordance with the notation in Sec. 1.2 (see also the Appendix), for $r = 2n$, *we have* $\widetilde{\mathbf{H}} = H^n$. *For* $r = 2n-1$, *the spaces* H_x^+, H_x^- *are reduced subspaces of the operator* $\mathrm{Re}\, q_n(x)$ *such that the following holds:*

on H_x^+ *the operator Re* $q_n(x)$ *is positive definite, while on* H_x^- *the operator* $-$*Re* $q_n(x)$ *is positive definite. The operators* P_x^+ *and* P_x^- *are the orthogonal projections onto* H_x^+ *and* H_x^- *respectively. If* $r = 2n-1$ *and if (2.5) holds,* H^+ *is an arbitrary Hilbert space such that* $\dim H^+ = \dim H_x^+$ *for all* $x \in (a,b)$*;* $V^\pm(x)$ *are isometric operators mapping* $H_x^\pm$ *onto* H^+*;* $\widetilde{\mathbf{H}} = H^+ \oplus H^{n-1}$*;* $0_\pm$ *is a zero-operator from* H *into* $H_0^\pm$*. For* $r = 2n-1 \geq 3$ *and* $\dim H = \infty$*,* $\widetilde{\mathbf{H}} = H^+ \oplus H^{n-1}$*;* H^+ *is an arbitrary separable Hilbert space,* $V_1^\pm(x)$ *are isometric operators mapping* $H_x^\pm \oplus H$ *onto* $H^+ \oplus H$*. The operators* $V^\pm(0)$ *and* $V_1^\pm(0)$ *are extended by zero to all of* $H_0^\mp$*.*

2) For all $h,\ g \in \widetilde{\mathbf{H}}^2$*, there exists* $y \in D(L^*)$ *such that*

$$\Gamma_1 y = h, \quad \Gamma_2 y = g.$$

Remark 5.1 Theorem 5.1 shows that the ordered triple $(\widetilde{\mathbf{H}}^2, \Gamma_1, \Gamma_2)$ is the space of boundary values of the minimal operator L (see the corresponding definition in [556] or in Bibliographical Comments to Appendix, Sec. A.4.1).

Remark 5.2 It can be checked directly that

$$P_1 \cdot (Sy)(0) = y^\wedge(0), \quad P_2 \cdot (Sy)(0) = y^\vee(0),$$

where $y^\wedge(0)$, $y^\vee(0)$ are defined by the formulas (2.8)–(2.10).

In subsequent proofs we will use various relations between the operators P_1 and P_2.

Lemma 5.2 *The following hold:*

$$P_1 = P_2 J; \quad P_2 = -P_1 J; \tag{5.31}$$

$$P_1^* P_1 + P_2^* P_2 = I_r; \quad P_2^* P_1 - P_1^* P_2 = J; \tag{5.32}$$

$$P_2 J P_1^* = -P_1 J P_2^* = I_{\widetilde{\mathbf{H}}}; \tag{5.33}$$

$$P_1 J P_1^* = P_2 J P_2^* = 0, \tag{5.34}$$

where J *is defined by the formula (5.12), and* $I_{\widetilde{\mathbf{H}}}$ *is the identity operator in* $\widetilde{\mathbf{H}}$.

Proof. For $r = 2n$ the Lemma follows by noting that

$$J = \begin{pmatrix} 0 & -I_n \\ I_n & 0 \end{pmatrix}.$$

For $r = 2n - 1$, by (5.6) and (5.12) we have

$$J = \left(\begin{array}{cc|c} 0 & -I_{n-1} & \\ I_{n-1} & 0 & \\ \hline & & i(P_0^+ - P_0^-) \end{array} \right). \tag{5.35}$$

Then (5.31) and (5.32) can be checked directly for $r = 2n - 1$. To prove (5.33) under the condition (2.5), it is enough to show that

$$\frac{1}{2}(V^+(0) - V^-(0))(P_0^+ - P_0^-)(V^{+*}(0) + V^{-*}(0)) = I_{H^+}, \tag{5.36}$$

i.e.

$$\frac{1}{2}V^+(0)P_0^+V^{+*}(0) + \frac{1}{2}V^-(0)P_0^-V^{-*}(0) = I_{H^+}.$$

This easily follows from

$$V^+(0)P_0^+V^{+*}(0) = V^-(0)P_0^-V^{-*}(0) = I_{H^+}.$$

Thus $P_2JP_1^* = I_{\tilde{\mathbf{H}}}$. The proof of (5.33) is completed by using (5.31). For $r \geq 3$ and $\dim H = \infty$, we can prove (5.33) analogously. Similarly, we can check (5.34), and the Lemma is proven. □

Proof of Theorem 5.1. Consider the problem on the half-line. For all y, $z \in \mathrm{D}(L^*)$, the Lagrange's formula holds:

$$\langle L^*y, z\rangle - \langle y, L^*z\rangle = \left[(y^\wedge, z^\vee)_{\tilde{H}} - (y^\vee, z^\wedge)_{\tilde{H}}\right]|_0^\infty. \tag{5.37}$$

On the other hand, by (5.29) and (5.32), we have

$$\begin{aligned}
&(\Gamma_1 y, \Gamma_2 z)_{\tilde{H}^2} - (\Gamma_2 y, \Gamma_1 z)_{\tilde{H}^2} \\
&= [(P_1(Sy)(x), P_2(Sz)(x))_{\tilde{H}} - (P_2(Sy)(x), P_1(Sz)(x))_{\tilde{H}}]|_0^\infty \\
&= ((P_2^*P_1 - P_1^*P_2)(Sy)(x), (Sz)(x))_{H^r}|_0^\infty = (J(Sy)(x), (Sz)(x))_{H^r}|_0^\infty.
\end{aligned}$$

Using (5.14) and (5.11), we get

$$\begin{aligned}
&(\Gamma_1 y, \Gamma_2 z)_{\tilde{H}^2} - (\Gamma_2 y, \Gamma_1 z)_{\tilde{H}^2} = (J(\widetilde{Y}_0)^{-1}\tilde{y}, (\widetilde{Y}_0)^{-1}\tilde{z})_{H^r}|_0^\infty \\
&= ((\widetilde{Y}_0)^{-1*} \cdot J \cdot (\widetilde{Y}_0)^{-1}\tilde{y}, \tilde{z})_{H^r}|_0^\infty = i(Q(x)\tilde{y}(x), \tilde{z}(x))_{H^r}|_0^\infty.
\end{aligned}$$

Since

$$W(y, z) = (y^\wedge, z^\vee)_{\tilde{H}} - (y^\vee, z^\wedge)_{\tilde{H}} = i(Q(x)\tilde{y}(x), \tilde{z}(x)) \tag{5.38}$$

(for $r = 2n$ this follows immediately from (5.3), (5.4), (2.8), and for $r = 2n-1$ this is checked similarly as in the Appendix), it follows that

$$(\Gamma_1 y, \Gamma_2 z)_{\widetilde{H}^2} - (\Gamma_2 y, \Gamma_1 z)_{\widetilde{H}^2} = (y^{\wedge}, z^{\vee})_{\widetilde{H}} - (y^{\vee}, z^{\wedge})_{\widetilde{H}} \tag{5.39}$$

Comparing (5.37) and (5.39), we get (5.28).

Let $h = h_1 \oplus h_2$, $g = g_1 \oplus g_2$ be arbitrary vectors from $\widetilde{\mathbf{H}}^2$. By $\phi(x, \lambda)$, $\psi(x, \lambda)$ we denote the vector-valued solutions of (2.1) satisfying the following initial conditions:

$$\phi^{\wedge}(0) = P_1(S\phi)(0) = -h_1, \quad \phi^{\vee}(0) = P_2(S\phi)(0) = g_1,$$

and

$$(S\psi)(\infty) = P_1^* h_2 + P_2^* g_2.$$

By Lemma 5.2 we have

$$P_1(S\psi)(\infty) = P_1 P_1^* h_2 + P_1 P_2^* g_2 = P_2 J P_1^* h_2 + P_2 J P_2^* g_2 = h_2,$$
$$P_2(S\psi)(\infty) = P_2 P_1^* h_2 + P_2 P_2^* g_2 = -P_1 J P_1^* h_2 - P_1 J P_2^* g_2 = g_2.$$

Let $0 \le f_1(t)$, $f_2 \in C_0^\infty$, $f_1(t) + f_2(t) = 1$ and let $f_2(t)$ be a compactly supported function equal to 1 in a neighborhood of 0. Let

$$y(x, \lambda) = f_1(x)\phi(x, \lambda) + f_2(x)\phi(x, \lambda).$$

Then $y \in \mathrm{D}(L^*)$, $\Gamma_1 y = h$ and $\Gamma_2 y = g$.

We analogously treat the problem on the whole real line. The Theorem is proven. □

Theorem 5.2 *In the absolutely indeterminate case and with conditions (A), (B), (C), every self-adjoint extension $\widetilde{L}$ of the minimal operator L is determined by the boundary conditions*

$$\cos \widetilde{A} \cdot \Gamma_1 y - \sin \widetilde{A} \cdot \Gamma_2 y = 0, \tag{5.40}$$

where $\widetilde{A}$ is a self-adjoint operator in $\widetilde{\mathbf{H}}^2$ and Γ_1, Γ_2 are defined by formulas (5.29), (5.30). Conversely, every boundary condition of the form (5.40) determines a self-adjoint extension of the operator L.

Proof. By Theorem 5.1 the triple $(\widetilde{\mathbf{H}}^2, \Gamma_1, \Gamma_2)$ is the space of boundary values of the minimal operator L; hence (see [773] and [529]), every self-adjoint extension $\widetilde{L}$ of the operator L determines a Hermitian relation θ in $\widetilde{\mathbf{H}}^2$:

$$(\Gamma_1 y)\theta(\Gamma_2 y) \Leftrightarrow y \in \mathrm{D}(\widetilde{L}).$$

Therefore, by the theorem on the general form of Hermitian relations (Theorem A.1), we have (5.40), where $\Gamma_1 y$, $\Gamma_2 y$ are determined by Theorem 5.1. This concludes the proof of the Theorem. □

Corollary 5.1 *If*

$$\widetilde{A} = \begin{pmatrix} A & 0 \\ 0 & B \end{pmatrix},$$

where A, B are self-adjoint operators in $\widetilde{\mathbf{H}}$, then we get separated boundary conditions. For example, for the problem on the half-line, we have:

$$\cos A \cdot y^{\wedge}(0) + \sin A \cdot y^{\vee}(0) = 0, \tag{5.41}$$

$$\cos B \cdot P_1(Sy)(\infty) - \sin B \cdot P_2(Sy)(\infty) = 0. \tag{5.42}$$

5.2 Parametrization of the Characteristic Operator Function

Let $\Phi(x,\lambda)$, $\Psi(x,\lambda) \in B(\widetilde{\mathbf{H}}, H)$ be the operator-valued solutions of the equation (2.1) with the initial conditions

$$\Phi^{\wedge}(0,\lambda) = \sin A; \quad \Phi^{\vee}(0,\lambda) = -\cos A; \tag{5.43}$$

$$\Psi^{\wedge}(0,\lambda) = \cos A; \quad \Phi^{\vee}(0,\lambda) = \sin A; \tag{5.44}$$

where the operators $\Phi^{\wedge}(x,\lambda)$, $\Phi^{\vee}(x,\lambda)$ are defined by formulas (2.25) and $\widetilde{\mathbf{H}}$ is defined in (2.15).

Theorem 5.3 *Under the hypotheses of Theorem 5.2, for all λ, $Im\ \lambda \neq 0$, the boundary condition (5.42) at the singular end-point is equivalent to the condition*

$$\begin{aligned} &\lim_{x\to\infty} [((\Psi^{\wedge})^*(x,\bar{\lambda}) + M^*(\bar{\lambda})(\Phi^{\wedge})^*(x,\bar{\lambda}))y^{\vee}(x) \\ &- ((\Psi^{\vee})^*(x,\bar{\lambda}) + M^*(\bar{\lambda})(\Phi^{\vee})^*(x,\bar{\lambda}))y^{\wedge}(x)] = 0, \end{aligned} \tag{5.45}$$

where the characteristic operator-valued function $M(\lambda)$ can be represented as

$$\begin{aligned} M(\lambda) = &-\left[(S\Psi)^*(\infty,\bar{\lambda})P_2^* \sin B - (S\Psi)^*(\infty,\bar{\lambda})P_1^* \cos B\right] \\ &\times \left[(S\Phi)^*(\infty,\bar{\lambda})P_2^* \sin B - (S\Phi)^*(\infty,\bar{\lambda})P_1^* \cos B\right]^{-1}. \end{aligned}$$

The function $M(\lambda)$ is analytic for all Im $\lambda \neq 0$ and satisfies the inequality

$$\frac{Im\ M(\lambda)}{Im\ \lambda} = \frac{M(\lambda) - M^*(\lambda)}{\lambda - \bar{\lambda}} \geq 0 \quad (Im\ \lambda \neq 0). \tag{5.46}$$

If $\dim H < \infty$, *then $M(\lambda)$ is meromorphic with the poles on the real line.*

As the operator-valued parameter B varies over the interval $-\frac{\pi}{2} I_{\widetilde{H}} < B \leq \frac{\pi}{2} I_{\widetilde{H}}$, the function $M(\lambda)$ ranges over all characteristic operator functions corresponding to the orthogonal spectral operator-valued functions with the boundary condition (5.41) at zero, i.e. for every λ, Im $\lambda \neq 0$, $M(\lambda)$ ranges over the operator circle (see [262] and [486]).

Proof. Consider the operator-valued solution

$$Y(x, \lambda) = \Phi(x, \lambda) P_1 + \Psi(x, \lambda) P_2 \colon H^r \to H \tag{5.47}$$

of the equation (2.1). To prove the completeness, it is enough to show that for any solution $y(x, \lambda)$ there exists $h \in H^r$ such that

$$Y^{\wedge}(0, \lambda) h = y_0^{\wedge}, \quad Y^{\vee}(0, \lambda) h = y_0^{\vee}, \tag{5.48}$$

where $y_0^{\wedge}$, $y_0^{\vee}$ are Cauchy data for the solution $y(x, \lambda)$: $y_0^{\wedge} = y^{\wedge}(0, \lambda)$, $y_0^{\vee} = y^{\vee}(0, \lambda)$. Using the initial conditions (5.43), (5.44), we can rewrite (5.48) as follows:

$$\begin{aligned} y_0^{\wedge} &= \sin A \cdot P_1 h + \cos A \cdot P_2 h, \\ y_0^{\vee} &= -\cos A \cdot P_1 h + \sin A \cdot P_2 h. \end{aligned}$$

Thus

$$P_1 h = \sin A \cdot y_0^{\wedge} - \cos A \cdot y_0^{\vee}, \quad P_2 h = \cos A \cdot y_0^{\wedge} + \sin A \cdot y_0^{\vee},$$

and by Lemma 5.2 we get

$$h = J\left[P_2^* (\sin A \cdot y_0^{\wedge} - \cos A \cdot y_0^{\vee}) - P_1^* (\cos A \cdot y_0^{\wedge} + \sin A \cdot y_0^{\vee})\right]. \tag{5.49}$$

Choosing h by the formula (5.49), we have (5.48), and hence the representation (5.9). Now that we have shown (5.10), we will prove that the solution $Y(x, \lambda)$ is fundamental.

Similarly as in (5.38), it is directly checked that the operator-valued solutions $Y(x, \lambda)$, $Z(x, \lambda)$ of the equation (2.1) satisfy the equality

$$W_{x,\lambda}\{Y^*, Z\} = i(\widetilde{Y})^*(x, \bar{\lambda}) \cdot Q(x) \cdot \widetilde{Z}(x, \bar{\lambda}), \tag{5.50}$$

where $W_{x,\lambda}\{Y^*, Z\}$ is defined by the formula (2.37). In particular, by (5.8) it follows that $W_{x,\lambda}\{Y^*, Z\}$ does not depend on x, and, from now on, we will omit the subscript x. Using (5.43), (5.44) and (5.47), we have from (5.50) with $Y = Z$:

$$i(\widetilde{Y})^*(x,\bar{\lambda})Q(x)\widetilde{Y}(x,\lambda) = [(\Phi^\vee P_1 + \Psi^\vee P_2)^*(\Phi^\wedge P_1 + \Psi^\wedge P_2)$$
$$- (\Phi^\wedge P_1 + \Psi^\wedge P_2)^*(\Phi^\vee P_1 + \Psi^\vee P_2)]\,|_{x=0} = (-P_1^* \cos A + P_2^* \sin A)$$
$$\cdot(\sin A \cdot P_1 + \cos A \cdot P_2) - (P_1^* \sin A + P_2^* \cos A)$$
$$\cdot(-\cos A \cdot P_1 + \sin A \cdot P_2) = P_2^* P_1 - P_1^* P_2.$$

By (5.22) we get

$$i(\widetilde{Y})^*(x,\bar{\lambda}) \cdot Q(x) \cdot \widetilde{Y}(x,\lambda) = J. \tag{5.51}$$

Therefore, $\widetilde{Y}(x,\lambda)$ has a bounded inverse for all x, λ, and (5.10) follows.

Define the operator $\chi(x,\lambda)$: $\widetilde{\mathbf{H}} \to H$ by the formula $\chi(x,\lambda)h = y_h(x,\lambda)$, where $y_h(x,\lambda)$ is a solution of the equation (2.1) satisfying the initial conditions at infinity

$$P_1(Sy_h)(\infty,\lambda) = \sin B \cdot h, \quad P_2(Sy_h)(\infty,\lambda) = \cos B \cdot h. \tag{5.52}$$

The operator $\chi(x,\lambda)$ is defined on the whole $\widetilde{\mathbf{H}}$. By the Closed Graph Theorem, to prove that $\chi(x,\lambda)$ is bounded for fixed x and λ, it is enough to show that $\chi(x,\lambda)$ is closed. Let $h_k \to 0$ as $k \to \infty$. Then by (5.52) we have $(S\chi h_k)(\infty,\lambda) \to 0$ as $k \to \infty$. Therefore, from (5.18) it follows that $(S\chi h_k)(x,\lambda) \to 0$. This and (5.14) imply $\widetilde{\chi}(x,\lambda)h_k \to 0$; hence, the operator $\chi(x,\lambda)$ is bounded. The operator-valued solution $\chi(x,\lambda)$ of the equation (2.1) satisfies the initial conditions at infinity:

$$P_1(S\chi)(\infty,\lambda) = \sin B, \quad P_2(S\chi)(\infty,\lambda) = \cos B. \tag{5.53}$$

Since $Y(x,\lambda)$ is a fundamental solution, we have

$$\chi(x,\lambda) = Y(x,\lambda) \cdot C(\lambda), \quad C(\lambda)\colon \widetilde{\mathbf{H}} \to H^r$$

$$\widetilde{\chi}(x,\lambda) = \widetilde{Y}(x,\lambda) \cdot C(\lambda).$$

This and (5.51) imply

$$C(\lambda) = (\widetilde{Y})^{-1}(x,\lambda)\widetilde{\chi}(x,\lambda) = -iJ(\widetilde{Y})^*(x,\bar{\lambda})Q(x)\widetilde{\chi}(x,\lambda). \tag{5.54}$$

(Recall that the first part of (5.54) does not depend on x due to (5.8).)

Setting

$$C_k(\lambda) = P_k \cdot C(\lambda), \quad k = 1, 2,$$

we have

$$\chi(x,\lambda) = \Phi(x,\lambda)C_1(\lambda) + \Psi(x,\lambda)\cdot C_2(\lambda),$$

and also

$$\widetilde{\chi}(x,\lambda) = \widetilde{\Phi}(x,\lambda)C_1(\lambda) + \widetilde{\Psi}(x,\lambda)\cdot C_2(\lambda),$$

where

$$C_k(\lambda) = -iP_kJ(\widetilde{Y})^*(x,\bar{\lambda})\cdot Q(x)\cdot\widetilde{\chi}(x,\lambda).$$

By (5.33) and (5.44) we have

$$Y^\wedge(0,\lambda)JP_1^* = (\Phi^\wedge(0,\lambda)P_1 + \Psi^\wedge(0,\lambda)P_2)JP_1^* = \Psi^\wedge(0,\lambda)$$

and

$$Y^\vee(0,\lambda)JP_1^* = \Psi^\vee(0,\lambda).$$

Therefore,

$$Y(x,\lambda)JP_1^* = \Psi(x,\lambda),$$

and, analogously,

$$Y(x,\lambda)JP_2^* = -\Phi(x,\lambda).$$

Thus, using (5.50), we get

$$C_1(\lambda) = i(\widetilde{\Psi})^*(x,\bar{\lambda})\cdot Q(x)\widetilde{\chi}(x,\lambda) = W_\lambda(\Psi^*,\chi),$$

$$C_2(\lambda) = -i(\widetilde{\Phi})^*(x,\bar{\lambda})\cdot Q(x)\widetilde{\chi}(x,\lambda) = -W_\lambda(\Phi^*,\chi).$$

Let

$$D_{x,\lambda}\{\Psi^*,\chi\} = (P_2S\Psi)^*(x,\bar{\lambda})\cdot(P_1S\chi)(x,\lambda) - (P_1S\Psi)^*(x,\bar{\lambda})\cdot(P_2S\chi)(x,\lambda).$$

Using (5.32) and (5.14), we get

$$\begin{aligned}D_{x,\lambda}\{\Psi^*,\chi\} &= (S\Psi)^*P_2^*P_1(S\chi) - (S\Psi)^*P_1^*P_2(S\chi)\\ &= (S\Psi)^*J(S\chi) = ((\widetilde{Y}_0)^{-1}\widetilde{\Psi})^*J((\widetilde{Y}_0)^{-1}\widetilde{\chi})\\ &= (\widetilde{\Psi})^*(\widetilde{Y}_0)^{-1*}J(\widetilde{Y}_0)^{-1}\widetilde{\chi} = i(\widetilde{\Psi})^*Q\widetilde{\chi} = W_\lambda(\Psi^*,\chi).\end{aligned}$$

Thus,

$$\begin{aligned}C_1(\lambda) &= W_\lambda\{\Psi^*,\chi\} = D_{\infty,\lambda}\{\Psi^*,\chi\}\\ &= (S\Psi)^*(\infty,\bar{\lambda})P_2^*\sin B - (S\Psi)^*(\infty,\bar{\lambda})P_1^*\cos B.\end{aligned}$$

By (5.19) the function $(S\Psi)(\infty,\lambda)$ is analytic in λ, and, hence, $C_1(\lambda)$ analytically depends on λ. Analogously, it can be shown that the following function is analytic:

$$C_2(\lambda) = -\left[(S\Phi)^*(\infty,\bar\lambda)P_2^* \sin B - (S\Phi)^*(\infty,\bar\lambda)P_1^* \cos B\right]; \tag{5.55}$$

hence, $\chi(x,\lambda)$ is also analytic.

By Lemma 2.4, the operator-valued function

$$\begin{aligned} M(\lambda) &= C_1(\lambda)\cdot C_2^{-1}(\lambda) \\ &= -\left[(S\Psi)^*(\infty,\bar\lambda)P_2^* \sin B - (S\Psi)^*(\infty,\bar\lambda)P_1^* \cos B\right] \\ &\quad \cdot \left[(S\Phi)^*(\infty,\bar\lambda)P_2^* \sin B - (S\Phi)^*(\infty,\bar\lambda)P_1^* \cos B\right]^{-1} \end{aligned}$$

is defined and analytic for all λ, $\operatorname{Im}\lambda \neq 0$, and if $\dim H < \infty$, $M(\lambda)$ is meromorphic with simple poles on the real axis. Besides,

$$\frac{M(\lambda)-M^*(\lambda)}{\lambda-\bar\lambda} = \frac{C_2^{-1*}(\lambda)[C_2^*(\lambda)C_1(\lambda) - C_1^*(\lambda)C_2(\lambda)]C_2^{-1}(\lambda)}{\lambda-\bar\lambda}. \tag{5.56}$$

Starting from $C_1(\lambda) = W_{0,\lambda}\{\Psi^*,\chi\}$, $C_2(\lambda) = -W_{0,\lambda}\{\Phi^*,\chi\}$, and using elementary transformations, we have

$$\begin{aligned} &C_2^*(\lambda)\cdot C_1(\lambda) - C_1^*(\lambda)\cdot C_2(\lambda) \\ &= (\chi^\vee)^*(0,\lambda)\chi^\wedge(0,\lambda) - (\chi^\wedge)^*(0,\lambda)\chi^\vee(0,\lambda). \end{aligned} \tag{5.57}$$

By formulas (5.37), (5.39), (5.29) and the initial condition (5.53), we get

$$\begin{aligned} &(\lambda-\bar\lambda)\int_0^\infty (\chi^\wedge)^*(x,\lambda)A(x)\chi^\wedge(x,\lambda)\,dx \\ &= (\chi^\vee)^*(0,\lambda)\cdot(\chi^\wedge)(0,\lambda) - (\chi^\wedge)^*(0,\lambda)\cdot(\chi^\vee)(0,\lambda). \end{aligned} \tag{5.58}$$

From (5.56), (5.57), (5.58), we get (5.46).

Let $F(x)\colon \widetilde{\mathbf{H}} \to H$ be an operator-valued function such that for all $h \in \widetilde{\mathbf{H}}$, we have $F(x)h \in \mathrm{D}(L^*)$. Then for all λ, $\operatorname{Im}\lambda \neq 0$ the following limit exists:

$$\begin{aligned} &\lim_{x\to\infty} W_{x,\lambda}\{[\Psi+\Phi M(\lambda)]^*, F\} \\ &= C_2^{-1*}\lim_{x\to\infty} W_{x,\lambda}\{\chi^*,F\} = C_2^{-1*}\lim_{x\to\infty} D_{x,\lambda}\{\chi^*,F\} \\ &= C_2^{-1*}\lim_{x\to\infty}[(P_2S\chi)^*(P_1SF) - (P_1S\chi)^*(P_2SF)] \\ &= C_2^{-1*}[\cos B\cdot P_1(SF)(\infty) - \sin B\cdot P_2(SF)(\infty)]. \end{aligned}$$

If for some $h \in \widetilde{\mathbf{H}}$, $y(x) = F(x)h \in \mathrm{D}(\widetilde{L})$, where $\widetilde{L}$ is the self-adjoint extension determined by the boundary conditions (5.41), (5.42), then we get (5.45). The Theorem is proven. □

5.3 Bibliographical Comments

The present chapter is based on the works by A. M. Khol'kin [541]–[543].

The description of all self-adjoint extensions of Sturm–Liouville operators on an unbounded interval and the parametrization of the characteristic function in the scalar case is due to M. G. Krein [588] and, later, to C. Fulton [438], and for Sturm–Liouville equations with operator potential in the absolutely indefinite case—to M. L. Gorbachuk [486]. For a scalar fourth order equation and a scalar fourth order operation with two terms in the quasi-regular case on the half-line, the description of self-adjoint extensions in terms of boundary conditions and the parametrization of the characteristic matrix was obtained by G. A. Mirzoev [680], [681] simultaneously and independently from [541]–[543]. A different description for all spectral functions of a scalar differential operator of even order with the help of characteristic matrices $M(\lambda)$ was given in the paper of A. V. Shtraus [850], whose results were generalized by V. M. Bruk [315], [316] to equations in a vector-function space. However, in those works, the question about the parametrization of $M(\lambda)$ in the case of a problem on the half-line was not considered.

The exposition of material in this chapter is based on the works of A. M. Khol'kin [541]–[543], where, in the absolutely indeterminate case, the above given results were generalized to symmetric differential operators of arbitrary order (both even and odd) with operator coefficients on an unbounded interval (the whole line or half-line).

In the case when the deficiency indices take intermediate values (between minimal and maximal possible values), a description of self-adjoint extensions of differential operators was given in the works of I.M. Guseinov and R. T. Pashaev [503], F. G. Maksudov and B. P. Allakhverdiev [645], B. P. Allakhverdiev [221], V. I. Mogilevskiy [691]–[693], M. M. Malamud and V. I. Mogilevskiy [650] and [651].

For Jacobi matrices with matrix-valued entries, some criteria for complete indefiniteness were established by A. G. Kostyuchenko–K. A. Mirzoev [575] and Yu. M. Dyukarev [390]. It is known that if one of the two deficiency indices of such matrices has maximal possible value, then so does the other one, and we have the absolutely indefinite case; see [560]. (For differential systems, cf. Theorem 5.6 below and [562, Theorems 4.2 and 4.3].)

5.3.1 *Everitt–Zettl Problem, Brusentsev's Example and Two Open Questions*

We note that in the case of a scalar minimal symmetric two-term differential operator of the simplest form in $L_2(a,\infty)$

$$Ly = (-1)^n y^{(2n)} + q(x)y, \quad a \le x < \infty, \tag{5.59}$$

the question of whether such an operator can be quasi-regular, i.e. can such an operator have deficiency indices $(2n, 2n)$, or, in different terminology, whether such an operator can be in the limit $2n$ case at infinity, is non-trivial for $n > 1$, i.e. for operators of order four or above. This question was considered for fourth order operators in W. N. Everitt's lecture at the conference in Uppsala in 1977 as well as in the works of some other mathematicians: R. M. Kaufman, T. T. Read and A. Zettl [537]; W. D. Evans and A. Zettl [405] and G. A. Mirzoev [681]. A. G. Brusentsev [324] obtained a solution of this problem (Brusentsev's result is also quoted in [785]) by constructing the following example.

Example 5.1 **(A. G. Brusentsev [324])** Let

$$L = l_1^+ l_1 \ge 0, \quad l_1 = (\exp v) l \exp(-v), \quad 1 \le x < \infty,$$

where

$$l = -\frac{d^2}{dx^2} - [2v'' + (v')^2], \quad v(x) = \alpha F(x^\beta) \in C^4,$$

where $F(x+1) = F(x) \not\equiv \text{const.}$, $\beta > 2$, $0 < |\alpha| < C_1(\beta)$, $\|F'\|_{L_1(0,1)} < C_2(\beta)\|F'\|_{L_2(0,1)}$.

As Brusentsev shows, for $F \in C^2$, all solutions of $l_1 u = 0$ belong to $L_2[1,\infty)$ and, hence, def $L = (4,4)$ by [415, Corollary 1] or [70, Sec. 6, Theorems 21 and 22]. Simultaneously, Brusentsev's example gives a negative answer to the question of whether the self-adjointness of a two-term fourth order differential operator in $L_2(-\infty,\infty)$ with a continuous potential follows from its positivity. On the contrary, for the Schrödinger operator, the analogous fact is true due to the well-known Povzner–Wienholtz theorem [752], [918] (see also [70]).

As for the two-term ordinary differential second order operators of the type (5.59) (with $n = 1$), it is well-known by Titchmarsh theorems (see [191, vol. 1] and also [70], [113], [114], [145], where other proofs are done, and [514], [770], [908], where non-monotone potentials are considered) that the case of Weyl limit-circle at infinity (i.e. limit-2 case) and simultaneously pure discreteness and non-semiboundedness of the spectrum may be obtained without difficulty if $q(x) \to -\infty$ quickly enough so that

$$\int^{\infty} |q(x)|^{-1/2}\, dx < \infty, \tag{5.60}$$

and under additional conditions of the monotonicity of $q < 0$, $q' < 0$, sign $q''(x) = \text{const}$ for large values of x, and[1]

$$q'(x) = O([-q(x)]^c), \quad 0 < c < 3/2, \quad x \to \infty.$$

(Under the same conditions, except (5.60), where now integral must be divergent, we have the case of limit-point at infinity and, simultaneously, the absolute continuity of the spectrum which now covers the whole real line.)

But for the two-term operator L in (5.59) of order $2n \geq 4$, the convergence or divergence of the integral

$$\int^{\infty} |q(x)|^{-1+1/(2n)}\, dx$$

[1]Note that just the condition (5.60), the monotonicity and the smoothness of the potential $q < 0$, without additional requirements, are not sufficient to guarantee the limit-circle case at infinity.

Example 5.2 Let $q(x) = -m_n^2$ at $2\pi n \leq x < 2\pi(n+1)$, $n = 0, 1, \ldots$, be a step-function, where $0 < m_n \leq m_{n+1}$ are integers such that $\sum_n (1/m_n) < \infty$, and m_n may tend to ∞ arbitrarily quickly. Then the equation $-y'' + q(x)y = 0$ has solution

$$y = c(x) = \cos(m_n x) \quad \text{at} \quad 2\pi n \leq x \leq 2\pi(n+1), \quad n = 0, 1, \ldots.$$

This solution does not belong to $L_2(0,\infty)$, and, therefore, this equation is in the limit-point case at ∞. If $\tilde{q}(x) \in C^\infty$ is sufficiently accurately smoothed $q(x)$, we obtain that $-y'' + \tilde{q}(x)y = 0$ is also in the limit-point case at ∞, and $\tilde{q}(x)$ may tend to $-\infty$ arbitrarily quickly. But in the considered case, there are potentials $Q(x) \geq q(x)$ for which we have the limit-circle case (unlike the case of potentials satisfying the Titchmarsh–Sears conditions). Therefore, the essential self-adjointness for the Sturm–Liouville equation with the potential of the type $q(x)$ or $\tilde{q}(x)$ in this example is called "unstable self-adjointness" in the sense of I. M. Gelfand. On the contrary, Titchmarsh–Sears conditions guarantee stable self-adjointness in the sense of I. M. Gelfand for Schrödinger equation.

under sufficiently regular $q(x) \to -\infty$ as $x \to \infty$ gives us, respectively, deficiency indices $(n+1, n+1)$ or (n, n), and, therefore, we cannot obtain quasi-regularity $(2n, 2n)$ in this way; see [145, Sec. 23, Theorem 4].

We have seen in Brusentsev's example that already in the fourth order case, the two-term quasi-regular operator is constructed in a much more complicated way than in the second order case.

In this regard, we would like formulate the following two open questions:

Open Question 5.1 Do two term quasi-regular operators of the type (5.59) of order six or greater exist?

Open Question 5.2 If quasi-regular two-term operators of the form (5.59) exist not for all n, then for what n they exist?

5.3.2 *On the Deficiency Indices of Scalar Symmetric Differential Operators of General Kind on the Half-Axis (Solved and Unsolved Questions)*

H. Weyl [916] was the first who established that a real second order symmetric operator (1.1) on the half-line (Sturm–Liouville operator) can have deficiency indices $(1, 1)$ or $(2, 2)$ (in terms of limit-point or limit-circle respectively) and, in these terms, he laid the foundation of the theory of extensions of singular differential operators (see the references to Sec. 5.2).

As is well-known (see, for example [6, Sec. 126] and [145, Sec. 17, no. 5]), the deficiency indices of the minimal symmetric differential operator of even order $r = 2n$ with real coefficients on the half-line (of the type (2.1) if the summation there is taken only over $k = 2j$, $j = 0, 1, \ldots, n$) lie within the limits n and $2n$, and the corresponding deficiency indices are equal. The question of what values (p, p) can be deficiency indices for such operators was unsolved for a long time. Moreover, for a number of decades there had been some controversy surrounding this question: it had been erroneously believed (W. Windau [919], D. Yu. Shin [845] and [846]) that a minimal symmetric (scalar) differential operator of order $2n$ can have deficiency indices (n, n) or $(2n, 2n)$ only, analogously to the well-known Weyl alternative "limit point–limit circle" for second order operators.

Theorem 5.4 **(I. M. Glazman [478], [479])** *If*

$$l_1 = -i(1+x)\frac{d}{dx}(1+x), \quad l_2 = \left(\frac{d}{dx}\right)^2 - 1,$$

then the operator L *of order* $2n$, *with minimal domain, which is generated by the differential expression*

$$l = (l_1)^{(p-n)}(l_2)^{(2n-p)}(l_1)^{(p-n)} \quad (n \le p \le 2n)$$

on the interval $(0, \infty)$, *has deficiency indices* (p, p), *and, therefore, deficiency indices* (p, p) *of minimal symmetric differential operators of even order* $2n$ *with real coefficients on the half-axis can actually take all the intermediate values* $n \le p \le 2n$.

Subsequently, new examples of operators with real coefficients whose deficiency indices take all possible values were constructed by a number of authors (S. A. Orlov [718], M. V. Fedoryuk [421], [422] and others) who used asymptotic methods (see [145], [164] and [195]) for the investigation of asymptotically Euler's equations.

The deficiency indices of symmetric singular differential operators with complex coefficients, in contrast to operators with real coefficients, may consist of pairs of unequal deficiency indices[2]. The simplest example is the minimal operator $i(d/dx)$ on the half-line, which has deficiency indices $(0, 1)$. For symmetric differential operators of even order ($r = 4$), the first example of an operator with deficiency index $(2, 3)$, i.e. with $N_+ \neq N_-$, is due to McLeod [671].

Remark 5.3 According to the general definition of deficiency indices of symmetric operators, the deficiency indices (m, s) of an arbitrary symmetric differential operator L in the space $L_2(a, b)$ consist of two numbers, where the first number, m, is equal to the maximal number of linearly independent square integrable on (a, b) solutions of the equation

$$L^* y = \lambda y,$$

where $\operatorname{Im} \lambda > 0$, and the second number, s, is the same for the case $\operatorname{Im} \lambda < 0$.

Theorem 5.5 (W. N. Everitt [406]; see also [6, vol. 2]) *The deficiency indices* (m, s) *of a minimal symmetric differential operator* L *of arbitrary order* $r = 2n$ *or* $r = 2n + 1$ *with complex coefficients on the half-axis satisfy the inequalities*

$$n \le m, s \le r; \quad m + s \ge r.$$

[2]or, more rigorously, "deficiency numbers"; deficiency index consists of a pair of deficiency numbers (N_+, N_-). But this terminology is not always used rigorously.

The following theorem for operators of even order was established by W. N. Everitt [408]. For operators of odd order, see V. I. Kogan and F. S. Rofe-Beketov [562] (see also [6, vol. 2]).

Theorem 5.6 **(Everitt–Kogan–Rofe-Beketov)** *If one of the deficiency indices (m, s) of a minimal symmetric differential operator L of order $r > 1$ with complex coefficients on the half-axis is equal to r, then the other is equal to r, i.e. $m = s = r$.*

(If $r = 1$, this theorem is false, as the above-mentioned example of the operator $i(d/dx)$ on the half-axis shows.)

Open Question 5.3 For symmetric differential operators with complex coefficients, can deficiency indices take any values that are compatible with the above mentioned Theorem 5.5 by Everitt and Theorem 5.6 by Everitt, Kogan and Rofe-Beketov?

As shown in [561], under additional restriction on admissible values of the deficiency indices, i.e. $|m - s| \leq 1$, the last question has an affirmative answer.

Theorem 5.7 **([561]; the formulation here is as in [562])** *For scalar minimal symmetric operators of arbitrary order (even or odd) with complex-valued coefficients on the half-line, any pair of deficiency indices (m, s) compatible with Theorems 5.5 and 5.6 and satisfying the additional restriction $|m - s| \leq 1$ is realizable.*

Corollary 5.2 *Among symmetric minimal differential operators of order $r \leq 4$ with complex coefficients on the half-line, there exist operators with all possible deficiency indices compatible with Theorems 5.5 and 5.6.*

Indeed, if $r \leq 4$, the deficiency indices of the corresponding operators cannot differ from each other by more than 1, by virtue of Theorems 5.5 and 5.6. All such possibilities are realizable by Theorem 5.7.

In [561] the authors obtained simple asymptotic formulas for solutions of a new class of linear homogeneous differential equations of arbitrary order with analytic coefficients having asymptotic expansions in a sector of the complex plane. Those asymptotic formulas involve the roots of certain characteristic polynomials whose coefficients are obtained from asymptotic expansions of coefficients of the differential equation. Unlike well-known asymptotic formulas for solutions of the class of differential equations with asymptotically constant coefficients and the class of "asymptotically Euler's equations" whose coefficients of $y^{(k)}$ have asymptotic behavior $a_k x^k$ as $x \to$

∞, $k = 0, 1, \ldots, r$ (r is the order of differential equation), the new class of equations is, in some sense, a hybrid of the above mentioned two classes. Let us consider the simplest example of an equation of the new class.

Example 5.3 (**[561]**) Let $0 < p < r$ and let a_0, a_1, $\ldots$, a_r be complex constants such that $a_p a_r \neq 0$. Then the differential equation

$$x^p \left(\sum_{k=p}^{r} a_k y^{(k)} \right) + \sum_{k=0}^{p-1} a_k x^k y^{(k)} = 0$$

has a fundamental system of solutions with the following asymptotic behavior at $x \to \infty$ (if the corresponding characteristic polynomials have only simple roots):

$$y_k = x^{\mu_k}(1 + o(1)), \qquad k = 1, \ldots, p;$$
$$y_k = \exp(\nu_k x)(1 + o(1)), \qquad k = p+1, \ldots, r.$$

Here, $\{\mu_k\}_{k=1}^{p}$ are the roots of the first characteristic polynomial

$$a_0 + a_1\mu + a_2\mu(\mu - 1) + \cdots + a_p\mu(\mu - 1)\cdots(\mu - p + 1) = 0,$$

and $\{\nu_k\}_{k=p+1}^{r}$ are the roots of the second characteristic polynomial

$$a_p + a_{p+1}\nu + a_{p+2}\nu^2 + \cdots + a_r\nu^{r-p} = 0.$$

(Differential equation of the considered type is reducible to a first order differential system with an irregular singularity at infinity.)

Using the obtained asymptotic formulas for solutions, examples of differential equations of arbitrary fixed order with all possible pairs of deficiency indices (m, s) compatible with Theorems 5.5 and 5.6, with $|m - s| \leq 1$, were constructed in [561].

Later, R. C. Gilbert [474] extended and developed those asymptotic formulas for a more general class of differential equations and proved the following Theorem.

Theorem 5.8 (**R. C. Gilbert [475]**) *For any given positive integer there exists a formally symmetric (scalar) ordinary differential operator whose deficiency indices differ by that positive integer.*

(See also the works [472] and [476] by the same author.)

Theorem of R. C. Gilbert is an essential step toward a solution of the problem posed in Question 5.3; however, this problem is not solved completely by Theorem 5.8. Indeed, this theorem guarantees the existence of a differential operator (possibly of sufficiently large order) with the prescribed difference $m - s$ of deficiency indices (m, s); however, we do not know whether or not differential operators of prescribed order can really have all pairs of deficiency indices that are compatible with Theorems 5.5 and 5.6.

At the same time, it is necessary to emphasize that Theorem 5.7 is not a corollary or a particular case of Theorem 5.8.

Open Question 5.4 For symmetric differential operators with complex coefficients, can deficiency indices (m, s) take any values that are compatible with Theorems 5.5 and 5.6 under additional restriction $|m - s| \leq p$, where $p = 2$ or $p = 3$, *etc.*?

For $p = 1$ the affirmative answer is given by Theorem 5.7. (This question is a subquestion of Question 5.3).

Below we will formulate V. I. Kogan–F. S. Rofe-Beketov's and R. C. Gilbert's results more completely.

5.3.3 *Kogan–Rofe-Beketov's Asymptotic Theorems and Deficiency Indices of Symmetric Differential Operators*

In [561] the authors distinguished a class of symmetric differential operators of arbitrary order (even or odd) with complex coefficients, and for this class of operators, deficiency indices turned out to be the numbers of roots (lying in a certain half-plane) of explicitly written polynomials.

The class, considered in [561] contains a sufficient set of differential operators for Theorem 5.7 to be verified. The corresponding work is based on asymptotic formulas derived in [561] for solutions of the differential equation

$$a_n(x)y^{(n)} + a_{n-1}(x)y^{(n-1)} + \cdots + a_0(x)y = 0, \qquad 0 < x_0 \leq x < \infty, \quad (5.61)$$

where the coefficients are holomorphic in some open sector S of the right half-plane (S contains the positive x-axis) and have asymptotic expansions in the form of power series as $x \to \infty$:

$$a_j(x) \sim x^{n-m} \left\{ a_{j,0} + \sum_{k=1}^{\infty} a_{j,k} x^{-k} \right\},$$
$$a_{n,0} \neq 0, \qquad j = n, n-1, \ldots, n-m+1; \quad (5.62)$$

$$a_j(x) \sim x^j \left\{ a_{j,0} + \sum_{k=1}^{\infty} a_{j,k} x^{-k} \right\},$$

$$a_{n-m,0} \neq 0, \qquad j = 0, 1, \ldots, n-m, \quad n \geq m \geq 0. \tag{5.63}$$

(The asymptotic formulas for the solution of (5.61) are also of independent interest.)

For the column vector-function

$$Y(x) := \text{col} \left\{ y(x), y'(x), \ldots, y^{(n-1)}(x) \right\}, \tag{5.64}$$

the equation (5.61) can be reduced in the usual way to the system

$$Y'(x) = A(x) Y(x) \tag{5.65}$$

with an irregular singular point at infinity. The principal term $A_0 = A(\infty)$ of the asymptotic expansion of the coefficient $A(x)$ as $x \to \infty$ may have multiple eigenvalues. The matrix $A(x)$ has the following entries:

$$A_{n,j}(x) = -\alpha_{j-1}(x), \qquad \alpha_j(x) := a_j(x)/a_n(x), \qquad j = 1, \ldots, n; \tag{5.66}$$

$$A_{k,k+1}(x) = 1, \qquad k = 1, \ldots, n-1; \tag{5.67}$$

and all other elements of the matrix $A(x)$ are equal to 0.

Thus, the following asymptotic expansions hold for $\alpha_j(x)$:

$$\alpha_j(x) \sim \alpha_{j,0} + \sum_{k=1}^{\infty} \alpha_{j,k} x^{-k}, \qquad j = n-1, \ldots, n-m; \tag{5.68}$$

$$\alpha_j(x) \sim x^{j+m-n} \left\{ \alpha_{j,0} + \sum_{k=1}^{\infty} \alpha_{j,k} x^{-k} \right\}, \qquad j = 0, 1, \ldots, n-m; \tag{5.69}$$

where

$$\alpha_{j,0} = \frac{a_{j,0}}{a_{n,0}}, \qquad \alpha_{n-m,0} \neq 0. \tag{5.70}$$

The first row of the fundamental matrix solution of the system (5.65) gives the fundamental system of solutions of the equation (5.61). We point out that the differential expression (5.61) is not assumed to be symmetric. In systems obtained from scalar differential equations it is often useful to apply the so-called "shearing transformation," which reduces the rank of the system (see [195]). But in the case considered in [561], in contrast to [195], "shearing transformation on a subspace" (in the terminology of [561]) was implemented, leading to the improvement of the structure of the system (5.65).

In this regard, the following substitution was used in [561]:

$$Y(x) = S(x)\widetilde{Y}(x), \tag{5.71}$$

where

$$S(x) = \operatorname{diag}\left\{x^{n-m}, x^{n-m-1}, \ldots, x, 1, \ldots, 1\right\}. \tag{5.72}$$

The system (5.65) is then transformed into the system

$$\widetilde{Y}'(x) = \widetilde{A}(x)\widetilde{Y}(x), \tag{5.73}$$

where $\widetilde{A}(x) = S(x)^{-1}A(x)S(x) - S(x)^{-1}S'(x)$ is a matrix holomorphic in the sector S for $x_0 \leq |x| < \infty$ and has in S an asymptotic power series representation

$$\widetilde{A}(x) \sim \sum_{r=0}^{\infty} \widetilde{A}_r x^{-r}, \qquad x \to \infty, \quad x \in S, \tag{5.74}$$

where the matrix $\widetilde{A}_0$ has the following entries:

$$\begin{aligned} &(\widetilde{A}_0)_{n,j} = -\alpha_{j-1,0}, \qquad j = 1, \ldots, n; \\ &(\widetilde{A}_0)_{k,k+1} = 1, \qquad k = n-m+1, \ldots, n-1; \end{aligned} \tag{5.75}$$

and all other elements of $\widetilde{A}_0$ are equal to 0.

Theorem 5.9 **([561])** *Let the coefficients $a_j(x)$, $j = 1, \ldots, n$ of the differential equation (5.61) be holomorphic in an open sector S with angle less than π and assume that they have uniform asymptotic expansions (5.62) and (5.63) in S. Assume that the roots μ_1, ..., μ_m of the equation*

$$M_m(\mu) := a_{n,0}\mu^m + a_{n-1,0}\mu^{m-1} + \cdots + a_{n-m,0} = 0 \tag{5.76}$$

are simple.

Then the equation (5.61) has a fundamental system of solutions $y_1(x), \ldots, y_n(x)$ such that m of these solutions admit in the sector S an asymptotic expansion of the form

$$y_j(x) \sim e^{\mu_j x} x^{z_j} \left(1 + \sum_{k=1}^{\infty} b_{j,k} x^{-k}\right)$$

$$\sim \exp\left\{\int^x \left(\mu_j + z_j x^{-1} + \sum_{k=2}^{\infty} c_{j,k} x^{-k}\right) dx\right\},$$
$$j = 1, \ldots, m, \tag{5.77}$$

where $\mu_j \neq 0$ are determined from (5.76), and

$$z_j = -\frac{\sum_{k=0}^{m} a_{n-k,1}\mu_j^{m-k+1} + a_{n-m-1,0}}{\mu_j M_m{}'(\mu_j)}, \quad j = 1, \ldots, m. \tag{5.78}$$

The coefficients $c_{j,k}$, where $k = 2, 3, \ldots$ and $j = 1, 2, \ldots, m$, can be determined by substituting the expansion (5.77) into the equation (5.61) and comparing like powers of x.

The remaining $n - m$ solutions have the asymptotic form

$$y_{j+m}(x) \sim x^{z_{j+m}} (\ln x)^{s_{j+m}} \left(1 + \sum_{k=1}^{\infty} b_{j+m,k} x^{-k}\right),$$
$$j = 1, \ldots, n-m, \tag{5.79}$$

where z_{j+m} are the roots of the equation

$$R(z) := \sum_{j=m}^{n-1} a_{n-j,0} \prod_{k=0}^{n-j-1} (z-k) + a_{0,0} = 0, \tag{5.80}$$

and s_{j+m} are non-negative integers, $0 \le s_{j+m} \le n - m - 1$.

If $R(z)$ does not have multiple roots or roots differing by an integer, then all $s_{j+m} = 0$, and the coefficients of the expansion (5.79) can be determined by substituting the expression (5.79) into the equation (5.61) and comparing like powers of x.

Furthermore, the asymptotic expansions can be differentiated as many times as we wish.

Corollary 5.3 ([561]) *The system (5.65) with coefficients (5.66)–(5.67), where $\alpha_j(x)$ have the form (5.68) and (5.69), and $\alpha_{n-m,0} \neq 0$ has a fundamental matrix–solution whose first row consists of m elements $y_j(x)$ of the form (5.77) and n-m elements of the form (5.79), and the remaining rows and their asymptotic expansions are obtained by successive differentiation of the first row or the asymptotic expansion of the first row respectively.*

(If we know only the coefficients $\alpha_{j,k}$, but not $a_{j,k}$, then we can put $a_{n,0} = 1$, $a_{n,k} = 0$ for $k < n$, and replace $a_{k,j}$ by $\alpha_{k,j}$ in the formulas (5.76), (5.78) and (5.80) that determine principal members of asymptotic expansions for $y_j(x)$, $j = 1, \ldots, n$.)

We now consider n-th order symmetric differential expressions $l[y]$ for which the equation

$$l[y] = \lambda y \tag{5.81}$$

turns out to be an equation of the form (5.61).

We will find the deficiency indices of minimal differential operators generated by the expression $l[y]$ (5.81) in the space $L_2[0,\infty)$. We assume that the expression $l[y]$ is written in the form analogous to (2.1), but with slightly changed notations and enumerations of coefficients (for convenience):

$$l[y] = \sum_{k=0}^{n} i^k l_k[y], \tag{5.82}$$

where

$${}_{2r}[y] = \left\{p_{2r}(x)y^{(r)}\right\}^{(r)} \tag{5.83}$$

and

$$l_{2r+1}[y] = (1/2)\left\{\left(p_{2r+1}(x)y^{(r)}\right)^{(r+1)} + \left(p_{2r+1}(x)y^{(r+1)}\right)^{(r)}\right\}. \tag{5.84}$$

The coefficients $p_j(x)$ are real-valued functions on the positive x-axis. We assume that $p_j(x)$ are holomorphic in an open sector S (with angle less than π) containing the positive x-axis, and we assume that $p_j(x)$ have in S uniform asymptotic expansions as $x \to \infty$ of the form

$$p_j(x) \sim x^{n-m+\nu} \sum_{k=0}^{\infty} p_{j,k} x^{-k}, \qquad p_{n,0} \neq 0,$$
$$j = n-m+1, \ldots, n; \tag{5.85}$$

$$p_j(x) \sim x^{j+\nu} \sum_{k=0}^{\infty} p_{j,k} x^{-k}, \qquad p_{n-m,0} \neq 0,$$
$$j = 0, 1, \ldots, n-m, \tag{5.86}$$

where ν is a non-negative integral parameter.

In order to apply Theorem 5.9 on the asymptotic behavior of solutions of (5.61), it is convenient, using the Leibnitz formula for derivative of the product, to write (5.81) in the following form:

$$\sum_{k=n-m-1}^{n} \left\{ i^k p_k(x) + i^{k+1}((k+1)/2)p'_{k+1}(x) + \dots \right\} y^{(k)} + \dots = \lambda y,$$
$$p_j(x) = 0, \qquad \text{for } j < 0 \text{ or } j > n. \tag{5.87}$$

This equation takes the form (5.61) when both sides are multiplied by $x^{-\nu}$. (Here, we write only those terms of the equation that determine the asymptotic behavior of solutions of the form (5.77).)

The characteristic equation (5.76) in this case takes the form

$$P_m(i\mu) = p_{n,0}(i\mu)^m + p_{n-1,0}(i\mu)^{m-1} + \dots + p_{n-m,0} = \lambda\delta_{0,n-m+\nu}, \tag{5.88}$$

and the formula (5.78) for the indices z_j ($j = 1, \dots, m$) reduces to the following expression:

$$z_j = -(n-m+\nu)/2 - i\lambda(i\mu_j)^{\nu-1}\left\{P_m{}'(i\mu_j)\right\}^{-1} \cdot \delta_{1,n-m+\nu}$$
$$+ \frac{\left\{\sum_{k=0}^{m} p_{n-k,1}(i\mu_j)^{m-k+1} + p_{n-m-1,0}\right\}}{\mu_j P_m{}'(i\mu_j)}, \tag{5.89}$$

where $\delta_{j,k}$ is the Kronecker symbol ($\delta_{j,k} = 1$ if $j = k$ and $\delta_{j,k} = 0$ if $j \neq k$).

After substitution

$$\zeta = z + \frac{1}{2}, \tag{5.90}$$

the characteristic equation (5.80) corresponding to solutions of the form (5.79) takes the following form:

$$F(\zeta, \nu) := \sum_{k=0}^{n-m} i^k p_{k,0} Q_k(\zeta, \nu) = \delta_{\nu,0}\lambda. \tag{5.91}$$

This follows immediately from (5.81) by virtue of (5.82)–(5.86). Here, Q_k are given by

$$Q_{2r}(\zeta, \nu) = \prod_{j=0}^{r-1} \left\{(\zeta + \nu/2)^2 - (j + (\nu+1)/2)^2\right\}, \qquad Q_0 := 1; \tag{5.92}$$

$$Q_{2r+1}(\zeta, \nu) = (\zeta + \nu/2) Q_{2r}(\zeta, \nu). \tag{5.93}$$

Let $N_1(\lambda, \nu)$ be the number of linearly independent solutions of the equation (5.81) that have the form (5.79) and belong to $L_2[0, \infty)$. Let

$N_2(\lambda, \nu)$ be the number of solutions of equation (5.81) that have the form (5.77) and belong to $L_2[0, \infty)$. Then

$$N_{\pm} = N_1(\lambda, \nu) + N_2(\lambda, \nu) \qquad (\text{Im } \lambda{>}0 \text{ or Im } \lambda{<}0) \tag{5.94}$$

determine the pair of deficiency numbers $\{N_+, N_-\}$ of the minimal operator generated by the expression $l[y]$ in $L_2[0, \infty)$. As is well known, $N_{\pm}$ do not depend on λ in the respective half-planes.

Lemma 5.3 ([561])

(1) The magnitudes of $N_1(\lambda, \nu)$ and $N_2(\lambda, \nu)$ do not depend on λ in each of the half-planes $Im\ \lambda > 0$ and $Im\ \lambda < 0$.

(2) The number $N_1(\lambda, \nu)$ is equal to the number of roots ζ of the equation (5.91) for which

$$Re\ \zeta = Re\ \{z + (1/2)\} < 0, \tag{5.95}$$

and, thus, for $\nu > 0$, $N_1(\lambda, \nu)$ does not depend on λ at all:

$$N_1(\lambda, \nu) = N_1(\overline{\lambda}, \nu) = N_1(0, \nu), \qquad (\nu > 0), \tag{5.96}$$

and[3]

$$[(n - m + 1)/2] \leq N_1(\lambda, \nu) \leq n - m, \qquad (\nu > 0), \tag{5.97}$$

and, in the indicated interval, $N_1(\lambda, \nu)$, where $\nu > 0$, can take any integral value depending on the real coefficients $p_{j,0}$ (where $j = 0, 1, \ldots, n - m$; $p_{n-m,0} \neq 0$) and on $\nu > 0$.

If $\nu = 0$, then

$$\begin{aligned} &N_1(\lambda, 0) = [(n - m)/2 + (1/4)\{1 - \text{sgn}\ (p_{n-m,0} Im\ \lambda)\}], \\ &\nu = 0, \qquad Im\ \lambda \neq 0, \end{aligned} \tag{5.98}$$

irrespective of the remaining coefficients $p_{j,k}$.

(3) The number $N_2(\lambda, \nu)$ can be expressed as a sum

$$N_2(\lambda, \nu) = \alpha(\lambda, \nu) + \beta(\lambda, \nu), \tag{5.99}$$

where $\alpha(\lambda, \nu)$ is the number of roots μ_j of (5.88) for which

$$Re\ \mu_j < 0, \tag{5.100}$$

[3] Here, $[x]$ denotes the greatest integer less than or equal to x.

and $\beta(\lambda,\nu)$ is the number of those purely imaginary roots μ_j of (5.88) for which the formula (5.89) gives

$$Re\ z_j < -\frac{1}{2} \tag{5.101}$$

(except in the case $n-m+\nu=0$, i.e. $n-m=\nu=0$, it is assumed that the polynomial $P_m(i\mu)$ (5.88) has only simple roots).
In the case $n=m$ and $\nu=0$,

$$N_2(\lambda,0) = \left[n/2 + (1/4)\left\{1 - \text{sgn}\ (p_{n,0} Im\ \lambda)\right\}\right], \quad Im\ \lambda \neq 0. \tag{5.102}$$

In the case $n-m+\nu>1$ $(n-m\geq 0,\ \nu\geq 0)$, the number $N_2(\lambda,\nu)$ does not depend on λ at all:

$$N_2(\lambda,\nu) = N_2(\overline{\lambda},\nu) = N_2(0,\nu), \qquad n-m+\nu>1. \tag{5.103}$$

Let r denote the number of purely imaginary roots μ_j of the equation

$$P_m(i\mu) := \sum_{j=0}^{m} p_{n-j,0}(i\mu)^{m-j} = 0, \tag{5.104}$$

i.e. r is the number of real roots $t_j = i\mu_j$ of the polynomial $P_m(t)$ in (5.104). Then

$$N_2(\lambda,\nu) = (m+r)/2, \qquad n-m+\nu>1. \tag{5.105}$$

Thus,

$$[(m+1)/2] \leq N_2(\lambda,\nu) \leq m, \qquad n-m+\nu>1, \tag{5.106}$$

and, in the indicated interval, $N_2(\lambda,\nu)$ can take (when $n-m+\nu>1$) any integral value depending on the real coefficients $p_{j,0}$, $j=n-m,\ldots,n$ (irrespective of λ and ν).
Finally, in the case $n-m+\nu=1$, we have
(a) $\nu=0$, $n-m=1$:

$$N_2(\lambda,\nu) = (n-1)/2 + (1/2)\text{sgn}\ (p_{1,0} Im\ \lambda) - (1/4)(1-(-1)^n)\text{sgn}\ (p_{n,0} Im\ \lambda), \qquad Im\ \lambda\neq 0; \tag{5.107}$$

(b) $\nu=1$, $n=m$:

$$N_2(\lambda,\nu) = \left[(n/2) + (1/4)(1 - \text{sgn}\ (p_{n,0} Im\ \lambda))\right], \quad Im\ \lambda\neq 0. \tag{5.108}$$

Theorem 5.10 ([561])

(1) Let L_0 be the minimal operator generated in $L_2[0,\infty)$ by n-th order symmetric differential expression $l[y]$ in (5.82). Assume that the coefficients of $l[y]$ are complex-valued and continuous for $0\leq x<\infty$, holomorphic for $0<x_0\leq|x|<\infty$ and have, in an open sector S of the right-half plane containing the positive x-axis, uniform asymptotic expansions of the form (5.85) and (5.86) as $x\to\infty$, where $\nu\geq 0$ and $m\in[0,n]$ are integers.
Further, let $p_n(x)\neq 0$, $0\leq x<\infty$, and let the roots of the polynomial $P_m(i\mu)$ in (5.88) be simple (except in the case $n-m+\nu=0$).
Then the deficiency indices $\{N_+,N_-\}$ of the operator L_0 in the case $n-m+\nu>1$ are determined (for $Im\ \lambda>0$ and $Im\ \lambda<0$ respectively) as the sum of the numbers of those roots μ_k and ζ_k of the equations (5.88) and (5.91) for which

$$Re\ \mu_k\leq 0, \qquad Re\ \zeta_k<0, \qquad n-m+\nu>1. \tag{5.109}$$

(2) Deficiency indices of the operator L_0 can have any equal values $N_+=N_-$ within the bounds

$$[(n+1)/2]\leq N_+=N_-\leq n, \tag{5.110}$$

and also any values differing by unity ($|N_+-N_-|=1$) within the bounds

$$[n/2]+1\leq\max\{N_+,N_-\}\leq\max\{1,n-1\}. \tag{5.111}$$

(3) Depending on the values of n, m and $\nu\geq 0$, the deficiency indices $\{N_+,N_-\}$ of the operator L_0 turn out to be as follows:
(a) if $0\leq n-m+\nu\leq 1$, $p_{n,0}>0$, then

$$\{N_+,N_-\}=\{[n/2],[(n+1)/2]\}; \tag{5.112}$$

(b) if $n-m+\nu>1$, $\nu>0$, then $N_+=N_-$, and deficiency indices may take any value within the bounds

$$[n/2+(1/4)(3-(-1)^m)]\leq N_+=N_-\leq n \tag{5.113}$$

depending on the coefficients $p_{j,0}$, $j=0,\ldots,n$;
(c) if $n-m+\nu>1$, $\nu=0$, i.e. $n-m>1$ and $\nu=0$, then

$$N_\pm=[(n+r)/2+(1/4)(1-\text{sgn}\ (p_{n-m,0}Im\ \lambda))],$$
$$Im\ \lambda>0 \text{ or } Im\ \lambda<0, \tag{5.114}$$

where r is the number of real roots of the polynomial $P_m(t)$ in (5.104). Thus, in this case,

$$N_+ - N_- = (1/2)((-1)^{n-m} - 1)\text{sgn }(p_{n-m,0}), \tag{5.115}$$

and any pairs $\{N_+, N_-\}$ satisfying (5.115) and

$$[n/2 + (1/4)(1 - \text{sgn }(p_{n,0}))] \leq N_+ \leq n-1,$$
$$[n/2 + (1/4)(1 + \text{sgn }(p_{n,0}))] \leq N_- \leq n-1$$

are possible. No other values for deficiency indices are possible in this case.

5.3.4 R. C. Gilbert's Class of Formally Self-Adjoint Ordinary Differential Operators Whose Deficiency Numbers Differ by an Arbitrary Pre-Assigned Positive Integer

Let S be an open sector of the complex plane, with vertex at the origin, containing the positive x-axis and has a positive central angle less than π.

Theorem 5.11 ([471]) *Let*

$$Ly := \sum_{r=0}^{n} a_r(x) y^{(r)}, \quad n \geq 2,$$

where $a_r(x)$ are holomorphic for $x \in S$, $0 < x_0 \leq |x| < \infty$, and $a_n(x) \equiv 1$.
Assume that for $r = 0, 1, \ldots, n$,

$$a_r(x) \sim \sum_{k=0}^{\infty} a_{r,k} x^{-k}, \qquad as\ x \to \infty$$

in every closed subsector of S. Let

$$I_h(\mu) = \sum_{r=0}^{n} a_{r,h} \mu^r, \qquad h = 0, 1, \ldots$$

Suppose that

$$I_0(\mu) = (\mu - b_1)^{m_1} (\mu - b_2)^{m_2} \cdots (\mu - b_\nu)^{m_\nu},$$

where b_1, b_2, ..., b_ν are distinct complex numbers; m_1, m_2, ..., m_ν are positive integers; and $m_1 + m_2 + \cdots + m_\nu = n$.

For $s = 1, 2, \ldots, \nu$, *let* $z_{s,p}$, $p = 1, 2, \ldots, m_s$, *be the roots of* $G_s(z) = 0$, *where*

$$G_s(z) := \left(I_0^{(m_s)}(b_s)/m_s!\right) z(z-1)\cdots[z-(m_s-1)]$$
$$+\left(I_1^{(m_s-1)}(b_s)/(m_s-1)!\right) z(z-1)\cdots[z-(m_s-2)] + \cdots$$
$$+I_{m_s-1}^{(1)}(b_s)z + I_{m_s}(b_s).$$

Suppose that for each s, $1\le s\le\nu$, *the following are true:*

(1) for $h = 0, 1, \ldots, m_s - 1$,

$$I_h^{(u)}(b_s) = 0 \qquad \textit{for } u = 0, 1, \ldots, m_s - h - 1;$$

(2) $z_{s,p}$, $1\le p\le m_s$, *are distinct and no two of them differ by a positive integer.*

Let T *be any closed subsector of* S. *Then* $Ly = 0$ *has a basis of solutions*

$$y_{s,p}(x), \quad 1\le s\le\nu, \quad 1\le p\le m_s,$$

of the form

$$y_{s,p}(x) = [1 + o(1)]x^{z_{s,p}}e^{b_s x}, \qquad \textit{as } x \to \infty, \qquad \textit{in } T.$$

Lemma 5.4 **([471])** *Let*

$$l[y] := \sum_{k=0}^{n} i^k l_k[y],$$

where $n = 2q + 1$, $q\ge 2$,

$$l_{2r}[y] = \left\{P_{2r}(x)y^{(r)}\right\}^{(r)},$$

$$l_{2r+1}[y] = \frac{1}{2}\left\{\left(P_{2r+1}(x)y^{(r)}\right)^{(r+1)} + \left(P_{2r+1}(x)y^{(r+1)}\right)^{(r)}\right\}.$$

Then

$$l[y] = \sum_{r=0}^{n} f_r(x)y^{(r)},$$

where

$$\begin{aligned}
&f_0(x) = i^0 P_0^{(0)} + iK_1 P_1^{(1)}, \\
&f_1(x) = iP_1^{(0)} + i^2 K_{1,1} P_2^{(1)} + i^3 K_{1,2} P_3^{(2)}, \\
&\cdots\cdots\cdots\cdots\cdots\cdots\cdots\cdots\cdots \\
&f_n(x) = i^n P_n^{(0)},
\end{aligned}$$

$K_{j,k}$ are some real constants, and

$$P_j^{(k)} = \frac{d^k P_j(x)}{dx^k}.$$

Theorem 5.12 ([471]) *Let $l[y]$ be as in Lemma 5.4, where $P_n(x) = x^3$ and $P_j(x)$ are holomorphic for $x \in S$, $0 < x_0 \leq |x| < \infty$, real-valued for real x and have the following asymptotic expansions as $x \to \infty$ in every closed subsector of S:*

$$\begin{aligned}
&P_r(x) \sim x^r \left(P_{r,0} + P_{r,1} x^{-1} + \cdots\right) \qquad \text{for } r = 0, 1, 2, \\
&P_r(x) \sim x^3 \left(P_{r,0} + P_{r,1} x^{-1} + \cdots\right) \qquad \text{for } 3 \leq r \leq n.
\end{aligned}$$

Let t_4, t_5, ..., t_n be real numbers (not necessarily distinct). Suppose that the coefficients $P_{j,k}$ satisfy the following conditions:

$$\begin{aligned}
&P_{n-1,0} = t_4 + \cdots + t_n; \\
&P_{n-2,0} = t_4 t_5 + t_4 t_6 + \cdots; \\
&P_{n-3,0} = t_4 t_5 t_6 + \cdots; \\
&\cdots\cdots\cdots\cdots\cdots\cdots \\
&P_{3,0} = t_4 \cdots t_n; \\
&P_{2,0} = 0, \qquad P_{1,0} = 6K_{1,2} P_{3,0}; \\
&P_{r,1} = 0, \qquad 3 \leq r \leq n-1; \\
&P_{2,1} = 6K_{2,2} P_{4,0}; \\
&P_{r,2} = 6K_{r,2} P_{r+2,0}, \qquad 3 \leq r \leq n-1.
\end{aligned}$$

Then, $l[y] - \lambda y = 0$ can be written in the form $Ly = 0$, where Ly is as in Theorem 5.11. Furthermore,

$$\begin{aligned}
&I_0(\mu) = \mu^3 (\mu - b_4) \cdots (\mu - b_n), \qquad \text{where } b_j = it_j; \\
&I_1(\mu) = (3/2) I_0^{(1)}(\mu), \qquad I_2(\mu) \equiv 0;
\end{aligned}$$

$$I_3(\mu) = \sum_{r=0}^{n-1} a_{r,3}\mu^r;$$

$$a_{0,3} = i^{-n}\left[i^0 P_{0,0} + iK_{0,1}P_{1,0} - \lambda\right].$$

Theorem 5.13 ([471]) *Let $w \geq 2$ be a positive integer and let $v = w - 1$. Let $r_1, \ldots, r_w, r_{w+1}, \ldots, r_{w+v}$ be real numbers with the following two properties:*

(1) $0 = r_1 < r_{w+1} < r_2 < r_{w+2} < \cdots < r_{w-1} < r_{2w-1} < r_w$;
(2) for $s, p = 1, 2, \ldots, w$, $s \neq p$,

$$D(s)/D(p) \neq 1,$$

where

$$D(k) := \prod_{\substack{1 \leq j \leq w \\ j \neq k}} (r_k - r_j)^3 \prod_{g=1}^{v} (r_k - r_{w+g}).$$

Let $t_{3u-2} = t_{3u-1} = t_{3u} = r_u$ for $2 \leq u \leq w$, and let $t_{3w+u} = r_{w+u}$ for $1 \leq u \leq v$. Let $l[y]$ be as in Theorem 5.12, where $n = 3w + v$. Then the deficiency numbers of the minimal operator in $L_2[0,\infty)$ corresponding to the differential expression $l[y]$ are

$$N_+ = w + v, \qquad N_- = 2w + v.$$

More detailed exposition and proofs of R. C. Gilbert's theorems can be found in his works [474]–[476].

5.3.5 *Deficiency Indices of Symmetric Differential Systems of the First and Arbitrary Orders and Some Open Questions*

In this section we first consider symmetric first-order systems

$$\frac{i}{2}\left\{\frac{d}{dx}(Q(x)y) + Q(x)\frac{dy}{dx}\right\} = (\lambda A(x) + B(x))y \tag{5.116}$$

on the half-axis $0 \leq x < \infty$ with a non-negative $(n \times n)$ matrix weight $A(x) \geq 0$. Under such conditions, the space $L_2(0,\infty; A(x)\,dx)$ is obviously not a Hilbert space, because some non-trivial vector-functions may have zero-norm in this metric. (To obtain a Hilbert-space, we must take the factor space of $L_2(0, \infty;\ A(x)\,dx)$ modulo the null-subspace consisting of all elements that have zero-norm in $L_2(0, \infty;\ A(x)\,dx)$. Therefore, we cannot say anything directly about the operator corresponding to the equation (5.116) in the Hilbert space $L_2(0, \infty;\ A(x)\,dx)$ and about deficiency indices of this operator. (Generally speaking, the equation (5.116) gives rise to a symmetric linear relation. This relation has deficiency indices connected with deficiency indices of the corresponding symmetric operator densely defined in some subspace; see, for example, [317] and [619]–[621] and the Appendix of this book. The equation (5.116) gives rise to an operator in the Hilbert space $L_2(0, \infty;\ A(x)\,dx)$ if $A(x) >> 0$ for almost all x.)

However, we may consider the number $N(\lambda)$ of linearly independent vector-solutions of the equation (5.116) that belong to space $L_2(0, \infty;\ A(x)\,dx)$.

Definition 5.2 **([562, Def. 1.2])** Formal deficiency number (f-deficiency number) $N(\lambda)$ of the system (5.116) is the maximum number of linearly independent A-square-integrable (on the considered interval; here $\mathbb{R}_+$) vector-solutions of the system (5.116) for the given $\lambda \in \mathbb{C}$.

Theorem 5.14 **([562, Theorem 2.1], [719])** *Formal deficiency numbers $N(\lambda)$ of the general symmetric first-order system (5.116), where $0 \leq x < \infty$, with a non-negative locally summable matrix weight $A(x) \geq 0$, $\det Q(x) \neq 0$, $Q(x) = Q^*(x)$ locally absolutely continuous and $B(x) \in L_{1,loc}(0, \infty)$, do not depend on λ in either Im $\lambda > 0$ or Im $\lambda < 0$.*

Definition 5.3 **([562, Def. 2.1])** Formal deficiency indices of the system (5.116) are the numbers (N_+, N_-), where

$$N_+ = N(\lambda) \quad (\text{Im}\ \lambda > 0), \qquad N_- = N(\lambda) \quad (\text{Im}\ \lambda < 0). \tag{5.117}$$

Theorem 5.15 **([562])** *Let* $\dim H = n < \infty$, *let the matrix coefficients of the first order system of the general type (5.116) be self-adjoint and locally integrable, let $Q(x)$ be locally absolutely continuous (element by element) and let*

$$\det Q(x) \neq 0, \qquad A(x) \geq 0 \qquad \textit{(in the matrix sense)}. \tag{5.118}$$

Additionally, let for some value of $\lambda = \lambda_0 \in \mathbb{C}$ all vector-solutions of the system (5.116) be square-integrable with the matrix weight $A(x)$ on the positive *half-axis. Then the system (5.116) is quasi-regular (i.e. all of its solutions are A-square-integrable on $\mathbb{R}_+$ at each $\lambda \in \mathbb{C}$) if and only if*

$$\inf_{0<x<\infty} \left\{ (\operatorname{sgn} \operatorname{Im} \lambda_0) \int_0^x \operatorname{tr}\,(Q(t)^{-1}A(t))\,dt \right\} > -\infty. \tag{5.119}$$

Remark 5.4 (**[562]**) If, in particular, $\operatorname{Im} \lambda_0 = 0$, then also

$$\operatorname{sgn} \operatorname{Im} \lambda_0 = 0,$$

and the condition (5.119) is automatically fulfilled.

Remark 5.5 In [562], the formulation of Theorem 5.15 contains only the sufficiency of the condition (5.119). However, the authors of [562] reviewed their paper [562] in *Referativnyi Zhurnal Matematika* (1976, No. 10, review No. B612)[4] and noted that the necessity of the condition (5.119) was also proven in [562, formula (3.4)]. This formula is as follows:

$$\begin{aligned} &\det G_b^{-1}\left\{U(\cdot,\overline{\lambda_0})\right\} \\ &= \exp\left\{2\cdot(\operatorname{Im} \lambda_0)\int_0^b \operatorname{tr}\,(Q(t)^{-1}A(t))\,dt\right\} \det G_b^{-1}\left\{U(\cdot,\lambda_0)\right\}, \end{aligned} \tag{5.120}$$

where $G_b\{Y\}$ is the Gram matrix of the matrix solution $Y(\cdot,\lambda)$ of (5.116), and $U(x,\lambda)$ is a matrix solution of (5.116) such that $U(0,\lambda) = I_n$:

$$G_b\{Y(\cdot,\lambda)\} := \int_0^b Y^*(t,\lambda)A(t)Y(t,\lambda)\,dt.$$

Remark 5.6 (**cf. [562, Theorem 1.2]**) When we count the number of A-square-integrable linearly independent vector-solutions of the equation (5.116) on the half-axis $(0,\infty)$ we can always suppose, without loss of generality, that

$$\det G_b\left\{U(\cdot,\lambda)\right\} > 0 \quad \text{at } b > 0$$

because we can always substitute $A(x) + I_n$ at $0 < x < 1$, instead of $A(x)$ in (5.116) without changing the number of A-square-integrable solutions on $0 < x < \infty$. (Here, we denote by I_n the $n \times n$ identity matrix.)

Since in the quasi-regular case the matrices $G_b^{-1}\left\{U(\cdot,\lambda_0)\right\}$ and $G_b^{-1}\left\{U(\cdot,\overline{\lambda_0})\right\}$ both monotonically decrease and tend to some positive non-degenerate matrices, we obtain the necessity of the condition (5.119) from

[4] In "review No. B612," letter "B" denotes the second letter of the Russian alphabet.

(5.120), when $b \to \infty$. Moreover, since in the quasi-regular case the condition (5.119) of Theorem 5.15 is necessary both for $\lambda = \lambda_0$ and for $\lambda = \overline{\lambda_0}$, this theorem can be formulated in the following equivalent form:

Theorem 5.15^{bis} ([562]) *For the system (5.116) with self-adjoint and locally-integrable $(n \times n)$-matrix coefficients, where $Q(x)$ is locally absolutely continuous and (5.118) holds, to be quasi-regular, it is necessary and sufficient that for some $\lambda = \lambda_0 \in \mathbb{C}$ all vector-solutions of (5.116) belong to $L_2(0,\infty;\, A(x)\,dx)$, and if Im $\lambda_0 \neq 0$, the following additional condition holds:*

$$\sup_{0<x<\infty} \left| \int_0^x \operatorname{tr} \left(Q(t)^{-1} A(t) \right) dt \right| < \infty. \tag{5.121}$$

Corollary 5.4 *For the canonical system*

$$J \frac{dy}{dx} = \lambda H(x) y \tag{5.122}$$

to be quasi-regular, where $(n \times n)$ matrix J is constant and satisfies the conditions

$$J^* = -J, \qquad J^2 = -I_n, \tag{5.123}$$

and locally summable $(n \times n)$ matrix $H(x)$ is non-negative

$$H(x) = H^*(x) \geq 0, \tag{5.124}$$

it is necessary and sufficient that

$$\int_0^\infty \operatorname{tr} H(x)\, dx < \infty. \tag{5.125}$$

Indeed, let in Theorem 5.15 (or Theorem 5.15^{bis}) $Q(x) = J$, $A(x) = H(x)$, $B(x) = 0$ and $\lambda_0 = 0$. Then for quasi-regularity of the system (5.122) it is necessary and sufficient that all vector-solutions of $J(dy/dx) = 0$ belong to the space $L_2(0,\infty;\, H(x)\,dx)$ or, equivalently, that

$$G_\infty \{U(\cdot, 0)\} = \int_0^\infty H(t)\, dt < I \cdot \infty \tag{5.126}$$

because, in the considered case, $U(x,0) \equiv I$.

But the conditions (5.125) and (5.126) are equivalent for non-negative matrices.[5]

Above we considered the quasi-regular case for the system (5.116), i.e. the case of maximum possible f-deficiency indices $N_+ = N_- = n$ for such system on the half-axis. We now consider the upper and lower bounds for f-deficiency indices of the system (5.116).

Theorem 5.16 ([562, Theorem 2.2], [13, Theorem 9.11.1 for canonical systems]) *Under the conditions of Theorem 5.14, the f-deficiency numbers of the general symmetric first-order system (5.116) on the half-axis $\mathbb{R}_+$ satisfy the inequalities*

$$\nu_-[Q]\leq N_+\leq n, \qquad \nu_+[Q]\leq N_-\leq n \tag{5.127}$$

and

$$N_+ + N_-\geq n, \qquad N_\pm\geq \dim N_0, \tag{5.128}$$

where $\nu_+[Q]$ and $\nu_-[Q]$ are, respectively, the numbers of positive and negative eigenvalues of the matrix $Q(x)$, and N_0 is the null manifold.

(It is well known that for a continuous non-singular matrix-function $Q(x)$, the numbers $\nu_\pm[Q]$ do not depend on x.)

We now consider deficiency indices of higher order symmetric systems.

Let the coefficients of the symmetric system (2.1) of order $r = 2n$ or $r = 2n - 1$, $n = 1, 2, \ldots$, be $(m \times m)$ matrix-functions on the semi-axis $\mathbb{R}_+$ satisfying the conditions from the beginning of Sec. 2.1 up to the formula (2.2). The only exception is the condition $W(x) >> 0$ (condition "$>>$") that may be replaced by a weaker condition $W(x)\geq 0$ (condition "$\geq$"). Under the condition "$>>$" we will talk about the deficiency indices of the equation (2.1) or operator (2.2), and under the condition "$\geq$" we will talk about the formal deficiency (f-deficiency) indices of the system (2.1). Under the condition "$>>$," the formal deficiency indices obviously coincide with the "usual" deficiency indices.

A recent monograph [43] deals with modern ideas and methods of the theory of order preserving random dynamical systems. Dynamical systems under noise perturbations are a possible area of application of this theory. The monograph [43] investigates "random attractors" and their properties and considers interesting examples and applications.

[5] Our Corollary 5.4 contains both Theorems 5.14 and 5.16 of [621] (see also [619] and [620]).

The following theorem holds in both even order ($r = 2n$) and odd order ($r = 2n - 1$) case.

Theorem 5.17 (**[562, Theorems 4.1–4.3]**) *By introducing additional unknowns in the form of quasi-derivatives (2.11) of the required solution (see (5.3) or (5.5) correspondingly), the formally symmetric system of differential equations of arbitrary order r, with a non-negative matrix weight $W(x)\geq 0$ (2.1), regarded as an ordinary differential equation for vector-functions when $(m \times m)$ matrix coefficients are sufficiently smooth (and as quasi-differential equation otherwise) and having non-degenerate leading coefficient ($p_n(x)$ if $r = 2n$ and Re $q_n(x)$ if $r = 2n - 1$), is equivalent to a symmetric first order system of the form (5.116) in the space of vector-functions (with mr components) having a non-negative matrix weight*

$$A(x) = W(x) \oplus 0_{(r-1)m\times(r-1)m}\geq 0, \tag{5.129}$$

and, consequently[6], *is equivalent to a canonical system of the form*

$$J\frac{dz}{dx} = \lambda H(x)z \tag{5.130}$$

with a non-negative $mr \times mr$ weight $H(x)\geq 0$.

The f-deficiency numbers $N_L(\lambda)$ of the system (2.1), defined as the maximal numbers of linearly independent W-square-integrable solutions of this system, are the same as f-deficiency numbers $N(\lambda)$ of the corresponding symmetric first-order system (5.130), and, therefore, they have constant values N_L in the half planes Im $\lambda > 0$ and Im $\lambda < 0$ respectively:

$$N_L(\lambda) = {N_L}^+ \quad (Im\ \lambda > 0); \qquad N_L(\lambda) = {N_L}^- \quad (Im\ \lambda < 0). \tag{5.131}$$

Moreover, for $r > 1$, ${N_L}^+ = mr$ if and only if ${N_L}^- = mr$, and if for $r > 1$ and some $\lambda_0\in\mathbb{C}$ we have $N_L(\lambda_0) = mr$, then the system (2.1) is quasi-regular, and, therefore, $N_L(\lambda) = mr$ for all $\lambda\in\mathbb{C}$.

Note that only one passage (5.3) to a new unknown vector-function with mr components (including quasi-derivatives, but without additional change[6] of the unknown) leads to the first order system (3.40) in the case $r = 2n$, and, in this case, $Q(x) = J/i$ and $\nu_\pm[Q] = mn$ and, therefore,

$$mn\leq {N_L}^\pm\leq 2mn, \quad (r = 2n). \tag{5.132}$$

[6]by additional change of the unknown vector-function by multiplying it by the corresponding chosen matrix-function (see [562, Lemma 1.1] or (5.11)–(5.16))

In the case $r = 2n - 1$, the analogous passage (5.5) to the first order system (5.116) (also without additional change[6] of the unknown) leads to the system (5.7), with $Q(x)$ defined by (5.6), and, in this case,

$$\nu_{\pm}[Q] = m(n-1) + \nu_{\pm}[\mathrm{Re}\; q_n] \qquad (r = 2n-1), \tag{5.133}$$

and, therefore, by Theorem 5.16

$$m(n-1) + \nu_{\mp}[\mathrm{Re}\; q_n] \leq N_L{}^{\pm} \leq m(2n-1) \qquad (r = 2n-1). \tag{5.134}$$

Note also that in both cases $r = 2n$ and $r = 2n - 1$, if $r > 1$, we have

$$Q^{-1}(x)A(x) = 0 \quad \text{and} \quad \mathrm{tr}\left\{Q^{-1}(x)A(x)\right\} = 0, \quad (r > 1), \tag{5.135}$$

and, therefore, the condition (5.117) of Theorem 5.14 is satisfied at $r > 1$ automatically for the equation (5.116) corresponding to the equation (2.1) by virtue of the corresponding changes (5.3) or (5.5).

Much less is known about realizability of r-th order $(m \times m)$ systems with prescribed f-deficiency indices on the half-axis than is known about the same question in the scalar case.

Theorem 5.18 ([562]) *For an arbitrary pair of integers $\{s, t\}$, where $0 \leq s, t \leq m$, there is an $(m \times m)$ first order differential system of the form (5.116) with weight $A(x) \equiv I_m$ such that the corresponding minimal differential operator on the half-axis has deficiency indices $\{s, t\}$:*

$$N_+ = s, \quad N_- = t.$$

This result is a simple corollary of the following fact noted in [561]: the first order scalar equation with $q(x) > 0$ and real $p(x)$

$$\frac{i}{2}\left\{(qy)' + qy'\right\} + p(x)y = \lambda y, \qquad x \in [0, \infty), \tag{5.136}$$

gives rise (on the half-axis) to a minimal operator with deficiency indices $\{1, 1\}$ if $q(x)^{-1} \in L_1(0, \infty)$, and deficiency indices $\{0, 1\}$ in the opposite case.

Indeed, we can construct a diagonal system of the type (5.116) with the prescribed deficiency indices $\{s, t\}$ using equations of the form (5.136) with the corresponding q_k and p_k for unknowns y_k $(k = 1, \ldots, m)$.

With regard to the second-order symmetric $(m \times m)$ systems,

$$-y'' + P(x)y = \lambda y, \qquad x \in (0, \infty), \tag{5.137}$$

V. B. Lidskiy [631] established sufficient conditions for the deficiency indices of such systems (in the case of real $P(x)$) to have extreme possible values, i.e. $\{m,m\}$ or $\{2m,2m\}$. Realizability of all other admissible values[7] of deficiency indices for second-order Hermitian systems is established by the following theorem.

Theorem 5.19 (Kalyabin [529]) *For an arbitrary pair of integers* $\{s,t\}$, *where* $m \leq s,t < 2m$ *or* $s = t = 2m$, *there is a Hermitian* $(m \times m)$ *matrix* $P(x)$ *such that the minimal differential operator corresponding to the second order differential system of on the half-axis (5.137) has the prescribed values* $\{s,t\}$ *of deficiency indices.*

In this case, the construction of the $(m \times m)$ matrix $P(x)$ for prescribed deficiency indices $\{s,t\}$ is much more complicated than the construction of the first order system with prescribed deficiency indices. For example, in the case $s = 3$ and $t = 2$ for (2×2) matrix $P(x)$, the following construction is given in [529]:

$$P(x) = (p_{j,k}(x)), \quad j,k = 1,2, \quad \text{where } p_{1,1}(x) = 0,$$
$$p_{1,2}(x) = \overline{p_{2,1}}(x) = \exp\left(x^2 - i\int_0^x \exp(2\xi^2)\,d\xi\right),$$
$$p_{2,2}(x) = \left((p_{1,2}(x))^{-1}\right)'' p_{1,2}(x) = -\exp(4x^2) + 4x^2 - 2.$$

Open Question 5.5 Obviously, in the case of symmetric differential systems with Hermitian matrix coefficients there are many more unsolved problems than in the case of scalar symmetric differential equations with complex coefficients. It is interesting whether in the case of systems, as well as in the scalar case, all admissible values of deficiency (or f-deficiency) indices $\{s,t\}$ can be realized for $(m \times m)$ systems of prescribed order r.

It is natural to try to solve this problem first in the scalar case. Since the problem has been solved affirmatively in the scalar case of order $r = 3$ and $r = 4$ (see Corollary 5.2), it is also natural to try to solve this problem for systems of order $r = 3$ and $r = 4$.

(By the way, asymptotic formulas for solutions of third order scalar differential equations with polynomial coefficients and deficiency indices for such type of operators were investigated in [473].)

What are the chances for the existence of affirmative solutions of such type of problems for all r in the vector or, at least, scalar case?

It is difficult to suggest an affirmative or negative answer. We can make some suggestions by comparison with more or less analogous problems. For

[7] A generalization of Lidskiy's minimality conditions for the deficiency (or f-deficiency) indices of the problem (2.1) with $r = 2$ and $W(x) \geq 0$ was obtained in [621].

example, in [583] I. V. Kovalishina and V. P. Potapov obtained an affirmative solution for a matrix analogue of the Nevanlinna–Pick problem— an interpolation problem for matrix-valued analytic Nevanlinna–Herglotz type functions. In this problem, a discrete sequence of $(m \times m)$ matrix balls (Weyl balls) with the so-called left and right matrix radii (sometimes they are called left and right semi-radii) arises. The considered problem has to do with possible values of the pair $\{s,t\}$ of ranks of the left and right radii of Weyl limit-balls. In [583] I. V. Kovalishina and V. P. Potapov obtained multiplicative formulas for the matrix radii of Weyl limit-balls. Using those formulas, they proposed a method for constructing values of the ranks of the left and right radii of the corresponding Weyl limit-balls for Nevanlinna–Pick problems with an arbitrarily prescribed admissible pair $\{s,t\}$, $0 \leq s, t \leq m-1$. As in the case of differential operators, maximal possible values $s = t = m$ can be reached only simultaneously, and the case $s = t = m$ for the Nevanlinna–Pick problem, is called (as in the case of differential operators) "absolutely (or completely) indefinite case." (Note that the work [582] by I. V. Kovalishina establishes a deep connection of Nevanlinna–Pick problem, Caratheodory problem and Hamburger's moment problem with the so-called J-theory[8] of V. P. Potapov and presents a uniform method for their solution. Additionally, the work [582] establishes criteria for completely indefinite cases for Nevanlinna–Pick and Hamburger matrix problems (analogous to Denjoy criterion for the scalar case), considers the theory of Weyl balls (circles) and describes a variety of solutions for these problems.) Also note a recent work by Yu. M. Dyukarev [391], where conditions for complete indefiniteness of the Nevanlinna–Pick problem for Stieltjes matrix-functions are obtained.

Based on results by I. V. Kovalishina and V. P. Potapov [583], V. K. Dubovoy [55], [385] proved that in the matrix interpolating Schur problem the pairs of ranks of the left and right (semi)-radii of the limit matrix ball can also achieve all admissible values.

However, unlike differential problems, interpolating problems are discrete, and, hence, can be related to difference operators. The latter operators may be unbounded only at infinity, unlike differential operators (that are locally unbounded).

H. Weyl [917] obtained a continuous analogue of the scalar Nevanlinna–Pick problem as a system of two linear first order equations with a linear fractional inclusion of the spectral parameter and constructed the limit–

[8] A systematic account of the principles of J-theory is given in the paper [393] and in the third section of the book [209].

point and limit–circle theory for this problem. Therefore, the Nevanlinna–Pick matrix problem is, in some sense, a discrete analogue of a first order differential system. Although a differential equation of arbitrary order can be reduced to a first order differential system, a possibility for realizing all admissible pairs of ranks of the left and right matrix (semi)-radii of the limit matrix ball for the Nevanlinna–Pick matrix problem may not guarantee that for differential equations (or systems) of arbitrary order all admissible pairs of deficiency numbers are realizable. Therefore, the last problem is open.

It is instructive to compare some aspects of this problem with the history of Hilbert's 21-st problem (see [32], [79], [80], [510] and [511]), which only recently received the negative answer in the works of A. A. Bolibrukh (see, for instance, [32]), after almost 100 years of numerous investigations by many mathematicians. For a long time it had been erroneously thought that in 1909 J. Plemelj (see [158] and [749]) obtained the affirmative solution of this problem. In this regard, our comparison will motivate some new versions of the Open Question 5.3.

5.3.6 *Some Comments on Hilbert's 21-st Problem and Bolibrukh Counterexample*

At the Second International Congress of Mathematicians in Paris on August 8, 1900 David Hilbert delivered his famous lecture "Mathematical Problems." One of those significant Problems, namely, the 21-st Problem, was still under Riemann's consideration. Therefore, Hilbert's 21-st Problem is also called the Riemann–Hilbert (RH) Problem.

In the 21-st Problem Hilbert proposed, "to show that there always exists a linear differential equation of Fuchsian type with prescribed singular points and prescribed monodromy group. Thus, the problem requires to find n functions of variable z that are regular on the whole complex z-plane, with a possible exception of given singular points—at those points the corresponding functions have only finite order poles, and by going around those points in the z-plane, the corresponding functions undergo a linear permutation."

We now give a more detailed explanation of the problem. Consider a first order linear homogeneous $n \times n$ differential system

$$\frac{dY}{dz} = R(z)Y \tag{5.138}$$

with rational coefficients, i.e. with the $n \times n$ matrix $R(z)$ with rational entries.

Let $a_1, \ldots, a_p$ be the poles of $R(z)$. Then solutions of the system (5.138) are analytic functions with singularities only at $a_1, \ldots, a_p$ and, possibly, at $a_0 = \infty$. Here, we consider the case of finite points a_j and suppose that the point $a_0 = \infty$ is a regular point[9] of the considered system (on the extended z-plane or on the corresponding Riemann sphere). A singular point a_j of the system (5.138) is called a regular singular point if every solution $Y(z)$ can increase as at most some (negative) power of $|z - a_j|$ when $z \to a_j$ through an arbitrary sectorial neighborhood with the vertex at the point a_j that does not coincide with the plane $\mathbb{C}$. A singular point a_j of the system (5.138) is called Fuchsian singular point if $R(z)$ has at $z = a_j$ only a first order pole. Every Fuchsian point is a regular singular point (see [46]) of the system (5.138). If all singular points of the system (5.138) are Fuchsian points, then this system is called Fuchsian system, and it can be written in the following form:

$$\frac{dY}{dz} = \left(\sum_{j=1}^{p} \frac{B_j}{z - a_j} \right) Y, \tag{5.139}$$

where B_j are constant matrices, and, by the residue theorem,

$$\sum_{j=1}^{p} B_j = 0 \tag{5.140}$$

(because ∞ is a regular point by assumption).

For the scalar n-th order linear differential equation

$$y^{(n)} + \sum_{k=0}^{n-1} b_k(z) y^{(k)} = 0, \tag{5.141}$$

the notion of a regular singular point is defined analogously to the case of a first order differential system. However, Fuchsian singular point a_j is now defined as a point near which

$$|b_k(z)| \leq C_k |z - a_j|^{k-n}, \quad k = 0, 1, \ldots, n-1; \qquad C_k > 0. \tag{5.142}$$

[9]The point $z = \infty$ is a regular point of the system (5.138) or a singular point of some kind of this system if, after substitution $z = \zeta^{-1}$, the system (5.138) turns into a system for which the point $\zeta = 0$ is a regular point or, correspondingly, a singular point of the same kind. Regular or singular points for a scalar differential equation at infinity are determined analogously.

It is known that for a scalar equation of the type (5.141), Fuchsian and regular singular points coincide (see [46, Chapter IV, Theorems 5.1 and 5.2]), unlike the case of a system of the type (5.138).

Let us now return to the system (5.138). Singular points $a_1, \ldots, a_p$ (and, possibly, a_0) of (5.138) can be branch points for solutions $Y(z)$ of this system. Therefore, after z goes around the singular point a_j, each fundamental system $Y(z)$ of solutions of (5.138) turns into another fundamental system, namely, into $Y(z)A_j$, where A_j are constant non-singular matrices, $j = 0, 1, \ldots, p$, depending on the choice of the fundamental system $Y(z)$ and the starting point z_0 of the paths around each singular point. (Evidently, z_0 is also the ending point of those paths.) The matrices A_j are called monodromy matrices, and A_j satisfy the so-called the Riemann relation

$$A_0 A_1 \cdots A_p = I. \tag{5.143}$$

Therefore, every fundamental system of solutions $Y(z)$ defines a homomorphism of the fundamental group of the space $\mathbb{C}P^1 \backslash \{a_0, a_1, \ldots, a_p\}$ with the designated point z_0 into the general linear group $GL(n, \mathbb{C})$ of $n \times n$ non-singular matrices with complex entries[10].

The above-mentioned homomorphism

$$\chi : \pi_1(\mathbb{C}P^1 \backslash \{a_0, a_1, \ldots, a_p\}, z_0) \longrightarrow GL(n, \mathbb{C}) \tag{5.144}$$

is called the monodromy representation or simply the monodromy of the system (5.138), and the group Ran χ (here, Ran χ denotes the range of the homomorphism χ) is called the monodromy group of this system. (In order to fix the choice of matrices A_j and the corresponding homomorphism, it is sufficient to fix $Y(z_0)$, $\det Y(z_0) \neq 0$, at the starting point z_0 of paths. Replacing the point z_0 by z_0' or changing the fundamental matrix $Y(z_0)$, the monodromy matrices A become $S^{-1}AS$, where S is a constant non-singular matrix. Therefore, the monodromy group of the system (5.138) is determined up to an equivalence.)

In 1857 B. Riemann [762] posed the inverse problem—given a homomorphism of the fundamental group of the space $\mathbb{C}P^1 \backslash \{a_0, a_1, \ldots, a_p\}$ with

[10] Here, $\mathbb{C}P^1$ denotes the extended complex-plane, and fundamental group refers to the group of homotopic classes of closed paths starting and ending at some (fixed for all paths) point $z_0 \in \mathbb{C}P^1 \backslash \{a_0, a_1, \ldots, a_p\}$. The product of two paths is defined as the path consisting of the given two paths traversed in the corresponding order; see, for example, [57] and [149]. The corresponding fundamental group is usually denoted by $\pi_1(\mathbb{C}P^1 \backslash \{a_0, a_1, \ldots, a_p\}, z_0)$.

prescribed points $a_0, a_1, \ldots, a_p$ into the group $GL(n, \mathbb{C})$, does there exist an $n \times n$ first order linear differential system of Fuchsian type that has a fundamental solution $Y(z)$ giving rise to the given homomorphism?

In the case $p = n = 2$, Riemann obtained the affirmative answer to this problem and explicitly found the corresponding system in [763].

H. Poincaré (1884) and, later, L. Schlesinger (1905-1906) considered RH-problem; however, their proofs contain some gaps and are not rigorous. In 1905, D. Hilbert considered the case $n = 2$ and of arbitrary p; see [512] and [513].

In 1908 J. Plemelj (see [158] and [749]) offered the proof that had later been considered flawless for many years; however, in early 1980s (see the works by A. T. Kohn [510] and V. I. Arnold and Yu. S. Ilyashenko [239]) gaps were found in J. Plemelj's proof[11]. As shown in [239], Plemelj's proof can be made rigorous, if one assumes additionally that one of the monodromy matrices A_j corresponding to going around the singular point a_j along a "small" loop is similar to a diagonal matrix. Also, J. Plemelj was the first to solve the problem analogous to RH-problem in the case of differential systems of the type (5.138) with regular singular points.

An effective solution of RH-problem for representations (5.144) that have all monodromy matrices A_j close to the identity matrix (corresponding to going around singular points a_j along "small" loops) was obtained by I. A. Lappo-Danilevskiy [107] in the 1920s with the help of the method of analytical functions of matrices also developed by Lappo-Danilevskiy.

In 1956 B. L. Krylov [600] constructed an effective solution of RH-problem for the case $n = 2$ and $p = 3$. For the case $n = 2$ and $p = 4$ an analogous problem was considered by N. P. Erugin [64], who showed, among other results, its connection with Painlevé equations.

An important new approach to the investigation of RH-problem is contained in the 1957 work by H. Röhrl [810] and is related to ideas of G. D. Birkhoff [273]. In his investigation of the RH-problem, H. Röhrl used topological methods (the theory of fiber bundles) and the theory of functions of several complex variables. H. Röhrl not only reproved Plemelj's results, but also proved solvability of the RH-problem on a non-compact Riemann surface as well as solvability of the problem analogous to the RH-problem for an arbitrary Riemann surface in the class of systems with regular singular points.

---

[11] Plemelj's method is based on reduction of the RH-problem to a certain homogeneous Hilbert's boundary problem in the theory of singular integral equations (see, for instance, [143] and [193]), and some inaccuracies were found in this reduction.

"Topological" approach was also used by Yu. S. Ilyashenko [239] in the investigation of a modified RH-problem. A rigorous proof of the solvability of RH-problem in the case of an arbitrary set of points $a_1, \dots, a_p$ and arbitrary 2-dimensional representation (5.144) follows from the works by W. Dekkers [52] and [352]. In those works, W. Dekkers used a result by P. Deligne [352] on the form of matrix coefficients of a system with regular singular points.

Some effective sufficient conditions for solvability of the inverse monodromy problem were established by A. R. Its and Yu. V. Novokshenov [522]. More detailed bibliography and history of RH-problem are contained in [9], [32], [294] and [295].

We will now formulate the main results of A. A. Bolibrukh [32]; see also [290]–[292].

Theorem 5.20 **(A. Bolibrukh)** *The RH-problem has the affirmative solution for every irreducible representation (5.144) of dimension $n = 3$.*

Theorem 5.21 **(A. Bolibrukh)** *For every $p > 3$, every set of points $a_1, \dots, a_p$ and every $n \geq 3$, there exists a representation (5.144) that does not correspond any Fuchsian system (5.139).*

This Theorem shows that RH-problem does not have the affirmative solution in the general case.

The following theorem generalizes the Theorem 5.20.

Theorem 5.22 **(A. Bolibrukh [293], V. Kostov [574])** *The RH-problem has the affirmative solution for every irreducible representation (5.144).*

With regard to the n-th order scalar Fuchsian equation (5.141), it is known that for $n > 1$ the RH-problem has the negative answer (see, for example, [717]) because the number of parameters of the corresponding equation is less than the set of equivalence classes of representations (5.144). (The corresponding calculation of parameters was initiated by H. Poincaré.)

As to the possibility of reducing the Fuchsian equation (5.141) to the Fuchsian system (5.139), the simplest way of introducing new unknown functions $y_1 = y$, $y_2 = y'$, ..., $y_n = y^{(n-1)}$ does not lead to the Fuchsian system. Another way, offered in [46], allows the construction of a Fuchsian system from an arbitrary equation which is Fuchsian in a neighborhood of a point a_j; however, this construction has essentially local nature.

Theorem 5.23 **(A. A Bolibrukh [32])** *For every Fuchsian equation*

(5.141) on $\mathbb{C}P^1$ it is possible to construct the Fuchsian system (5.139) with the same singular points and the same monodromy.

(This Theorem answers a question of N. P. Erugin [64].)

Let us now return to the case of RH-problem for Fuchsian systems and consider the Bolibrukh counterexample.

Bolibrukh Counterexample ([32]). The monodromy representation χ of the three-dimensional system (5.138) with four regular singular points with the matrix coefficient $R(z)$ of the form

$$R(z) = \frac{R_1(z)}{z^2} + \frac{R_2(z)}{6(z+1)} + \frac{R_3(z)}{2(z-1)} + \frac{R_4(z)}{3(z-1/2)} \tag{5.145}$$

cannot be realized as the monodromy representation (5.144) of the Fuchsian system (5.139) in the case when the matrices $R_1(z), \ldots, R_4(z)$ are

$$R_1(z) = \begin{pmatrix} 0 & 1 & 0 \\ 0 & z & 0 \\ 0 & 0 & -z \end{pmatrix}, \qquad R_2(z) = \begin{pmatrix} 0 & 6 & 0 \\ 0 & -1 & 1 \\ 0 & -1 & 1 \end{pmatrix},$$
$$R_3(z) = \begin{pmatrix} 0 & 0 & 2 \\ 0 & -1 & -1 \\ 0 & 1 & 1 \end{pmatrix}, \qquad R_4(z) = \begin{pmatrix} 0 & -3 & -3 \\ 0 & -1 & 1 \\ 0 & -1 & 1 \end{pmatrix}. \tag{5.146}$$

Thus, we see that the representation for which the corresponding Fuchsian system does not exist is given indirectly.

It can also be seen that singular points -1, 1, $1/2$ of the system (5.145) are Fuchsian points; the point ∞ is a holomorphic point for the system (5.145) because the residue of $R(z)$ at ∞ is zero, and the point $z = 0$ is the regular singular point for the system (5.145), (5.146).

As shown by A. A. Bolibrukh [32], "non-correspondence to a Fuchsian system" property of the representation χ (5.144) is unstable under small deformations of the set of singular points.

* * *

Let us now return to Open Questions in the Deficiency Indices Problem. Here, we will formulate some other versions of these questions. Of course, the corresponding questions could be formulated independently of the history of Hilbert's 21-st problem; nevertheless, they now seem more natural.

Open Questions 5.6 To consider previous Questions 5.3, 5.4 and 5.5 for differential operators with analytic coefficients belonging to certain special classes, such as polynomial or rational coefficients, or entire or meromorphic coefficients, or with coefficients depending on the prescribed number of parameters for differential operators of a given order. (See, for example, [227], [472], [473] and [859].) We can also consider operators whose coefficients are analytic in some sector containing the positive half-axes in the complex plane and have asymptotic expansions there (as, for example, in Secs. 5.3.3 and 5.3.4). One of the questions is as follows: for which of those classes is the answer to Questions 5.3, 5.4 and 5.5 negative?

If we consider symmetric differential operators with complex-valued coefficients belonging to a sufficiently restricted class, we can, of course, obtain the negative answer to the Question 5.3 for this class of operators. However, it would be interesting to obtain the negative answer for some not very restricted classes of considered operators. Another question is as follows: what are the deficiency indices of the corresponding class of differential operators at various values of numerical or functional parameters?

It would be very interesting to see what results could be obtained by using operator theory methods (that turned out to be very successful in the case of differential operators with real-valued coefficients; see Sec. 5.3.2, Theorem 5.4).

5.3.7 *Some Comments on Section 5.2*

Regarding Sec. 5.2, we will mention a series of works dealing with the notion of Weyl–Titchmarsh characteristic function [916], [191] and their generalizations and applications, especially concerning the construction of Weyl solutions and the investigation of their characteristic properties: [656][12], and the monographs [2], [5], [24], [46], [55], [96], [112], [121], [191, Volume 1], and [203]. The last monograph contains the theory of regular matrix functions of Schur and Caratheodory class in the unit disk (they are obtained from matrix functions of Nevanlinna–Herglotz class with the help of Cayley transform), the theory of matrix balls of Weyl–Titchmarsh type and their applications as well as an extensive bibliography. We shall also mention the papers [233], [242], [262], [272], [308], [310], [322], [341], [354]–[357], [442], [459], [467], [470], [480], [541]–[543], [552], [588], [616], [633], [646]–[648], [650], [652], [656], [680], [688], [719], [751], [771], [789], [828],

[12] The account of corresponding results of V. A. Marchenko [656] will be given later in these Bibliographical Comments (Sec. 5.3.8).

[839], [916], [923] and [924]. In [552] V. I. Khrabustovskiy obtained, among other results, the necessary and sufficient conditions for the Weyl characteristic function corresponding to separated boundary conditions (for the formulation of this result, see Bibliographical Comments to Appendix).

The concept of an abstract Weyl function corresponding to a self-adjoint extension of a symmetric operator was introduced in the works of V. A. Derkach and M. M. Malamud; see [354], [359], [360], [361], [646] and also V. A. Derkach–M. M. Malamud–H. S. V. de Snoo–S. Hassi [358], [362]. Here, the method of abstract boundary conditions plays a very useful role. The concept of an abstract Weyl function plays the role similar to that of the Weyl–Titchmarsh function for the Sturm–Liouville operator on a half-axis. More precisely, the abstract Weyl function for the Sturm–Liouville operator coincides with the classical Weyl–Titchmarsh function, and, for an arbitrary symmetric operator, it coincides with the so-called M. Krein's "Q-function." (See the works by M. G. Krein and H. Langer [592], [593]; M. G. Krein and I. E. Ovcharenko [594]–[598]; and H. Langer and B. Textorius [615]. In some cases, for the so-called dual pair of bounded operators, Q-function coincides with the transfer function of a linear dynamical system from D. Z. Arov's work [241].) In this regard, note the monographs [171] and [172] by L. A Sakhnovich where the representation of the fundamental solution in the form of transfer function is applied in various spectral and interpolation problems.

For non-self-adjoint Sturm–Liouville operators, the Weyl function of classical type is introduced by V. B. Lidskiy [633]. For non-self-adjoint (finite) systems, see E. S. Birger and G. A. Kalyabin [272].

For other generalizations of the Weyl function or matrix for some classes of non-self-adjoint differential operators, see the works by V. A. Yurko [202]–[205] and [923]–[925].

5.3.8 *Characteristic Properties of Weyl Solutions for the Sturm–Liouville and Dirac Equations. V. A. Marchenko's Theorems*

Let us consider the Sturm–Liouville (scalar) equation

$$L[y] := -y'' + q(x)y = \lambda y; \quad \lambda = k^2, \tag{5.147}$$

with the real-valued potential $q \in L^2_{\rm loc}(-\infty, \infty)$.

Denote by $c(k, x)$ and $s(k, x)$ the solutions of (5.147) with the initial data:

$$c(k,0) = s'(k,0) = 1, \qquad c'(k,0) = s(k,0) = 0.$$

By the Weyl's [916] classical theorem, there exist functions $m^{\pm}(\lambda)$, holomorphic in $\lambda \in \mathbb{C}\backslash\mathbb{R}$, such that the following solutions $\psi^{\pm}$ of (5.147) belong, respectively, to $L^2(0,+\infty)$ or $L^2(-\infty,0)$:

$$\psi^+(\lambda,x) = c(k,x) + m^+(\lambda)s(k,x) \in L^2(0,+\infty) \tag{5.148}$$

$$\psi^-(\lambda,x) = c(k,x) - m^-(\lambda)s(k,x) \in L^2(-\infty,0). \tag{5.149}$$

These solutions are called Weyl solutions, and the functions $m^{\pm}(\lambda)$ are called Weyl (or Weyl–Titchmarsh) functions of (5.147) on $\mathbb{R}^{\pm}$ respectively. The following equalities are known for the $L^2(\mathbb{R}^{\pm})$ norms of $\psi^{\pm}$:

$$\|\psi^{\pm}(\lambda,x)\|^2 = \frac{\text{Im } (m^{\pm}(\lambda))}{\text{Im } \lambda}.$$

In the case of the Weyl limit-point at $+\infty$ or $-\infty$, the corresponding Weyl functions $m^{\pm}(\lambda)$ are defined uniquely. In the case of the Weyl limit-circle at $+\infty$ or $-\infty$, to each point of these circles there corresponds its own Weyl function. In the case of semi-boundedness of the minimal operator corresponding to (5.147) on the half-axis $\mathbb{R}^+$ or $\mathbb{R}^-$, we have the limit-point case for the corresponding half-axis. Denote by $\mu^{\pm}$ the infimum of corresponding operators on the corresponding half-axis.

We have:

$$\mu^{\pm} = \inf \int_{\mathbb{R}^{\pm}} (|y'(x)|^2 + q(x)|y(x)|^2)\, dx,$$

where the infimum is taken over the set of the $L^2(\mathbb{R}^{\pm})$-normalized functions $y(x) \in C^2$ with compact support (its own support for each function) and satisfying the boundary condition $y(0) = 0$.

In accordance with what we said above, if $\mu^+ > -\infty$ then the Weyl function $m^+(\lambda)$ is uniquely defined, and $m^-(\lambda)$ is uniquely defined when $\mu^- > -\infty$. Additionally, in these cases, the functions $m^+(\lambda)$ and $m^-(\lambda)$, respectively, are holomorphic everywhere in $\mathbb{C}\backslash[\mu^{\pm},\infty)$. If μ^+ or μ^- is equal to $-\infty$, the negative λ-half-axis contains an unbounded set of singularities of the functions $m^{\pm}(\lambda)$ respectively.

Thus, we see that the Weyl solutions for the semi-bounded Sturm–Liouville equations are defined uniquely (up to a multiplicative constant) by the conditions (5.148), (5.149) respectively, i.e. by their behavior with respect to the variable x, when x tends to $+\infty$ or $-\infty$.

As V. A. Marchenko [656] has shown, the Weyl solutions of semi-bounded equations (5.147) are also uniquely defined by their behavior with respect to the variable λ when $|\lambda| \to \infty$. Following V. A. Marchenko, we replace the pair $\psi^{\pm}(\lambda, x)$ of Weyl solutions by one solution, as in the following formula, where $\lambda = k^2$:

$$\psi(k,x) = \begin{cases} \psi^+(\lambda, x) & \text{if } \operatorname{Im} k > 0, \\ \psi^-(\lambda, x) & \text{if } \operatorname{Im} k < 0. \end{cases}$$

This solution will also be called the Weyl solution of (5.147). The solution $\psi(k, x)$ is defined and holomorphic in k in the complex k-plane everywhere outside the set

$$\sigma = \mathbb{R} \cup [-i\gamma^-, i\gamma^+],$$

where $\gamma^{\pm} = [\max(0, -\mu^{\pm})]^{1/2}$ (i.e. outside the real axis $\mathbb{R}$ and bounded or unbounded segment $[-i\gamma^-, i\gamma^+]$ of the imaginary axis).

We have that $\psi(k, x) \in L^2(\mathbb{R}^{\pm})$ with respect to x when k lies in the upper or, correspondingly, lower complex half-plane.

In what follows, by $K(\epsilon)$ we denote the complement of an ϵ-neighborhood of the set σ. It is evident, that for each positive ϵ, the Weyl solution $\psi(k, x)$ of (5.147) is holomorphic in the domain $K(\epsilon)$. In the case of the null potential $q = 0$, we have for the corresponding Weyl solution of (5.147):

$$\psi_0(k,x) = e^{ikx}.$$

As V. A. Marchenko showed in the following theorem for an arbitrary equation (5.147), the principal term of the asymptotics for the Weyl solution when $|k| \to \infty$ in the domain $K(\epsilon)$ is $\psi_0(k, x)$ for all $\epsilon > 0$.

Theorem 5.24 **(V. A. Marchenko [656])** *The Weyl solution $\psi(k, x)$ of the equation (5.147) can be represented in the form*

$$\psi(k,x) = e^{ikx}(1 + u(k,x)),$$

when the function $u(k, x)$ satisfies the following condition in the domain $K(\epsilon)$ for all $\epsilon > 0$ and $N > 0$:

$$\lim_{|k|\to\infty} \sup_{|x|<N} \left(|u(k,x)| + \left| k^{-1} \frac{du}{dx}(k,x) \right| \right) = 0.$$

This asymptotic formula allows a unique characterization of the Weyl solution by its behavior with respect to the variable $\lambda = k^2$.

Theorem 5.25 **(V. A. Marchenko [656])** *Let the operator L in (5.147) be semi-bounded on the right (left) half-axis. If some solution $\phi(k,x)$ of the equation (5.147) depends on k analytically in the half-plane Im $k > c$ (respectively, Im $k < c$) for some constant c, and, for some $\theta < 1$,*

$$\lim_{|k|\to\infty} e^{-|k|^\theta} e^{-ikx}\phi(k,x) = 0,$$

then, in the corresponding half-plane,

$$\phi(k,x) = \phi(k,0)\psi(k,x),$$

where $\psi(k,x)$ is the Weyl solution of the equation (5.147).

Those results, after comparison with V. A. Marchenko [655], lead their author to the following theorem:

Theorem 5.26 **(V. A. Marchenko [656])** *Suppose that the equation (5.147) has solution $\phi(k,x)$ which possesses the logarithmic derivative $\frac{\phi'}{\phi}$, allowing expansions in Laurent series in the domain $|k| > r$ for each $x \in (-\infty, \infty)$:*

$$\frac{d\phi(k,x)/dx}{\phi(k,x)} = ik + \sum_{n=1}^{\infty} a_n(x)(2ik)^{-n}.$$

Then: 1) The operator L is semi-bounded on both half-axes $\mathbb{R}^\pm$ and

$$\phi(k,x) = \phi(k,0)\psi(k,x),$$

where $\psi(k,x)$ is Weyl solution of the equation (5.147).
2) The potential $q(x)$ is infinitely differentiable and satisfies the inequalities:

$$-(4r^2 - a_1(0)) \le q(x) \le r^2,$$
$$\sup_{-\infty<x<\infty} |q^{(n)}(x)| \le 2\mu^{n+2}(n+1)!,$$

where

$$\mu = \frac{1}{2}(10r^2 - 2a_1(0))^{1/2}.$$

In addition to Weyl solutions for the Sturm–Liouville equation (5.147), V. A. Marchenko [656] considered vector-valued Weyl solutions $\Psi^\pm(\lambda,x)$ for the Dirac equation

$$D[Y] \equiv J\frac{dY}{dx} + V(x)Y = \lambda Y, \tag{5.150}$$

where $J \equiv i\sigma_3$, $\sigma_3 = \begin{pmatrix} 1 & 0 \\ 0 & -1 \end{pmatrix}$ is one of the Pauli matrices and

$$V(x) = \begin{pmatrix} 0 & \nu(x) \\ \overline{\nu(x)} & 0 \end{pmatrix} \in L^2_{\text{loc}}(-\infty, \infty).$$

For $\lambda \in \mathbb{C}\backslash\mathbb{R}$ the Weyl functions $M^{\pm}(\lambda)$ are holomorphic and are uniquely defined by the conditions

$$\Psi^{\pm}(\lambda, x) \equiv F(\lambda, x) \pm M^{\pm}(\lambda)G(\lambda, x) \in L^2(\mathbb{R}^{\pm}), \tag{5.151}$$

where only upper or only lower signs are taken; $F(\lambda, x)$, $G(\lambda, x)$ are the solutions of the equation (5.150) with the initial conditions

$$F(\lambda, 0) = \begin{pmatrix} 1 \\ i \end{pmatrix}, \quad G(\lambda, 0) = \begin{pmatrix} i \\ 1 \end{pmatrix}.$$

The functions and Weyl solutions are uniquely defined because for the Dirac equation we have the Weyl limit point case, as shown by B. M. Levitan (see [113] and [114]), and, independently, by V. V. Martynov [663].

Here we will give a simple proof of this fact from the paper [656] by V. A. Marchenko. Multiply both sides of (5.150) on the left by $Y^*(x)$, integrate both sides from 0 to t, and take the imaginary part of both sides. Integrating by parts the left hand side of the obtained equation, we get the following equation for the solution $Y = G(\lambda, x) = \text{col}\ (G_1, G_2)$:

$$|G_1(\lambda, t)|^2 - |G_2(\lambda, t)|^2 = 2\text{Im}\ \lambda \int_0^t \{|\bar{G_1}(\lambda, x)|^2 + |\bar{G_2}(\lambda, x)|^2\}\, dx.$$

Since for non-real λ the right hand side (and, hence, the left hand side) of the obtained equation is a strictly monotone function of t, vanishing at $t = 0$, it follows that the left hand side of the equation is not integrable on any of the t-half-axes $\mathbb{R}^{\pm}$. Therefore, for Im $\lambda \neq 0$, the solution $G(\lambda, t)$ does not belong to any of the spaces $L^2(\mathbb{R}^{\pm})$, and, hence, the Weyl functions $M^{\pm}(\lambda)$ for the Dirac equation (5.150) are uniquely determined by the condition (5.151). Thus, for the Dirac equation, we always have the limit point case, and this concludes the proof.

The following equalities hold for the $L^2(\mathbb{R}^{\pm})$ norms of Weyl solutions $\Psi^{\pm}(\lambda, x)$:

$$\|\Psi^{\pm}(\lambda, x)\|^2 = 2\frac{\text{Im}\ M^{\pm}(\lambda)}{\text{Im}\ \lambda}.$$

Following [656], we denote

$$\Psi(\lambda, x) = \begin{cases} \Psi^+(\lambda, x) & \text{if Im } \lambda > 0, \\ \Psi^-(\lambda, x) & \text{if Im } \lambda < 0, \end{cases}$$

and call these functions Weyl solutions of the equation (5.150).

We have $\Psi(\lambda, x) \in L^2(\mathbb{R}^\pm)$ for $\pm\text{Im } \lambda > 0$ respectively.

When the potential $\nu = 0$, and, hence, $V = 0$, Weyl solution of (5.150) has the form

$$\Psi_0(\lambda, x) = e^{i\lambda x} \begin{pmatrix} 0 \\ 2i \end{pmatrix}.$$

In the general case, $\Psi_0(\lambda, x)$ is the main term, as $|\lambda| \to \infty$, in the asymptotics of the Weyl solution of the Dirac equation (5.150) in the set $\Delta(\epsilon) = \{\lambda\colon |\text{Im } \lambda| > \epsilon\}$.

Theorem 5.27 (V. A. Marchenko [656] and, with stronger assumptions, T. Misyura [688]) *Weyl solution $\Psi(\lambda, x)$ of (5.150) can be represented as*

$$\Psi(\lambda, x) = e^{i\lambda x} \begin{pmatrix} u_1(\lambda, x) \\ 2i + u_2(\lambda, x) \end{pmatrix},$$

where, for all $\epsilon > 0$ and $N > 0$, the functions $u_1(\lambda, x)$ and $u_2(\lambda, x)$ satisfy the following condition in the set $\Delta(\epsilon)$:

$$\lim_{|\lambda| \to \infty} \sup_{|x| < N} (|u_1(\lambda, x)| + |u_2(\lambda, x)|) = 0.$$

Obtained asymptotics allows us to uniquely characterize Weyl solutions of the Dirac equation (5.150) by their behavior with respect to λ, as opposed to the earlier known characterization of these solutions by their behavior with respect to x as $x \to \pm\infty$.

Theorem 5.28 (V. A. Marchenko [656]) *If the solution*

$$\Phi(\lambda, x) = \begin{pmatrix} \Phi_1(\lambda, x) \\ \Phi_2(\lambda, x) \end{pmatrix}$$

of (5.150) depends analytically on λ in the half-planes $|Im\ \lambda| > c$, and for some $\theta < 1$,

$$\lim_{|\lambda| \to \infty} e^{-|\lambda|^\theta} e^{-i\lambda x} \Phi(\lambda, x) = 0,$$

then, in those half-planes,

$$\Phi(\lambda, x) = \frac{1}{2}(\Phi_1(\lambda, 0) - i\Phi_2(\lambda, 0))\Psi(\lambda, x),$$

where $\Psi(\lambda, x)$ *is the Weyl solution of (5.150).*

Comparing these theorems with the work of V. A. Marchenko in the monograph [19], the following theorem follows:

Theorem 5.29 (V. A. Marchenko [656]) *Assume that the equation (5.150) has a solution*

$$\Phi(\lambda, x) = e^{i\lambda x}\begin{pmatrix} u_1(\lambda, x) \\ u_2(\lambda, x) \end{pmatrix}$$

such that, in the domain $|\lambda| > r$, *the functions* $u_1(\lambda, x)$, $u_2(\lambda, x)$ *for all* $x \in (-\infty, \infty)$ *have Laurent expansions*

$$u_1(\lambda, x) = \sum_{n=1}^{\infty} \lambda^{-n}\alpha_n(x), \quad u_2(\lambda, x) = \beta + \sum_{n=1}^{\infty} \lambda^{-n}\beta_n(x).$$

Then:

1) $\Phi(\lambda, x) = \frac{1}{2}(u_1(\lambda, 0) - iu_2(\lambda, 0))\Psi(\lambda, x)$, *where* $\Psi(\lambda, x)$ *is the Weyl solution of (5.150).*

2) the potential $\nu(x)$ *is infinitely differentiable and satisfies the inequalities*

$$\sup_{-\infty < x < \infty} \left|\nu^{(k)}(x)\right| \leq \frac{1}{4}(4\omega)^{k+1}k!, \quad k = 0, 1, 2, \ldots,$$

where $\omega = \max(r, |\lambda_j|)$ *and* λ_j *are zeros of the function* $u_1(\lambda, 0) - iu_2(\lambda, 0)$ *in the set* $|\lambda| > r$.

(When there are no zeros λ_j in the domain $|\lambda| > r$, we have $\omega = r$; the set of zeros λ_j is bounded for $\beta \neq 0$, or, generally, when $u_1(\lambda, 0) - iu_2(\lambda, 0) \not\equiv 0$.)

Chapter 6

Discrete Levels in Spectral Gaps of Perturbed Schrödinger and Hill Operators

In this chapter we assume that $\dim H < \infty$. Our approach to the investigation of discrete levels in spectral gaps is different from the approach used in Sec. 4.4. If the reader is interested only in the case of a scalar equation, she/he may omit Sec. 6.1.

6.1 Factorized Phase Matrix of the Perturbed System and the Discrete Spectrum in Gaps of the Continuous Spectrum

Consider the Sturm–Liouville system

$$l_1[u] \equiv -u'' + q(t)u = \lambda u \tag{6.1}$$

and the perturbed system

$$l_2[y] \equiv -y'' + [q(t) + p(t)]y = \lambda y \tag{6.2}$$

with Hermitian $(n \times n)$ matrix-valued coefficients $q(t) = q^*(t) \in L_{1,\text{loc}}$, $p(t) = p^*(t) \in L_{1,\text{loc}}$. By $U_1(t,\lambda)$, $U_2(t,\lambda)$ we denote $(n \times n)$ matrix-valued solutions of (6.1) such that the matrix

$$U(t,\lambda) = \begin{pmatrix} U_1(t,\lambda) & U_2(t,\lambda) \\ U_1'(t,\lambda) & U_2'(t,\lambda) \end{pmatrix} \tag{6.3}$$

is a J-unitary Wronski matrix for all $\operatorname{Im}\lambda = 0$, i.e.

$$U^*(t,\lambda)JU(t,\lambda) = J = \begin{pmatrix} 0_n & -I_n \\ I_n & 0_n \end{pmatrix}, \tag{6.4}$$

where I_n is the $n \times n$ identity matrix, 0_n is the $n \times n$ zero matrix. The condition (6.4) can be replaced by an equivalent condition $U(t,\lambda)JU^*(t,\lambda) = J$.

The Wronskian $W\{U_1^*, U_2\} \equiv (U_1^* U_2' - U_1^{*\prime} U_2)(t,\lambda)$ is independent of t. Since (6.4) holds for $t = 0$, it follows that (6.4) holds for all t. In particular, (6.4) holds if $U_1(t,\lambda)$ and $U_2(t,\lambda)$ are given by the initial conditions

$$\begin{aligned} U_1(0,\lambda) &= U_2'(0,\lambda) = \cos A, \\ U_1'(0,\lambda) &= -U_2(0,\lambda) = -\sin A, \end{aligned} \tag{6.5}$$

where $A = A^*$ is a Hermitian $(n \times n)$-matrix which may depend on λ.

Consider the matrix θ_U—phase matrix of the pair U_1, U_2:

$$\theta_U(t,\lambda) = (U_1 - iU_2)^{-1}(U_1 + iU_2). \tag{6.6}$$

Under the above assumptions, it is known (both in the real case [632] and in the general Hermitian case [546]) that $\theta_U(t,\lambda)$ is a unitary matrix, and for all f, $g \in \mathbb{C}^n$,

$$U_1 f = 0 \Leftrightarrow \theta_U f = -f, \quad U_2 g = 0 \Leftrightarrow \theta_U g = g, \tag{6.7}$$

$$\text{nul } U_2(t,\lambda) = \text{nul } [\theta_U(t,\lambda) - I]$$

$$\frac{\partial}{\partial t}\theta_U(t,\lambda) = i r_U(t,\lambda)\theta_U(t,\lambda), \tag{6.8}$$

where

$$r_U = 2(U_1 - iU_2)^{-1}(U_1^* + iU_2^*)^{-1} >> 0 \tag{6.9}$$

is a strictly positive matrix coefficient. Now it follows that, as t increases, the eigenvalues of $\theta_U(t,\lambda)$ move strictly monotonely along the unit circle in the counterclockwise direction (see [13] and [632]). Here nul $U = \dim\ker U$ and hence nul $U =$ def U. (Unlike (1.12), it is not necessary to prove the last equality since $\dim H < \infty$.)

The construction analogous to (6.6) can be carried out for the case of the perturbed equation (6.2). However, if we want to take into account the smallness of perturbation $p(t)$ or use any information about the spectrum of the unperturbed equation (for example, the presence of a gap in the spectrum), then it is more convenient to proceed in a different way.

Let $Y(t, \lambda)$ be any $(n \times n)$ matrix-valued solution of the system (6.2) satisfying the following self-adjoint conditions for $\operatorname{Im} \lambda = 0$:

$$W\{Y^*, Y\} = (Y^* Y' - Y^{*\prime} Y)(t, \lambda) = 0, \tag{6.10}$$

and the non-singularity condition

$$(Y'^* Y' + Y^* Y)(t, \lambda) >> 0. \tag{6.11}$$

Each of these conditions is satisfied for all t if it is satisfied for some $t = t_0$. The conditions (6.10), (6.11) together are equivalent to the requirements[1]

$$\det Y(t, \lambda) \not\equiv 0 \tag{6.12}$$

and

$$\cos A(t, \lambda) \cdot Y'(t, \lambda) - \sin A(t, \lambda) \cdot Y(t, \lambda) = 0, \tag{6.13}$$

where $A(t, \lambda) = A^*(t, \lambda)$ is a Hermitian matrix. The condition (6.13) is satisfied for all t if it is satisfied for some value of $t \in \mathbb{R}$.

It is proven in [546, Lemma 1.4] that (6.12) holds if (6.10) and (6.11) are satisfied. Assuming (6.10) and (6.11), we can prove (6.13) as follows. First, note that the linear relation between h, $h' \in \mathbb{C}^n$ determined by the equation

$$Y^*(t_0, \lambda) h' - Y'^*(t_0, \lambda) h = 0, \quad h, h' \in \mathbb{C}^n, \tag{6.14}$$

is a Hermitian relation due to (6.10), (6.11) (see Corollary A.1). By Lemma A.2 the coefficients of the equation (6.14) can be represented as

$$Y'(t_0, \lambda) = \sin A \cdot K, \quad Y(t_0, \lambda) = \cos A \cdot K, \tag{6.15}$$

where $A = A^*$ and $\det K \neq 0$, so that (6.13) holds.

Thus, we will look for a solution $Y(t, \lambda)$ of the system (6.2) in the form

$$Y(t, \lambda) = U_1(t, \lambda) C_1(t, \lambda) + U_2(t, \lambda) C_2(t, \lambda), \tag{6.16}$$

---

[1] Note that from the condition (6.11) alone, (6.12) does not follow. This can be seen on the example of identically singular matrix $Y(t) = \begin{pmatrix} 1 + \cos t & \sin t \\ \sin t & 1 - \cos t \end{pmatrix}$ with nowhere singular derivative $Y'(t)$ so that $Y(t)$ satisfies (6.11) but not (6.12). The matrix $Y(t)$ is not a solution of any equation of the form (6.1) or (6.2) since $Y''(t)$ is not singular when $Y(t)$ is singular.

where $C_1(t,\lambda)$, $C_2(t,\lambda)$ are $(n \times n)$ matrices, by using the variation of parameters method, i.e. by requiring that

$$\begin{pmatrix} Y \\ Y' \end{pmatrix} = U(t,\lambda) \begin{pmatrix} C_1(t,\lambda) \\ C_2(t,\lambda) \end{pmatrix}. \tag{6.17}$$

Here U is the Wronski matrix (6.3) of the system (6.1). We will write the systems (6.1) and (6.2) in the form

$$J\frac{d}{dt}U = \begin{pmatrix} \lambda - q & 0 \\ 0 & I \end{pmatrix} U \tag{6.18}$$

and

$$J\frac{d}{dt}\begin{pmatrix} Y \\ Y' \end{pmatrix} = \begin{pmatrix} \lambda - q - p & 0 \\ 0 & I \end{pmatrix} \begin{pmatrix} Y \\ Y' \end{pmatrix}, \tag{6.19}$$

and substitute (6.17) into (6.19). Then multiplying by U^* on the left and using (6.4) and (6.18), we get

$$J\frac{d}{dt}\begin{pmatrix} C_1 \\ C_2 \end{pmatrix} = -U^*(t,\lambda) \begin{pmatrix} p(t) & 0 \\ 0 & 0 \end{pmatrix} U(t,\lambda) \begin{pmatrix} C_1 \\ C_2 \end{pmatrix}. \tag{6.20}$$

From (6.17) we find the initial values $C_1(0,\lambda)$, $C_2(0,\lambda)$:

$$\begin{pmatrix} C_1(0,\lambda) \\ C_2(0,\lambda) \end{pmatrix} = -JU^*(0,\lambda)J \begin{pmatrix} Y(0,\lambda) \\ Y'(0,\lambda) \end{pmatrix}. \tag{6.21}$$

If, in particular, $U(0,\lambda) = I$, then

$$C_1(0,\lambda) = Y(0,\lambda), \quad C_2(0,\lambda) = Y'(0,\lambda).$$

Lemma 6.1 *The matrix*

$$\theta_C(t,\lambda) = (C_2 + iC_1)(C_2 - iC_1)^{-1} \tag{6.22}$$

is unitary for all t, $\lambda \in \mathbb{R}$ *and satisfies the equation*

$$\frac{d}{dt}\theta_C(t,\lambda) = i\theta_C(t,\lambda) r_C(t,\lambda), \tag{6.23}$$

where

$$r_C(t,\lambda) = -\frac{1}{2}(\theta_U\theta_C - I)^*(U_1 - iU_2)^* \cdot p(t)(U_1 - iU_2)(\theta_U\theta_C - I), \tag{6.24}$$

and $\theta_C(0,\lambda)$, *together with the matrix* A *in the representation (6.15), is defined by the self-adjoint boundary condition for* $Y(t,\lambda)$ *at* $t = 0$ *for the given*

$\lambda \in \mathbb{R}$, and it does not depend on K in (6.15) due to the formulas (6.21) and (6.22).

Proof. By the conditions (6.10) and (6.11), the matrix $\begin{pmatrix} Y(0,\lambda) \\ Y'(0,\lambda) \end{pmatrix}$ can be complemented to J-unitary $(2n \times 2n)$ matrix which, using (6.15) with $t_0 = 0$, can be represented as

$$\widetilde{Y}(0,\lambda) = \begin{pmatrix} Y(0,\lambda) & -(\sin A)K^{*-1} \\ Y'(0,\lambda) & (\cos A)K^{*-1} \end{pmatrix}, \tag{6.25}$$

as noted in [546]; see [922, Lemma 1.2] for the real case.

Since $U(0,\lambda)$ and $\widetilde{Y}(0,\lambda)$ are J-unitary, it follows that the $(2n \times n)$ matrix on the left hand side of (6.21) can also be complemented to $(2n \times 2n)$ J-unitary matrix $\widetilde{C}(0,\lambda)$:

$$\widetilde{C}(0,\lambda) = -JU^*(0,\lambda)J\widetilde{Y}(0,\lambda), \tag{6.26}$$

and $(2n \times n)$ matrix-valued solution of the system (6.20), (6.21) can be complemented to J-unitary $(2n \times 2n)$ matrix-valued solution $\widetilde{C}(0,\lambda)$ of the equation (6.20) with the initial conditions (6.26).

This means (see [13, Theorem 10.2.2] and [546, Lemma 1.2]) that the matrix $\theta_C(t,\lambda)$ in (6.22) is unitary, and it satisfies the differential equation (6.23), where, in accordance with [546], and by (6.20),

$$r_C(t,\lambda) = -2(C_2^* + iC_1^*)^{-1}(U_1C_1 + U_2C_2)^*p(U_1C_1 + U_2C_2)(C_2 - iC_1)^{-1}.$$

Now (6.24) follows immediately from the equality

$$U_1C_1 + U_2C_2 = \frac{1}{2i}(U_1 - iU_2)(\theta_U\theta_C - I)(C_2 - iC_1), \tag{6.27}$$

which can be checked directly using (6.6) and (6.22). This concludes the proof of the Lemma. □

Corollary 6.1 *The matrices $C_1(t,\lambda)$ and $C_2(t,\lambda)$ cannot both annihilate the same vector.*

Indeed, this is so because by (6.26) we have $|\det \widetilde{C}(t,\lambda)| = 1 \neq 0$, or, differently, if $C_1h = C_2h = 0$ for some $t = t_0$, $0 \neq h \in \mathbb{C}^n$, then $Yh = Y'h = 0$, and $\det Y(t,\lambda) \equiv 0$, contrary to (6.12).

It is well-known (see [13] and [794][2]) that the matrix

$$\theta(t,\lambda) = (Y' + iY)(Y' - iY)^{-1}, \tag{6.28}$$

called *phase matrix*, corresponding to the solution $Y(t,\lambda)$ of the system (6.2) satisfying (6.10), (6.11), is unitary,

$$\text{nul } Y(t,\lambda) = \text{nul } (\theta(t,\lambda) - I), \tag{6.29}$$

and, as t increases, the eigenvalues of $\theta(t,\lambda)$ can pass through the point $+1$ on the unit circle only in the counterclockwise direction. We will show that the matrix $\Theta(t,\lambda)$:

$$\Theta(t,\lambda) \equiv \theta_U(t,\lambda)\theta_C(t,\lambda) \tag{6.30}$$

satisfies the same conditions.[3] At the same time, only the factor $\theta_C(t,\lambda)$ depends on the perturbation $p(t)$, and this dependence is to a large extent explicit due to (6.23) and (6.24). The factor $\theta_U(t,\lambda)$ depends only on the unperturbed equation (6.1) and carries the information about spectral properties of (6.1).

Lemma 6.2 *Let $Y(t,\lambda)$ be a solution of the system (6.2) satisfying the conditions (6.10), (6.11). Then*

$$\textit{nul}\, Y(t,\lambda) = \textit{nul}\, (\Theta(t,\lambda) - I), \tag{6.31}$$

where $\Theta(t,\lambda) = \theta_U\theta_C$, and the unitary matrices $\theta_U(t,\lambda)$, $\theta_C(t,\lambda)$ are defined by the formulas (6.6) and (6.22). Moving along the unit circle as t increases, the eigenvalues of the unitary matrix $\Theta(t,\lambda)$ can pass the point $+1$ only in the counterclockwise direction.

Proof. By (6.16), (6.27) and (6.30) we have

$$Y(t,\lambda) = \frac{1}{2i}(U_1 - iU_2)(\Theta(t,\lambda) - I)(C_2 - iC_1). \tag{6.32}$$

Since the matrices $U(t,\lambda)$ and $\widetilde{C}(t,\lambda)$ are J-unitary, we have $\det[(U_1 - iU_2)(C_2 - iC_1)] \neq 0$. This and (6.32) imply (6.31).

We will now prove the second assertion of the Lemma. Note that by (6.30), (6.23)–(6.24) and (6.8)–(6.9), we have

$$\Theta_t{}'(t,\lambda) = ir_U(t,\lambda)\Theta + i\Theta r_C(t,\lambda). \tag{6.33}$$

[2]See Lemmas 1.1 and 1.2 (Sec. 1b) from Sec. 1.1 of the present book and the formula (1.24).

[3]Therefore, the matrix $\Theta(t,\lambda)$ is a variant of the phase matrix.

Using formula (1.30) or, similarly, [13, (V.6.1)], we introduce the Hermitian matrix

$$A(t,\lambda) = i(e^{i\alpha}I + \Theta)(e^{i\alpha}I - \Theta)^{-1}, \tag{6.34}$$

where $e^{i\alpha}$ is chosen outside the spectrum of $\Theta(t,\lambda)$ for the given λ and t (here α is real). By (6.33) we have

$$\begin{aligned} A'(t,\lambda) &= 2(e^{i\alpha}I - \Theta)^{-1} r_U(t,\lambda)(e^{i\alpha}I - \Theta)^{*-1} \\ &+2(e^{i\alpha}I - \Theta)^{*-1} r_C(t,\lambda)(e^{i\alpha}I - \Theta)^{-1}. \end{aligned} \tag{6.35}$$

By (6.24) and (6.30), for all vectors $h \in \mathbb{C}^n$, the equation $\Theta(t,\lambda)h = h$ implies $r_C(t,\lambda)h = 0$ for the same values of t and λ. Thus, in this case, from (6.35) it follows that for all $h \neq 0$

$$(A'(t,\lambda)h, h) = 2|e^{i\alpha} - 1|^{-2}(r_U(t,\lambda)h, h) > 0,$$

due to (6.9). Therefore, the eigenvalues of $A(t,\lambda)$ corresponding by (6.34) to the eigenvalue $+1$ of $\Theta(t,\lambda)$ increase with respect to t for the given λ (see Lemma 1.3). Hence, the eigenvalues of $\Theta(t,\lambda)$ pass through the point $+1$ on the unit circle in the counterclockwise direction; compare with the proof of Lemma 1.2 (part 1b) starting from the formula (1.30). The Lemma is proven. □

By $N_1(\mu,\nu)$, $N_1^{\pm}(\mu,\nu)$ we denote the number of points in the interval (μ,ν) (counting multiplicities) of the spectrum of the self-adjoint boundary-value problem for the equation (6.1) on the real line or on the half-lines $\mathbb{R}^+$, $\mathbb{R}^-$ respectively.[4] The notations $N_2(\mu,\nu)$, $N_2^{\pm}(\mu,\nu)$ have the same meaning in the case of the perturbed problem (6.2). Assume that $U(0,\lambda)$ and the boundary conditions do not depend on λ.

Theorem 6.1 *If $N_1^+(\mu,\nu) < \infty$, then $N_2^+(\mu,\nu) < \infty$ if and only if*

$$\overline{\lim_{t\to\infty}} \left|\ln\det[\theta_C(t,\nu)\theta_C^{-1}(t,\mu)]\right| < \infty. \tag{6.36}$$

In this case, the following estimate holds (the upper signs must be taken into consideration):

$$\begin{aligned} \overline{\lim_{t\to\pm\infty}} \Big| &N_2^{\pm}(\mu,\nu) - N_1^{\pm}(\mu,\nu) \\ &\mp \frac{1}{2\pi i}\ln\det[\theta_C(t,\nu)\theta_C^{-1}(t,\mu)]\Big| \le 6n. \end{aligned} \tag{6.37}$$

[4]The existence of self-adjoint extensions for the operators (6.1) and (6.2) is guaranteed in the presence of spectral gaps or when q and p are real.

The same statement is true for $N_1^-(\mu,\nu)$, $N_2^-(\mu,\nu)$ *if* $t \to -\infty$ *in (6.36), (6.37). If* $N_1(\mu,\nu) < \infty$, *then* $N_2(\mu,\nu) < \infty$ *if and only if*

$$\sup_{-\infty<t<\infty} \left|\ln\det[\theta_C(t,\nu)\theta_C^{-1}(t,\mu)]\right| < \infty. \tag{6.38}$$

In this case[5]

$$|N_j(\mu,\nu) - [N_j^+(\mu,\nu) + N_j^-(\mu,\nu)]| \le 2n, \quad (j = 1,2). \tag{6.39}$$

Remark 6.1 In the formulas (6.36)–(6.38),

$$\ln\det[\theta_C(t,\nu)\theta_C^{-1}(t,\mu)] = i\int_0^t \mathrm{tr}\ [r_c(s,\nu) - r_c(s,\mu)]\, ds. \tag{6.40}$$

The equality (6.40) follows from

$$\det\theta_c(t,\lambda) = \exp\left\{i\int_0^t \mathrm{tr}\ r_c(s,\lambda)\, ds\right\}\det\theta_c(0,\lambda),$$

where $\theta_c(0,\lambda)$ does not depend on λ by the assumptions made before the formulation of Theorem 6.1. The last equality follows from (6.23) using Liouville–Ostrogradsky Theorem.

Proof. By $N_{j,b}(\mu,\nu)$, $(j = 1,2)$ we denote the number of eigenvalues in the interval (μ,ν) of the boundary-value problems (6.1), (6.2) respectively, on the interval $(0,b)$ with the self-adjoint boundary conditions at $t = 0$ and $t = b$ (naturally, independent of λ). By $L_{j,b}^{\min}$ we denote the minimal differential operators with respect to the right end-point (with the self-adjoint boundary condition at the left end-point $t = 0$). Since the deficiency numbers def $L_{j,b}^{\min} \le n$, $(0 < b \le \infty)$, it follows that for all $0 < b < b' \le \infty$ (see [70, No. 9, p. 58])

$$N_{j,b'}(\mu,\nu) \ge N_{j,b}(\mu,\nu) - n. \tag{6.41}$$

Furthermore, by refining the arguments from [70, No. 9, p. 59], which are based on the fact that the minimal operator represents the closure of the corresponding differential expression from the set of compactly supported functions and on the above estimate for deficiency numbers of the minimal

[5] In (6.37), (6.39), it is assumed that if boundary conditions are required at both ends of $\mathbb{R}$ or $\mathbb{R}^\pm$, then the conditions are separated. It is also assumed that the boundary conditions at $+\infty$ for the problems on $\mathbb{R}$ and $\mathbb{R}^+$ coincide. The analogous assumption is made at $-\infty$ for the problems on $\mathbb{R}$ and $\mathbb{R}^-$.

operator, we get[6] the following: if $N_j^+(\mu,\nu) \equiv N_{j,\infty}(\mu,\nu) < \infty$, then

$$\overline{\lim_{b\to\infty}} \left|N_j^+(\mu,\nu) - N_{j,b}(\mu,\nu)\right| \le n, \tag{6.42}$$

and if $N_j^+(\mu,\nu) = \infty$, then

$$\lim_{b\to\infty} N_{j,b}(\mu,\nu) = \infty. \tag{6.43}$$

From (6.43) and (6.41) with $b' = \infty$ it follows that

$$N_j^+(\mu,\nu) < \infty \Leftrightarrow \overline{\lim_{b\to\infty}} N_{j,b}(\mu,\nu) < \infty. \tag{6.44}$$

The relations (6.41)–(6.44) hold for any choice of self-adjoint conditions at $t = b$, in particular, if the condition $y(b) = 0$ holds. Denoting the corresponding values $N_{j,b}^0(\mu,\nu)$ by index ^{0}, and using the oscillatory theorem for systems (Theorem 1.1), we have

$$N_{j,b}^0(\mu,\nu) = \sum_{0<t<b} \text{nul } Y_j(t,\nu) - \sum_{0<t\le b} \text{nul } Y_j(t,\mu), \tag{6.45}$$

where $Y_j(t,\lambda)$ is a matrix-valued solution of the system (6.j), $j = 1, 2$ with the given self-adjoint boundary condition for $t = 0$, $\det Y_j(t,\lambda) \not\equiv 0$. By Lemma 6.2, from (6.45) for $j = 2$ (see also (6.40)), we have

$$\left|N_{2,b}^0(\mu,\nu) - \frac{1}{2\pi i} \ln\det[\Theta(b,\nu)\Theta^{-1}(b,\mu)]\right| \le 2n. \tag{6.46}$$

Analogously, from (6.7)–(6.9) we have

$$\left|N_{1,b}^0(\mu,\nu) - \frac{1}{2\pi i} \ln\det[\theta_U(b,\nu)\theta_U^{-1}(b,\mu)]\right| \le 2n, \tag{6.47}$$

where $N_{1,b}^0(\mu,\nu)$ satisfies the same boundary condition at $t = 0$ for the equation (6.1) as does the solution $U_2(t,\lambda)$ of (6.1). (In the notation of (6.45), the solution $U_2(t,\lambda)$ is denoted by $Y_1(t,\lambda)$.) Using (6.30), from (6.46) and (6.47) we obtain

$$\left|N_{2,b}^0(\mu,\nu) - N_{1,b}^0(\mu,\nu) - \frac{1}{2\pi i} \ln\det[\theta_C(b,\nu)\theta_C^{-1}(b,\mu)]\right| \le 4n. \tag{6.48}$$

[6] Indeed, if $N_j^+(\mu,\nu) = m \ge n$, then there exists a system of $m-n$ linearly independent compactly supported functions $\phi_k \in \mathrm{D}\,(L_{j,\infty}^{\min})$ such that $\left\|l_j[\phi_k] - \frac{\mu+\nu}{2}\phi_k\right\| < \frac{\nu-\mu}{2}\|\phi_k\|$. Since for all compactly supported functions $\phi_k \in \mathrm{D}\,(L_{j,\infty}^{\min})$, for all sufficiently large $b > 0$, we have $\phi_k \in \mathrm{D}\,(L_{j,b}^{\min})$, the equation (6.42) follows, and, similarly, we get (6.43). (In the latter case, we use the possibility of choosing m as large as we want.)

Finally, by (6.42) with $j = 1$ and $j = 2$ and with $N_{j,b}(\mu, \nu)$ replaced by $N^0_{j,b}(\mu, \nu)$, we get (6.37) with the top signs.

The first assertion of the Theorem, related to condition (6.36), follows from (6.44) with $j = 1$ and $j = 2$ due to (6.48). The corresponding results for $N_1^-(\mu, \nu)$ and $N_2^-(\mu, \nu)$, as $t \to -\infty$, are proven similarly. The assertions related to relations (6.1), (6.2) on the real line, in conjunction with (6.38) and (6.39), can be derived from the previous results by using the decomposition method for differential operators (see [70] and [6]) on the real line at $t = 0$. This concludes the proof of the Theorem. □

Remark 6.2 The results given above can be generalized to perturbations of self-adjoint Sturm–Liouville systems in the general form

$$-[P(t)u' + R(t)u]' + R^*(t)u' + Q(t)u = \lambda W(t)u$$

and broad classes of canonical systems.

6.2 Generalization of D'Alembert–Liouville–Ostrogradsky Formula and Its Application to Growth Estimates for Solutions of Canonical Almost Periodic Systems

In what follows, the estimate of growth of solutions of differential equations and systems is used in the investigation of discrete levels in spectral gaps.

In this section, we consider two types of systems of differential equations: canonical Hermitian system

$$U' = A(t)U + B(t)V, \quad V' = C(t)U - A^*(t)V, \tag{6.49}$$

where $B(t) = B^*(t)$, $C(t) = C^*(t)$ and $A(t)$ are $n \times n$ matrix-valued functions (Hamiltonian system); and non-self-adjoint (generally speaking) system of two scalar equations

$$u' = a(t)u + b(t)v, \quad v' = c(t)u + d(t)v, \tag{6.50}$$

where the coefficients $a(t)$, $b(t)$, $c(t)$, $d(t)$ are not assumed to be real, $-\infty < t < \infty$. As a consequence of the results proven in this section, we also get the results for Sturm–Liouville type systems.

For the system (6.49), an important role is played by solutions $(U(t), V(t))$, where U, V are $n \times n$ matrix-valued functions satisfying so called self-adjointness conditions and non-singularity conditions:

$$U^*V - UV^* = 0, \tag{6.51}$$

$$\det(U^*U + V^*V) \neq 0. \tag{6.52}$$

By (6.49), the conditions (6.51), (6.52) are satisfied for all $t \in (-\infty, +\infty)$ if they are satisfied for one value $t = t_0$. The condition (6.52) means that the matrix-valued solution $(U(t), V(t))$ consisting of two $n \times n$ matrices represents n linearly independent vector-valued solutions of the system (6.49). Here, the n linearly independent vector-valued solutions are just pairs of column-vectors with equal column index, belonging to matrices $U(t)$ and $V(t)$ respectively. This is clear if we note that the system (6.49) in canonical notation has the form

$$J\frac{d}{dt}\begin{pmatrix} U \\ V \end{pmatrix} = \begin{pmatrix} -C & A^* \\ A & B \end{pmatrix}\begin{pmatrix} U \\ V \end{pmatrix}, \quad J = \begin{pmatrix} 0 & -I_n \\ I_n & 0 \end{pmatrix}. \tag{6.53}$$

By $C(t, s)$ we will denote the Cauchy matrix of the system (6.49), i.e. $2n \times 2n$ matrix whose columns are vector-valued solutions of (6.53) and $C(s, s) = I_{2n}$. In what follows, whenever we talk about almost periodic functions, we mean functions periodic in the sense of Bohr, i.e. uniformly almost periodic functions. The notation I_n denotes the $n \times n$ identity matrix.

Theorem 6.2 *Assume that the coefficients of the Hermitian canonical system (6.49) are $(n \times n)$ matrix-valued almost periodic functions. Assume that the system (6.49) has a matrix-valued almost periodic* [7] *solution $(U(t), V(t))$ satisfying the conditions (6.51), (6.52), where U, V are $n \times n$ matrices. Then the remaining n linearly independent vector-valued solutions of the system (6.49) grow at most linearly as $|t| \to \infty$, and are represented by a matrix-valued solution $(Y(t), Z(t))$ of (6.49) with the following asymptotics (with normalization condition $U^*Z - V^*Y = I_n$):*

$$\begin{aligned} Y(t) &= U(t)[K + o_n(1)]t - V(t)G(t), \\ Z(t) &= V(t)[K + o_n(1)]t + U(t)G(t), \end{aligned} \tag{6.54}$$

where $G(t) > 0$ is an almost periodic Hermitian matrix-valued function

$$G(t) = (U^*(t)U(t) + V^*(t)V(t))^{-1}, \tag{6.55}$$

$K = K^*$ is a constant matrix

$$K = \lim_{|t| \to \infty} \frac{1}{t} F(t), \tag{6.56}$$

[7] See Remark 6.3 for a possibility to weaken this requirement.

$$F(t) = \int_0^t G\{U^*(B+C)U - V^*(B+C)V - 4Re\,[U^*(Re\;A)V]\}\,G\,d\tau, \tag{6.57}$$

$o_n(1)$ is an $n \times n$ matrix with elements of order $o(1)$ in both formulas (6.54), where $|t| \to \infty$, $Re\;A := \frac{1}{2}(A + A^*)$.

The proofs of Theorems 6.2–6.4 are given at the end of the Section and are based on Theorems 6.5 and 6.6.

Theorem 6.3 *Under the hypotheses of Theorem 6.2, the Cauchy matrix of the systems (6.49), (6.54) has the asymptotics*

$$C(t,s) = \begin{pmatrix} U(t) \\ V(t) \end{pmatrix} [K + o_n(1)](-V^*(s), U^*(s))(t-s) + O_{2n}(1), \tag{6.58}$$

where, as $|t-s| \to \infty$, the matrix $o_n(1)$ tends to 0 uniformly in s, and $O_{2n}(1)$ is a $2n \times 2n$ matrix with bounded elements.

Remark 6.3 In the hypotheses of Theorems 6.2 and 6.3, we can omit the almost periodicity requirement for $V(t)$ if the matrix $B^{-1}(t)$ exists and is bounded on the entire real line, which is equivalent to

$$\inf_t |\det B(t)| > 0, \tag{6.59}$$

and if the derivatives $B'(t)$, $A'(t)$ exist and are bounded on the real line because, in this case, almost periodicity of $V(t)$ follows from almost periodicity of $U(t)$.

Similarly, if $C^{-1}(t)$ exists and is bounded together with the derivatives $C'(t)$, $A'(t)$, then, in the hypotheses of Theorems 6.2 and 6.3, we can omit the almost periodicity requirement for $U(t)$ because, in this case, almost periodicity of $U(t)$ follows[8] from almost periodicity of $V(t)$.

Finally, if we additionally assume (6.59) but do not assume differentiability of coefficients, the almost periodicity requirement for $V(t)$ in Theorems 6.2 and 6.3 can be replaced by the boundedness requirement for $V(t)$ or $U'(t)$ since, in this case, $V(t)$ is almost periodic if $U(t)$ is almost periodic.

Proof of Remark 6.3. If $\det B(t) \neq 0$, then from (6.49) it follows that

$$V(t) = B^{-1}(t)U'(t) - B^{-1}(t)A(t)U(t), \tag{6.60}$$

[8] Clearly, we must not omit the almost periodicity requirement for both $U(t)$ and $V(t)$ even when the conditions allowing any of the omissions of almost periodicity requirement in Remark 6.3 hold simultaneously.

$$(B^{-1}U')' - (B^{-1}AU)' = V' = CU - A^*B^{-1}U' + A^*B^{-1}AU, \qquad (6.61)$$

$$\begin{aligned} &U'' - (B'B^{-1} + A - BA^*B^{-1})U' \\ &-(C + BA^*B^{-1}A + A' - B'B^{-1}A)U = 0. \end{aligned} \qquad (6.62)$$

From (6.59) and boundedness of A', B', it follows that the coefficients of (6.62) are bounded, and, hence, from boundedness of $U(t)$, by a generalized version of Esclangon's Lemma (see [115, Chapter VI, Sec. 5, Proposition 4]), it follows that U' and U'' are bounded. Boundedness of U'' implies continuity of U' on the real line. This together with almost periodicity of U guarantees (see [115, Chapter I, Sec. 1, Property V]) almost periodicity of U'. Hence by (6.60), it follows that $V(t)$ is almost periodic. If, in addition, the coefficients of the system satisfy (6.59), then from the boundedness assumption on $V(t)$, it follows by (6.49) that V' is bounded, and this implies that V is uniformly continuous. Again by (6.49), since V is uniformly continuous, it follows that U' is uniformly continuous, and hence U' is almost periodic. Now by (6.60) it follows that $V(t)$ is almost periodic. We have proven two cases of the Remark. The other cases can be treated analogously. □

Corollary 6.2 *Assume that the Sturm–Liouville type system*

$$-(P(t)X' + R(t)X)' + R^*(t)X' + Q(t)X = \lambda W(t)X \qquad (6.63)$$

has $n \times n$ matrix-valued almost periodic coefficients, and

$$P^*(t) = P(t), \quad Q^*(t) = Q(t), \quad W^*(t) = W(t), \qquad (6.64)$$

$$\inf_{-\infty<t<\infty} |\det P(t)| > 0. \qquad (6.65)$$

Assume also that for some real $\lambda = \lambda_0$ the equation (6.63) has an almost periodic $n \times n$ matrix-valued solution $X(t)$ satisfying the conditions

$$X^*X^{[1]} - X^{[1]*}X \equiv X^*PX' - X'^*PX + X^*(R - R^*)X = 0, \qquad (6.66)$$

$$\det\left(X^{[1]*}X^{[1]} + X^*X\right) \neq 0, \qquad (6.67)$$

where

$$X^{[1]} = PX' + RX \qquad (6.68)$$

is a quasi-derivative corresponding to (6.63). Then, if there exist bounded derivatives $P'(t)$ and $R'(t)$ and if $X'(t)$ is bounded, which is equivalent to boundedness of $X^{[1]}(t)$, then for the same value $\lambda = \lambda_0$, the equation (6.63) has a matrix-valued solution $Y(t)$ satisfying the condition $X^*Y^{[1]} - X^{[1]*}Y \equiv I$. The solution $Y(t)$, together with its quasi-derivative $Y^{[1]}(t)$, has the following asymptotics as $|t| \to \infty$:

$$\begin{aligned} Y(t) &= X(t)[K + o_n(1)]t - X^{[1]}(t)G(t), \\ Y^{[1]}(t) &= X^{[1]}(t)[K + o_n(1)]t + X(t)G(t), \end{aligned} \tag{6.69}$$

where $G(t)$ and K are defined by the formulas (6.55), (6.56), (6.57) with the substitution

$$U(t) = X(t), \quad V(t) = X^{[1]}(t), \tag{6.70}$$

and also

$$\begin{aligned} B(t) &= P^{-1}(t), \quad A(t) = -P^{-1}(t)R(t), \\ C(t) &= Q(t) - \lambda_0 W(t) - R^*(t)P^{-1}(t)R(t). \end{aligned} \tag{6.71}$$

The substitutions (6.70), (6.71) in the formulas (6.58), (6.56), (6.57) give the asymptotics for the Cauchy matrix of the system (6.63) as $|t-s| \to \infty$.

Proof. It suffices to note that using substitutions (6.70) and (6.71), the system (6.63) can be reduced to the system (6.49). □

Theorem 6.4 *Assume that the coefficients of the non-self-adjoint (generally speaking) system of two scalar equations (6.50) are almost periodic functions, and that the function $\exp\left\{\int_0^t (a+d)\,d\xi\right\}$ is also almost periodic. Then, if the system (6.50) has an almost periodic non-trivial solution $(u(t), v(t))$, then the solution $(y(t), z(t))$ of (6.50), which is linearly independent with $(u(t), v(t))$ and satisfies the condition $(uz - vy)|_{t=0} = 1$, has the following asymptotics as $|t| \to \infty$:*

$$\begin{aligned} y(t) &= u(t)[k + o(1)]t - v(t)\Delta_\phi^{-1}(t)\exp\left\{\int_0^t (a+d)\,d\xi\right\}, \\ z(t) &= v(t)[k + o(1)]t + \phi(t)u(t)\Delta_\phi^{-1}(t)\exp\left\{\int_0^t (a+d)\,d\xi\right\}, \end{aligned} \tag{6.72}$$

where

$$\Delta_\phi(t) = \phi(t)|u(t)|^2 + |v(t)|^2, \tag{6.73}$$

and $\phi(t)$ is an arbitrary almost periodic function such that $\phi'(t)$ is also almost periodic (in particular, a constant function), and

$$\inf_t |\phi(t)| > 0, \quad |arg\ \phi(t)| < \pi - \epsilon, \quad \epsilon > 0, \tag{6.74}$$

where k is a constant independent of the choice of $\phi(t)$,

$$k = \lim_{|t|\to\infty} \frac{1}{t} f(t),$$

$$\begin{aligned} f(t) = \int_0^t \Delta_\phi^{-2}(\tau)\{\phi\bar{u}^2(b\phi + \bar{c}) - \bar{v}^2(\bar{b}\phi + c) \\ -2\phi\bar{u}\bar{v} Re\ (a-d) - \phi'\bar{u}\bar{v}\} \exp\left\{\int_0^\tau (a+d)\, d\xi\right\} d\tau. \end{aligned} \tag{6.75}$$

As $|t-s| \to \infty$, the Cauchy matrix $c(t,s)$ of the system (6.50) has uniform asymptotics with respect to s:

$$c(t,s) = [k + o(1)] \begin{pmatrix} -u(t)v(s) & u(t)u(s) \\ -v(t)v(s) & v(t)u(s) \end{pmatrix} (t-s) + O_2(1). \tag{6.76}$$

Roughly, (6.72) gives the asymptotics in the following simpler form:

$$\begin{aligned} y(t) &= u(t)[k + o(1)]t + O(1), \\ z(t) &= v(t)[k + o(1)] + O(1). \end{aligned} \tag{6.77}$$

Remark 6.4 If $\inf |b(t)| > 0$, then in the hypotheses of Theorem 6.4, we can replace the almost periodicity requirement on $v(t)$ by the boundedness requirement on $v(t)$ or on $u'(t)$. If, in addition, $b'(t)$ and $a'(t)$ exist and are bounded, then no restrictions on $v(t)$ or on $u'(t)$ are required.

The proof of Remark 6.4 is similar to the proof of Remark 6.3.

Corollary 6.3 *Assume that the scalar non-self-adjoint Sturm–Liouville type equation*

$$-(p(t)x' + r(t)x)' + s(t)x' + q(t)x = \lambda w(t)x, \tag{6.78}$$

has almost periodic coefficients, where $\inf|p(t)| > 0$, *and the function* $\exp\left\{\int_0^t (s-r)p^{-1}\,d\xi\right\}$ *is almost periodic. Assume that for some* $\lambda \in \mathbb{C}$ *the equation (6.78) has a nontrivial almost periodic solution* $x(t)$. *Assume also that the derivatives* $p'(t)$, $r'(t)$ *exist and are bounded, or that* $x'(t)$ *is bounded. Then for the same value of* λ, *the equation (6.78) has a solution* $y(t)$ *satisfying the condition* $p(xy' - x'y)|_{t=0} = 1$. *Moreover,* $y(t)$ *and its quasi-derivative* $y^{[1]} = py' + ry$ *have asymptotics (6.72) with the following substitutions in (6.72), (6.73), (6.75):*

$$u(t) = x(t),\ v(t) = x^{[1]}(t) \equiv px' + rx,\ z(t) = y^{[1]}(t), \tag{6.79}$$

$$\begin{aligned} &b(t) = p^{-1}(t),\ a(t) = -p^{-1}(t)r(t),\ d(t) = p^{-1}(t)s(t), \\ &c(t) = q(t) - \lambda w(t) - p^{-1}(t)r(t)s(t). \end{aligned} \tag{6.80}$$

Proof. The proof of the Corollary follows from Theorem 6.4 by using the substitutions (6.79), (6.80) reducing the equation (6.78) to the system (6.50).

Before proving Theorems 6.2–6.4, we will establish the formulas enabling us to construct the solutions $(Y(t), Z(t))$ of the systems (6.49), (6.50) from the given solutions $(U(t), V(t))$ of (6.49), (6.50) such that $(Y(t), Z(t))$ and $(U(t), V(t))$ are linearly independent, without "sewing together" or analytic continuation (if the latter one is possible) of the constructed solutions at the zeros of $\det U(t)$. In fact, for some systems of the form (6.49), (6.50), it may happen that $U(t) \equiv 0$ when $\det V(t) \neq 0$. The formulas that we construct are generalizations of the well-known D'Alembert–Liouville–Ostrogradsky formula

$$y(t) = u(t)\int_0^t u^{-2}(\xi)\,d\xi. \tag{6.81}$$

Given a solution $u(t)$ of the Sturm–Liouville equation

$$-u'' + q(t)u = \lambda u, \tag{6.82}$$

the formula (6.81) enables us to construct a solution $y(t)$ of (6.82) so that $y(t)$, $u(t)$ are linearly independent. The formulas that we construct also represent generalizations of the matrix generalization of (6.81); see formula (10.20) in [78, Chapter XI]. The advantage of our formulas, in comparison with the above described known formulas, is that the integrands in our formulas are continuous on the whole real line, including the points where

$\det U(t) = 0$ or $u(\xi) = 0$, unlike (6.81) or the above mentioned formula (10.20) in [78, Chapter XI] for solutions of the canonical system. The introduction of functional or numerical parameter $\phi(t)$ into our formulas allows us to obtain the expressions of the form (6.81) by passing to the limit with respect to the parameter. In the following theorems, we do not assume that the coefficients of the systems (6.49), (6.50) are almost periodic.

Theorem 6.5 *Assume that the coefficients of the Hermitian system (6.49) are matrix-valued locally integrable functions. Let $(U(t), V(t))$ be some matrix solution of (6.49) satisfying the conditions (6.51), (6.52). Then the pair $(Y(t), Z(t))$, where*

$$\begin{aligned} Y(t) &= U(t)F(t) - V(t)G(t), \\ Z(t) &= V(t)F(t) + U(t)G(t), \end{aligned} \tag{6.83}$$

with $G(t)$, $F(t)$ defined by the formulas (6.55), (6.57), represents the second matrix-valued solution of the system (6.49) satisfying the condition $U^*(t)Z(t) - V^*(t)Y(t) \equiv I$ and also the conditions of the form (6.51), (6.52).

Proof. The proof is a modification of the variation of parameters method. We will look for functions F, G in (6.83) so that

$$Y' = A(t)Y + B(t)Z, \quad Z' = C(t)Y - A^*(t)Z, \tag{6.84}$$

and other conditions for (Y, Z) are satisfied. Differentiate each term in the equalities (6.83) and equate the obtained expressions for Y', Z' with the corresponding right hand sides of (6.84). In the obtained equations, instead of Y and Z, we substitute the corresponding expressions (6.83), and, instead of U' and V', we substitute the expressions (6.49). Simplifying, we get

$$\begin{aligned} UF' - VG' &= [(B + C)U - 2(\text{Re } A)V]G, \\ VF' + UG' &= -[(B + C)V + 2(\text{Re } A)U]G. \end{aligned} \tag{6.85}$$

Multiplying from the left both sides of the first equation in (6.85) by U^*, and the second by V^*, adding both equations, and using (6.51), we get

$$\begin{aligned} &(U^*U + V^*V)F' \\ &= \{U^*(B + C)U - V^*(B + C)V - 4\text{Re } [U^*(\text{Re } A)V]\} G. \end{aligned} \tag{6.86}$$

Similarly, from (6.85) we can obtain a differential equation for G (linear, homogeneous and not containing F) defining G up to a matrix factor such that this equation is satisfied by G given in (6.55). But we can find G in a simpler way (and uniquely) from (6.83) by multiplying the first equation from the left by $(-V)^*$, the second by U^*, and adding both equations. As a result, using the hypotheses of the Theorem, we get

$$I \equiv U^*Z - V^*Y = (U^*U + V^*V)G(t), \tag{6.87}$$

implying (6.55) for G. Now we see that F in (6.57) satisfies the equation (6.86), and hence Y, Z in (6.83) with F, G as specified in the Theorem, satisfy the system (6.49). Also by (6.87), Y and Z satisfy the other assertions of the Theorem. It remains to check the validity of (6.51) for the pair Y, Z, i.e. the equality $Y^*Z - Z^*Y = 0$. But this equality follows directly from (6.83) and (6.51) using (6.55) when $F = F^*$. (The calculation is particularly simple for $t = 0$ when $F(0) = 0$, which corresponds to (6.57).) The Theorem is proven. □

Remark 6.5 The normalization condition $F(0) = 0$ in (6.57) is equivalent to the condition

$$\{(Y^*Y + Z^*Z)(U^*U + V^*V)\}\,|_{t=0} = I,$$

since

$$Y^*Y + Z^*Z = F^*G^{-1}F + G$$

by (6.55), (6.83) and (6.51).

Theorem 6.6 *Let $(u(t), v(t))$ be an arbitrary non-trivial solution of the system of two scalar equations (6.50) with locally integrable complex-valued coefficients. Then the pair $(y(t), z(t))$,*

$$\begin{aligned} y(t) &= u(t)f(t) - \bar{v}(t)g(t), \\ z(t) &= v(t)f(t) + \phi(t)\bar{u}(t)g(t), \end{aligned} \tag{6.88}$$

where

$$g(t) = \Delta_\phi^{-1} \exp\left\{\int_0^t (a + d)\, d\xi\right\}, \tag{6.89}$$

$f(t)$, $\Delta_\phi(t)$ are defined by the formulas (6.75), (6.73), $\phi(t) \neq 0$ is an arbitrary continuous complex-valued function such that $\phi'(t)$ is also continuous, and $\phi(t)$ is never strictly negative, represents the solution of the system (6.50). The solution $(y(t), z(t))$ is linearly independent with $(u(t), v(t))$

and satisfies the following conditions:

$$(uz - vy)|_{t=0} = 1, \quad \{(\phi|y|^2 + |z|^2)(\phi|u|^2 + |v|^2)\}_{t=0} = \phi(0). \tag{6.90}$$

(If Im $\phi(0) = 0$, then the normalization condition $f(0) = 0$ used in (6.75) is uniquely determined by (6.90).)

Proof. The proof is similar to the proof of Theorem 6.5, so we omit it. We only need to keep in mind that for all solutions (u, v), (y, z) of the system (6.50), we have

$$(uz - vy)(t) = (uz - vy)(0) \exp\left\{\int_0^t (a + d)\, d\xi\right\}$$

by Liouville–Ostrogradsky Theorem. □

Corollary 6.4 *Under the hypotheses of Theorem 6.5,* $(2n \times 2n)$ *Cauchy matrix of the systems (6.49), (6.53) can be represented as*

$$C(t, s) = \begin{pmatrix} U(t) \\ V(t) \end{pmatrix} [F(t) - F(s)](-V^*(s), U^*(s))$$
$$+ \begin{pmatrix} U(t) \\ V(t) \end{pmatrix} G(s)(U^*(s), V^*(s))$$
$$+ \begin{pmatrix} V(t) \\ -U(t) \end{pmatrix} G(t)(V^*(s), -U^*(s)), \tag{6.91}$$

and under the hypotheses of Theorem 6.6, (2×2) *Cauchy matrix of the system (6.50) can be represented as*

$$c(t, s) = \begin{pmatrix} u(t) \\ v(t) \end{pmatrix} (-v(s), u(s))[f(t) - f(s)]$$
$$+ \begin{pmatrix} u(t) \\ v(t) \end{pmatrix} (\phi(s)\bar{u}(s), \bar{v}(s))\Delta_\phi^{-1}(s)$$
$$+ \begin{pmatrix} \bar{v}(t) \\ -\phi(t)\bar{u}(t) \end{pmatrix} (\bar{v}(s), -u(s))\Delta_\phi^{-1}(t) \exp\left\{\int_s^t (a + d)\, d\xi\right\}. \tag{6.92}$$

Proof of Theorem 6.2. By Theorem 6.5 it suffices to show that $G(t)$ in (6.55) is almost periodic (since almost periodicity of $G(t)$ implies that the function $F'(t)$, representing the integrand in (6.57), is almost periodic) and to apply the mean value theorem for an almost periodic function (see [115, Chapter II, Sec. 3]). To prove that $G(t)$ in (6.55) is almost periodic, it suffices to show that

$$\inf_t \det[U^*(t)U(t) + V^*(t)V(t)] > 0. \tag{6.93}$$

Assume the opposite, i.e. assume that there exists a sequence $t_j \in \mathbb{R}$ and a sequence of unit vectors $h_j \in \mathbb{C}^n$ such that

$$h_j^*[U^*(t_j)U(t_j) + V^*(t_j)V(t_j)]h_j \to 0.$$

Then we get

$$U(t_j)h_j \to 0, \quad V(t_j)h_j \to 0.$$

Since the unit sphere in $\mathbb{C}^n$ is compact, we can choose a subsequence of $\{h_j\}_{j=1}^{\infty}$, also denoted by h_j, such that $h_j \to h$, where $h \in \mathbb{C}^n$ and $|h| = 1$. Then we have

$$U(t_j)h_j \to 0, \quad V(t_j)h_j \to 0.$$

Since the coefficients of the system (6.49) are bounded, then there exists a sequence $N_j \to \infty$ such that[9]

$$\sup_{|t-t_j|<N_j} \{|U(t)h| + |V(t)h|\} \to 0.$$

But this is impossible for a non-zero scalar almost periodic function $|U(t)h| + |V(t)h|$. This conclude s the proof of the inequality (6.93) and Theorem 6.2. □

Proof of Theorem 6.3. The proof follows from the representation (6.91) for the Cauchy matrix of the systems (6.49), (6.53), from almost periodicity of $G(t)$ and $F'(t)$ established in the proof of Theorem 6.2, and from the stronger form[10] of the mean value theorem for an almost periodic function which implies

$$F(t) - F(s) = (t-s)\left\{\frac{1}{t-s}\int_s^t F'(\xi)\,d\xi\right\} = (t-s)\{K + o_n(1)\},$$

uniformly in s as $|t-s| \to \infty$. This concludes the proof of the Theorem. □

[9]Indeed, the solution of the system (6.49) with zero Cauchy data at the point t_j is identically equal to zero. If the Cauchy vector data $(U(t_j)h, V(t_j)h)$ are sufficiently small, then the solution of (6.49) is small in the interval $|t-t_j| < N_j$. (The estimates of the solution of the Cauchy problem are uniform with respect to translations since the coefficients of (6.49) are uniformly bounded on the entire real axis.)

[10]See [115, Chapter II, Sec. 3].

Proof of Theorem 6.4. The proof follows from Theorem 6.6 and (6.92) similarly as the proofs of Theorems 6.2 and 6.3. We note only that k does not depend on the choice of $\phi(t)$ because the solution (y, z) is determined by the first condition in (6.90) with accuracy up to a term of the form $(u, v) \cdot \text{const}$. This corresponds to the definition of $f(t)$ with accuracy up to an additive constant not influencing the value of k due to the first formula in (6.75). □

Remark 6.6 Using substitutions (6.70), (6.71) or (6.79), (6.80), Theorems 6.5 and 6.6 can be reformulated for self-adjoint systems (6.63) and second order non-self-adjoint equations (6.78).

We now formulate those results.

Theorem 6.7 *Assume that the coefficients of the self-adjoint Sturm–Liouville-type system (6.63) are $n \times n$ matrix-valued locally integrable functions satisfying (6.64). Assume that the functions $A(t)$, $B(t)$, $C(t)$ in (6.71) are locally integrable.*

Let $X(t)$ be an $n \times n$ matrix-valued solution of (6.63) for some real $\lambda = \lambda_0$, satisfying the conditions (6.66), (6.67). Then for the same value $\lambda = \lambda_0$, the equation (6.63) has a solution $Y(t)$ defined, together with its quasi-derivative $Y^{[1]}(t)$ in (6.68), by the formulas

$$\begin{aligned} Y(t) &= X(t)F(t) - X^{[1]}(t)G(t), \\ Y^{[1]}(t) &= X^{[1]}(t)F(t) + X(t)G(t), \end{aligned} \tag{6.94}$$

where $G(t)$ and $F(t)$ are defined by the formulas (6.55) and (6.57) with the substitutions (6.70) and (6.71). This solution satisfies the conditions of the form (6.66), (6.67), and the corresponding Wronskian with $X(t)$ is the identity matrix, i.e.

$$X^*(t)Y^{[1]}(t) - X^{[1]*}(t)Y(t) \equiv I_n.$$

Theorem 6.8 *Assume that the coefficients of the scalar non-self-adjoint Sturm–Liouville type equation (6.78) and the expressions (6.80) are locally integrable complex-valued functions. Let $x(t)$ be a non-trivial solution of the equation (6.78) for some $\lambda \in \mathbb{C}$. Then for the same value of λ, the equation (6.78) has a solution $y(t)$ defined, together with its quasi-derivative $y^{[1]}(t) := py' + ry$, by the formulas*

$$\begin{aligned} y(t) &= x(t)f(t) - \overline{x^{[1]}(t)}g(t), \\ y^{[1]}(t) &= x^{[1]}(t)f(t) + \phi(t)\overline{x(t)}g(t), \end{aligned} \tag{6.95}$$

where $g(t)$ and $f(t)$ are defined by the formulas (6.95), (6.75), (6.73) with the substitutions (6.79), (6.80), and $\phi(t) \neq 0$ is an arbitrary continuous complex-valued function such that $\phi'(t)$ is continuous and $\phi(t)$ is never strictly negative. The solutions $y(t)$ and $x(t)$ are linearly independent and

$$(xy^{[1]} - x^{[1]}y)|_{t=0} = 1,$$

$$\{(\phi|y|^2 + |y^{[1]}|^2)(\phi|x|^2 + |x^{[1]}|^2)\}_{t=0} = \phi(0). \tag{6.96}$$

6.3 A Reduction of the Problem about the Number of Discrete Levels in a Finite Spectral Gap of One-Dimensional Schrödinger Operator to a Similar Problem for a Semi-Infinite Gap of the Transformed Schrödinger Operator

Continuing the previous considerations (see Sec. 6.1), we are able, in the scalar case ($\dim H = 1$), to reduce the study of the discrete spectrum of the perturbed problem (6.2) on the real line or half-line, which is introduced by the perturbation into a finite spectral gap of the starting equation (6.1), to the study of properties of negative spectrum of equations of the form (6.1) with some potential $q^\wedge(s) \in L_{1,\text{loc}}$ for which explicit formulas are specified.[11] Since the latter question has been studied far more thoroughly than the former one (see, for example, [35], [70], [78] and [187]), we are interested in the above mentioned reduction. In what follows, we assume that $p, q \in L^1_{\text{loc}}$.

We fix the same value of $\lambda \in \mathbb{R}$ in (6.1)–(6.2). The formulas (6.16), (6.17), (6.22) and (6.23) can now be written for real u_1, u_2, $y(t,\lambda)$ and with $u_1u_2' - u_1'u_2 \equiv 1$ in the form

$$y(t) = a(t)[u_1(t)\cos\gamma(t) + u_2(t)\sin\gamma(t)], \tag{6.97}$$

$$y'(t) = a(t)[u_1'(t)\cos\gamma(t) + u_2'(t)\sin\gamma(t)], \tag{6.98}$$

where $c_1 = a\cos\gamma$, $c_2 = a\sin\gamma$, $\theta_c = -e^{-2i\gamma}$ and[12]

$$\gamma'(t) = p(t)[u_1(t)\cos\gamma(t) + u_2(t)\sin\gamma(t)]^2, \tag{6.99}$$

[11] Those explicit formulas express $q^\wedge$ in terms of q, p and a solution of the unperturbed equation (6.1) for the value of λ coinciding with the prescribed end-point of the spectral gap under consideration (see (6.114)).

[12] The equation (6.99) can be obtained directly in our case by differentiating the fraction θ_c (6.22) and substituting instead of c_1', c_2' the corresponding expressions from (6.20).

$\gamma(0)$ is defined by the following boundary condition for $y(t)$ at $t = 0$: if $y'(0) = hy(0)$, then

$$\tan\gamma(0) = -(u_1'(0) - hu_1(0))(u_2'(0) - hu_2(0))^{-1}. \tag{6.100}$$

The representation (6.97)–(6.98) and the phase equation (6.99) are well-known in physics literature (see, for example, [16] and [34]), where those equations are considered in the case of potentials with special form, when $q(t) \to 0$ as $t \to \infty$, and the solutions $u_1(t)$, $u_2(t)$ have explicit formulas and posses trigonometric asymptotics. We will not impose these restrictions by using a well-known representation which works in the general case:

$$u_1(t) = r(t)\sin\alpha(t), \quad u_2(t) = r(t)\cos\alpha(t), \tag{6.101}$$

where

$$\alpha'(t) = -\frac{1}{r^2(t)}, \quad r(t) = (u_1^2 + u_2^2)^{1/2}. \tag{6.102}$$

Then (6.97)–(6.99) give

$$y(t) = a(t)r(t)\sin[\alpha(t) + \gamma(t)], \tag{6.103}$$

$$\frac{y'}{y} = \frac{r'}{r} + \alpha'\cot(\alpha+\gamma) = \frac{r'}{r} - \frac{1}{r^2}\cot(\alpha+\gamma). \tag{6.104}$$

In some sense, the given construction is a generalization of Prüfer's transform [192], [78]:

$$\begin{aligned} y = \rho(t)\sin\phi(t), \quad y' &= \rho\cos\phi = y\cot\phi, \\ \phi' &= \cos^2\phi + [\lambda - q - p]\sin^2\phi. \end{aligned} \tag{6.105}$$

The phase equation (6.99) has advantage over (6.105) since (6.99) enables us to bring out the influence of perturbation on the phase in a more explicit form. From the formulas (6.97)–(6.104) we have the following Lemma.

Lemma 6.3 *A non-trivial real solution $y(t)$ of the equation (6.1) has roots precisely where there is a "full phase"*

$$\psi(t) \equiv \alpha(t) + \gamma(t) = n\pi, \quad n = 0, \pm1, \pm2, \ldots \tag{6.106}$$

When passing through each of the roots $t = t_n$, the function $\psi(t)$ is decreasing (as can be seen from (6.104)). Therefore, $y(t)$ oscillates as $t \to \infty$ if and only if $\psi(t) \to -\infty$ as $t \to \infty$.

If necessary, we will denote the dependence of the considered quantities on λ by $\psi(t,\lambda)$, $\gamma(t,\lambda)$, *etc.* The notations $N_j(\mu,\nu), N_j^{\pm}$ are the same as in Sec. 6.1 (see the paragraph preceding Theorem 6.1).

Theorem 6.9 *If $N_1^+(\mu,\nu)<\infty$, then $N_2^+(\mu,\nu)<\infty$, $(=\infty)$ if and only if for some (and, hence, for any) real $\gamma(0)$ the following condition holds due to (6.99):*

$$\overline{\lim_{t\to\infty}}|\gamma(t,\mu)-\gamma(t,\nu)|<\infty\,(\textit{respectively}\ =\infty). \tag{6.107}$$

In this case, the following estimate holds under conditions (6.100) and $\gamma(0,\nu)=\gamma(0,\mu)$:

$$\overline{\lim_{t\to+\infty}}\left|N_2^+(\mu,\nu)-N_1^+(\mu,\nu)+\frac{1}{\pi}(\gamma(t,\nu)-\gamma(t,\mu))\right|\le 6.$$

The same holds for N_1^-, N_2^- if in (6.107) we have $t\to-\infty$. If $N_1(\mu,\nu)<\infty$, then $N_2(\mu,\nu)<\infty$ if and only if

$$\sup_{-\infty<t<\infty}|\gamma(t,\mu)-\gamma(t,\nu)|<\infty. \tag{6.108}$$

Proof. By Lemma 6.3, the proof is completely analogous to the proof of Theorem 6.1. If we note that, by the formulas (6.6), (6.22), (6.30)

$$\theta_u=-e^{-2i\alpha},\quad \theta_c=-e^{-2i\gamma},\Theta=-e^{-2i\psi}, \tag{6.109}$$

then Theorem 6.9 is a consequence of Theorem 6.1. □

We now transform the equation (6.99) into Riccati equation, from which we then pass to Sturm–Liouville equation. Setting

$$\tan\gamma(t)=Q(t)z(t), \tag{6.110}$$

where

$$Q(t)=\exp\left\{2\int_0^t p(s)u_1(s)u_2(s)\,ds\right\}, \tag{6.111}$$

from (6.110) and (6.99) we get

$$Q'z+Qz'=\frac{\gamma'}{\cos^2\gamma}=p[u_1+u_2zQ]^2,$$

and this implies

$$z'(t)=p(t)[u_1^2(t)Q^{-1}(t)+u_2^2(t)Q(t)z^2(t)]. \tag{6.112}$$

The solution $z(t)$ of the equation (6.112) can be continuously continued as $t \to \infty$ (starting from some value of t) if and only if $\overline{\lim}|\gamma(t)| < \infty$. Furthermore, setting

$$t = t(s),\ \tilde{z}(s) = z(t(s)),\ s = s(t) = \int_0^t |p(\xi)|u_2^2(\xi)Q(\xi)\,d\xi, \qquad (6.113)$$

we have

$$\tilde{z}'(s) = [\tilde{z}^2(s) + q^\wedge(s)]\text{sign}\ p(t(s)),$$

where

$$q^\wedge(s) = \frac{u_1^2(t(s))}{u_2^2(t(s))}Q^{-2}(t(s)). \qquad (6.114)$$

If $p(t)$ does not change its sign, then the substitution

$$\tilde{z} = -\frac{\eta'(s)}{\eta(s)}\text{sign}\ p(t_0), \quad (\exists t_0 \colon p(t_0) \neq 0) \qquad (6.115)$$

leads to the equation

$$\eta'' + q^\wedge(s)\eta = 0, \quad 0 \leq s < s(\infty) \leq \infty. \qquad (6.116)$$

Lemma 6.4 *The function $q^\wedge(s)$ in (6.114) is locally integrable.*

Proof. The proof follows from the equation

$$q^\wedge(s)\,ds = |p(t)|u_1^2Q^{-1}(t)\,dt. \qquad (6.117)$$

□

Note also that the intervals where $p(t) \equiv 0$ are constancy intervals of $s(t)$ in (6.113) and for $z(t)$ in (6.112), and, hence, although the corresponding values of s are the discontinuity points of $t(s)$, the function $\tilde{z}(s) = z(t(s))$ preserves continuity. Since $s(t)$ is locally absolutely continuous, then the Lebesgue measure on the s-axis of the image of the t-set where $s'(t) = \infty$, or $s'(t)$ does not exist, or where $s'(t)$ is not equal to the integrand in (6.113), is equal to zero.

The following Theorem follows from our constructions.

Theorem 6.10 *Assume that $p(t)$ does not change its sign. Then, in accordance with (6.99), $\overline{\lim}_{t\to\infty}|\gamma(t)| < \infty$ $(= \infty)$ if and only if the equation (6.116) is non-oscillatory (oscillatory) as $s \nearrow s(\infty)$.*

Theorem 6.11 *Assume that perturbation $p(t)$ has non-constant sign. Assume that the two equations of the type (6.116) constructed for perturbations $\pm|p(t)|$, respectively, or for perturbations $p_\pm(t) := \max\{\pm p(t), 0\} \geq 0$ ($p = p_+ - p_-$), respectively, are non-oscillatory as $s \nearrow s(\infty)$. Then (6.99) implies $\overline{\lim_{t\to\infty}}|\gamma(t)| < \infty$.*

Proof. The proof follows by using Theorem 6.10 and by applying a theorem about a differential inequality (see [41] and [42]) to the equation (6.99) and to the solutions $\gamma^\vee(t)$ and $\gamma^\wedge(t)$ of analogous equations with $p_+(t)$ and $-p_-(t)$ instead of $p(t)$ respectively. This gives the inequality $\gamma^\wedge(t) \leq \gamma(t) \leq \gamma^\vee(t)$ for $t > 0$, provided $\gamma^\wedge(0) \leq \gamma(0) \leq \gamma^\vee(0)$. (See the Bibliographical Comments.) □

6.4 Applications to Perturbations of Hill Operators and Schrödinger Operators with an Almost Periodic Potential. Kneser-Type Constants and Effective Masses

We will demonstrate the application of Theorems 6.10 and 6.11 to the equations (6.1)–(6.2) in the case of a real-valued $q \in L_{1,\text{loc}}$ or uniformly almost periodic (a.p.) $q(t)$, i.e. almost periodic in the sense of Bohr [31], [111] with a real-valued perturbation $p(t) \in L_{1,\text{loc}}$ which is small at infinity in the following sense:

$$\lim_{x\to\pm\infty} \int_x^{x+1} |p(t)|\, dt = 0. \tag{6.118}$$

It is well-known [276] (see also [70, pp. 171–174]) that, in this case, the essential spectra of operators L_1 and L_2 coincide: $\sigma_e(L_2) = \sigma_e(L_1)$, and, under the influence of perturbation, in each of the spectral gaps of the operator L_1, may appear finitely or infinitely many discrete levels. These levels may be condensed only to the end-points of gaps. (In the presence of $q(t) \in L_{1,\text{loc}}$ which is not assumed to be semi-bounded below, the proofs from [276] and [70] work in our case if we assume that the operator $-\frac{1}{2}\frac{d^2}{dt^2} + q(t)$ on the real line is semi-bounded below. In our case, the latter condition is obviously satisfied.) It is also known (see, for example [524] and [835]) that $\sigma(L_1) = \sigma_e(L_1)$, i.e. L_1 *does not have isolated eigenvalues.* The necessary and sufficient conditions for the zonal spectrum[13] $\sigma(L_1)$ to

[13] i.e. the spectrum consisting of a sequence of segments that belong to the essential spectrum, tend to infinity and are separated by a sequence of gaps (see the bibliographical comments at the end of Chapter 6, formula (6.192) and Lemma 6.7).

correspond to the purely periodic potential $q(t)$ belonging to the given Sobolev space $W^m_{2,\mathrm{loc}}$ are given by Marchenko–Ostrovskiy Theorem [658] (see also [121] and the Bibliographical Comments for this Chapter). We will find precise estimates for perturbations guaranteeing the finiteness or the infinitude of the number of discrete levels which are brought into the given gap in a neighborhood of its boundary. These results are new not only for almost periodic (a.p.) but also for the periodic $q(t)$.

Lemma 6.5 *If $q(t)$ is a real-valued a.p. function and for some $\lambda \in \mathbb{R}$, the equation (6.1) has a non-trivial a.p. real-valued solution $u_1 = e(t, \lambda)$, then the solution $u_2 = \mathcal{E}(t, \lambda)$, linearly independent with u_1, has the form*

$$\mathcal{E}(t, \lambda) = [A_\lambda + o(1)]te(t, \lambda) + O(1),\ |t| \to \infty, \tag{6.119}$$

where, under the condition $W\{e, \mathcal{E}\} \equiv e\mathcal{E}' - e'\mathcal{E} = 1$,

$$A_\lambda = M_t\{F_k(t, \lambda|e)\} \equiv \lim_{b-a\to\infty} \frac{1}{b-a}\int_a^b F_k\,dt, \tag{6.120}$$

with

$$F_k(t, \lambda|u) = (q + 1 - \lambda^{1-k})(\lambda^k u^2 - u'^2)(\lambda^k u^2 + u'^2)^{-2}, \tag{6.121}$$

$k = 0, 1$ (F_0 is defined for all $\lambda \in \mathbb{R}$, and for $\lambda > 0$, both F_0 and F_1 are well-defined). If $\lambda = \lambda_0 = \inf \sigma(L_1)$, then the following equation also holds:

$$A_{\lambda_0} = M_t\left\{e^{-2}(t, \lambda_0)\right\}. \tag{6.122}$$

Lemma 6.5 follows from the generalization of D'Alembert–Liouville–Ostrogradsky formula $u_2 = u_1 \int u_1^{-2}\,dt$ given in Lemma 6.6. This generalization is applicable to the case of oscillating solutions and does not require any "sewing together" for $u_2(t)$ at the zeros of $u_1(t)$.

Lemma 6.6 *Let u_1 and u_2 be real-valued solutions of the equation (6.1) with an arbitrary real $q(t) \in L^1_{\mathrm{loc}}$ (we do not assume that $q(t)$ is periodic or a.p.) such that $W\{u_1, u_2\} = 1$. Then the following equality holds for $k = 0, 1$ with $\lambda > 0$ (or for $k = 0$ with any $\lambda \in \mathbb{R}$):*

$$u_2(t, \lambda) = u_1(t, \lambda) \int F_k(t, \lambda|u_1)\,dt - \frac{u_1'(t, \lambda)}{\lambda^k u_1^2(t, \lambda) + u_1'^2(t, \lambda)}, \tag{6.123}$$

where $F_k(t, \lambda|u)$ is defined by the formula (6.121). (These formulas can be modified with $|u_1|^2 + |u_1'|^2$ in the denominator, so that they can be applied to *the non-real case [785], and can also be generalized to systems; see Sec. 6.2 and the Bibliographical Comments.)*

In the following theorem, we will assume, for simplicity, that

$$p(t) \in L_1(-\infty, \infty). \tag{6.124}$$

Theorem 6.12 *Assume that $q(t)$ is a.p. function (or a periodic $L_{1,loc}$ function). Assume also that $p(t)$ satisfies (6.124), and let $(\lambda_{2\nu-1}, \lambda_{2\nu})$ be a gap*[14] *in $\sigma_e(L_1)$. Suppose that (6.1) has a.p. solution $u = e(t, \lambda_j)$, where $j = 2\nu - 1$ or $j = 2\nu$.*

1) If for some $N > 0$, $\epsilon > 0$,

$$(-1)^{j+1} p(t) \geq [|K_j| + \epsilon] t^{-2}, \quad t > N \text{ or } t < -N, \tag{6.125}$$

then there is an infinite number of eigenvalues of the operator L_2 in (6.2) between λ_j and $(\lambda_{2\nu-1} + \lambda_{2\nu})/2$.

2) If for some $N > 0$, $\epsilon > 0$,

$$(-1)^{j+1} p(t) \geq [|K_j| - \epsilon] t^{-2}, \quad |t| > N, \tag{6.126}$$

where

$$K_j = (4 A_{\lambda_j} M_t \{e^2(t, \lambda_j)\})^{-1}; \quad (-1)^j K_j > 0, \tag{6.127}$$

and $A_{\lambda_j} \neq 0$, then there is at most finite number of eigenvalues of the operator L_2 (6.2) between λ_j and $(\lambda_{2\nu-1} + \lambda_{2\nu})/2$. (The conditions 1) and 2) as well as the requirement on the existence of a.p. solution $e(t, \lambda_j)$ are fulfilled in the case of a periodic $q(t)$ (then $e(t, \lambda_j)$ is even periodic or antiperiodic), and, in the case of a.p. $q(t)$, it is proven[15] *below that the above conditions are fulfilled for zonal potentials in the sense of B. M. Levitan (i.e. the potential corresponding to the Schrödinger operator having the zonal spectrum after B. Levitan; see the formulation of Lemma 6.7).*

Proof of Lemma 6.6. For $k = 0, 1$, using (6.1) and (6.121), we get

$$\frac{d}{dt} \frac{u_2 + \frac{u_1'}{\lambda^k u_1^2 + {u_1'}^2}}{u_1} = F_k(t, \lambda | u_1), \tag{6.128}$$

[14] Note that although the gaps cannot always be enumerated in the order from left to right (for example, if the gaps are complementary intervals of the Cantor set), the set of gaps and their end-points is at most countable, and, hence, the gaps can be enumerated so that $(\lambda_{2\nu-1}, \lambda_{2\nu})$ is the ν-th gap $\nu = 0, 1, 2, \ldots$, where $\nu = 0$ corresponds to the semi-infinite gap $(\lambda_{-1} = -\infty, \lambda_0)$.

[15] See Lemma 6.7 and the formula (6.166).

where the expression under d/dt on the left hand side of (6.128) is continuous at the roots of $u_1(t,\lambda)$, taking values $u_2'(t,\lambda)/u_1'(t,\lambda)$. This is seen by representing the expression under d/dt on the left hand side of (6.128) in the form

$$[\lambda^k u_1 u_2 + u_1' u_2' + (1 - W\{u_1, u_2\}) u_1' u_1^{-1}](\lambda^k u_1^2 + {u_1'}^2)^{-1},$$

and the Lemma is proven. □

Proof of Lemma 6.5. The proof follows from the formula (6.123) by setting $u_1(t,\lambda) = e(t,\lambda)$, $u_2(t,\lambda) = \mathcal{E}(t,\lambda)$ since, in this case, $F_k(t,\lambda|e)$ is u. a. p. function,

$$\begin{aligned}\mathcal{E}(t,\lambda) &= e(t,\lambda)\int_1^t F_k(t,\lambda|e)\,dt - \frac{e'(t,\lambda)}{\lambda^k e^2(t,\lambda) + e'^2(t,\lambda)}\\ &= e(t,\lambda)\int_1^t (F_k(t) - M\{F_k\})\,dt + A_\lambda t e(t,\lambda) + O(1)\\ &= [A_\lambda + o(1)]te(t,\lambda) + O(1),\end{aligned}$$

and (6.119) is proven. The equation (6.122) follows similarly from D'Alembert–Liouville–Ostrogradsky formula. This concludes the proof of Lemma 6.5. □

Proof of Theorem 6.12. In (6.2) we set $p(t) = \beta t^{-2}$ for $t > 1$ (and, analogously, for $t < -1$)[16], and we apply Theorem 6.10. For $k = 0$ or $k = 1$, we denote

$$\zeta(t) = \frac{1}{t}\int_1^t (F_k(t) - A_\lambda)\,dt \to 0, \quad \text{as } t \to \infty. \tag{6.129}$$

Then $\mathcal{E}(t,\lambda) = [A_\lambda + \zeta(t)]te(t,\lambda) + O(1)$.
In accordance with Sec. 6.3, we have

$$\begin{aligned}Q(t) &= \exp\left\{2\int_1^t p(s)e(s)\mathcal{E}(s)\,ds\right\}\\ &= \exp\left\{2\beta\int_1^t \frac{e^2(s)}{s}(A_\lambda + \zeta(s))\,ds + C_{1\lambda} + O\left(\frac{1}{t}\right)\right\} \qquad (6.130)\\ &= \exp\left\{2\beta[A_\lambda(M\{e^2\} + \delta(t))\ln t + \alpha_1(t)\ln t] + C_{1\lambda} + o(1)\right\}\\ &= C_\lambda t^{2\beta A_\lambda M\{e^2\} + \alpha(t)} \cdot \{1 + o(1)\}, \quad (t \to \infty),\end{aligned}$$

[16] In physics, β is called the coupling constant. From Theorem 6.12 we see that Kneser-type constants K_j are critical values of the coupling constant.

where $C_\lambda = \text{const}$,

$$\delta(t) = \frac{1}{\ln t}\int_1^t \frac{\epsilon(s)}{s}\,ds \to 0, \quad (t\to\infty), \tag{6.131}$$

$$\epsilon(t) = \frac{1}{t}\int_1^t (e^2(s) - M\{e^2\})\,ds \to 0, \quad (t\to\infty), \tag{6.132}$$

$$\alpha_1(t) = \frac{1}{\ln t}\int_1^t \frac{e^2(s)\zeta(s)}{s}\,ds \to 0, \quad (t\to\infty), \tag{6.133}$$

$$\alpha(t) = 2\beta[A_\lambda\delta(t) + \alpha_1(t)] \to 0, \quad \alpha'(t)t\ln t \to 0, \tag{6.134}$$

as $t\to\infty$.

Furthermore, (6.134) allows us to use the properties of proximate orders[17], and we get

$$\begin{aligned} s(t) &= \int_1^t |p(\tau)|\mathcal{E}^2(\tau)Q(\tau)\,d\tau \\ &= C_\lambda|\beta|A_\lambda^2\int_1^t \tau^{2\beta A_\lambda M\{e^2\}+\alpha(\tau)}\cdot[e^2(\tau)+o(1)]\,d\tau \\ &= C_\lambda|\beta|A_\lambda^2\int_1^t \tau^{2\beta A_\lambda M\{e^2\}+\alpha(\tau)}\cdot[M\{e^2\} + (e^2(\tau) - M\{e^2\}) + o(1)]\,d\tau \\ &= C_\lambda|\beta|A_\lambda^2 M\{e^2\}(2\beta A_\lambda M\{e^2\}+1)^{-1}\cdot t^{2\beta A_\lambda M\{e^2\}+1+\alpha(t)}\cdot(1+o(1)), \end{aligned} \tag{6.135}$$

where, while integrating by parts, we used the equality $e^2(t) - M\{e^2\} = (t\epsilon(t))'$ following from (6.132), and β in (6.135) and below satisfies the condition

$$2\beta A_\lambda M\{e^2\} > -1. \tag{6.136}$$

Similarly, by (6.117), under the condition (6.136), we have

$$\begin{aligned} \int_{s(t)}^{s(\infty)} q^\wedge(s)\,ds &= \int_{s(t)}^{\infty} q^\wedge(s)\,ds \\ &= C_\lambda^{-1}|\beta|\int_t^\infty e^2(\tau)\tau^{-2(1+\beta A_\lambda M\{e^2\})-\alpha(\tau)}\cdot\{1+o(1)\}\,d\tau \\ &= C_\lambda^{-1}|\beta|M\{e^2\}\cdot(1+2\beta A_\lambda M\{e^2\})^{-1}t^{-1-2\beta A_\lambda M\{e^2\}-\alpha(t)}(1+o(1)). \end{aligned} \tag{6.137}$$

[17] See, for example, [110, formulas (1.53), (1.53′)], [71, Chapter II, Sec. 2], and also the bibliographical comments at the end of Chapter 6 of this book.

Therefore,

$$\lim_{t\to\infty} s(t)\int_{s(t)}^{\infty} q^{\wedge}(s)\,ds = \frac{\beta^2 A_\lambda^2 M^2\{e^2\}}{(2\beta A_\lambda M\{e^2\}+1)^2}. \tag{6.138}$$

Starting from (6.138) and using Hille–Wintner integral comparison theorem [187, Theorem 2.12], we will find a condition for the auxiliary equation (6.116) corresponding to perturbation $p(t) = \beta t^{-2}$ to be oscillatory or non-oscillatory. For convenience, we recall Hille–Wintner integral comparison theorem: if for the equations $v'' + q_1(s)v = 0$ and $v'' + q_2(s)v = 0$, the inequality

$$0 \le \int_t^{\infty} q_1(s)\,ds \le \int_t^{\infty} q_2(s)\,ds, \tag{6.139}$$

holds for all $t \ge a \ge 0$ and the integrals in (6.139) converge (at least conditionally), then if the second equation is not oscillatory as $t \to \infty$, the first equation is also not oscillatory. (The continuity requirement on q_1, q_2 in [187] is not needed for the proof of the above theorem. The proof also works under the assumption $q_1, q_2 \in L_{1,\mathrm{loc}}$. Further generalizations of Hille–Wintner integral comparison theorem are given in [610] and [187].)

Let us compare the equation (6.116) with the equation $v'' + \frac{c}{s^2}v = 0$ which is non-oscillatory for $c = 1/4$ and oscillatory for $c > 1/4$, where $\int_t^{\infty} cs^{-2}\,ds = ct^{-1}$. Setting first $q_1(s) = q^{\wedge}(s)$, $q_2(s) = \frac{1}{4}s^{-2}$ and then $q_1(s) = \frac{1+\epsilon}{4}s^{-2}$, $\epsilon > 0$, $q_2(s) = q^{\wedge}(s)$, we see, using (6.137)–(6.139), (6.120) and Proposition 6.1 below, that the assertions of Theorem 6.12 hold for perturbations of the form $p_\beta(t) = \beta t^{-2}$, $(|t| > 1)$.

Proposition 6.1 *Assume that $p(t) \in L_1(-\infty,\infty)$ and $q(t)$ is u. a. p function. Then*

$$\sup_{t\in\mathbb{R}} \left|\gamma\left(t, \frac{\lambda_{2\nu-1}+\lambda_{2\nu}}{2}\right)\right| < \infty.$$

Now for general perturbations $p(t) \in L_1$, the Theorem follows form a theorem about a differential inequality[18] applied to phase equations of the form (6.99) with $p(t)$ and $p_\beta(t)$ respectively. It remains to prove Proposition 6.1.

Proof of Proposition 6.1. By Theorems 6.10 and 6.11 it is enough to show that the equations of the type (6.116) are non-oscillatory as $s \to s(\pm\infty)$ when λ belongs to a spectral gap (for example, $\lambda = \frac{1}{2}(\lambda_{2\nu-1}+\lambda_{2\nu})$).

[18]See the bibliographical comments pertaining to Sec. 6.3 of this book.

We estimate $Q(t)$ in (6.111), taking as $u_{1,2}(t,\lambda)$ Weyl solutions $\psi_\pm(t,\lambda) \in L^2(0,\infty)$ of the equation (6.1) with the normalization condition given by the identity Wronskian (see (6.150)). From (6.111) we get

$$Q(t) = \exp\left\{-\int_1^t p(s)\frac{\psi_+(s,\lambda)\psi_-(s,\lambda)}{m_+(\lambda)+m_-(\lambda)}\,ds\right\}$$
$$= \exp\left\{\int_1^t p(s)G(s,s,\lambda)\,ds\right\}.$$

Since the Green's function $G(t,t,\lambda)$ is almost periodic on the diagonal for all λ in the resolvent set of (6.1) with u. a. p. potential (see [727, Theorems 3.1 and 3.2]), we have, using (6.124),

$$|Q(t)| + |Q^{-1}(t)| < C = C(\lambda).$$

Therefore, by (6.113) and (6.117), for all $t > 0$, we have

$$a(t) := s(t)\int_{s(t)}^{s(\infty)} |q^\wedge(s)|\,ds$$
$$\le \int_0^t |p(\xi)|\psi_-^2(\xi)Q(\xi)\,d\xi \int_t^\infty |p(s)|\psi_+^2(s)Q^{-1}(s)\,ds$$
$$\le \frac{C^2(\lambda)}{(m_+(\lambda)+m_-(\lambda))^2}\int_0^t |p(\xi)|\psi_-^2(\xi)\,d\xi \int_t^\infty |p(s)|\psi_+^2(s)\,ds.$$

As shown in [70, Chapter II, Sec. 27, formula (17)] [19],

$$\int_t^\infty |p(s)|\psi_+^2(s)\,ds \le 2\eta(t)\int_t^\infty ({\psi'}_+^2(s)+\psi_+^2(s))\,ds,$$

where $\eta(t) \to 0$ as $t \to +\infty$.

Hence, by the inequality

$$\|y'\| \le C_{12}\|y\|^{1/2}\|y''\|^{1/2},$$

in $L^2(0,\infty)$ (see [70, Chapter II, Sec. 26] or [643]), and by (6.1), we have

$$\|{\psi'}_+\|^2 \le C_{12}^2\|\psi_+\|\cdot\|(q-\lambda)\psi_+\| \le C_{12}(\lambda)\|\psi_+\|^2,$$

and, hence,

$$\int_t^\infty |p(s)|\psi_+^2(s,\lambda)\,ds \le C_\lambda\eta(t)\int_t^\infty \psi_+^2(s,\lambda)\,ds.$$

[19] In the proof of a theorem from [276].

Similarly, for all $t > 0$,

$$\int_0^t |p(\xi)|\psi_-^2(\xi,\lambda)\,d\xi \le C_\lambda \int_{-\infty}^t \psi_-^2(\xi,\lambda)\,d\xi,$$

and, hence, as $t \to +\infty$, we have

$$\begin{aligned} a(t) &\le \frac{C_+(\lambda)}{(m_+(\lambda)+m_-(\lambda))^2}\eta(t)\int_{-\infty}^t \psi_-^2(\xi,\lambda)\,d\xi \int_t^\infty \psi_+^2(s,\lambda)\,ds \\ &\le \frac{C_+(\lambda)}{(m_+(\lambda)+m_-(\lambda))^2}\eta(t)\sup_\alpha\left\{\int_{-\infty}^\alpha \psi_-^2(\xi,\lambda)\,d\xi \int_\alpha^\infty \psi_+^2(s,\lambda)\,ds\right\} \to 0 \end{aligned}$$

because

$$\frac{1}{(m_+(\lambda)+m_-(\lambda))^2}\sup_\alpha\left\{\int_{-\infty}^\alpha \psi_-^2(\xi,\lambda)\,d\xi \int_\alpha^\infty \psi_+^2(s,\lambda)\,ds\right\} \le \|R_\lambda\|^2 < \infty,$$

where R_λ is the resolvent of the operator (6.1) determined by the Green's function

$$G(t,s,\lambda) = \begin{cases} -\frac{1}{m_+(\lambda)+m_-(\lambda)}\psi_+(t,\lambda)\psi_-(s,\lambda), & t > s, \\ \\ -\frac{1}{m_+(\lambda)+m_-(\lambda)}\psi_-(t,\lambda)\psi_+(s,\lambda), & t < s, \end{cases}$$

where λ belongs to a spectral gap.

The last inequality is an inequality of Mackenhoupt type [700] (see also [151, Chapter II, Sec. 7]), and can be proven by noting that its left hand side does not exceed

$$\sup_\alpha \frac{\|R_\lambda f_\alpha\|}{\|f_\alpha\|}, \text{ where } f_\alpha(t) = \begin{cases} \psi_+(t,\lambda), & \alpha < t < \infty, \\ 0, & -\infty < t < \alpha. \end{cases}$$

Therefore, $a(t) \to 0$ as $t \to +\infty$, and, hence, (6.116) is non-oscillatory for $s \to s(+\infty)$ (by the Hille–Wintner comparison theorem [187] or [70, Theorem 8, p. 30]). It can be proven similarly that an equation of the type (6.116) is non-oscillatory for $s \to s(-\infty)$. Proposition 6.1 and Theorem 6.12 are proven. □

Corollary 6.5 *If $q(t) \equiv 0$, we have $\lambda_0 = 0$ and $K_0 = \frac{1}{4}$, and Theorem 6.12 is a spectral reformulation of classical Kneser's Theorem (see [185] and [58]), which has an equivalent formulation in terms of oscillation of solutions. (Therefore, K_j are called Kneser-type constants.)*

Remark 6.7 We always have $0 < K_0 \leq 1/4$, with the equality sign possible only when $q(t) = \text{const}$, as is easily seen from (6.122) and (6.127) for λ_0 using Cauchy–Schwarz–Bunyakovskiy inequality.

Example 6.1 Consider the Lamé's equation

$$-u'' + n(n+1)k^2 \text{sn}\,^2 t \cdot u = \lambda u \tag{6.140}$$

with $0 < k < 1$.

For $n = 1$, (in other words, in the one-gap case), we have

$$\begin{gathered} \lambda_0 = k^2,\ \lambda_1 = 1,\ \lambda_2 = 1 + k^2 \\ e(t, \lambda_0) = \text{dn}\ t, \quad e(t, \lambda_1) = \text{cn}\ t, \quad e(t, \lambda_2) = \text{sn}\ t \end{gathered} \tag{6.141}$$

and the corresponding Kneser-type constants can be expressed via complete elliptic integrals K and E of the 1-st and 2-nd kind:

$$\begin{gathered} 0 < K_0 = 4^{-1} k'^2 K^2 E^{-2} < 1/4, \quad k'^2 = 1 - k^2, \\ K_1 = -4^{-1} k^2 k'^2 K^2 (E - k'^2 K)^2,\ K_2 = 4^{-1} k^2 K^2 (E - K)^2. \end{gathered} \tag{6.142}$$

For $n = 2$ (in other words, in the two-gap case), we have

$$\begin{aligned} &\lambda_0 = 2\{1 + k^2 - \sqrt{1 - k^2 + k^4}\}, \\ &\lambda_1 = 1 + k^2, \quad \lambda_2 = 1 + 4k^2, \\ &\lambda_3 = 4 + k^2, \quad \lambda_4 = 2\{1 + k^2 + \sqrt{1 - k^2 + k^4}\}, \end{aligned} \tag{6.143}$$

and the corresponding periodic (or anti-periodic) solutions have the following form:

$$\begin{aligned} &e(t, \lambda_0) = 3\text{dn}^2 t + k^2 - 2 + \sqrt{1 - k^2 + k^4}, \\ &e(t, \lambda_1) = \text{cn}\ t \cdot \text{dn}\ t, \quad e(t, \lambda_2) = \text{sn}\ t \cdot \text{dn}\ t, \\ &e(t, \lambda_3) = \text{sn}\ t \cdot \text{cn}\ t, \quad e(t, \lambda_4) = 3\text{dn}^2 t + k^2 - 2 - \sqrt{1 - k^2 + k^4}. \end{aligned} \tag{6.144}$$

Formulas (6.143) are found according to the formulas for spectral zones of the potential $6\wp(t)$ given in [206, Chapter II, Sec. 10.3]. In the case of the three-gap potential, the spectrum is found by solving three quadratic equations; see [4, Sec. V] and [6, vol. 2, p. 273–274 in Russian version]. (See also [198, Chapter XXIII] and [63]. The books [3], [82], [106] and [109] are devoted to the study of elliptic functions.) The formulas (6.143)–(6.144) enable us to explicitly calculate Kneser-type constants for the two-gap Lamé's equation using Theorem 6.12. We will not go over this calculation here.

Corollary 6.6 *Assume that $|p(t)| \leq Ct^{-2}$ ($C > 0$, $|t| > N$). Then every sufficiently distant spectral gap of the problem (6.1) contains at most finite number of eigenvalues of the perturbed problem (6.2). If $|t^2 p(t)| \to \infty$ and sign $p(t) = const$ as $t \to \infty$ (or $t \to -\infty$), then every finite spectral gap contains infinitely many eigenvalues of the problem (6.2).*

To prove this, it suffices to note that by (6.120), (6.121) with $k = 1$ and (6.127), we have

$$|K_j| \to \infty, \quad \text{as } j \to \infty. \tag{6.145}$$

The following lemma shows that the class of u. a. p. zonal potentials [112] generates the equations with the following property: to the ends of spectral gaps correspond u. a. p. solutions. This is needed to apply Theorem 6.12 to perturbations of such potentials.

Lemma 6.7 *Assume that the spectrum of the operator L_1 in (6.1) has zonal structure: $\sigma(L_1) = \bigcup_{\nu=0}^{\infty} [\lambda_{2\nu}, \lambda_{2\nu+1}]$, where $-\infty < \lambda_0 < \lambda_1 < \lambda_2 < \cdots < \lambda_{2\nu-1} < \lambda_{2\nu} < \cdots$, and*

$$\exists C > 0,\ \delta > 0 : \lambda_{n+2} - \lambda_n > Cn^{\delta},\ n = 0, 1, 2, \ldots \tag{6.146}$$

and

$$\exists p > 1 : \quad \sum_{\nu=1}^{\infty} \lambda_{2\nu}^{p} (\lambda_{2\nu} - \lambda_{2\nu-1}) < \infty. \tag{6.147}$$

Then for $\lambda = \lambda_j$ the equation (6.1) has a non-trivial u. a. p. solution.

Proof. By $\theta(t, \lambda)$, $\phi(t, \lambda)$ we denote the solutions to (6.1) with the initial data

$$\theta(0, \lambda) = \phi'(0, \lambda) = 1, \quad \theta'(0, \lambda) = \phi(0, \lambda) = 0. \tag{6.148}$$

Let $m_{\pm}(\lambda)$ denote the Weyl–Titchmarsh functions of the equation (6.1) corresponding to $t \to \pm\infty$ respectively. The functions $m_{\pm}(\lambda)$ are chosen so that

$$\text{Im } m_{\pm}(\lambda) > 0, \quad \text{for all Im } \lambda > 0. \tag{6.149}$$

Then, outside the spectrum of (6.1), Weyl solutions (in the case of the Hill operator, Weyl solutions are Floquet–Bloch solutions) have the following form:

$$\psi_{\pm}(t,\lambda) = \theta(t,\lambda) \pm m_{\pm}(\lambda)\phi(t,\lambda) \in L_2(\mathbb{R}^{\pm}). \tag{6.150}$$

The latter is well-known in the case Im $\lambda \neq 0$, but it also holds in the spectral gaps on the real line because in the inequality

$$\pm \int_0^{\pm\infty} |\psi_{\pm}(t,\lambda)|^2 \, dt \leq \frac{\text{Im } m_{\pm}(\lambda)}{\text{Im } \lambda},$$

where $\lambda = u + iv$ and u belongs to a spectral gap, we can pass to the limit as $v \to 0$. The limit on the right hand side (and, hence, on the left hand side) is finite due to the representation of $m_{\pm}(\lambda)$ as a Stieltjes transform of the spectral function of the problem on the corresponding half-line.

We now note that (as shown in [112, Chapter 9, Sec. 1])

$$m_{\pm}(\lambda) = \lim_{N\to\infty} m_{\pm}^N(\lambda),$$

where $m_{\pm}^N(\lambda)$ are Weyl–Titchmarsh functions for the N-zone u. a. p. potentials $q_N(t)$ corresponding to the first N gaps (and spectral parameters $\xi_j(0)$, $\sigma_j(0) = \pm$) of the infinite-zone u. a. p. potentials with the rest of the gaps deleted. In this case, we have the locally uniform convergence of potentials $q_N(t) \to q(t)$ and solutions of the equation (6.1) $\theta_N(t,\lambda) \to \theta(t,\lambda)$, $\phi_N(t,\lambda) \to \phi(t,\lambda)$, and hence $\psi_{\pm}^N(t,\lambda) = \theta_N \pm m_{\pm}^N \phi_N \to \theta \pm m_{\pm}\phi = \psi_{\pm}(t,\lambda)$. This allows us to pass to the limit as $N \to \infty$ on both sides of the well-known formula[20]

$$\psi_{+}^N(t,\lambda)\psi_{-}^N(t,\lambda) = \frac{\prod_{j=1}^{N}(\lambda - \xi_j^N(t))}{\prod_{j=1}^{N}(\lambda - \xi_j(0))}, \tag{6.151}$$

where $\lambda \neq \xi_j(0)$, $j = 1, 2, \ldots, \xi_j^N(t) \in [\lambda_{2j-1}, \lambda_{2j}]$ are spectral parameters of the potential $q_N(x+t)$, by using the locally uniform estimate given

[20] For the case of a finite-zone potential, see [112, Sec. 8.2.10–8.2.14]. For the case of an even u. a. p. potential, see [6, p. 261 (α)]. For the case of a periodic potential, see [206, Chapter II, Sec. 4, (15)].

in [112, Proof of Theorem 9.2.2]

$$\sum_{j=1}^{N} |\xi_j(t) - \xi_j^N(t)| \le C \sum_{j=N+1}^{\infty} (\lambda_{2j} - \lambda_{2j-1}),$$

where C depends on the interval.

As a result, we get

$$\psi_+(t,\lambda)\psi_-(t,\lambda) = \prod_{j=1}^{\infty} [(\lambda - \xi_j(t))/(\lambda - \xi_j(0))], \tag{6.152}$$

where $\xi_j(t)$ are the spectral parameters of the potential $q(x+t)$ and are u. a. p. functions (see [112, Chapter XI, Sec. 4])[21]. Besides, due to a special form of $m_\pm(\lambda)$ (see formula (9.1.10) from [112]), we have $m_+(\lambda_j) = -m_-(\lambda_j)$, and, hence, $\psi_+(t,\lambda_j) = \psi_-(t,\lambda_j) =: e(t,\lambda_j)$. (Note that by a suitable choice of coordinates, we can always achieve $m_\pm(\lambda_j) \ne \infty$ or $\lambda_{2j-1} \ne \xi_j(0) \ne \lambda_{2j}$, which we henceforth assume for $j = 1, 2, \ldots$) It follows that the square of the real-valued solution $e(t,\lambda_j) = \psi_\pm(t,\lambda_j)$ is almost periodic. Since $e^2(t,\lambda_j)$ is almost periodic, the following Lemma shows that $e(t,\lambda_j)$ is also almost periodic.

Lemma 6.8 *If the equation (6.1) with real bounded potential $q(t)$ has a real solution $u(t,\lambda)$ such that $u^2(t,\lambda)$ is u. a. p. for some $\lambda \in \mathbb{R}$, then $u(t,\lambda)$ is u. a. p. Moreover, there exists $\epsilon > 0$ such that*

$$u^2(t,\lambda) + u'^2(t,\lambda) \ge \epsilon > 0, \quad \forall t \in \mathbb{R}. \tag{6.153}$$

Proof. Since $u^2(t,\lambda)$ is u. a. p, it follows that $u^2(t,\lambda) \not\to 0$ as $t \to \pm\infty$. Thus there exists $\epsilon_1 > 0$ such that if $u^2(t',\lambda) < \epsilon_1$ at the local extremum point t', then $u(t',\lambda) = 0$. Otherwise, there exists a sequence $t_n \to \infty$ of local extremum points of $u^2(t,\lambda)$ such that $u(t_n,\lambda) \to 0$, $u(t_n,\lambda) \ne 0$. It follows that $u'(t_n,\lambda) = 0$ since $(u^2(t_n,\lambda))' = 0$. Now since $q(t)$ is bounded, it follows that if, contrary to (6.153), $|u(t_n,\lambda)| + |u'(t_n,\lambda)| \to 0$, then for every $\delta > 0$ there exist arbitrarily long t-intervals on which $|u(t,\lambda)| < \delta$. But this is impossible in the case of u. a. p. $u(t,\lambda) \not\equiv 0$. Therefore, if $u(t',\lambda) = 0$ and τ is the ϵ'-almost-period for $u^2(t,\lambda)$, where $\epsilon' < \epsilon_1$, then

[21] Thus, in the case of zonal u. a. p. potentials, we have obtained a proof, independent of [835], of almost periodicity of the Green's function $G(t,t,\lambda) = -\psi_+(t,\lambda)\psi_-(t,\lambda)/(m_+(\lambda) + m_-(\lambda))$ outside the spectrum due to (6.152). (Here the poles of the numerator and the denominator of $G(t,t,\lambda)$ cancel each other out for $\lambda = \xi_j(0)$, and $m_+(\lambda) + m_-(\lambda) \ne 0$ outside the essential spectrum due to the absence of isolated eigenvalues).

$u(t' + \tau, \lambda) = 0$ and both t' and $t' + \tau$ are points at which the sign of $u(t, \lambda)$ changes. Thus $u(t, \lambda)$ and $u^2(t, \lambda)$ are simultaneously u. a. p., and Lemmas 6.8 and 6.7 are proven. □

Note that in the case of u. a. p. $q(t)$, without any additional requirements, it may happen that there are no almost periodic solutions at the end-points of the spectral gaps of (6.1); see [835, p. 597].

Note that Kneser constants K_n are related in a simple way to the so called effective masses $æ_n$, which play an important role in physical processes[22]. In the case of a periodic potential, the effective masses are known:

$$æ_n := \left[(d^2\lambda/dk^2)|_{\lambda=\lambda_n}\right]^{-1}, \tag{6.154}$$

where $k = k(\lambda)$ is the quasi-impuls (see (6.157) and below). This definition of the effective mass can also be carried over to the case of u. a. p. zone-type potentials. Let us clarify what we have just said. It is well-known that we can associate to the Hill operator (6.1) the monodromy matrix

$$T(\lambda) = \begin{pmatrix} \theta(\omega, \lambda) & \phi(\omega, \lambda) \\ \theta'(\omega, \lambda) & \phi'(\omega, \lambda) \end{pmatrix}, \tag{6.155}$$

where ω is the period of the potential: $q(t + \omega) = q(t)$. The characteristic numbers $\rho_1(\lambda)$, $\rho_2(\lambda)$ of $T(\lambda)$, called the multipliers of the Hill equation (6.1), are the solutions of the quadratic equation[23]

$$\rho^2 - [\theta(\omega, \lambda) + \phi'(\omega, \lambda)]\rho + 1 = 0, \tag{6.156}$$

where $\rho_1^{-1}(\lambda) = \rho_2(\lambda) \equiv \rho(\lambda)$ are chosen so that $|\rho(\lambda)| \geq 1$, and the equality is attained precisely on the spectrum. The functions $\rho_j(\lambda) \neq 0$ are regular and single-valued in the whole λ-plane with cuts along the spectrum. For $\rho(\lambda) \neq \pm 1$, the equation (6.1) has the Floquet–Bloch fundamental system of solutions

$$\begin{aligned} \psi_1(t, \lambda) &= e^{ik(\lambda)t}\chi_1(t, \lambda), \\ \psi_2(t, \lambda) &= e^{-ik(\lambda)t}\chi_2(t, \lambda), \end{aligned} \tag{6.157}$$

where

$$-ik(\lambda) = \omega^{-1} \ln \rho(\lambda) \equiv \alpha(\lambda) - i\beta(\lambda), \tag{6.158}$$

[22]The concept of effective mass originated in the theory of crystals; see, for example, [208, Chapter II, Sec. 7].

[23]$\det T(\lambda)$ is the Wronski determinant, and, in our case, $\det T(\lambda) = 1$.

and $\chi_j(t,\lambda)$, $j = 1,2$, are ω-periodic functions[24] with respect to t (multiple-valued with respect to λ). The function $k(\lambda)$ in (6.157), normalized by the asymptotics

$$k(\lambda) \sim \sqrt{\lambda}, \quad (\text{Re } k(\lambda) \geq 0 \text{ for Im } \lambda \geq 0; \tag{6.159}$$

$$\text{Im } k(\lambda) \geq 0 \text{ for } \lambda \in \mathbb{C} \backslash [\lambda_0, \infty)),$$

as $\lambda \to \infty$, in the λ-plane with $[\lambda_0, \infty)$ deleted, is called the *quasi-impuls of the Hill equation* (6.1); see, for example, [121], [191, vol. 2], [206], [253], [426], [603], [658], [786] and [858].

Let us consider (6.1) with u. a. p. $q(t)$. Corresponding to such an equation, in [524], the rotation number[25]

$$\beta(\lambda) := \lim_{t\to\infty} \frac{1}{t} \arg\, [\phi'(t,\lambda) + i\phi(t,\lambda)],\ \lambda \in \mathbb{R} \tag{6.160}$$

was introduced. Moreover, it was proven that $\beta(t)$ exists, is continuous, non-decreasing with respect to λ and constant in the spectral gaps. It was also proven that $\beta(\lambda)/\pi$ is equal to the average number of zeros of $\phi(t,\lambda)$ with respect to the unit length on the t-axis[26], and, finally,

$$\beta(\lambda) = \text{Im } W(\lambda + i0) = \text{Re } k(\lambda + i0), \quad \lambda \in \mathbb{R}, \tag{6.161}$$

where $W(z)$ is an analytic function in $\mathbb{C}\backslash[\lambda_0, \infty)$ given by

$$W(z) = M_t \left\{ \frac{-1}{2G(t,t;z)} \right\} =: ik(z), \tag{6.162}$$

where $G(t,t;z)$ is the Green's function for (6.1) on the diagonal. Thus, $-iW(\lambda)$ coincides with the quasi-impuls in the case of the Hill operators. Therefore, it is natural to take the function $-iW(z)$ in (6.162) as the definition of the *quasi-impuls for the u. a. p. equation* (6.1)[27]. Subsequently,

---

[24] It is easy to see that the vectors col $\{\psi_j(0,\lambda),\ \psi_j'(0,\lambda)\}$, $j = 1,2$, are eigenvectors of the monodromy matrix (6.155).

[25] In [524], this number is denoted by $\alpha(\lambda)$. There is also a similar result in [258] and [730].

[26] These facts allow us to identify $\beta(\lambda)$ in (6.160) and (6.161) with Re $k(\lambda)$ in (6.157), i.e. with $\beta(\lambda)$ in (6.158), and put $ik(z)$ on the left hand side of (6.162).

[27] The function $W(z)$ is sometimes called the rotation number function. The necessary and sufficient conditions for the uniqueness of reconstruction of the rotation number function $W(z) = W(z,q)$ (6.162) by the countable set of polynomial conservation laws $\{\sigma_n(q)\}_{n=0}^{\infty}$ in terms of growth of the sequence $\{\sigma_n(q)\}$ were obtained in [712]. (This problem was posed by V. A. Marchenko.)

the definition of the effective mass in (6.154), which is known in the purely periodic case, is obviously transferred to the zonal u. a. p. case.

Indeed, in the case of a u. a. p. zonal potential, the limiting values of the function $-iW(\lambda+i0) = k(\lambda+i0)$ on the real axis are infinitely differentiable everywhere except for the end-points of spectral gaps, where the limiting values have singularities of the form

$$k - k_n = [2æ_n(\lambda - \lambda_n)]^{1/2} + O(\lambda - \lambda_n). \tag{6.163}$$

This shows that the definition of the effective mass $æ_n$ is correct in the u. a. p. case. Moreover, since $k(\lambda)$ maps the upper λ-half-plane into the first quadrant, whereby the real axis is mapped to a "comb with vertical teeth," where the "teeth" correspond to spectral gaps on the real λ-axis, the following relations hold:

$$æ_n \neq 0, \text{ sign } æ_n = (-1)^n, \; n = 0, 1, 2, \ldots \tag{6.164}$$

In each finite spectral gap $(\lambda_{2\nu-1}, \lambda_{2\nu})$, there is a point η_ν such that $k'(\lambda)|_{\lambda=\eta_\nu} = 0$. These points are mapped to the upper end-points of the "teeth" of the "comb."

In the case of the Hill operator, the expressions for the effective masses are obtained in [733]. Comparing those expressions with the formulas (6.119), (6.127) for Kneser-type constants, we get

$$K_n = (8æ_n)^{-1}, \quad n = 0, 1, \ldots \tag{6.165}$$

By (6.127) and (6.164) we obtain a well-known formula for the Hill operator:

$$A_{\lambda_j} \neq 0, \quad j = 0, 1, \ldots \tag{6.166}$$

where λ_j is an end-point of the non-degenerate gap (gap with non-zero length).

In the case of zone-type potentials, a modification of the construction used in [733] shows that the formulas (6.165) and (6.166) still hold. In the zonal u. a. p. case (see [786]), the quantities $æ_n$ can be represented as infinite products analogously as in the formulas in [428] corresponding to the Hill operator. (In [428], the connection between $æ_n$ and K_n was not considered. The product was expressed in terms of λ_0, λ_1, $\lambda_2, \ldots$ and the roots of $F'(\lambda) = 0$, where

$$F(\lambda) = F(\lambda; \omega) = \frac{1}{2}(\theta(\omega, \lambda) + \phi'(\omega, \lambda)) \tag{6.167}$$

is Lyapunov function, or, said differently, Hill's discriminant.)

As shown in [112], in the case of infinite-zone potentials, the conditions

$$\int_{\lambda_{2\nu-1}}^{\lambda_{2\nu}} \left[f^{-1/2}(z) \prod_{k=1}^{\infty} \frac{\eta_k - z}{\lambda_{2k}} \right] dz = 0, \quad \nu = 1, 2, \ldots,$$

where $f(z) = \prod_{k=1}^{\infty}(1 - z/\lambda_k)$, uniquely determine the numbers $\eta_k \in (\lambda_{2k-1}, \lambda_{2k})$ (here, for the simplicity of notation, we put $\lambda_0 = 0$).

Proposition 6.2 *Let $q(t)$ be u. a. p. potential of zonal type as in Lemma 6.7. Then the effective masses $æ_n$ can be represented as*

$$æ_n = \frac{1}{2} \left\{ \frac{\lambda - \lambda_n}{\lambda - \lambda_0} \prod_{j=1}^{\infty} \frac{(\lambda - \eta_j)^2}{(\lambda - \lambda_{2j-1})(\lambda - \lambda_{2j})} \right\} \Bigg|_{\lambda = \lambda_n},$$

and by (6.165), the Kneser-type constants K_n can also be represented as an infinite product.

Corollary 6.7 *In the above considered case of zone-type u. a. p. potential, we have*

$$\sum_{n=0}^{\infty} æ_n = \frac{1}{2}, \quad \sum_{n=0}^{\infty} K_n^{-1} = 4.$$

(In the periodic case, the first equation was obtained in [428].)

We also note that in the case of a *periodic* potential $q(t + \omega) = q(x)$, the Kneser-type constants, and hence by (6.165), the effective masses are expressed in terms of the Hill discriminant $F(\lambda)$:

$$K_j = -\frac{\omega^2}{8|F|'(\lambda_j)}, \quad æ_j = -\frac{1}{\omega^2}|F|'(\lambda_j).$$

(Here, $|F|'(\lambda)$ denotes the derivative of $|F|(\lambda)$ along the real λ-axis.)

6.5 Discrete Levels in Spectral Gaps of Perturbed Hill Operators

We consider the scalar Hill operator (6.1) and its perturbation (6.2). The functions p, $q \in L_{1,\text{loc}}$ are real-valued.

Theorem 6.13 **([769])** *Assume that*

$$\int_{-\infty}^{\infty} (1 + |t|)|p(t)|\, dt < \infty. \tag{6.168}$$

Then in each gap of the continuous spectrum of the self-adjoint problem (6.2), where $q(t)$ is a periodic function, there are at most finitely many eigenvalues, whereby each of the sufficiently distant spectral gaps contains at most two eigenvalues.

The eigenvalues cannot lay on the continuous spectrum, and, in particular, the end-points of the spectral gaps are not eigenvalues.

Remark 6.8 Under (more general than (6.168)) condition

$$\lim_{x\to\pm\infty} x \int_{x}^{\pm\infty} |p(t)|\, dt = 0, \tag{6.169}$$

the number of discrete levels introduced by the perturbation into a semi-infinite gap of the Hill operator is finite; see M. Sh. Birman [276, Theorem 4.11] and [70, p. 56].

Proof of Theorem 6.13. Assume that the period $\omega = 1$ and $\rho(\lambda)$ is defined by (6.156). For $\rho(\lambda) \neq \pm 1$, we write the Floquet–Bloch fundamental system of solutions $e_j(t, \lambda)$ in the form[28]

$$e_j(t,\lambda) = \phi(1,\lambda)\theta(t,\lambda) + [\rho_j(\lambda) - \theta(1,\lambda)]\phi(t,\lambda), \tag{6.170}$$

where $j = 1, 2$, and $e_j(t + 1, \lambda) = \rho_j(\lambda)e_j(t, \lambda)$, $|\rho_1(j)| \leq 1$, $|\rho_2(\lambda)| \geq 1$. (When $\phi(1, \lambda) = 0$, one of the solutions $e_j(t, \lambda)$ is identically equal to 0 with respect to t, and its normalization has to be changed in this case.) For $\rho(\lambda) = \pm 1$, we have $e_1(t, \lambda) = e_2(t, \lambda)$, and $e_j(t, \lambda)$, $j = 1, 2$ are periodic, while the other solutions grow at most linearly with respect to t.

Lemma 6.9 *If the matrices $A = (a_{ik})_{i,k=1}^2$ and $B = \begin{pmatrix} p & r \\ o & s \end{pmatrix}$ are unitarily equivalent, then*

$$|r| \leq |a_{12}| + |a_{21}|. \tag{6.171}$$

Proof. Let $A = UBU^*$, where $U = (u_{ik})$ is a unitary matrix. Then

$$a_{12} = pu_{11}\bar{u}_{21} + ru_{11}\bar{u}_{22} + su_{12}\bar{u}_{22},$$
$$a_{21} = p\bar{u}_{11}u_{21} + r\bar{u}_{12}u_{21} + s\bar{u}_{12}u_{22},$$

[28]These are concrete representatives of solutions of type ψ_j in (6.157).

or, since $u_{11}\bar{u}_{21} = -u_{12}\bar{u}_{22}$ (because U is unitary), we have

$$a_{12} = [ru_{11} + (s-p)u_{12}]\bar{u}_{22},$$
$$a_{21} = [ru_{21} + (s-p)u_{22}]\bar{u}_{12}.$$

If $u_{11}u_{22} \neq 0$, we have $a_{12}\frac{u_{22}}{\bar{u}_{22}} - a_{21}\frac{u_{12}}{\bar{u}_{12}} = r \det U$, and (6.171) holds. On the other hand, if $u_{11}u_{22} = 0$, then it is easy to see that the equality holds in (6.171). The Lemma is proven. □

Lemma 6.9 may be useful when we want to define more precisely the values of constants in our estimates.

Lemma 6.10 *For any solution $u(t,\lambda)$ of the equation (6.1) with periodic potential $q(t) = q(t+1)$, the following estimate holds for complex λ and all real t and s:*

$$|u(t,\lambda)| + \frac{|u'(t,\lambda)|}{1+|\sqrt{\lambda}|}$$
$$\leq Ce^{\alpha(\lambda)\cdot|t-s|} \cdot (1+|t-s|)\left(|u(s,\lambda)| + \frac{|u'(s,\lambda)|}{1+|\sqrt{\lambda}|}\right), \quad (6.172)$$

where $\alpha(\lambda) = \ln|\rho(\lambda)| \geq 0$. (The Lemma also holds in the case of complex potentials.)

Proof. Let $\vec{h}(t,\lambda) = \left\{u(t,\lambda); \frac{|u'(t,\lambda)|}{1+|\sqrt{\lambda}|}\right\}$ be a column vector. Then $\vec{h}(n,\lambda) = B^n(\lambda)\vec{h}(0,\lambda)$, where $n = 0,1,2,\ldots$, and the matrix

$$B(\lambda) = \begin{pmatrix} \theta(1,\lambda) & (1+|\sqrt{\lambda}|)\phi(1,\lambda) \\ (1+|\sqrt{\lambda}|)^{-1}\theta'(1,\lambda) & \phi'(1,\lambda) \end{pmatrix}$$

has the eigenvalues $\rho(\lambda)$ and $\rho^{-1}(\lambda)$. Using the same unitary transformation for all n, the matrices $B^n(\lambda)$ can be reduced to the form

$$B_n(\lambda) := \begin{pmatrix} \rho^n(\lambda) & r_n(\lambda) \\ 0 & \rho^{-n}(\lambda) \end{pmatrix},$$

and by Lemma 6.9, for $n = 1$, we get the estimate

$$|r_n(\lambda)| \leq n \cdot |\rho(\lambda)|^{n-1}\left\{\frac{|\theta'(1,\lambda)|}{1+|\sqrt{\lambda}|} + (1+|\sqrt{\lambda}|)|\phi(1,\lambda)|\right\}$$
$$\leq C \cdot n \cdot |\rho(\lambda)|^n,$$

which, using induction and the equality $B_n(\lambda) = (B_1(\lambda))^n$, holds for all n. Therefore, $|\vec{h}(n,\lambda)| \leq C\cdot(n+1)\cdot|\rho(\lambda)|^n \cdot |\vec{h}(0,\lambda)|$. Using the asymptotics

$\rho(\lambda) \sim e^{-i\sqrt{\lambda}}$, as $|\lambda| \to \infty$, it is not hard to extend the above estimate to intervals of t-values. In the same manner, we get the estimate for $t < 0$. Finally, since $q(t)$ is periodic, the constant C in the formula (6.172) can be chosen independently of t, and the Lemma is proven. □

Lemma 6.11 *Under the condition (6.168), the perturbed equation (6.2) has Jost-type solutions:*

$$E_1(t,\lambda) = e_1(t,\lambda) - \int_t^\infty K(t,s,\lambda)p(s)e_1(s,\lambda)\,ds,$$
$$E_2(t,\lambda) = e_2(t,\lambda) + \int_{-\infty}^t K(t,s,\lambda)p(s)e_2(s,\lambda)\,ds. \qquad (6.173)$$

For all t, the functions $E_j(t,\lambda)$ are analytic with respect to λ in the plane which is cut along the spectrum of (6.1) and are continuous in the whole complex plane including the points of the cuts (at these points the continuity is understood in the one-sided sense). The integral kernel $K(t,s,\lambda)$ is an entire function of λ for every pair of values t, s. Moreover, $K(t,s,\lambda)$ admits the estimate

$$|K(t,s,\lambda)| \le C \cdot \frac{1+|t-s|}{1+|\sqrt{\lambda}|} \cdot e^{\alpha(\lambda)\cdot|t-s|} \qquad (6.174)$$

and represents the Cauchy function of the equation (6.2), i.e. satisfies (6.2) with respect to t, $-\infty < t < \infty$, where $K(t,t,\lambda) = 0$, $K'_t(t,s,\lambda)|_{t=s} = 1$. (The Lemma also holds for complex-valued $q(t)$ and $p(t)$.)

Proof. The integral kernel $K(t,s,\lambda)$ is related to the Cauchy function $C(t,s,\lambda)$ of the unperturbed equation (6.1) by

$$K(t,s,\lambda) = C(t,s,\lambda) + \int_s^t C(t,\xi,\lambda)p(\xi)K(\xi,s,\lambda)\,d\xi. \qquad (6.175)$$

Since $C(t,s,\lambda)$ satisfies (6.1) and the initial conditions $C(t,t,\lambda) = 0$, $C'_t(t,s,\lambda)|_{t=s} = 1$, we can apply the estimate (6.172) to $C(t,s,\lambda)$. Then, solving the equation (6.175) by the method of successive approximations, we get the estimate (6.174) for $K(t,s,\lambda)$ and we also check the other assertions of the Lemma. This concludes the proof of the Lemma. □

To conclude the proof of the Theorem, we use the decomposition method (see [70] or [6, p. 31]) and some additional arguments. Consider the auxiliary operators L_1, L_2, L_3 determined by the equation (6.2) on the intervals $(-\infty,-M')$, $(-M',M)$, (M,∞) with the boundary conditions

$y(-M') = y(M) = 0$. The continuous spectrum of the operators L_1, L_3 is the same as in the problems (6.1) and (6.2). If an interval $\Delta = (\mu - \delta, \mu + \delta)$ on the λ-axis contains at most n points of the spectrum of L_1, L_2 and L_3 (counting multiplicities), then the problem (6.2) has at most $n+2$ eigenvalues in the interval Δ. (Otherwise, from the corresponding eigenfunctions of (6.2), it would be possible to construct a linear manifold of dimension greater than n of functions vanishing at $-M'$ and M such that

$$\int_{-\infty}^{\infty} |l[y] - \mu y|^2 \, dt < \delta^2 \int_{-\infty}^{\infty} |y|^2 \, dt.$$

This would mean that the operator $L \equiv L_1 \oplus L_2 \oplus L_3$ has more than n eigenvalues $\lambda_i \in \Delta$, contradicting the assumption.)

For each spectral gap, we choose its own splitting points $-M'$, M of the differential operator under consideration. To this end, using the asymptotics of the solutions $e_j(t,\lambda)$ in (6.170) and some additional facts about the motion of the roots of $e_j(t,\lambda)$ as λ changes, we will show that for every gap $(\lambda_{2k-1}, \lambda_{2k})$, there exist numbers ϵ_{jk}, $0 \le \epsilon_{jk} \le 1$, such that for some value of γ, fixed for all gaps, we have

$$|e_j(\epsilon_{jk}, \lambda)| \ge \gamma \cdot \max_{0 \le t \le 1} |e_j(t,\lambda)| > 0, \tag{6.176}$$

where $j = 1, 2$; $\lambda_{2k-1} \le \lambda \le \lambda_{2k}$; $k = 0, 1, 2 \ldots$

(For the semi-infinite gap $-\infty < \lambda \le \lambda_0$, i.e. for $k = 0$, we need to put $\epsilon_{1,0} = 0$, $\epsilon_{2,0} = 1$ in (6.176).) This and Lemma 6.11 show that for a fixed sufficiently large integer N, the functions $E_1(N + \epsilon_{1k}, \lambda)$ and $E_2(-N + \epsilon_{2k}, \lambda)$ do not have roots in the corresponding gaps $\lambda_{2k-1} \le \lambda \le \lambda_{2k}$, where $k = 1, 2, \ldots$, i.e. these gaps do not contain any eigenvalues of auxiliary problems for the intervals $(N + \epsilon_{1k}, \infty)$, $(-\infty, -N + \epsilon_{2k})$ on the t-axis. Indeed, by (6.173), (6.174) and (6.176), we have

$$\begin{aligned}
&|E_1(N + \epsilon_{1,k}, \lambda)| \ge |e_1(N + \epsilon_{1,k}, \lambda)| \\
&- \int_N^{\infty} |K(N + \epsilon_{1,k}, s, \lambda) p(s) e_1(s, \lambda)| \, ds \\
&\ge e^{-\alpha(\lambda) N} |e_1(\epsilon_{1,k}, \lambda)| \left\{ 1 - \frac{C e^{\alpha(\lambda)}}{\gamma(1 + |\sqrt{\lambda}|)} \int_N^{\infty} (1+s)|p(s)| \, ds \right\} > 0
\end{aligned}$$

for all finite gaps, provided that

$$\int_N^{\infty} (1+s)|p(s)| \, ds < \frac{\gamma}{C} \exp\left\{ - \max_{\lambda_1 \le \lambda < \infty} \alpha(\lambda) \right\},$$

where $C > 0$ is the same as in (6.174). Similarly, we consider $E_2(-N + \epsilon_{2,k}, \lambda)$. Finally, each gap will contain at most finite number of eigenvalues λ'_m of the regular problem for the equation (6.2) on the interval $(-N + \epsilon_{2k}, N + \epsilon_{1k})$. The length of the gaps tends to zero as $\lambda \to +\infty$; see [191]. Therefore, by carefully varying the numbers ϵ_{jk}, it is possible to fully free each of sufficiently distant gaps of the eigenvalues of the corresponding three auxiliary problems.[29] This will conclude the proof of the part of the theorem pertaining to the discrete levels in the spectral gaps.

Lemma 6.12 *Let $(\lambda_{2k-1}, \lambda_{2k})$ be a gap in $\sigma(L)$. The zeros $t'_{\nu,k}$ and $t''_{\nu,k}$ of the solutions $e_j(t, \lambda_{2k-1})$ and $e_j(t, \lambda_{2k})$ are intertwined and independent of $j = 1, 2$. Let*

$$\cdots < t'_{\nu,k} < t''_{\nu,k} < t'_{\nu+1,k} < t''_{\nu+1,k} < \cdots, \qquad (\nu = 0, \pm 1, \pm 2, \ldots).$$

As λ increases from λ_{2k-1} to λ_{2k}, all t-zeros of $e_1(t, \lambda)$ move strictly monotonely in one direction, while t-zeros of $e_2(t, \lambda)$ move in the opposite direction and alternate with the zeros of $e_1(t, \lambda)$. The increasing with respect to λ zeros run through the intervals $(t'_{\nu,k}, t''_{\nu,k})$, while decreasing with respect to λ zeros run through the intervals $(t''_{\nu,k}, t'_{\nu+1,k})$.

Proof of Lemma 6.12. No point $t_0 \in \mathbb{R}$ can simultaneously be a solution of $e_j(t, \lambda')$ and $e_n(t, \lambda'')$, where $\lambda_{2k-1} \le \lambda' \ne \lambda'' \le \lambda_{2k}$, $j, n = 1, 2$ ($j = n$ or $j \ne n$). Otherwise, if $e_j(t_0, \lambda') = e_n(t_0, \lambda'') = 0$, then λ' and λ'' are eigenvalues of the problem

$$y'' + [\lambda - q(t)]y = 0,\ y(t_0) = y(t_0 + 1) = 0,\ (t_0 \le t \le t_0 + 1), \qquad (6.177)$$

which is impossible because, $[\lambda_{2k-1}, \lambda_{2k}]$ contains (see [113] and [46]) exactly one eigenvalue of (6.177). As λ changes, all t-zeros of $e_j(t, \lambda)$ move in one direction due to (6.157), where $\chi_j(t + 1, \lambda) = \chi_j(t)$, and by the Sturm comparison theorem. Indeed, if, as λ increases, two t-zeros of $e_j(t, \lambda)$ move in the opposite direction, then among the t-zeros of $e_j(t, \lambda)$, there exists a pair of neighbors that moved in the opposite direction. This contradicts the Sturm comparison theorem, and the Lemma is proven. □

Now we can take ϵ_{1k} in (6.176) to be any number from $[0, 1)$ belonging to the middle third of the interval covered by the zeros of $e_2(t, \lambda)$, and, hence,

---

[29] We choose ϵ_{jk} in (6.176) with accuracy up to $\pm\frac{c}{\sqrt{\lambda}}$ for some $c > 0$ (as shown after Lemma 6.12). Due to the asymptotics $\lambda'_m = \left(\frac{m\pi}{l}\right)^2 + O(1)$, where $l = 2N + \epsilon_{1k} - \epsilon_{2k}$, we have the variation of eigenvalues $\lambda'_m \pm \frac{2\pi cm}{l^2} + O(1)$, allowing us to eliminate the eigenvalues from the distant gap $(\lambda_{2k-1}, \lambda_{2k})$.

not covered by the zeros of $e_1(t,\lambda)$ for $\lambda_{2k-1} \le \lambda \le \lambda_{2k}$. The existence of the common $\gamma > 0$ for all spectral gaps is guaranteed by the asymptotics

$$e_1(t,\lambda) = A_1(\lambda)[\sin(t\sqrt{\lambda} + \eta_1(\lambda)) + O(\lambda^{-1/2})],\ (\lambda \to +\infty) \tag{6.178}$$

and by the orthogonality (on the period) of $e_1(t,\lambda_{2k-1})$ and $e_1(t,\lambda_{2k})$ as eigenfunctions of either periodic or anti-periodic problem (6.1). Indeed, for γ from distant gaps, where the asymptotics (6.178) is sufficiently accurate, the zeros $t'_{\nu,k}$, $t''_{\nu,k} \in [0,1]$ of the solutions $e_1(t,\lambda_{2k-1})$ and $e_2(t,\lambda_{2k})$, respectively, are close to the zeros of the curves $\sin(t\sqrt{\lambda_{2k-1}} + \eta_1(\lambda_{2k-1}))$ and $\sin\left(t\sqrt{\lambda_{2k-1}} + \eta_1(\lambda_{2k-1}) \pm \frac{\pi}{2}\right)$ because $\lambda_{2k} - \lambda_{2k-1} \to 0$ as $k \to \infty$ and $\eta_1(\lambda_{2k}) \approx \eta_1(\lambda_{2k-1}) \pm \frac{\pi}{2}$ due to the orthogonality of solutions on the period and the asymptotics (6.157). Suppose, for definiteness, that the t-zeros of $e_2(t,\lambda)$ are increasing with respect to λ in the gap $\lambda_{2k-1} < \lambda < \lambda_{2k}$. Then by Lemma 6.12, the intervals $(t'_{\nu,k}, t''_{\nu,k})$, containing approximately one quarter of the sine wave in (6.178), do not contain any zeros of $e_1(t,\lambda)$. If $\epsilon_{1k} \in [0,1]$ is chosen to belong to the middle third of the interval $(t'_{\nu,k}, t''_{\nu,k})$ for some ν, then the distance between ϵ_{1k} and the closest t-zeros of $e_1(t,\lambda)$, for all $\lambda \in [\lambda_{2k-1}, \lambda_{2k}]$, is at least $1/12$ of the length of the sine wave in (6.178). Hence, for $e_1(\epsilon_{1k},\lambda)$ in (6.176), for large k, we can take $\gamma \approx \sin(\pi/6) = 1/2$ or, with the superfluous guarantee, $\gamma = 1/4$.

As for the finite number M of gaps with indices k for which the asymptotics in (6.178) is insufficiently accurate to use $\gamma = 1/4$ in (6.176), for each of those gaps, for all $j = 1,2$, there exists a value $\gamma = \gamma_{jk} > 0$ such that (6.176) holds. (Note that for $\lambda \in (-\infty, \lambda_0]$, the solutions $e_j(t,\lambda)$ do not oscillate and do not have zeros on the t-axis.) Finally, we can chose γ in (6.176) for all j and k as follows:

$$\gamma = \min_{\substack{0 \le k \le M \\ j = 1,2}} \left\{\gamma_{jk}, \frac{1}{4}\right\} > 0.$$

The inequality (6.176) is proven. The absence of eigenvalues in the continuous spectrum follows from the asymptotics of solutions of the perturbed problem; see (6.173), [769] and [780]. This concludes the proof of the Theorem. □

Theorem 6.13 allows the appearance of at most two eigenvalues in each of the distant spectral gaps. The following theorem addresses the question whether there exist discrete levels in the distant[30] spectral gaps of the

[30] We emphasize that we deal with non-degenerate spectral gaps, i.e. those that have not degenerated to a point. Degenerate spectral gaps, strictly speaking, are not spectral

perturbed problem (6.2).

Theorem 6.14 *Under the hypotheses of Theorem 6.13, let*

$$\int_{-\infty}^{\infty} p(t)\,dt \neq 0. \tag{6.179}$$

Then in every sufficiently distant spectral gap of the perturbed Hill operator, there exists exactly one eigenvalue.

Proof. The eigenvalues of (6.2) are the zeros of the Wronski determinant

$$W(\lambda) = E_1(t,\lambda)E'_2(t,\lambda) - E'_1(t,\lambda)E_2(t,\lambda). \tag{6.180}$$

When $\phi(1,\lambda) = 0$, then one of the unperturbed solutions $e_j(t,\lambda)$, as well as $E_j(t,\lambda)$, where $j = 1, 2$, is identically equal to 0 for all t, (and, instead, we should take solutions with different normalization condition), and, hence, as easily seen, the eigenvalues of the problem (6.2) in the gap $(\lambda_{2k-1}, \lambda_{2k})$, where $k = 0, 1, 2, \ldots$, are precisely those values of λ such that

$$\widetilde{W} := \frac{1}{\phi(1,\lambda)} W(\lambda) = 0, \quad \lambda \in (\lambda_{2k-1}, \lambda_{2k}). \tag{6.181}$$

Here we assume that $\phi(1, \lambda_j) \neq 0$, where λ_j is an arbitrary point in $(\lambda_{2k-1}, \lambda_{2k})$. This can always be achieved by the appropriate choice of the origin of the coordinate system on the t-axis for all spectral gaps. It suffices to take the origin to be a point on the real t-axis not belonging to the countable set of t-zeros of the solutions $e(t, \lambda_j)$, $(j = 0, 1, 2, \ldots)$[31]. From [46, Chapter VIII, Sec. 3] (see also [936]), it follows that in each spectral gap there exists precisely one simple[32] zero of $\phi(1,\lambda)$. Hence

$$\text{sign}\ \phi(1, \lambda_{2k-1}) = -\text{sign}\ \phi(1, \lambda_{2k}). \tag{6.182}$$

By Lemma 6.11 we have

$$K_t{}'(t,s,\lambda)E_j(t,\lambda) - K_t(t,s,\lambda)E_j{}'(t,\lambda) = E_j(s,\lambda),\ (j = 1, 2). \tag{6.183}$$

The left hand side of (6.183) is a Wronskian, and, hence, is independent of t. Therefore, by (6.173) it follows that

$$(e_1E_2{}' - e_1{}'E_2)(t,\lambda) = w(\lambda) + \int_{-\infty}^{t} p(s)e_1(s,\lambda)E_2(s,\lambda)\,ds, \tag{6.184}$$

gaps; they belong to the essential spectrum.

[31] In [934] it was shown that it is possible to do this.

[32] Since $\frac{d}{dt}(\phi' \cdot \dot{\phi} - \dot{\phi}' \cdot \phi) = \phi^2$, we have $\dot{\phi}(1,\lambda) = \int_0^1 \phi^2(t,\lambda)\,dt$ when $\phi(1,\lambda) = 0$. Thus, in this case, $\dot{\phi}(1,\lambda) \neq 0$.

where

$$w(\lambda) = (e_1 e_2{}' - e_1{}' e_2)(t,\lambda); \quad w(\lambda_k) = 0, \quad k = 0,1,2,\dots \tag{6.185}$$

This and (6.180) show that

$$\begin{aligned} W(\lambda) &= (e_1 E_2{}' - e_1{}' E_2)(t,\lambda) - \int_t^{\infty} [K(t,s,\lambda)E_2{}'(t,\lambda) \\ &\quad - K_t{}'(t,s,\lambda)E_2(t,\lambda)]p(s)e_1(s,\lambda)\,ds \\ &= w(\lambda) + \int_{-\infty}^{\infty} p(s)e_1(s,\lambda)E_2(s,\lambda)\,ds \equiv w(\lambda) + I(\lambda), \end{aligned} \tag{6.186}$$

where

$$\begin{aligned} I(\lambda) &= \int_{-\infty}^{\infty} p(t)e_1(t,\lambda)e_2(t,\lambda)\,dt \\ &\quad + \int_{-\infty}^{\infty} p(t)e_1(t,\lambda)\,dt \int_{-\infty}^{t} K(t,s,\lambda)p(s)e_2(s,\lambda)\,ds \\ &\equiv I_1(\lambda) + I_2(\lambda). \end{aligned} \tag{6.187}$$

Using the asymptotics (6.178) of solutions as $\lambda_k \to \infty$, Riemann–Lebesgue theorem, the estimate (6.174) for $K(t,s,\lambda)$, and the relation

$$e_1(t,\lambda_k)e_2(t,\lambda_k) = \frac{1}{2}A_1^2(\lambda_k)\left\{1 - \cos(2(t\sqrt{\lambda_k} + \eta_1(\lambda_k)))\right\} + O(\lambda_k{}^{-1/2}),$$

we have

$$I_1(\lambda_k) = \left\{\max_t |e_1(t,\lambda_k)|^2\right\}\left\{\frac{1}{2}\int_{-\infty}^{\infty} p(t)\,dt + o(1)\right\},$$

$$\begin{aligned} |I_2(\lambda_k)| \le \frac{C}{\sqrt{|\lambda_k|}} &\left\{\max_t |e_1(t,\lambda_k)|^2\right\}\left\{\int_{-\infty}^{\infty} |p(t)|\,dt\right\} \\ &\times \left\{\int_{-\infty}^{\infty} (1+|t|)|p(t)|\,dt\right\}. \end{aligned}$$

Hence, for sufficiently large λ_k, using (6.179), (6.185)–(6.187) and (6.181)–(6.182), we have:

$$\operatorname{sign} \widetilde{W}(\lambda_{2k-1}) = -\operatorname{sign} \widetilde{W}(\lambda_{2k}), \quad (\lambda_k \to \infty). \tag{6.188}$$

Thus, there are eigenvalues in each spectral gap (but no more than two by Theorem 6.13). We will show that the function $\widetilde{W}(\lambda)$, regular in each spectral gap, can have only simple zeros in the spectral gaps. By standard arguments, it is easy to show[33] that

$$E_1E_2 = \frac{d}{dt}\left(E_1'\dot{E}_2 - E_1\dot{E}_2'\right) = \frac{d}{dt}\left(E_2'\dot{E}_1 - E_2\dot{E}_1'\right), \qquad (6.189)$$

where the *dot* denotes the differentiation with respect to λ. Hence, if $\widetilde{W}(\widetilde{\lambda}) = 0$, i.e. $\widetilde{\lambda}$ is an eigenvalue and the solutions $E_1(t, \widetilde{\lambda})$, $E_2(t, \widetilde{\lambda})$ are linearly dependent, we get:

$$\int_{-\infty}^{0} + \int_{0}^{\infty} (E_1E_2)(t, \widetilde{\lambda})\, dt = \left(E_1'\dot{E}_2 - E_1\dot{E}_2'\right)(0, \widetilde{\lambda}) - \left(E_2'\dot{E}_1 - E_2\dot{E}_1'\right)(0, \widetilde{\lambda}),$$

i.e. $\int_{-\infty}^{\infty}(E_1E_2)(t, \widetilde{\lambda})\, dt = -\dot{W}(\lambda)$, or, if we change the normalization condition, which does not influence the validity of

$$\dot{\widetilde{W}}(\widetilde{\lambda}) = -\int_{-\infty}^{\infty} \left\{\frac{(E_1E_2)(t, \lambda)}{\phi(1, \lambda)}\right\}_{\lambda = \widetilde{\lambda}} dt \neq 0.$$

Therefore, the zeros of $\widetilde{W}(\lambda)$ are simple. This concludes the proof of the Theorem. □

Remark 6.9 If, under the hypotheses of the Theorem 6.14, $p(t)$ does not change its sign, and the integral (6.168) is sufficiently small, then in every finite spectral gap, there exists precisely one eigenvalue.

6.6 Bibliographical Comments

The main results of this chapter are due to F. S. Rofe-Beketov; see [769], [780], [781], [785], [786], [790] and [791].

6.6.1 *Some Comments on Section 6.1*

Regarding Sec. 6.1, the proofs of Lemma 6.1 and Lemma 6.2 use the well-known technique developed in [13], [632], [794] and [922]; however, this technique is used in a different situation—for a factorized matrix or for one of the matrix factors entering the factorized matrix. In the scalar case, this approach was suggested in [781] (see also Sec. 6.3 of this book). In the

[33] Indeed, differentiating parentheses in (6.189) with respect to t and replacing the second derivatives with respect to t by using $-E''_j + (q+p)E_j = \lambda E_j$ and $-\dot{E}_j'' + (q+p)\dot{E}_j = \lambda\dot{E}_j + E_j$, we get the desired equalities (6.189).

proof of Theorem 6.1, we use the oscillatory theorem for a matrix Sturm–Liouville problem, proven by the authors in [794] (see also Chapter 1 of this book).

6.6.2 *Some Comments on Section 6.2*

Regarding Sec. 6.2, the transformation of matrix solutions of canonical systems and systems of Sturm–Liouville-type equations was introduced in [790], and for a scalar Sturm–Liouville equation—in [781] and [786]. Since those transformations map one solution of an equation or system into another solution of the same system, they proved to be useful in the questions treated in this book. (Recently, the corresponding formulas were generalized for some new classes of differential systems; see [811].)

Darboux transforms [349] and their generalizations (see [133], [304], [657]), mapping solutions of equations into solutions of other equations or systems, play an important role in questions of integration of non-linear evolution equations.

Darboux transforms represent one of two well-known methods for getting explicit solutions of both linear equations (Schrödinger-type equations with a spectral parameter) and non-linear (evolution) equations. While the inverse problem method reduces the question of solving a non-linear evolution equation to the question of solving a linear integral equation of the inverse problem, which is explicitly solvable only in special cases, Darboux transforms enable us to explicitly get a certain class of solutions of non-linear equations, including solitons, bypassing integral equations.

Note that Darboux transform depends on two numerical parameters, and its reiteration depends on any *finite* number of parameters. An analogue of Darboux transform depending on a functional parameter (in other words, continuum of parameters) was constructed in [304] and [657]. Besides, in those works, a generalization of Darboux transform was constructed for equations in operator-valued functions with operator coefficients of arbitrary order. It is technically convenient that Darboux transform is based on taking derivatives. The above mentioned works [304] and [657] show why it is plausible to consider analogous equations and transforms with respect to generalized derivatives. All these questions are, in essence, of algebraic nature; here, only algebraic structure is important (see [122]). We will also mention the following works on related topics: [580], [826], [827] and [829]. Transformation operators of special type for matrix Sturm–Liouville equations, representing the solutions of such systems for all λ in terms of linear combinations of solutions (and their derivatives) of systems with another matrix potential with the same λ, are studied in [115, Chapter VI]. The coefficients of these linear combinations are fractional linear functions of λ. The paper [898] is devoted to application of the Darboux transform to the investigation of cosmological models.

An estimate of the Cauchy function of a self-adjoint scalar equation (6.82) with a finite-zone periodic potential (a uniform estimate in λ on the spectrum) was obtained in [629]. In a much simpler case of the Hill operator, this estimate was obtained in [769]. The formulas of type (6.69) and (6.77) for the scalar equation (6.82) were announced in [786].

6.6.3 *Some Comments on Section 6.3 and Differential Inequalities*

Section 6.3 is based on [780], [781], [785] and [786].

In [157, p. 54, Exercise 7], a theorem about differential inequalities (see also [41], [42] and [743]) was stated in the form used in this book.

Theorem 6.15 (about differential inequalities) *Assume that for all* $t > t_0$,

$$\begin{aligned} y'(t) &= f(t,y), \\ u'(t) &> f(t,u), \quad v'(t) \geq f(t,v), \end{aligned} \tag{6.190}$$

where y, u, v *are given continuously differentiable functions,* $y(t_0) = u(t_0) = v(t_0)$, *and the function* $f(t,y)$ *is continuous. Then for all* $t > t_0$, *we have* $u(t) > y(t)$. *If, in addition, through every point* $(t, y(t))$, $t \geq t_0$, *passes a unique integral curve* $y = y(t)$ *of the equation (6.190), then* $v(t) \geq y(t)$ *for all* $t \geq t_0$.

This theorem is the simplest case of general investigations on differential inequalities of arbitrary order due to S. A. Chaplygin [41] and [42]; see also M. Petrovisch [743] and A. D. Myshkis [702].

In order to apply the above formulated theorem to equations of the form (6.99) with $p(t) \in L_{1,\mathrm{loc}}$ and $\pm p^{\pm}(t)$, we approximate $p(t)$ in L_1 by a sequence of continuous functions $p_n(t)$, and $\pm p^{\pm}(t)$ by sequences $\pm p_n^{\pm}(t)$ respectively. Using the theorem about differential inequalities, we obtain the following inequality for the solutions γ_n, $\gamma_n^{\vee}$, $\gamma_n^{\wedge}$ of the corresponding equations with $t \geq t_0$ and with the initial data $\gamma_n(t_0) = \gamma_n^{\vee}(t_0) = \gamma_n^{\wedge}(t_0) =$

$\gamma(t_0)$:

$$\gamma_n^{\wedge}(t) \le \gamma(t) \le \gamma_n^{\vee}(t_0).$$

Letting $n \to \infty$, we get

$$\gamma^{\wedge}(t) \le \gamma(t) \le \gamma^{\vee}(t_0).$$

(It is easy to see that using the method of successive approximations for all t_0, $\gamma(t_0) \in \mathbb{R}$, the equation (6.99) has a unique solution on the real line.)

Note that Chaplygin's theorem about differential inequalities inspired many investigations and generalizations and generated a broad literature on the topic. In addition to the works mentioned in the bibliographical comments in [41] and in [42], we also mention [20], [101], [127], [638], [764] and references there.

Note, by the way, that the spectrum of second order nonlinear boundary problems on a finite interval with separated boundary conditions may behave completely differently from the spectrum of the linear regular Sturm–Liouville problem.

Example 6.2 As shown in [764], there exists a positive number $32/(9e) \le \lambda_0 \le 32/9$ for the non-linear boundary problem

$$y'' y^{1/2} = \lambda e^x, \qquad y(0) = y(1) = 1, \tag{6.191}$$

such that for $\lambda \le \lambda_0$ the problem (6.191) is solvable, but for $\lambda > \lambda_0$ this problem is not solvable.

In E. Kamke's handbook on differential equations [86, Equation 6.242], it is noted that our equation (6.191) is the equation of the spatial current of displacement in a cylindrical condenser.

On the other hand, there are many non-linear problems whose spectrum and eigenfunctions are in many aspects analogous to the ones for the linear case. For example, a recent paper [644] constructs a system of eigenfunctions of a non-linear Sturm–Liouville problem forming a basis in $L_p(0,1)$.

6.6.4 *The Spectrum of Self-Adjoint Periodic and Almost Periodic Operators. Marchenko–Ostrovskiy and Pastur–Tkachenko Theorems.*

Regarding Sec. 6.4, in the case of a real periodic potential $q(t+\omega) = q(t) \in L_{1,\mathrm{loc}}$, it is known (see [60] and [191, Volume II]) that the spectrum of the

Hill operator is absolutely continuous and consists of a sequence of zones $[\lambda_{2k}, \lambda_{2k+1}] \subset \mathbb{R}$ extending to $+\infty$ and separated by the gaps $(\lambda_{2k-1}, \lambda_{2k})$:

$$-\infty < \lambda_0 < \lambda_1 \leq \lambda_2 < \lambda_3 \leq \lambda_4 < \cdots \tag{6.192}$$

If $\lambda_{2k-1} = \lambda_{2k}$, the corresponding gap is degenerate. The points λ_{4k-1} and λ_{4k} form the spectrum of the periodic problem: $u(\omega, \lambda) = u(0, \lambda)$, $u'(\omega, \lambda) = u'(0, \lambda)$, while the points λ_{4k+1} and λ_{4k+2} form the spectrum of the anti-periodic problem: $u(\omega, \lambda) = -u(0, \lambda)$, $u'(\omega, \lambda) = -u'(0, \lambda)$ for the equation (6.1) on the period ω.

Full characterization of the spectrum of the Hill operator in terms of free parameters is given by Marchenko–Ostrovskiy theorem (see [121] and [658]).

Theorem 6.16 (V. A. Marchenko–I. V. Ostrovskiy [121], [658]) *The sequence (6.192) corresponds (in the sense indicated above) to the spectrum of the Hill operator (6.1) with a real-valued potential* $q(t + \omega) = q(t) \in W^n_{2,loc}$ *(Sobolev class) if and only if there exists a sequence* $h_k \geq 0$, $k = 1, 2, \ldots$ *such that*

$$\sum_{k=1}^{\infty} \left(k^{n+1} h_k\right)^2 < \infty,$$

and

$$\sqrt{\lambda_{2k-1} - \lambda_0} = z(\pi k - 0), \quad \sqrt{\lambda_{2k} - \lambda_0} = z(\pi k + 0),$$

where $z(\zeta) = z(\zeta; \omega)$ *conformally maps the domain ("comb")*

$$\Pi_+ := \{\zeta\colon Im\, \zeta > 0\} \setminus \bigcup_{k=1}^{\infty} \{\zeta\colon Re\, \zeta = \pm k\pi,\, 0 \leq Im\, \zeta \leq h_k\} \tag{6.193}$$

onto the upper half-plane such that $z(0) = 0$, $z(i\eta; \omega)(i\eta)^{-1} \to \omega^{-1}$ *as* $\eta \to +\infty$.

It is well-known that in the closure of each finite spectral gap $[\lambda_{2k-1}, \lambda_{2k}]$ of the Hill operator (including degenerate gaps), there is precisely one eigenvalue ν_k of the Dirichlet problem for the same equation (6.1) on the period:

$$u(0, \nu_k) = u(\omega, \nu_k) = 0. \tag{6.194}$$

To find all Hill equations (6.1) with the prescribed spectrum (6.192), one must (in accordance with V. Marchenko and I. Ostrovskiy [658], [659] and the book [121]; see also [112]) add to the fixed data (6.192) all possible sequences $\{\nu_k\}_{k=1}^{\infty}$, where $\nu_k \in [\lambda_{2k-1}, \lambda_{2k}]$ are arbitrary, and all possible sequences of signs + or − for $k = 1, 2, \dots$ (except for degenerate gaps). Each pair of the above-mentioned sequences together with the spectrum (6.192) defines uniquely the potential $q(t)$, and, therefore, the Hill operator (6.1). In other words, V. Marchenko–I. Ostrovskiy's works [658], [659] (and [121]) contained both full characterization of the spectrum of Hill operators (see their Theorem above) and the comprehensive and effective solution of the inverse spectral problem, including necessary and sufficient conditions and a complete set of independent spectral data that determine potential uniquely. This set can be presented as a sequence of complex numbers $\theta = \{\theta_k\}_{k=0}^{\infty}$ uniquely determined by the following conditions:

$$\begin{aligned}
&\theta_0 = \lambda_0, \\
&|\theta_k| \equiv h_k = \text{Im}\ \zeta(\gamma_k; \omega), \\
&\text{Im}\ \theta_k = (-1)^k \cdot \text{Im}\ \zeta(\beta_k; \omega) \cdot \text{sign}\ G(\nu_k; \omega), \\
&\text{sign Re}\ \theta_k = \text{sign}\ (\beta_k - \gamma_k), \qquad k = 1, 2, \dots
\end{aligned} \tag{6.195}$$

Here, $\beta_k = (\nu_k - \lambda_0)^{1/2} > 0$, and $\{\gamma_k\}_{k=1}^{\infty}$ is the sequence of positive roots of the equation

$$\dot{F}(\gamma^2 + \lambda_0; \omega) = 0,$$

$$G(\lambda; \omega) := \frac{1}{2}(\theta(\omega, \lambda) - \phi'(\omega, \lambda)), \tag{6.196}$$

and $\{\nu_k\}_{k=1}^{\infty}$ are determined by (6.194). ([659]; see also [730].)

The function $\zeta(z; \omega)$ maps the upper half-plane $C_+ = \{z : \text{Im}\ z > 0\}$ conformally onto the domain Π_+ (6.193). Therefore, $\zeta(z; \omega)$ is a function of Nevanlinna–Herglotz type and, simultaneously, it is the inverse function of $z(\zeta) = z(\zeta; \omega)$ from the Marchenko–Ostrovskiy Theorem.

It is easy to see that $z(\zeta; \omega) = \omega^{-1} z(\zeta; 1)$.

If $q \in L_{2,\text{loc}}$ then, in accordance with the Marchenko–Ostrovskiy Theorem, the sequence θ belongs to the space w_1^2 of complex-valued sequences with the norm

$$\|\theta\|_\omega = |\theta_0| + \left(\frac{\pi}{\omega}\right)^2 \left(\sum_{k=1}^{\infty} k^2 |\theta_k|^2\right)^{1/2}.$$

The spectrum σ of the Hill operator with potential $q(t+\omega) = q(t)$ belonging to $L_{2,\text{loc}}$, which is uniquely determined by each sequence θ belonging to w_1^2, is the following set:

$$\sigma = \{\lambda \geq \lambda_0 : \text{Im}\, \zeta((\lambda - \lambda_0)^{1/2}; \omega) = 0\}. \tag{6.197}$$

An analogue of Marchenko–Ostrovskiy Theorem for a periodic Dirac operator was obtained by T. V. Misyura in [686]. The asymptotics and some other properties of the spectrum of operators of this type were considered in [687] and [688] by T. V. Misyura and in more recent works by P. Djakov and B. S. Mityagin—[372], [375], [376] and [380]. Convergence of spectral expansions of periodic Dirac operators was investigated by B. S. Mityagin in [689] and [690].

In the case of an almost periodic potential, the spectra of operators of the form (6.1) exhibit a much greater variety than in the periodic case. The spectra can fail to be absolutely continuous, can be pure point spectra [695], can be concentrated on Cantor-type sets [630] remaining absolutely continuous, *etc.* In the case of an almost periodic potential, the absence of *isolated* eigenvalues and the semi-boundedness below are preserved.

Among differential operators with almost periodic coefficients, spectral properties of operators with almost periodic zone-type potentials, investigated in the monograph by B. M. Levitan [112] and in the papers [623]–[625], are more analogous to spectral properties of periodic differential equations than those of almost periodic differential equations of the general form. The description of zone-type potentials after B. M. Levitan is given in the formulation of Lemma 6.7 in the present book.

In [729] and [730], L. A. Pastur and V. A. Tkachenko developed the spectral theory for a one-dimensional Schrödinger operator with a potential from the space of almost periodic Stepanov functions, which can be approximated in the metric of this space by periodic functions with a period of infinite growth, so that the speed of approximation is greater than exponential. L. A. Pastur and V. A. Tkachenko also gave a description, similar to the one in Marchenko–Ostrovskiy Theorem, of independent spectral data for reconstructing the potential uniquely. For limit-periodic potentials [729], [730], the spectrum is absolutely continuous, nowhere dense, and is of multiplicity two, and generalized eigenfunctions are of "quasi-Bloch" (in other words, Floquet) type—similar to the periodic case [446].

Below we give precise formulation of L. A. Pastur–V. A. Tkachenko results, following their works [729], [730] and [153].

Following L. A. Pastur and V. A, Tkachenko, let us consider the space

S^2 of Stepanov almost periodic functions with the norm

$$\|f\|_{S^2} = \sup_x \left(\int_x^{x+1} |f(t)|^2 \, dt \right)^{1/2} \tag{6.198}$$

and denote by $A = \{a_n\}_{n=1}^{\infty}$ a (strictly) increasing sequence of positive numbers such that a_{n+1}/a_n are integers. Let R_n denote the set of all real numbers of the form $r = \pi k / a_n$ for $\pm k = 0, 1, 2, ...$, and let $R(A) = \bigcup_{n=1}^{\infty} R_n$. (It is easy to see that R_n is contained in R_{n+1} for each n.) It is known (see the book [111]) that to each element q of S^2 there corresponds a countable set of Fourier coefficients $\{\hat{q}_r\}$, $r \in \mathbb{R}$, and for this set, the norm

$$\left(\sum_{r \in \mathbb{R}} |\hat{q}_r|^2 \right)^{1/2} < \infty.$$

The Pastur–Tkachenko class $Q_\infty(A)$ is now defined as a subset of the set of all functions $q \in S^2$ for which all non-zero coefficients $\hat{q}_r$ belong to $R(A)$ and satisfy the condition

$$\lim_{n \to \infty} \left(\exp(b a_{n+1}) \sum_{r \in R(A) \setminus R_n} |\hat{q}_r|^2 \right) = 0, \qquad \text{for all } b > 0. \tag{6.199}$$

To every element $q \in Q_\infty(A)$ corresponds a sequence of a_n-periodic functions

$$q_n(x) = \sum_{r \in R_n} \hat{q}_r \cdot \exp(2irx),$$

and this sequence converges in S^2 to $q(x)$ with the rate corresponding to (6.199), i.e.

$$\lim_{n \to \infty} (\exp(b a_{n+1}) \|q - q_n\|_{S^2}) = 0, \qquad \text{for all } b > 0. \tag{6.200}$$

We now introduce the space $K(A)$ of complex-valued functions $\kappa(r) := \kappa_r$, where $r \in R(A)$, with the norm of w_1^2-type:

$$\|\kappa\|_{K(A)} := \sup_{n \geq 1} \|\kappa\|_{w_1^2(n)} = \lim_{n \to \infty} \|\kappa\|_{w_1^2(n)}, \tag{6.201}$$

where

$$\|\kappa\|_{w_1^2(n)} := |\kappa_0| + \left(\sum_{r\in R_n} r^2|\kappa_r|^2\right)^{1/2}.$$

Following [730], let us select a subset $K_\infty(A)$ of $K(A)$ consisting of functions κ_r such that $\operatorname{Im}\kappa_0 = 0$, $\kappa_r = \kappa_{-r}$, and

$$\lim_{n\to\infty}\left(\exp(ba_{n+1})\sum_{r\in R(A)\setminus R_n} r^2|\kappa_r|^2\right) = 0, \qquad \text{for all } b > 0. \tag{6.202}$$

Let, for some fixed potential $q\in Q_\infty(A)$, a sequence of a_n-periodic functions $\{q_n(x)\}_{n=1}^\infty$ satisfy (6.200), and let $\{\theta_k^{(n)}\}_{k=0}^\infty$ be the complete set of spectral data for the Hill equation with the potential $q_n(x)$ in the sense of Marchenko and Ostrovskiy determined by (6.195) and (6.196). Then the function $\kappa^{(n)}$, which is given by the following equalities, belongs to $K_\infty(A)$:

$$\kappa_r^{(n)} = \begin{cases} \theta_0^{(n)} & \text{when } r = 0, \\ (\pi/a_n)\theta_{|k|}^{(n)} & \text{when } r = (\pi/a_n)k\in R_n,\ \pm k = 1,2,..., \\ 0 & \text{when } r\notin R_n. \end{cases} \tag{6.203}$$

The first Pastur–Tkachenko Theorem gives a parametric description of the class of potentials $Q_\infty(A)$ in terms of the spectrum, analogous to the description known in the periodic case (see Marchenko–Ostrovskiy Theorem above, their works [658], [659], and the book [121]), and, in particular, it solves the inverse problem of spectral analysis for the class $Q_\infty(A)$.

Theorem 6.17 (L. A. Pastur–V. A. Tkachenko [729], [730] and [153]) *The sequence of functions $\kappa^{(n)} = \{\kappa_r^{(n)}\}\in K_\infty(A)$, arising from a potential $q\in Q_\infty(A)$, has a limit κ in the metric (6.201) for $K(A)$, and this limit belongs to $K_\infty(A)$.*

Such mappings of potentials $q\in Q_\infty(A)$ to functions $\kappa\in K_\infty(A)$ give rise to a one-to-one mapping T_∞ of $Q_\infty(A)$ onto $K_\infty(A)$. (The map T_∞ turns into a homeomorphism between the spaces $K_\infty(A)$ and $Q_\infty(A)$ when they are made into countably normed spaces in a suitable way.)

Thus, the function $\kappa\in K_\infty(A)$ represents a complete set of independent spectral data corresponding to the limit-periodic potential $q\in Q_\infty(A)$.

The parametric description of the Hill operators, as we have seen above, is closely related to the Nevanlinna–Herglotz function $\zeta(z;\omega)$ (that coincides with a quasi-impulse). An analogous analytic object that plays an important role in spectral theory also exists in the Pastur–Tkachenko theory of the limit-periodic Schrödinger operators:

Theorem 6.18 (L. A. Pastur–V. A. Tkachenko [729], [730] and [153]) *If $q \in Q_\infty(A)$, $q_n(x)$ are a_n-periodic potentials satisfying (6.200) and $\zeta_n(\mu; a_n)$ is a sequence of Nevanlinna–Herglotz functions arising from the Hill operators with potentials q_n, then the sequence $\kappa_n(\mu) = (\pi/a_n)\zeta_n(\mu; a_n)$ converges uniformly in the closed half-plane $\mathbb{C}_+$ to a Nevanlinna–Herglotz function $\kappa(\mu)$. The limit function $\kappa(\mu)$ maps $\mathbb{C}_+$ conformally onto the domain*

$$\Pi^+_\infty(A) = \mathbb{C}_+ \setminus \bigcup_{r \in R(A)\setminus\{0\}} \{\mu\colon Re\ \mu = r,\ 0 \le Im\ \mu \le |\kappa_r|\},$$

in such a way that $\kappa(0) = 0$ and

$$\lim_{\eta \to +\infty} ((i\eta)^{-1}\kappa(i\eta)) = \pi;$$

it is absolutely continuous on the real axis and has derivative $\kappa' \in L_{p,loc}(\mathbb{R})$, where $p \in (1, 2)$.

Theorem 6.19 (L. A. Pastur–V. A. Tkachenko [729], [730] and [153]) *A necessary and sufficient condition for a closed set $\sigma \subset \mathbb{R}$ to be the spectrum of a Schrödinger operator of the form (6.1) with the limit-periodic potential $q \in Q_\infty(A)$ is that this set can be represented in the form*

$$\sigma = \{\lambda \ge \lambda_0 : Im\ \kappa((\lambda - \lambda_0)^{1/2}) = 0\}, \tag{6.204}$$

where λ_0 is a real number, and the function $\kappa(\mu)$ gives a conformal mapping of the upper half-plane $\mathbb{C}_+$ onto the domain $\Pi^+_\infty(A)$. The function $\kappa \in K_\infty(A)$ is uniquely determined by the conditions $\kappa(0) = 0$ and

$$\lim_{\eta \to +\infty} (i\eta)^{-1}\kappa(i\eta) = \pi.$$

Theorem 6.19 enables us to construct limit-periodic Schrödinger operators whose spectra have prescribed geometric properties.

For example, if a function $\kappa \in K_\infty(A)$ has nonzero values at all points of the set $R(A) \cap \Delta$, where Δ is an interval, then the corresponding Schrödinger operator L_q (i.e. operator of the form (6.1) with potential q, corresponding to the function κ) has a Cantor (or Cantor-type, i.e. nowhere dense) spectrum on the interval

$$\kappa^{-1}(\Delta)^2 + \lambda_0.$$

Examples of such potentials were given by J. Moser [699] and J. Avron and B. Simon [245]. In addition, to the set of potentials $q \in Q_\infty(A)$ with a Cantor spectrum there corresponds the set of precisely those functions $\kappa \in K_\infty(A)$ that have a dense support in $R(A)$. This set of functions κ is a dense G_δ-set in $K_\infty(A)$.

As noted in [730], the function $\pi^{-1}\text{Re}\ \kappa((\lambda-\lambda_0)^{1/2})$ coincides with the normalized distribution function $N(\lambda)$ of eigenvalues of L_q, which can be defined for any almost periodic potential and even for a random metrically transitive potential $q(x)$. Hence, the spectrum of L_q can be identified, as shown in the work by L. A. Pastur [728], with the set of points of growth of Re $\kappa((\lambda-\lambda_0)^{1/2})$. Also, the values of $\pi N(\lambda)$ at the lacunae of the spectrum are constant and belong to $R(A)$. This fact is consistent with the theorem of Johnson and Moser [524], according to which these values for an arbitrary Bohr almost periodic potential belong to the module of its frequencies[34]. In our case, the set of values of $N(\lambda)$ at the lacunae coincides with the support of the function $\kappa \in K_\infty(A)$ parametrizing the potential $q(x)$.

Finally, the quantity Im $\kappa((\lambda-\lambda_0)^{1/2})$ is the leading Lyapunov index $\gamma(\lambda)$ of the equation $l_q[y]=\lambda y$ (of the form (6.1)), and, as noted in [730], in accordance with the general result for metrically transitive potentials [576], the fact that $\gamma(\lambda)=0$ for $\lambda \in \sigma$ means that any set of positive Lebesgue measure in the spectrum contains an absolutely continuous component. However, in the Pastur–Tkachenko case a stronger assertion holds:

Theorem 6.20 **(L. A. Pastur–V. A. Tkachenko [729], [730] and [153])** *The spectrum of the operator L_q (of the form (6.1) on the real axis) with a limit-periodic potential $q \in Q_\infty(A)$ is absolutely continuous and has multiplicity two.*

The above formulated Theorem shows that some important properties of the spectrum of the operator (6.1) with the limit periodic potential q of Pastur–Tkachenko class $Q_\infty(A)$ coincide with the analogous properties of the Hill equation. This Theorem follows from the following theorem which is due to the same authors and states the existence of Floquet–Bloch type solutions of the considered equations. These solutions are also analogous to solutions of the corresponding type of the Hill equation (6.157).

Theorem 6.21 **(L. A. Pastur–V. A. Tkachenko [729], [730] and [153])** *If $q \in Q_\infty(A)$, then for almost all (with respect to Lebesgue measure)*

[34] Some other results of [524] are quoted in our book; see Sec. 6.4, formulas (6.160)–(6.162) and also two footnotes there.

values $\lambda \in \sigma(L_q) \in \mathbb{R}$, *where* $\lambda = \kappa_0 + \mu^2$, *the equation (6.1) has linearly independent solutions* $\psi^{\pm}(x,\mu)$, $\psi^{+} = \overline{\psi^{-}}$, *of the form*

$$\psi^{\pm}(x,\mu) = \exp(\pm i\kappa(\mu)x)\xi^{\pm}(x,\mu),$$

where $\xi^{\pm}(\cdot,\mu) \in S^2$, *and there exists a sequence* $\xi_n^{\pm}(x,\mu)$ *of* a_n*-periodic functions such that for any* $\Delta \subset \mathbb{R}_+$ *and any* $b > 0$,

$$\lim_{n\to\infty} \exp(ba_{n+1}) \int_{\Delta} \|\xi^{\pm}(\cdot,\mu) - \xi_n^{\pm}\|_{S^2}^{p}\, d\mu = 0 \tag{6.205}$$

for any $p \in (1,2)$.

The set of functions $\psi^{\pm}(x,(\lambda-\kappa_0)^{1/2})$ *for Lebesgue- almost all values of* λ *belonging to the spectrum* σ *of* L_q *forms a complete system of generalized eigenfunctions for it, so that the corresponding resolution of the Dirac* δ*-function has the form*

$$\delta(x-t) = (2\pi)^{-1} \int_{\sigma} Re\, \left(\psi^{+}(x,(\lambda-\kappa_0)^{1/2})\psi^{-}(t,(\lambda-\kappa_0)^{1/2})\right) d\lambda.$$

More precisely, the resolution of the identity E_s *of* L_q *has the form*

$$\begin{aligned}(E_s f)(x) = (2\pi)^{-1} \int_{\sigma\cap[\kappa_0,s]} &(\Psi_f^{+}((\lambda-\kappa_0)^{1/2})\psi^{-}(x,(\lambda-\kappa_0)^{1/2}) \\ &+ \Psi_f^{-}((\lambda-\kappa_0)^{1/2})\psi^{+}(x,(\lambda-\kappa_0)^{1/2}))\, d\lambda/2,\end{aligned}$$

where $\Psi_f^{\pm}(\mu)$ *is the generalized Fourier transform of the function* $f \in L_2(\mathbb{R})$ *with respect to the functions* $\psi^{\pm}(x,\mu)$:

$$\Psi_f^{\pm}(\mu) = \int_{\mathbb{R}} \psi^{\pm}(x,\mu) f(x)\, dx.$$

It is interesting to note that S. A. Molchanov and V. A. Chulaevskiy [695] have constructed examples of limit-periodic potentials that can be approximated with less than exponential rate and lead to pure point spectrum. In view of this result, the conditions of Theorems 6.20 and 6.21 are the best possible.

It is also necessary to note some earlier works on spectral analysis of one-dimensional Schrödinger operators with almost-periodic coefficients, such as works by J. Moser [699], V. A. Chulaevskiy [340], J. Avron and B. Simon [245]. In those works limit-periodic potentials are also considered, but, unlike in the above given account of L. Pastur–V. Tkachenko results, the corresponding potentials belong to the space of limit-periodic functions in the sense of H. Bohr. The works [245], [340] and [699] establish some important results concerning direct spectral analysis of considered operators and density of the set of considered potentials with nowhere dense spectrum in the considered metric space with the topology of uniform convergence on the whole axis[35].

Introduction of the Stepanov class of almost-periodic functions and limit-periodic potentials allowed Pastur and Tkachenko (see [729], [730] and the book [153]) to obtain more explicit description of a larger set of limit-periodic potentials with prescribed properties of the spectrum (including Cantor-type spectra with positive Lebesgue measure) and to solve the inverse spectral problem by means of "completion" of the spectral theory of the Hill operator, which was given by V. Marchenko and I. Ostrovskiy (see [658], [659] and the book [121]).

We may say that Pastur–Tkachenko theory of limit-periodic one-dimensional Schrödinger operators now has a completed form as Marchenko–Ostrovskiy spectral theory of Hill operators.

An analogue of Pastur–Tkachenko theory for limit-periodic Dirac operators is constructed by I. E. Egorova [394], [396] (see also [399]). For limit-periodic Jacobi matrices, see I. E. Egorova [395]. Almost-periodic solutions with nowhere dense spectrum of non-linear evolutionary equations are investigated in the works by I. E. Egorova [397], [398] and A. Boutet de Monvel–I .E. Egorova [298].

* * *

Similar to the role of polynomials among entire functions, the role of finite-zone potentials is especially important among both periodic and almost periodic potentials.

In the non-periodic case, finite-zone potentials were first considered by N. I. Akhiezer [216], where consideration was restricted to even potentials. As noted in [112, p. 172], this restriction was probably caused by the fact that at that time (before 1961), the inverse problem on the real line with respect to the spectral matrix-function had not yet been studied thoroughly enough. (We note that the solution of the mentioned inverse problem was

---

[35] In [245] it was proven, among other results, that for a massive set of limit-periodic real-valued potentials, the spectra of the corresponding one-dimensional Schrödinger operators are 0-dimensional topological subspaces of the complex plane $\mathbb{C}$, lying, evidently, on the real axis $\mathbb{R}$. This result was generalized in [929] for limit-periodic complex-valued potentials of Stepanov class S^2 (6.198). In this case, the spectra are also 0-dimensional topological subspaces of $\mathbb{C}$; however, now they are not contained in $\mathbb{R}$.

obtained only later in [771][36] and was effectively applied in the papers on inverse problems for periodic and almost periodic potentials on the real line, such as [112], [625], *etc.*) However, the work of N. I. Akhiezer [216] is interesting most importantly because it was the first work establishing the connection between the inverse problems of spectral analysis for finite-zone potentials and methods of algebraic geometry (Jacobi's problem of inverting hyper-elliptic integrals); see also [5] and [6].

In the case of finite-zone almost periodic potentials, explicit solutions to Schrödinger equations were constructed by A. P. Its and V. B. Matveev [520], [521], who used the above mentioned connection discovered in [216]. Regarding finite-zone Dirac operators, see [687].

Another approach is due to S. P. Novikov [706] and is based on the fact that Korteweg-de Vries equation gives rise to a completely integrable Hamiltonian system [439], [926]; see also [30] and [189].

The set of finite-zone potentials is dense in both the set of periodic zone-type potentials and the set of almost periodic zone-type potentials, as shown in [659] and [624] respectively; see also [112]. The closure of the set of finite-zone potentials, leading to a more general form of infinite-zone potentials with finite cluster points of spectral gaps, was considered in [626].

The reference [254] (see also [22]) is devoted to the question of under what specification of parameters, the solutions of integrable non-linear equations can be expressed in terms of Abelian functions of lower genus—in particular, elliptic functions.

It is interesting to note a new result by V. Batchenko and F. Gesztesy [249] about the spectrum of *non-self-adjoint* Schrödinger operators with complex-valued finite-zone almost periodic potentials. In this case, spectral properties of such operators are in many respects similar to spectral properties of the non-self-adjoint Hill operators with finite-zone potentials (see some properties below), though the proofs are essentially more complicated.

R. N. Davydov [350] obtained an explicit expression for the spectral matrix for the scalar equation (6.1) on the axis with a bounded, generally complex-valued potential satisfying the n-th analogue of the stationary Korteweg-de Vries equation. In the case of non-real $q(t)$, the spectral matrix entries are treated as Marchenko-type generalized functions (i.e. dis-

---

[36]See also [112], [459], [536], [552], [554], [636], [778], [789], and the last section of Bibliographical Comments to Chapter 6 "The inverse Sturm–Liouville Problem for the Spectral Matrix on the Whole Axis." Note that the inverse scattering problem on the axis was solved by L. D. Faddeev [417], [418] and V. A. Marchenko [121].

tributions, see [121]) acting on a main space of entire functions which are exponential-type with respect to $\lambda^{1/2}$. As known, potentials of such type are finite-gap and any n-gap periodic potential satisfies one of stationary higher Korteweg-de Vries equations; see [206] and [249]. Thus, the necessary and sufficient conditions for the form of the spectral matrix (at least for the real finite-gap Hill operator) follow from [350]. Explicit form of the spectral matrix and its necessity and sufficiency for any real finite-gap Hill operator can also be obtained directly from spectral data by comparing Marchenko–Ostrovskiy theorem (see [121] and [658]) with the studies of B. M. Levitan [112].

A new class of periodic potentials, which can be called "semi-finite-zone" or "semi-infinite-zone" potentials, was introduced and investigated by P. Djakov and B. Mityagin in [372], [377] and [378]. Such potentials have only finite number of simple (i.e. non-degenerate) eigenvalues for the periodic boundary problem and infinite number of simple eigenvalues for the anti-periodic boundary problem (or *vice versa*).

Some works by M. V. Novitskiy (see, for example, [714]–[716]) are devoted to spectral asymptotics of Hill operators with quasi-analytic potentials. In those works, Novitskiy proved the McKean–Trubowitz conjecture and obtained a description of periodic potentials of the Carleman class in terms of decay of the gap lengths (an extension of Trubowitz theorem).

Spectral theory and properties such as stability, dichotomy, construction of solutions, perturbations of periodic, almost periodic differential and difference operators, and also stochastic operators, whose properties are in many ways close to those of almost periodic operators, are the main theme of a large number of works. We will mention the following monographs (or parts of monographs): [6], [21], [30], [39], [50], [51], [60], [100], [112], [113], [115], [117], [121], [132], [138], [153], [154], [167], [176], [186], [191], [196], [199], [200], and [206]; the papers: [216], [249], [253], [258], [261], [276]–[282], [299], [302], [312], [328]–[330], [332], [344], [347], [348], [366], [369]–[371], [373], [374], [387]–[389], [400], [401], [424]–[428], [443], [446], [450], [454], [459]–[464], [468], [469], [507], [508], [516], [519]–[521], [524], [530], [531], [545]–[552], [554], [555], [571], [572], [603], [604], [622]–[627], [629], [639]–[641], [658], [670], [676]–[678], [694], [697], [703], [706], [712]–[716], [722]–[727], [729], [765], [769], [775], [780], [781], [785], [786], [790], [803]–[805], [824], [825], [831], [835]–[838], [843], [844], [858], [862], [864], [867], [868], [870]–[872], [882], [883], [886]–[889], [892], [911]–[913], [928], [930], [934], [936], and [939].

Beside periodic equations, so-called Wannier–Stark operator

$$-\frac{d^2}{dx^2} - Fx + q(x), \quad 0 \leq x < \infty,$$

with a real periodic function $q(x)$ and constant $F > 0$, attracts attention of many physicists and mathematicians.

Such operators describe a periodic crystal lattice in homogeneous electric field. The case of a singular periodic potential q plays an essential role here. Accelerating properties of linear potential compete here with closing properties of the singular lattice. In this regard, we will mention the following works: [228], [244], [260], [331], [353], [416], [616], and [754]–[756]. A simple method of localization of the absolutely continuous spectrum, based on the use of the spectral "B–F" (Buslaev–Faddeev) identities [333], is offered in [351].

Problems on a finite interval for differential operators with a singular potential are investigated in the works [832]–[834]. Periodic and antiperiodic problems in $L_2(0,1)$ with singular potentials are considered in [676]–[678] and [697].

* * *

Without aiming to give a comprehensive review of the theory of stability of differential equations with operator-valued coefficients (there are many excellent references for this topic), we briefly discuss the results of the papers [803]–[805] and the book [51] devoted to a generalization of the well-known Lyapunov's stability criterion for operator-valued Hill equation. In the considered case, those results demonstrate a specific effect of the influence of integrability type of the operator-valued coefficient $P(t)$ on stability property of the equation.

In the monograph of Yu. L. Daletskiy and M. G. Krein [51, Chapter 5, Problem 8] and, independently, in the papers [803], [804], and later in [805], the criteria for stable boundedness for solutions of Hill's equation

$$y'' + \lambda P(t) y = 0, \qquad -\infty < t < +\infty, \tag{6.206}$$

with an operator coefficient

$$P(t) = P(t+T) = P^*(t) \in B(H) \tag{6.207}$$

with a non-negative average value

$$P_{av} := \frac{1}{T} \int_0^T P(t)\, dt \geq 0, \tag{6.208}$$

obtained in the scalar case by A. M. Lyapunov [120] and in the finite-dimensional case by M. G. Krein [590], are transferred to the case of a separable Hilbert space H.[37] (Here, $\lambda \neq 0$ belongs to a so called central stability zone (if $P_{\mathrm{av}} = 0$) and right one-sided stability zone (if $P_{\mathrm{av}} \geq 0$).)

Solutions of (6.206) are said to be (see [591]) stable bounded[38] (on the real line) if they are bounded for all $-\infty < t < +\infty$ together with any solutions of the perturbed equation

$$y'' + \lambda \widetilde{P}(t) y = 0, \qquad \widetilde{P}(t) = \widetilde{P}(t+T) = \widetilde{P}^*(t) \in B(H),$$

when the coefficient $\widetilde{P}(t)$ is sufficiently close to $P(t)$ in integral sense:

$$\int_0^T \|P(t) - \widetilde{P}(t)\| < \delta$$

(here, $\delta > 0$ is sufficiently small).

Variant 1° of Theorem 1 from [804] gives a stability criterion that contains as special cases the criteria of A. M. Lyapunov [120] and M. G. Krein [590]. Furthermore, as shown in [805], the mentioned criterion from [804] is more general than the stability criterion from the monograph of Yu. L. Daletskiy and M. G. Krein [51]. Indeed, if any variant of conditions in [51] guaranteeing the stability is satisfied, then the variant 1° of Theorem 1 from [804] guaranteeing the stability is also satisfied. However, the fulfilment of conditions in variant 1° of Theorem 1 from [804] does not imply the fulfillment of conditions guaranteeing the stability in [51]. The latter can be seen on an example of the equation (6.206) with 2×2 matrix coefficient

$$P(t) = \begin{pmatrix} 1 & 0 \\ 0 & \sin t \end{pmatrix}.$$

The corresponding equation is stable by Theorem 1 from [804], but it does not satisfy the above mentioned stability conditions from [51].

As shown in Theorem 1 from [803] and in Theorem 1(1°) from [804], in the case $\dim H = \infty$, we cannot replace the Bochner integrability condition in the uniform operator topology of the operator coefficient $P(t)$ (see (6.207) and (6.208)) in those theorems by Bochner integrability condition on the period $[0, T]$ of the vector-valued function $P(t)f$ for all $f \in H$ and the condition $\|P(t)\| \in L_{1,\mathrm{loc}}$. This can be seen on an example from [803]

[37] This problem was posed by M. G. Krein [591].

[38] Sometimes one talks about strong stability of the equation (6.206); see, for example, [51].

and [804] when the coefficient $P(t)$ admits in some orthonormal basis a diagonal matrix representation

$$P(t) = \text{diag}\,(\epsilon_1 + \sin t, \epsilon_2 + \sin 2t, \dots, \epsilon_k + \sin kt, \dots),$$

where $0 \leq \epsilon_k \to 0$ as $k \to \infty$.

If uniform Bochner integrability of the coefficient $P(t)$ is not assumed, the stability criterion for the equation (6.206) is given in Theorem 1(2°) from [804] and Theorem 2 from [805]. (The latter theorem shows that in the second variant the above mentioned Yu. L. Daletskiy–M. G. Krein stability criterion [51, Chapter 5, Problem 8], the uniform Bochner integrability requirement on $P(t)$ can be replaced by strong Bochner integrability requirement on $P(t)$ and the requirement $\|P(t)\| \in L_{1,\text{loc}}$.)

6.6.5 *The Spectrum of Non-Self-Adjoint Operators. Batchenko–Gesztesy and Rofe-Beketov Theorems*

The works [441], [667], [668], [669], [731], [732], [767], [831], [843], [889], [890], [891], and [892] are devoted to a non-self-adjoint Hill operator.

The reference [930] is devoted to properties of the essential spectra of equations and systems with sufficiently slowly changing, generally speaking, complex-valued coefficients (in particular, periodic or almost periodic).

Here we quote some results from [767] on the spectrum of non-self-adjoint differential operators $L[y]$ of arbitrary order with periodic complex-valued coefficients; see also [667].

The question on topological properties of the spectrum of differential equations with complex-valued periodic coefficients was posed by I. M. Gel'fand at the Differential Equations Conference in 1957 in Kharkov.

Theorem 6.22 **([767])** *Let $L[y]$ be a closed operator generated in the space $L_2(-\infty, \infty)$ by the differential expression*

$$l[y] = p_0(x)y^{(n)} + p_1(x)y^{(n-1)} + \cdots + p_n(x)y, \tag{6.209}$$

where $p_j(x + \omega) = p_j(x)$, $j = 0, 1, \dots n$, $\omega > 0$, $p_0(x) \neq 0$ are sufficiently smooth. Then the spectrum of the operator L

(1) is purely continuous, i.e. essential without any eigenvalues and without residual spectrum.

(2) coincides with the conditional stability set Λ *of* $l[y]$, *i.e. with the set of such values of* $\lambda \in \mathbb{C}$ *for which the equation* $l[y] = \lambda y$ *has at least one nontrivial solution which is bounded on the whole axis* $-\infty < x < +\infty$.

(3) satisfies one of the following:

(a) it consists of infinite number (in general) or finite number of analytic arcs

(b) it fills the whole complex plane

(c) it is empty.

(The possibilities (3) (b) and (3) (c) are excluded if $p_0(x) = const.$*)*

Here we briefly outline the proof. If λ belongs to the conditional stability set, then we can construct a sequence of test functions $f_m(x, \lambda)$ (using the existence of a nontrivial bounded solution[39] of $l[y] = \lambda y$). Therefore, by H. Weyl's theorem, λ belongs to the essential spectrum of the operator L in (6.209), and since L has neither eigenvalues nor residual spectrum, we see that λ is a point of the purely continuous spectrum of L. Conversely, if λ does not belong to the conditional stability set, then λ is a regular point of the operator L in (6.209), i.e. λ belongs to the resolvent set of L, as we can see from the following Lemma:

Lemma 6.13 (**[767]**) *If* $l[y] = \lambda y$ *has no nontrivial bounded solutions, then the operator* L *in (6.209) has the resolvent* $R_\lambda = (L - \lambda I)^{-1}$, *which is a bounded integral operator with the Carleman-type kernel* $G(x, t, \lambda)$ *such that* $G(x + \omega, t + \omega, \lambda) = G(x, t, \lambda)$ *and*

$$|G(x, t, \lambda)| \le C e^{-\alpha|x-t|}(1 + |x - t|)^{m-1},$$

where $C = C(\lambda)$; $\alpha = \omega^{-1}\min_j |\ln|\rho_j(\lambda)||$; *and* m *is the maximal exponent of elementary divisors of the monodromy matrix:* $m = m(\lambda) \le n$.

(Note that the kernel $G(x, t, \lambda)$ *may be triangular.)*

Such resolvent kernel $G(x, t, \lambda)$ may be constructed by a generalization of the method offered by M. Burnat [328] for the case of second order self-adjoint operators. Namely, if $l[y] = \lambda y$ has no nontrivial bounded solutions,

---

[39] Indeed, if $y_0(x, \lambda)$ is such a solution, then we can put

$$f_m(x, \lambda) = \frac{y_0(x, \lambda) h_m(x)}{\|y_0(\cdot, \lambda) h_m(\cdot)\|},$$

where $h_m \in C_0^\infty(\mathbb{R})$, $0 \le h_m(x) \le 1$, $h_m(x) = 1$ for $a_m < x < b_m$; $h_m(x) = 0$ in $\mathbb{R} \setminus (a_m - 1, b_m + 1)$, $a_m < b_m < a_{m+1} - 2$, $a_m \operatorname{sign} m \to +\infty$ and $b_m - a_m \to \infty$ as $m \to \pm\infty$ (cf. [70]).

then this equation does not have multipliers with the unit modulus, and, therefore, it has a fundamental system of n solutions (Floquet-type solutions) such that each of these solutions belongs to $L_2(\mathbb{R}_+)$ or $L_2(\mathbb{R}_-)$ (any nontrivial solution cannot belong to $L_2(\mathbb{R})$). From such system of solutions we can construct Green function (and, hence, the kernel $G(x,t,\lambda)$ of the resolvent for such λ) analogously to the case of a self-adjoint singular differential expression (see, for example, [6], [58] and [145]). Using the Floquet form of solutions, one can check that this kernel admits the estimate indicated in Lemma 6.13 and, therefore, gives rise to a bounded integral operator of Carleman type. (Analogously to the case of Hill equation (cf. (6.155) and (6.156)), multipliers of the considered equation are the roots $\rho_j(\lambda)$, $j = 1, \dots, n$, of the characteristic equation

$$\det(y_j^{(k-1)}(\omega,\lambda) - \rho\delta_{j,k})_{j,k=1}^n = 0,$$

for the monodromy matrix $(y_j^{(k-1)}(\omega,\lambda))_{k,j=1}^n$, where

$$l[y_j(x,\lambda)] = \lambda y_j(x,\lambda), \quad y_j^{(k-1)}(0,\lambda) = \delta_{j,k}, \quad j,k = 1,\dots,n,$$

and $\delta_{j,k}$ is the Kronecker symbol: $\delta_{j,j} = 1$; $\delta_{j,k} = 0$ if $j \neq k$.)

Finally, (3a) follows from the assertion (2) because, in accordance with I. M. Gelfand's method [446], we obtained that Λ coincides with the union over all $t \in [0, 2\pi]$ of sets of all λ-roots $\lambda = \lambda(t)$ of the equation

$$\det(y_j^{(k-1)}(\omega,\lambda) - e^{it}\delta_{j,k})_{j,k=1}^n = 0.$$

The cases (3) (b) and (3) (c) of the Theorem are impossible if $p_0(x) = \text{const}$, by the known asymptotic behavior of solutions of $l[y] = \lambda y$ in (6.209) when $|\lambda| \to \infty$ and by asymptotic behavior of multipliers which can be obtained from the characteristic equation using the known asymptotic behavior of its coefficients.

Corollary 6.8 ([767]) *For all λ from a connected component of the resolvent set of the operator L in (6.209), the number of multipliers with modulus < 1 is constant.*

Example 6.3 ([767]) The problem

$$e^{-2ix}(-y'' + iy') = \lambda y, \quad -\infty < x < +\infty,$$

has continuous spectrum, which fills whole the plane, because the considered equation has bounded solutions at each λ:

$$y = \exp(\pm\lambda^{1/2}e^{ix}) \quad \text{if } \lambda \neq 0,$$

$$y = 1 \quad \text{and} \quad y = e^{ix} \quad \text{if } \lambda = 0.$$

Here, the whole λ-plane belongs not only to the conditional stability set, but also to the stability set of the differential operator.

Example 6.4 ([767]) The problem obtained from Example 6.3 by substitution $z = e^{-x}y$, i.e.

$$e^{-2ix}(-z'' + (i-2)z' + (i-1)z) = \lambda z, \quad -\infty < x < +\infty,$$

does not have spectrum because the corresponding equation does not have any bounded nontrivial solution $z(x)$.

Example 6.5 ([767]) This example deals with the problems for the same equation as in Example 6.4, but on the left or right half-axis (instead of the whole x-axis) with the boundary condition $z(0) = 0$. For the problem on the right half-axis $0 \le x < +\infty$, each point of λ-plane is an eigenvalue, but for the same problem on the left half-axis $-\infty < x < 0$, the whole λ-plane belongs to the residual spectrum.

Corollary 6.9 ([767], [843]) *From Theorem 6.22, for the second-order equation with complex-valued coefficients with period* $\omega > 0$

$$-y'' + p_1(x)y' + p_2(x)y = \lambda y, \qquad -\infty < x < \infty, \tag{6.210}$$

we have the following more concrete properties of the spectrum:

(1) If Re $P = 0$*, where*

$$P = \int_0^\omega p_1(x)\,dx,$$

then the spectrum has multiplicity two and coincides with the closure of the stability set (not only conditional stability). In this case, the spectrum consists of a finite or infinite number of analytic arcs contained in some strip $|Im\ \lambda| <$ *const and it extends to the right up to infinity.*

(2) In the opposite case, if Re $P \neq 0$*, then the spectrum has multiplicity 1 because, in this case, both multipliers of the equation (6.210) cannot have unit modulus simultaneously. The spectrum necessarily contains precisely one unbounded (connected) component, and this component approaches asymptotically the parabola*

$$\lambda = \left(t - \gamma\omega^{-1}\right)^2 + C, \qquad -\infty < t < \infty,$$

where C is a complex constant (depending on p_1, p_2) and $\gamma = P/(2i)$. In this case, the spectrum can have additionally at most finite number of simple closed analytical curves. No spectrum component can be contained inside other components. (If we consider an unbounded spectral component, we assume as the inner domain the one that contains $+\infty$.) No two components of the spectrum can have more than one common point, and if there is a common point, then it is a corner point for each of these components[40].

Some results on the properties of conditional stability set of the second order equation (6.210) are obtained by M. I. Serov [843]. By Theorem 6.22, the same properties hold for the spectrum of (6.210); see Corollary 6.9. The paper [843] contains an unnecessary condition on boundedness of those components that cannot contain other components.

Theorem 6.23 ([767]) *The resolvent set of the equation (6.210) is connected if and only if Re $P = 0$.*

Proof. By virtue of Corollary 6.9, it is sufficient to consider only the case Re $P = 0$, i.e. the case Im $\gamma = 0$. In this case, instead of (6.156) with θ and ϕ determined by (6.1) and (6.148), we have the following equation for multipliers $\rho(\lambda)$:

$$\rho^2 - (\theta(\omega, \lambda) + \phi'(\omega, \lambda))\rho + e^{2i\gamma} = 0, \tag{6.211}$$

where θ and ϕ are now solutions of the equation (6.210) instead of (6.1), but with the same initial data (6.148) as before.

To obtain the spectrum from (6.211), we must find all values of λ for which this equation has at least one solution ρ with the unit modulus, i.e. we must define the λ's from (6.211) when there is substitution $\rho = e^{it}$ at all real t or (which suffices) at $-\pi \le t \le \pi$. After such substitution, we obtain from (6.211) the following equation to determine the spectrum of (6.210):

$$\Psi(\lambda) := (\theta(\omega, \lambda) + \phi'(\omega, \lambda))e^{-i\gamma} = 2\cos(t - \gamma), \tag{6.212}$$

where $-\pi \le t \le \pi$.

From (6.212) we see that, on the spectrum, Im $\Psi(\lambda) = 0$, and, therefore, the spectrum cannot contain any closed curve, because $\Psi(\lambda)$ is an entire

[40]Two or more components of the spectrum that have one common point can be considered as one smooth analytic self-intersecting curve (for instance, as the figure of eight).

analytic function differing from a constant, and Im $\Psi(\lambda)$ is a harmonic function satisfying the maximum principle.

The possibility that the spectrum has two connected branches extending to infinity is also excluded by the asymptotic behavior of $\Psi(\lambda)$ at $|\lambda| \to \infty$. Therefore, the resolvent set of (6.210) is connected if Re $P = 0$. The theorem is proven. □

Theorem 6.24 ([**767**]) *The problem for the equation (6.210) on the half-axis* $0 \le x < \infty$ *with the boundary condition* $y'(0) - hy(0) = 0$ *or* $y(0) = 0$ *has the same continuous spectrum as the equation (6.210) on the whole axis.*

Additionally, if Re $P < 0$, *then inner components of the resolvent set of the problem (6.210) on the whole axis are filled in by eigenvalues of the problem (6.210) (with boundary conditions at zero) on the half-axis, and finite number of such eigenvalues may lie down on the continuous spectrum.*

In the case Re $P > 0$, *the same sets are covered by the residual spectrum. Additionally, there can possibly occur only isolated eigenvalues and isolated points of the residual spectrum, and the number of such points is at most finite if Re* $P \neq 0$. *If Re* $P = 0$, *then eigenvalues and residual spectrum cannot lie over the continuous spectrum.*

It is necessary to mention important results by M. Gasymov [441] and L. Pastur–V. Tkachenko [731] on the spectral analysis of non-self adjoint Hill operators (6.1) with potentials of the form

$$q(t) = \sum_{n=0}^{\infty} c_n e^{int},$$

with Fourier series expansion containing only non-negative n in the exponents e^{int} (direct and inverse problems).

For non-self-adjoint Hill equations (6.1) with complex-valued potentials $q(t)$ of general type, L. Pastur and V. Tkachenko [732] and V. Tkachenko [890]–[892] obtained important results on geometry of the spectrum, including necessary and sufficient conditions. Those works are based on profound investigation of analytic properties of the Hill discriminant (Lyapunov function) for Hill equations in the non-self-adjoint case.

Passing to almost-periodic non-self-adjoint Schrödinger operators, here we briefly explain the above mentioned interesting recent results by V. Batchenko and F. Gesztesy [249]. They considered and investigated the spectrum of Schrödinger operators l_1 in (6.1) with complex-valued quasi-periodic algebro-geometric KdV potentials $q(t)$, where $-\infty < t < \infty$, (i.e. with potentials q satisfying one (and, therefore, infinitely many) equation(s) of the stationary Korteweg-de Vries (KdV) hierarchy) associated with non-singular hyperelliptic curves in [249], and, in more details, in [250]. As in the case of a periodic potential (see Theorems 6.22, 6.23 and Corollary 6.9 in the present book), it turns out that the spectrum of l_1 coincides with the conditional stability set of l_1. However, in the non-periodic case, it is not possible to describe the conditional stability set by means of investigating the multipliers' modulus. As shown in [249] and [250], it turns out that the spectrum can be explicitly described in terms of the mean value $W(z) := ik(z)$ (6.162) of the inverse of diagonal Green's function of l_1. Arguing heuristically, is natural to expect that on the continuous spectrum the quasi-impuls $k(z)$ is real-valued, and, therefore, the spectrum of l_1 coincides with the set Re $W(z) = 0$ in the complex z-plane. As a result, the spectrum of l_1 consists of finitely many simple analytic arcs and one semi-infinite simple analytic arc in the complex plane. The spectrum cannot contain a closed loop. Crossings as well as confluences of spectral arcs are possible and discussed as well in [249] and [250], where these results are obtained not only in the space $L_2(\mathbb{R}, dx)$, but are also extended to $L_p(\mathbb{R}, dx)$ setting for $p \in [1, +\infty)$.

We now precisely formulate the above mentioned results following the **work of V. Batchenko and F. Gesztesy [249]**. Throughout this formulation, we assume that

$$q \in C^\infty(\mathbb{R}). \tag{6.213}$$

By L we denote the (scalar) differential expression (6.1) with $q = q(x)$ instead of $q(t)$, in accordance with KdV-theory and with [249]:

$$L = -\frac{d^2}{dx^2} + q. \tag{6.214}$$

To construct the KdV hierarchy, we need a second differential expression P_{2n+1} of order $2n+1$, where $n \in \mathbb{N}_0 = \{0, 1, 2, ...\}$, defined recursively as follows. First, define $\{f_m\}_{m \in \mathbb{N}_0}$ recursively by

$$\begin{aligned} &f_0 = 1, \\ &f_{m,x} = -\frac{1}{4} f_{m-1,xxx} + q f_{m-1,x} + \frac{1}{2} q_x f_{m-1}, \quad m \in \mathbb{N} = \{1, 2, 3, ...\}, \end{aligned} \tag{6.215}$$

where $\{c_k\}_{k \in \mathbb{N}} \subset \mathbb{C}$ denote integration constants, which naturally arise when solving (6.215).

The differential expressions P_{2n+1} of order $2n+1$ are then defined by

$$P_{2n+1} = \sum_{m=0}^{n} \left(f_{n-m} \frac{d}{dx} - \frac{1}{2} f_{n-m,x} \right) L^m, \qquad n \in \mathbb{N}_0. \tag{6.216}$$

Using the recursion (6.215), the commutator of P_{2n+1} and L can be explicitly computed, and one obtains

$$[P_{2n+1}, L] = 2 f_{n+1,x}, \qquad n \in \mathbb{N}_0. \tag{6.217}$$

In particular, (L, P_{2n+1}) represents the celebrated Lax pair of the KdV hierarchy. Varying $n \in \mathbb{N}_0$, the stationary KdV hierarchy is then defined in terms of vanishing of the commutator of P_{2n+1} and L in (6.217) by

$$-[P_{2n+1}, L] = -2 f_{n+1,x}(q) = \text{s-KdV}_n(q) = 0, \qquad n \in \mathbb{N}_0. \tag{6.218}$$

Here, $\text{s-KdV}_n(q) = 0$ denotes the stationary n-th KdV equation for potential q of the KdV hierarchy corresponding to the Lax pair (L, P_{2n+1}), where L is the Schrödinger operator (6.214) with potential q, and P_{2n+1} is defined by (6.216) and (6.215):

$$\begin{aligned} P_1 &= \frac{d}{dx}, \\ P_3 &= \left(\frac{1}{2} q + c_1 \right) \frac{d}{dx} - \frac{1}{4} q_x + \frac{d}{dx} \left(-\frac{d^2}{dx^2} + q \right) \\ &= -\frac{d^3}{dx^3} + \left(\frac{3}{2} q + c_1 \right) \frac{d}{dx} + \frac{3}{4} q_x, \end{aligned}$$

and so on. The notation $f_m(q)$ denotes the functional dependence of f_m on q.

Here, we used (see (6.215))

$$f_0 = 1, \qquad \frac{df_1}{dx} = \frac{1}{2} \frac{dq}{dx},$$

and, hence,

$$f_1(q) = \frac{1}{2} q + c_1.$$

Therefore,

$$\frac{df_2}{dx} = -\frac{1}{8} \frac{d^3 q}{dx^3} + \frac{1}{2} q \frac{dq}{dx} + \frac{1}{2} \left(\frac{1}{2} q + c_1 \right) \frac{dq}{dx},$$

and

$$f_2(q) = -\frac{1}{8}\frac{d^2 q}{dx^2} + \frac{3}{8}q^2 + \frac{1}{2}c_1 q + c_2,$$

and so on.

Therefore, by virtue of (6.218),

$$\text{s-KdV}_0(q) := -\frac{dq}{dx};$$

$$\text{s-KdV}_1(q) = -2\frac{df_2(q)}{dx} = \frac{1}{4}\frac{d^3 q}{dx^3} - \frac{3}{2}q\frac{dq}{dx} - c_1\frac{dq}{dx};$$

and so on.

By definition, the set of solutions of (6.218) with $n \in \mathbb{N}_0$ and $c_k \in \mathbb{C}$, $k \in \mathbb{N}$, represents a class of algebro-geometric KdV solutions. Sometimes algebro-geometric stationary KdV solutions q are simply called KdV potentials.

When we assume that q satisfies the n-th stationary KdV equation, we mean that q satisfies one of the n-th stationary KdV equations after a particular choice of integration constants $c_k \in \mathbb{C}$, $k = 1, 2, \ldots, n$, $n \in \mathbb{N}$, has been made.

Next, following [249], we introduce a polynomial F_n of degree n with respect to the spectral parameter $z \in \mathbb{C}$:

$$F_n(z, x) = \sum_{m=0}^{n} f_{n-m}(x) z^m. \tag{6.219}$$

The recursion relation (6.215) and the equation (6.218) imply that

$$F_{n,xxx} - 4(q - z)F_{n,x} - 2q_x F_n = 0. \tag{6.220}$$

Multiplying (6.220) by F_n, a subsequent integration with respect to x results in

$$\frac{1}{2}F_{n,xx}F_n - \frac{1}{4}F_{n,x}^2 - (q - z)F_n^2 = R_{2n+1}, \tag{6.221}$$

where R_{2n+1} is a monic polynomial of degree $2n + 1$. We denote its roots by $\{E_m\}_{m=0}^{2n}$, and, hence, write

$$R_{2n+1}(z) = \prod_{m=0}^{2n} (z - E_m), \qquad \{E_m\}_{m=0}^{2n} \in \mathbb{C}. \tag{6.222}$$

One can show that the equation (6.221) leads to an explicit determination of the integration constants $c_1, \ldots, c_n$ in

$$\text{s-KdV}_n(q) = -2f_{n+1,x}(q) = 0 \tag{6.223}$$

in terms of the zeros $E_0, \ldots, E_{2n}$ of the associated polynomial R_{2n+1} in (6.222). In fact, one can prove that

$$c_k = c_k(\widehat{E}), \qquad \widehat{E} := \{E_0, E_1, \ldots, E_{2n}\}, \qquad k = 1, \ldots, n, \tag{6.224}$$

where

$$c_k(\widehat{E}) = -2^{-2k} \times \sum_{j_0+\cdots+j_{2n}=k} \frac{(2j_0)!\cdots(2j_{2n})!E_0^{j_0}\cdots E_{2n}^{j_{2n}}}{(j_0!)^2\cdots(j_{2n}!)^2(2j_0-1)\cdots(2j_{2n}-1)}. \tag{6.225}$$

Note that, without loss of generality, we may assume that solutions of $\text{s-KdV}_n(q) = 0$ satisfy (6.213). Indeed, if $q \in C^{2n+1}(\mathbb{R})$ satisfies the n-th stationary KdV equation (6.223) (for a given set of integration constants c_k, $k = 1, 2, \ldots, n$), then we obtain the equation (6.221), where F_n is introduced according to (6.219). But the equation (6.221), in turn, as shown inductively in [465, Proposition 2.1] yields (6.213) and $f_m \in C^\infty(\mathbb{R})$, $m = 0, 1, \ldots, n$.

In accordance with [249], let us denote by $CP(\mathbb{R})$ and $QP(\mathbb{R})$ the sets of continuous periodic and quasi-periodic functions on $\mathbb{R}$ respectively. A function f is called quasi-periodic with fundamental periods $(\omega_1, \omega_2, \ldots, \omega_N) \in (0, +\infty)^N$ if the frequencies $2\pi/\omega_1, \ldots, 2\pi/\omega_N$ are linearly independent over a field $\mathbb{Q}$ of rational numbers and if there exists a continuous function $F \in C(\mathbb{R}^N)$, periodic with period 1 in each variable

$$F(x_1, \ldots, x_j + 1, \ldots, x_n) = F(x_1, \ldots, x_j, \ldots, x_n), \tag{6.226}$$

where $x_j \in \mathbb{R}$, $j = 1, 2, \ldots, N$, such that

$$f(x) = F(x/\omega_1, \ldots, x/\omega_N), \qquad x \in \mathbb{R}. \tag{6.227}$$

It is well-known that such quasi-periodic function f is the Bohr almost-periodic function; see, for instance, B. M. Levitan [111].

Beside the notation $M_t\{f\}$ or $M_x\{f\}$ for the mean value of an almost periodic function f, as in formulas (6.120) and (6.162) of the present book, we introduce for simplicity the notation $\langle f \rangle$ as in [249], when this cannot cause confusion.

For the rest of our quotation of results [249], it will be convenient to introduce the following hypothesis due to V. Batchenko and F. Gesztesy:

Hypothesis (B-G) Assume that the roots E_0, ..., E_{2n} of the polynomial $R_{2n+1}(z)$ are simple:

$$E_m \neq E_h \qquad \text{for } m \neq h, \qquad m, h = 0, 1, ..., 2n; \tag{6.228}$$

and, moreover, suppose that $q \in C^\infty(\mathbb{R}) \cap QP(\mathbb{R})$ satisfies the n-th stationary KdV equation (6.218) on $\mathbb{R}$.

Assuming the Hypothesis (B-G), let us now consider L in (6.214) not as a differential expression, but as a closed operator in $L_2(\mathbb{R}, dx)$ with

$$\mathrm{D}(L) := \mathrm{Dom}(L) = W^{2,2}(\mathbb{R}) := H^{2,2}(\mathbb{R}) \text{ (Sobolev space)}.$$

One can now introduce the set $\Sigma \subset \mathbb{C}$ by

$$\Sigma = \{\lambda \in \mathbb{C}: \mathrm{Re}\ (\langle g(\lambda, \cdot)^{-1} \rangle) = 0\}. \tag{6.229}$$

Here, $g(\lambda, x)$ denotes the diagonal Green's function of L, i.e. $G(x, x; \lambda)$, and $\langle f(\cdot) \rangle = M\{f(\cdot)\}$ is a mean value of $f(x)$; cf. (6.162) and (6.120). But, unlike the self-adjoint case, the λ-domain, free of the spectrum $\sigma(L)$, is more complicated than the one in the self-adjoint case: $\mathbb{C} \backslash [\lambda_0, +\infty)$. The general theory says only that $\sigma(L) \subset W(L)^{\sim}$, i.e. the spectrum belongs to the closure of the numerical range $W(L)$ of the operator L (see Sec. A.1.2 in the Appendix, or Kato [87, p. 269]). Following [249], we introduce an appropriate set of (nonintersecting) cuts S_j—joining the points $E_{m(j)}$ and $E_{h(j)}$, $j = 1, \ldots, n$, and S_{n+1}—joining E_{2n} and ∞, and denote

$$S = \bigcup_{j=1}^{n+1} S_j, \qquad S_j \cap S_k = \emptyset, \quad j \neq k. \tag{6.230}$$

Following [249] we define the corresponding cut plane by $\Pi := \mathbb{C} \backslash S$, and introduce the holomorphic function $\Pi \to \mathbb{C}$ (see (6.222)):

$$R_{2n+1}(z)^{1/2} = \left(\prod_{m=0}^{2n} (z - E_m) \right)^{1/2} \tag{6.231}$$

with an appropriate choice of the square root branch in (6.231). For definiteness, we choose the branch of the square root in (6.231) so that

$$(-1)^n \mathrm{Im}\ (R_{2n+1}(z)^{1/2}) > 0, \qquad \text{as } z \to -\infty, \tag{6.232}$$

and the cut S_{n+1}, joining E_{2n} and $+\infty$. By (6.229) the definition of the set Σ is independent of the choice of the branch of $g(\lambda, \cdot)$ and the choice of the system of cuts S_j.

The construction of Green's function uses the stationary Baker–Akhiezer functions—Weyl solutions $\psi(\cdot; x, x_0)$ of the equation $L\psi = \lambda\psi$ with the quasi-periodic algebro-geometric potential $q(x)$ (complex-valued in the case under consideration and satisfying s-KdV equations (6.218)). Here, λ belongs to the two-sheeted hyperelliptic Riemann surface[41] of genus n (each sheet is identical to the cut plane Π) corresponding to the hyperelliptic curve K_n:

$$\Phi_n(z, y) := y^2 - R_{2n+1}(z) = 0. \tag{6.233}$$

The functions $\psi(z; x, x_0)$ are normalized by their exponential relative to $z^{1/2}$ asymptotics at $z \to \infty$ on K_n and by the condition $\psi(z; x_0, x_0) = 1$. These functions turn into the Floquet–Bloch solutions in the case of a periodic potential q.

We can write:

$$\psi(z; x, x_0) = \exp\left(\int_{x_0}^{x} \phi(z, x')\, dx'\right), \qquad z \in K_n, \quad (x, x_0) \in \mathbb{R}^2. \tag{6.234}$$

More precisely, we denote the points on K_n by $P = (z, y) \in K_n$. Then for ϕ in (6.234) we can write:

$$\phi(P, x) = \frac{iy + (1/2)F_{n,x}(z, x)}{F_n(z, x)} \tag{6.235}$$

($\phi(\cdot, x)$ is the so-called fundamental meromorphic function on K_n).

When the sheet of Riemann's surface is fixed, one can write z instead of P.

Following [249], we denote the zeros of the polynomial $F_n(\cdot, x)$ in (6.219) by $\{\mu_j(x)\}_{j=1}^{n}$ and, hence, write

$$F_n(z, x) = \prod_{j=1}^{n} (z - \mu_j(x)). \tag{6.236}$$

The considered conditions imply

$$\mu_j \in C(\mathbb{R}), \qquad |\mu_j(x)| \le C, \quad j = 1, \ldots, n, \quad x \in \mathbb{R}, \quad C > 0. \tag{6.237}$$

[41] The main information about Riemann surfaces can be found, for example, in the books [56], [75], [82], [109] and [206].

Let us return to the diagonal Green's function of L in (6.214). We have

$$g(P;x) = \frac{\psi(P;x,x_0)\psi(P^*;x,x_0)}{W\{\psi(P;.,x_0),\psi(P^*;.,x_0)\}}, \qquad P\in K_n, \qquad x,x_0\in\mathbb{R}, \quad (6.238)$$

where $P=(z,y)$, $P^*=(z,-y)$, and $W\{u,v\}=uv'-u'v$ is the Wronskian. We can show that

$$g(P;x) = i\frac{F_n(z,x)}{2y}, \qquad P=(z,y)\in K_n, \quad x\in\mathbb{R}. \quad (6.239)$$

Together with $g(P;x)$ we also introduce its branch $g_+(z,x)$, defined on the fixed by (6.232) sheet $\Pi_+\subset K_n$ and denote this branch also by $g(z;x)$:

$$g(z;x) = \frac{i}{2}F_n(z;x)R_{2n+1}(z)^{-1/2}, \qquad z\in\Pi, \quad x\in\mathbb{R}. \quad (6.240)$$

Under the Hypothesis (B-G), it is possible to see that $q^{(k)}$ and f_k, where $k\in\mathbb{N}$, and, hence, all x and z-derivatives of $F_n(z,\cdot)$, $z\in\mathbb{C}$, and $g(z,\cdot)$, $z\in\Pi$, are quasi-periodic. Moreover, taking limits to points on S, the last results extends to both sides of the cuts in the set $S\setminus\cup_{m=0}^{2n}E_m$ (cf. (6.230)) by continuity with respect to z.

We are interested in the construction of $\langle g(z,\cdot)^{-1}\rangle$. For this purpose, note that the expression of the type (6.238) is a special case of the following expression $G(z;x)$:

$$G(z;x) := \frac{u_1(z,x)u_2(z,x)}{W\{u_1(z,\cdot),u_2(z,\cdot)\}}, \quad (6.241)$$

where $u_1(z,x)$ and $u_2(z,x)$ are arbitrary solutions of the equation $Lu=zu$, and L is the expression (6.214) with arbitrary $q\in L_{1,\mathrm{loc}}(\mathbb{R})$. It is known that (see, for instance, [447]),

$$-(G^{-1})_z = 2G - G_{xxz} + [G^{-1}G_xG_z]_x. \quad (6.242)$$

Substituting $G=g$ and integrating, we have the following Lemma.

Lemma 6.14 **([249])** *Assume the Hypothesis (B-G) and let $z,z_0\in\Pi$. Then*

$$\langle g(z,\cdot)^{-1}\rangle = -2\int_{z_0}^{z}\langle g(z',.)\rangle\,dz' + \langle g(z_0,.)^{-1}\rangle, \quad (6.243)$$

where the path connecting z_0 and z is assumed to lie in the cut plane Π. Here, both parts of the equality are understood directly when they exist or as an analytic continuation on z, $z_0\in\Pi$ in the other case. Moreover, by *taking limits to points on S (6.230) in (6.243), the result (6.243) extends to either side of the cuts in the set S by continuity with respect to z.*

Remark 6.10 Following [249] we introduce the set Π_C such that for $z\in\Pi_C$, $g(z,\cdot)^{-1}$ is quasi-periodic and, hence, $\langle g(z,\cdot)^{-1}\rangle$ is well defined. We put

$$\Pi_C = \Pi\backslash\Big(\{z\in\mathbb{C}: |z|\leq C+1\}\cup\{z\in\mathbb{C}: \mathrm{Re}(z)\geq \min_{m=0,\ldots,2n}[\mathrm{Re}(E_m)]-1,$$

$$\min_{m=0,\ldots,2n}[\mathrm{Im}(E_m)]-1\leq\mathrm{Im}(z)\leq \max_{m=0,\ldots,2n}[\mathrm{Im}(E_m)]+1\}\Big), \tag{6.244}$$

where $C>0$ is the constant in (6.237).

Moreover, without the loss of generality, we may assume that Π_C contains no cuts, i.e.

$$\Pi_C\cap S=\emptyset. \tag{6.245}$$

If one analytically continues $g(z,x)$ with respect to z, the function $g(z,x)$ will acquire zeros for some $x\in\mathbb{R}$ and, hence, $g(z,\cdot)^{-1}\notin QP(\mathbb{R})$.

Nevertheless, as shown by the right-hand side of (6.243), $\langle g(z,\cdot)^{-1}\rangle$ admits an analytic continuation in z from Π_C to all of Π, and from now on, $\langle g(z,\cdot)^{-1}\rangle$, $z\in\Pi$, always denotes this analytic continuation. From (6.239) we also have

$$\mathrm{Re}(\langle g(P,\cdot)^{-1}\rangle) = -2\mathrm{Im}(y\langle F_n(z,\cdot)^{-1}\rangle); \tag{6.246}$$

compare the last equality with (6.162).

Lemma 6.15 *Assume the Hypothesis (B-G). Then Σ in (6.229) coincides with the conditional stability set of L, i.e.*

$$\Sigma=\Lambda. \tag{6.247}$$

(For the definition of the set Λ, see the formulation of (2) in Theorem 6.22.)

The proof is based on the formula for the Baker–Akhiezer solutions of $L\psi = z\psi(P;\cdot,x_0)$, namely

$$\begin{aligned}\psi(P;x,x_0) &= \Theta(P;x,x_0)\exp\{(x-x_0)\\ &\qquad\cdot\big[(1/2)\mathrm{Re}\ \langle g(P,\cdot)^{-1}\rangle + iT\big]\},\end{aligned} \tag{6.248}$$

where

$$T\in\mathbb{R},\quad P\in K_n\backslash\Big\{P_\infty\bigcup\{\hat{\mu}_j(x_0)\}_{j=1}^n\Big\}.$$

Here, $\widehat{\mu}_j(x_0)$ is point $\mu_j(x_0)$ "lifted" from $\mathbb{C}$ to the curve K_n, and P_∞ is the point at infinity which compactifies the curve K_n.

The formula (6.248) is a direct corollary of the formulas (3.29) and (3.32) of [250] and, as shown there,

$$\Theta(P;\cdot,x_0)\in L_\infty(\mathbb{R};\,dx), \qquad P\in K_n\backslash\{\widehat{\mu}_j(x_0)\}_{j=1}^n. \tag{6.249}$$

(Note that the formula (6.248) is heuristically connected with (6.162).)

Theorem 6.25 **(V. Batchenko–F. Gesztesy [249])** *Assume the Hypothesis (B-G). Then the point spectrum and residual spectrum of L in (6.214) are empty, and, therefore, the spectrum of L is purely continuous.*

Here we emphasize that, unlike in the case of a non-self-adjoint periodic operator, this theorem is not valid in the case of an arbitrary almost periodic potential q.

Theorem 6.26 **(V. Batchenko–F. Gesztesy [249])** *Assume the Hypothesis (B-G). Then the spectrum of L coincides with Σ and, therefore, with Λ—the conditional stability set of L. In particular,*

$$\{E_m\}_{m=0}^{2n}\subset\sigma(L), \tag{6.250}$$

and $\sigma(L)$ has no isolated points.

In the special self-adjoint case (i.e. real-valued q), the equality $\sigma(L)=\Sigma$ (see (6.229)) is equivalent to vanishing of the Lyapunov exponent of L which characterizes the purely absolutely continuous (in that case) spectrum of L, as discussed by Kotani; see [576]–[579] (see also [37]), Pastur [727], and Pastur and Figotin [154], [155]. In the case when q is periodic and complex-valued, this has also been studied by Kotani; see [579].

The explicit formula for Σ in (6.229) permits a qualitative description of the spectrum of L as follows. Differentiating (6.243) we obtain for some constants $\{\eta_j\}_{j=1}^n\in\mathbb{C}$:

$$\begin{aligned}\frac{d}{dz}\langle g(z,\cdot)^{-1}\rangle &= -2\langle g(z,\cdot)\rangle \\ &= -i\left[\prod_{j=1}^{n}(z-\eta_j)\right]\left(\prod_{m=0}^{2n}(z-E_m)\right)^{-1/2}, z\in\Pi.\end{aligned} \tag{6.251}$$

The relation (6.251) may be extended to either side of the cuts in S by continuity relative to z.

Theorem 6.27 **(V. Batchenko–F. Gesztesy [249])** *Assume the Hypothesis (B-G). Then the spectrum $\sigma(L)$ of L has the following properties:*

(1) $\sigma(L)$ is contained (as it is clear from considering the numerical range) in the semi-strip

$$\sigma(L)\subset\{z\in\mathbb{C}: Im(z)\in[M_1, M_2],\ Re(z)\geq M_3\}, \tag{6.252}$$

where

$$M_1 = \inf_{x\in\mathbb{R}}[Im(q(x))], \qquad M_2 = \sup_{x\in\mathbb{R}}[Im(q(x))],$$
$$M_3 = \inf_{x\in\mathbb{R}}[Re(q(x))]. \tag{6.253}$$

(2) $\sigma(L)$ consists of finitely many simple analytic arcs and one simple semi-infinite arc. The analytic arcs may only end at the points $\eta_1, \ldots, \eta_n$; $E_0, \ldots, E_{2n}$; and at infinity. The semi-infinite arc σ_∞ asymptotically approaches the half-line in the following sense: asymptotically, σ_∞ can be parametrized by

$$\sigma_\infty = \{z\in\mathbb{C}: z = R + iIm(\langle q\rangle) + O(R^{-1/2}) \text{ as } R\to+\infty\}. \tag{6.254}$$

(3) Each E_m, where $m = 0, \ldots, 2n$, is met by at least one of the arcs. More precisely, a particular E_{m_0} is hit by precisely $2N_0 + 1$ analytic arcs, where $N_0\in\{0, \ldots, n\}$ denotes the number of η_j that coincide with E_{m_0}. Adjacent arcs meet at an angle $2\pi/(2N_0 + 1)$ at E_{m_0}. (Thus, generically, $N_0 = 0$ and precisely one arc hits E_{m_0}.)

(4) Crossings of spectral arcs are permitted. This phenomenon takes place precisely when for a particular $j_0\in\{1, \ldots, n\}$, $\eta_{j_0}\in\sigma(L)$ is such that

$$Re\left(\langle g(\eta_{j_0}, \cdot)^{-1}\rangle\right) = 0, \tag{6.255}$$

for some $j_0\in\{1, \ldots, n\}$ with $\eta_{j_0}\notin\{E_m\}_{m=0}^{2n}$.
In this case, $2M_0 + 2$ analytic arcs converge toward η_{j_0}, where $M_0\in\{1, \ldots, n\}$ denotes the number of η_j that coincide with η_{j_0}. Adjacent arcs meet at an angle $\pi/(M_0 + 1)$ at η_{j_0}. (Thus, generically, $M_0 = 1$ and two arcs cross at a right angle.)

(5) The resolvent set $\mathbb{C}\backslash\sigma(L)$ of L is path-connected.

Of course, these results can be applied to the special case of algebro-geometric complex-valued periodic potentials (see the references [283], [284] [914] and [915] the results of which must be partially contained in some theorems concerning arbitrary periodic complex-valued potentials. The corresponding connection between the algebro-geometric approach and standard Floquet theory is briefly pointed out in Appendix C of [250]). But even in this special case, items (3) and (4) of Theorem 6.27 provide additional new details on the nature of the spectrum of L in (6.214).

The methods of papers [249] and [250] extend, as noted by their authors, to the case of algebro-geometric non-self-adjoint second order finite difference (Jacobi) operators associated with the Toda lattice hierarchy and to the case of Dirac-type operators related to the focusing nonlinear Schrödinger hierarchy. Moreover, they extend to the infinite genus limit $n \to \infty$ using the approach in [456].

In conclusion, note that non-self-adjoint differential operators considered above (in Sec. 6.6.5) can, generally speaking, fail to be spectral operators in Dunford sense [59] in periodic case as well as in almost-periodic case.

* * *

Theorem 6.12, together with the basic ideas of the proof, appeared in [786]. (The case of a perturbed Hill operator with some gap between the constants in the sufficient and necessary conditions was considered in [781].)

The case $\epsilon = 0$ in Theorem 6.12 was considered in [838] with the help of the standard perturbation of the form

$$|K_j|t^{-2} + \beta(t \ln t)^{-2}.$$

We also mention the works [464] and [865] containing the improvements of Kneser's theorem.

We now make a few remarks regarding the proof of Theorem 6.12. *Proximate order* is a function $\rho(t)$ satisfying $\lim_{t\to\infty} \rho(t) = \rho$ (i.e. the limit exists and is equal to some $\rho \in \mathbb{R}$) and $\lim_{t\to\infty} t\rho'(t) \ln t = 0$; see [110] and [71].

We will specify the formulas (1.53) and (1.53′) from [110] or the corresponding formulas from [71, Chapter 2, Sec. 2]: for $\lambda < \rho + 1$ and as $r \to \infty$,

$$\int_\alpha^r t^{\rho(t)-\lambda}\, dt = \frac{1}{\rho + 1 - \lambda} r^{\rho(r)+1-\lambda} + o(r^{\rho(r)+1-\lambda}),$$

and for $\lambda > \rho + 1$ and as $r \to \infty$,

$$\int_r^\infty t^{\rho(t)-\lambda}\, dt = \frac{1}{\lambda - \rho - 1} r^{\rho(r)-\lambda+1} + o(r^{\rho(r)+1-\lambda}).$$

6.6.6 *The Spectrum of Perturbed Self-Adjoint and Non-Self-Adjoint Operators*

The assertions given in Corollary 6.6 for periodic potentials were presented by one of the authors at the Second Summer Mathematical School in Baku in 1975; see [781], [780]. Under the conditions $c_1|t|^{-\alpha} < |p(t)| < c_2|t|^{-\alpha}$, $0 < \alpha < 1$, $p(t) \nearrow 0$, as $|t| \to \infty$ monotonely on $\mathbb{R}_\pm$, the infinitude of the number of discrete levels introduced into each spectral gap was established independently by L. B. Zelenko [928]. Moreover, in [928], asymptotic formulas for the distribution of discrete levels were given. In particular, for $\lambda_{2\nu-1} < \lambda \nearrow \lambda_{2\nu}$, Zelenko's formula gives

$$N(\lambda_{2\nu-1}, \lambda) \sim (\pi\omega)^{-1}|2F'(\lambda_{2\nu})|^{1/2} \int_{p(t)<\lambda-\lambda_{2\nu}} (\lambda - \lambda_{2\nu} - p(t))^{1/2}\, dt,$$

where $F(\lambda) = \frac{1}{2}[\theta(\omega, \lambda) + \phi'(\omega, \lambda)]$ is Hill's discriminant. (See also more recent works of A. V. Sobolev [867], [868] and K. M. Schmidt [838]–[841].)

The first theorem giving a sufficient condition for the infinitude of the number of discrete levels in a spectral gap of the Dirac operator was proven by B. Ya. Skachek [863].

For the canonical system with the so-called pseudo-exponential potential, bound states and scattering problems are investigated in [481] and [482].

Interesting facts about multi-dimensional Schrödinger operators with spherically symmetric potentials depending on the radius were discovered in [312], [507], [508], [555] and [837]. For example (see [507]), it turns out that in the two-dimensional case, finite spectral gaps of the corresponding one-dimensional Hill operator are densely filled by eigenvalues, while in a semi-infinite gap there is an infinite number of eigenvalues which converge to the infimum of the essential spectrum [837].

In contrast to the Schrödinger operator, the spectrum of the Dirac operator (if the potential $V = V(r)$ is spherically symmetric and tends to infinity "in a regular way" as $r \to \infty$ and the mass $m(r) = m$ is constant) is, as shown in [836], the whole real line. If $m(r) = V(r)$ is of "regular growth" at infinity, then the absolutely continuous spectrum is the negative half-line, while there is a purely discrete spectrum on the positive half-axis; see [842]. The work [528] by H. Kalf, T. Okaji and O. Yamada investigates non-existence of eigenvalues of Dirac operators with variable mass $m(r)$ and 4×4 matrix-valued potential $V(x)$ which is not assumed to be spherically symmetric and is not necessarily smooth or Hermitian, and $V(x)$ may decay or diverge at infinity. As a result, in [528] it is shown that there are no eigenvalues for a large class of $m(r)$ and $V(x)$ such that $|m(r)| << |V(x)| \to \infty$ and $V(x)$ is positive or negative definite at infinity.

Regarding Sec. 6.5, Theorem 6.13, together with the main ideas of the proof, was published in [769]. Theorem 6.13 is the first result in the direction of the solution to the question about discrete levels in finite spectral gaps, which was posed in 1961 in the lecture by M. Sh. Birman and I. M. Glazman at the Fourth Soviet Mathematical Congress (see also [70, Chapter V, p. 56]). The proof of Lemma 6.12 appeared in [780].

Theorem 2 in [780], whose generalization is given below by the author of the theorem, is devoted to the question about eigenvalues in the continuous spectrum of a perturbed operator of a more general form.

Theorem 6.28 *Let $q_j(t+\omega_j) = q_j(t) \in L_{1,loc}$, $j = 1, 2$ be two periodic functions (possibly complex-valued) and let*

$$q(t) = \begin{cases} q_1(t), & 0 < t < \infty \\ q_2(t), & -\infty < t < 0. \end{cases} \tag{6.256}$$

Then the essential spectrum of the operator L_q in (6.1) satisfies the equality

$$\sigma_e(L_q) = \sigma(L_{q_1}) \cup \sigma(L_{q_2}), \tag{6.257}$$

where L_{q_j} are Hill operators with the potentials q_j on the real line.

If

$$p \in L_1(-\infty, \infty), \tag{6.258}$$

then the essential spectrum of the perturbed operator (6.2) satisfies

$$\sigma_e(L_q + p) = \sigma_e(L_q) \tag{6.259}$$

and does not carry any eigenvalues, with a possible exception of those $\lambda \in \sigma_e(L_q)$ for which (6.1) has unbounded solutions on both half-lines $\mathbb{R}_\pm$. If (6.168) holds, then $\sigma_e(L_q+p)$ is completely free of any eigenvalues. (Theorem also holds for a problem on the half-line with a boundary condition at $t = 0$.)

Proof. If for some $\lambda = \lambda'$, all solutions of (6.1) are bounded on some half-line, say for $0 < t < \infty$, then none of the solutions, except for the trivial one, belongs to $L_2(0, \infty)$.

Now we need the following Lemma.

Lemma 6.16 **([770, Part 2 of Lemma 2])** *If $u_1(t)$, $u_2(t)$ are solutions of (6.1) such that $u_1u'_2 - u'_1u_2 \equiv 1$ and for some λ,*

$$\int_0^\infty r^2(t)|p(t)|\,dt < \infty,$$

where $r^2 = |u_1|^2 + |u_2|^2$, then for any non-trivial solution $y(t)$ of the equation (6.2), the following asymptotic relation holds for the same λ and for some C_1 and C_2:

$$y(t) = C_1u_1(t) + C_2u_2(t) + O\{r(t)\int_t^\infty r^2(s)|p(s)|\,ds\}, \tag{6.260}$$

where $|C_1| + |C_2| \neq 0$.

By Lemma 6.16 and (6.258) we see that if $\lambda = \lambda'$, the solutions of (6.2) have the same asymptotics as the solutions of (6.1) as $t \to \infty$. Hence, λ' is not an eigenvalue of (6.2). Now suppose that the condition (6.168) holds. If (6.1) has unbounded solutions for $\lambda = \lambda' \in \sigma_e(L_q)$ (for example, as $t \to +\infty$), then one of the following two conditions holds:

a) on the interval $(0, \infty)$, the equation (6.1) has a fundamental system of solutions of the form

$$e_1(t, \lambda'), \quad g_1(t, \lambda') = \alpha_1(\lambda')te_1(t, \lambda') + f_1(t, \lambda'), \tag{6.261}$$

where both $e_1(t, \lambda')$, $f_1(t, \lambda')$ are either ω_1 periodic or ω_1-anti-periodic ($2\omega_1$-periodic), $\alpha_1(\lambda') \neq 0$;

b) on the interval $(0, \infty)$, the equation (6.1) has a fundamental system of solutions consisting of an exponentially growing and an exponentially decaying solution. But, then, on the negative half-axis $(-\infty, 0)$, the equation (6.1) cannot have an analogously behaving fundamental solution. Otherwise, λ' would not belong to $\sigma_e(L_q)$ and would be either an isolated eigenvalue of the operator L_q or a regular point, and this can be shown by constructing Green's function for (6.1); compare with [767]. Thus in case b), $\lambda' \in \sigma(L_{q_2}) = \sigma_e(L_{q_2})$.

In the latter case, either all solutions of (6.1) are bounded as $t \to -\infty$ and do not belong to $L_2(-\infty, 0)$, and, hence, as shown at the beginning of the proof, (when $t \to +\infty$ and when (6.258) holds), the solutions of (6.2) behave the same, and λ' is not an eigenvalue of (6.2) or (6.1) has a fundamental system on $-\infty < t < 0$ analogous to (6.261):

$$e_2(t, \lambda'), \quad g_2(t, \lambda') = \alpha_2(\lambda')te_2(t, \lambda') + f_2(t, \lambda'), \tag{6.262}$$

where both $e_2(t,\lambda')$, $f_2(t,\lambda')$ are either ω_2 periodic or ω_2-anti-periodic ($2\omega_2$-periodic), $\alpha_2(\lambda') \neq 0$.

We will show that under the condition (6.168), in the case (6.262), as in the case (6.261), the perturbed equation (6.2) has, as $t \to +\infty$ and $t \to -\infty$ respectively, fundamental systems with asymptotics (6.261) and (6.262), and, hence, λ' is not an eigenvalue of L_q. (Analogously as in a) and b), we consider the case when (6.1) has unbounded solutions as $t \to -\infty$.)

Consider (6.261) when the condition (6.168) holds.

In this case, (6.2) has a fundamental system of solutions $E_1(t,\lambda')$, $G(t,\lambda')$, where $E_1(t,\lambda')$ is defined by the formula (6.173) and, as $t \to +\infty$, due to (6.174), has asymptotics of $e_1(t,\lambda')$, and $G_1(t,\lambda')$ is determined from the equation

$$G_1(t,\lambda') = g_1(t,\lambda') - g_1(t,\lambda')w^{-1}(\lambda') \int_t^{\infty} p(s)e_1(s,\lambda')G_1(s,\lambda')\,ds$$
$$-e_1(t,\lambda')w^{-1}(\lambda') \int_N^t p(s)g_1(s,\lambda')G_1(s,\lambda')\,ds, \qquad (6.263)$$

where $w(\lambda')$ is the Wronskian of solutions $e_1(t,\lambda')$ and $g_1(t,\lambda')$, which converges, by the method of successive approximations, if $N > 0$ is chosen so that the integral $\int_N^{\infty}(t+1)|p(t)|\,dt$ is sufficiently small (cf. [1, Theorem 1.5.1]). From (6.263) we see that $G_1(t,\lambda')$ has the asymptotics of $g_1(t,\lambda')$ as $t \to +\infty$, as needed. The case (6.262) is analogous.

It remains to check the equality (6.259) when the condition (6.258) holds.

If $\lambda \in \sigma_e(L_q)\backslash(\{\lambda\colon \rho_1(\lambda) = \pm 1\} \cup \{\lambda\colon \rho_2(\lambda) = \pm 1\})$, where $\rho_j(\lambda)$ are multipliers of $L_j - \lambda$, then, as shown above, under the condition (6.258), the equation (6.2) has bounded solutions on at least one of the half-lines $\mathbb{R}_{\pm}$ (not belonging to L^2 on the corresponding half-line). By constructing a sequence of compactly supported test "almost eigenfunctions" with non-overlapping supports[42], it follows that $\lambda \in \sigma_e(L_q + p)$. Since the essential spectrum is closed, it follows that $\sigma_e(L_q) \subseteq \sigma_e(L_q + p)$.

If $\lambda \notin \sigma_e(L_q)$, then none of the multipliers $\rho_j(\lambda)$ of $L_j - \lambda$, $j = 1, 2$, has the absolute value 1; hence, (6.1) has exponentially decreasing solutions $e_1(t,\lambda)$, $(t \to +\infty)$, and $e_2(t,\lambda)$, $(t \to -\infty)$. In this case, the formulas (6.173) give the solutions $E_1(t,\lambda)$, $E_2(t,\lambda)$ with respective asymptotics $e_1(t,\lambda)$, $(t \to +\infty)$, and $e_2(t,\lambda)$, $(t \to -\infty)$ under the condition (6.258) instead of (6.168). This is easy to see since *outside neighborhoods of the*

[42]Compare with the construction of test functions for $l[y] = \lambda y$ in (6.209), especially the footnote below Theorem 6.22.

points where $\rho(\lambda) = \pm 1$, it is possible to omit the term $|t-s|$ in the numerator of (6.174). The same holds for the estimate (6.172) because on the complement of the set of points where $\rho(\lambda) = \pm 1$, the monodromy matrix is diagonalizable. The above shown presence of exponentially decaying solutions $E_1(t,\lambda)$ $(t \to +\infty)$ and $E_2(t,\lambda)$ $(t \to -\infty)$ shows that $\lambda \notin \sigma_e(L_q + p)$; however, λ is a regular point of (6.2) if the Wronskian $W\{E_1, E_2\} \neq 0$ (because, in this case, Green's function exists and generates a bounded operator) or λ is an isolated eigenvalue of (6.2) if $W\{E_1, E_2\} = 0$. Therefore, (6.259) and the theorem are proven. □

Special cases of the above theorem are contained in [769], [780] and [935]. In [780], the case $q_2(t) = q_1(t-a)$ was considered, corresponding to the perturbation of the potential $q_1(t)$ of a one-dimensional crystal lattice composed of a phase-shift of a periodic potential on the negative part of the real axis ("partial shift") and a small at infinity potential $p(t)$ corresponding to a local impurity in the crystal ($p(t)$ is the so-called potential of local impurity). The case considered in Theorem 6.28 corresponds to two one-dimensional half-spaces filled with two different crystal lattices plus a local impurity with the potential $p(t)$ on the background of the potential of the two lattices.

Here, we call attention to the works by I. A. Anders and V. P. Kotlyarov [224], E. Ya. Khruslov and V. P. Kotlyarov [553] and V. P. Kotlyarov [581] in which the problems for step-type data (6.256), where q_j = const and $j = 1, 2$, with perturbation are considered, and some other works of the same authors: [223], [300], [301] and [303].

Also note that one-dimensional Schrödinger equation with an almost-periodic potential can be considered as a one-dimensional model of the so-called quasi-crystal (see, for example, Supplement 3 and the Appendix to Supplement 3 in the book by S. P. Novikov and A. T. Fomenko [149]).

The following theorem is concerned with the eigenvalues of a perturbed non-self-adjoint Hill operator.

Theorem 6.29 **(V. A. Zheludev [935])** *The eigenvalues of a perturbed Hill operator with complex-valued continuous $q(t)$ and $p(t)$ do not have finitely spaced accumulation points if for some $\delta > 0$*

$$|p(t)| < C \exp\{-\delta |t|^{1/2}\}.$$

This theorem generalizes a well-known result of B. S. Pavlov [734], [735] and [736]. (Zheludev's result turns into Pavlov's result if $q(t) \equiv 0$.) In [736], there is a result showing that it is not possible to weaken the requirement on the perturbation:

Theorem 6.30 **(B. S. Pavlov [736, II])** *For every* $\beta \in (0, \frac{1}{2})$ *and* $\lambda_0 \in (0, \infty)$, *there exists a differential operator* l_h *of the form (6.1) on the half-line* $0 < t < \infty$ *with the boundary condition* $u'(0) = hu(0)$ *such that for some complex* h, *the operator* l_h *has infinitely many eigenvalues with the accumulation point* λ_0. *The potential of this operator is a real-valued infinitely differentiable function satisfying the condition*

$$|q(t)| < C \exp\{-\delta t^{\beta}\}, \quad C > 0, \delta > 0, 0 < t < \infty.$$

In the form given here, Theorem 6.14, was first published by F. S. Rofe-Beketov in [775] (see also [776]). Independently, but with an additional requirement on the finiteness of the second moment in the perturbation, Theorem 6.14 was proven by N. E. Firsova [425], who was able to remove this extra requirement later in [427].

Starting from the results of [769], requiring (6.179) and $t^2 p(t) \in L^1(-\infty, \infty)$, as a requirement in addition to the hypotheses of Theorems 6.13 and 6.14, V. A. Zheludev [934], [936] proved the appearance of eigenvalues in every sufficiently distant spectral gap of the perturbed Hill operator. Zheludev showed that if $p(t)$ does not change its sign and $|p(t)| <$ const, then every sufficiently distant spectral gap contains precisely one eigenvalue. Theorem 6.14 is based on two conditions: Zheludev's condition (6.179) and Rofe-Beketov's condition (6.168). Neither of these two conditions can be completely removed. For example, the function $p(t) = -(1 + t^2)^{-2/3}$ satisfies the condition (6.179); however, as easily seen from Theorem 6.12, $p(t)$ introduces an infinite number of eigenvalues into every spectral gap of (6.2).

Regarding Remark 6.8, Zheludev proved that if the second moment of the perturbation $p(t)$ is small and does not change its sign, then precisely one eigenvalue appears in every finite spectral gap.

More delicate and more complicated question about discrete levels in spectral gaps of a perturbed Hill operator when $\int_{-\infty}^{\infty} p(t) = 0$, *a priori* 0, 1 or 2 levels in each distant gap (due to [769]), was considered by N. E. Firsova in [429] and [430]. Perturbations of periodic systems of arbitrary order affecting all coefficients were considered by V. I. Khrabustovskiy [545]–[547].

6.6.7 *Matrix Operators or Differential-Algebraic Operators and Related Topics*

More recently, so-called matrix operators or differential-algebraic operators, including periodic systems of differential equations with a singular leading coefficient, have been studied; see [210], [243], [664], [673], [848] and [921]. These operators are of interest in physics; for example, in linear magnetic hydrodynamics, in the description of the oscillation of plasma in a periodic magnetic field [116]. In connection with mathematical aspects of these topics, we mention the monographs [65], [174], [179], [180] and the following monographs that investigate applied problems by methods of operator theory: [17], [67], [68], [91], [92], [93], [104] and [144]. More specifically, those books apply methods of functional analysis to the study of small movements and normal oscillations of hydromechanical systems having cavities filled with either ideal or viscous fluids and to other important problems.

In particular, in [67] the stationary wave motion with harmonic dependence on time is considered, and [68] is devoted to the investigations of time-dependent problems of the theory of internal waves in stratified and rotating fluids. Here, mathematical aspects of the study of well-posed solvability in classical and generalized sense as well as the construction of explicit solutions for some practically important problems are considered. Additionally, the results of the investigation of limiting behavior of solutions at large values of time, stabilization and transition to stationary oscillation stage are presented. The main results for the problems considered in [67] and [68] were obtained by the authors of those books.

The monograph [144] (and, partially, [17]) studies fluids with a free surface considered under zero-gravity or low-gravity conditions; it also studies equilibrium states, their stability conditions and their bifurcations. The second part of [144] is devoted to small oscillations in ideal and viscous fluids and the conditions of the onset of self-gravitational and thermocapillary convection. Since low-gravity fluid mechanics, which is of great importance in space technology, is hard to test in laboratory, a significant part of the text of [144] is devoted to problems of computer simulation. The material in [144] is presented in a qualitative and quantitative way that does not require an advanced level of mathematics. The authors also produce in [144] rigorous mathematical results on stability and bifurcation of ordinary and partial differential equations. Many numerical results are collected in tables and graphs.

The monograph [91] presents evolutionary and spectral problems arising from the study of small motions and eigenoscillations of an ideal fluid. These problems are self-adjoint in a suitably chosen Hilbert space or in a Pontryagin space. The monograph [92] is a collection of non-self-adjoint problems on small motions and normal oscillations of a viscous fluid filling a bounded region (container). The monograph [92] is also concerned with the study of small oscillations of complex hydrodynamic systems. In Chapter 10 of [92], partially dissipative hydrosystems are studied for the first time. The contents of [91] and [92] are based on the authors' and their students' works.

The book [105], containing lectures given by its author at the University of Rome, is devoted to problems of the theory of attractors for semigroups and applications to evolution equations arising from physics, and is based on its author's results obtained in the period from 1975 to 1990.

The book [95] is devoted to differential equations with operator coefficients, where the coefficients are bounded as operators from one to another suitable Hilbert space (for example, the coefficients themselves may be some differential operators from one to another Sobolev space). The authors of [95] investigate many applications of such equations to boundary value problems for partial differential equations.

A recent monograph [43] deals with modern ideas and methods of the theory of order preserving random dynamical systems. This theory is applicable to dynamics of systems under noise perturbations. The monograph [43] investigates properties of an important concept of "random attractors" and considers many interesting examples and applications; see also the works [758], [759] and [760] devoted to the description of asymptotic dynamics of reaction-diffusion equations in a thin two-layer domain with a junction condition on the interface. Basic objects of study in [758], [759] and [760] are global attractors and invariant manifolds of the dissipative compact semigroup corresponding to the problem. Those works contain first results on asymptotic dynamics for the corresponding problem in composite domains. The works [437], [601] and [602] are devoted to coupled thermo-mechanical problems and two-fluid model in magnetohydrodynamics.

Last decades have been characterized by implementation of various functional analysis methods in the field of continuous medium mechanics. Important pioneering work in this area was done by K. O. Friedrichs, O. Ladyzhenskaya, J. Leray, S. L. Sobolev, and H. Weyl. Later on, many mathematicians and researchers in mechanics from all over the world contributed jointly to the development of this area.

The problems considered in the above mentioned books are described or reduced to differential equations with operator-valued coefficients, where the leading coefficient is either an identically degenerated operator or a compact operator (and, therefore, does not have a bounded inverse in an infinite dimensional space). Such problem was considered by S. L. Sobolev [869] in connection with the investigation of oscillations of a fluid in a container under weightlessness conditions. This problem is described by a system of partial differential equations allowing reduction to an equation of the form:

$$A\frac{du}{dt} + Bu = f(t)$$

in a Hilbert space. Equations of such kind are sometimes called "abstract Sobolev equations," or "implicit" or "degenerate operator-differential equations."

In connection with the magnetohydrodynamic theory of processes in the nuclear matter, in addition the above-mentioned book [116], we note the papers [214] and [215]. In these papers, small oscillations of density and magnetization of the neutron matter are investigated near equilibrium values defined by spontaneous magnetization of the neutron matter. The spectra of oscillations are found. Special attention is paid to investigation of resonance phenomena between acoustic and magnetic branches of oscillations and, also, between properly electro-magnetic and magnetic oscillatory branches. Here we also note the recent paper [812].

We also mention the works [211]–[213] where the theory of propagation of a chain nuclear reaction in multiplicating medium is studied in diffusion approximation and the propagation velocity of a nuclear reaction is obtained in the overcritical regime. The problem of realizing slow nuclear burning regime is of current concern, and proposals that use slow intermediate reactions with fast neutrons are offered in [423], [483], [484] and [885], allowing many orders decrease[43] of effective velocity of nuclear reaction propagation in a matter. In many aspects, this problem is analogous to the biological problem of the extension velocity of a steadily progressive wave of gene increase due to the local establishment of a favorable mutation, for the case of a uniform linearly distributed population; see the works [66], [431] and [566]. The results of [566] and recent developments are reviewed in [173] and [740]; see also [346].

Methods of the theory of analytic functions, spectral theory of non-self-adjoint operators and operator pencils and Maslov's operator method

[43]i.e. decrease 10^n times, where $n >> 1$.

were applied in the investigation of the first or higher-order linear operator-differential (or abstract) equations (that are not always solved with respect to the highest-order derivative) in the classical works [219], [381], [611] [612] and [665] as well as in more recent works [445], [488], [745]–[747], [818]–[820], and [904]–[906]. Those methods were also applied to semilinear equations

$$A\frac{du}{dt} + Bu = f(t, u)$$

in the works [419], [420], [525] [526], [538], [821], [822], [823] and [907]. Applications to the problems of electric circuits, electrodynamics, filtration of fluids in non-homogeneous media and other degenerate problems for partial differential equations of the non-Kovalevskaya type were considered in the works [420], [538], [818], [821]–[823] and [905]. In the above-mentioned type of problems, describing the manifold of admissible initial data and obtaining criteria for correctness of the Cauchy problem are theoretically difficult and, at the same time, are of interest in applications.

In connection with the Boltzmann equation, Povzner equation and other equations in kinetic theory of gases, we note the works [38], [230], [287]–[289], [336], [368], [452], [453], [497]–[500], [741], [742], and [753].

In conclusion of this Subsection, we note recent publications [564] and [565] by A. N. Kolmogorov, where some problems on turbulent motion of an incompressible viscous fluid are posed, and the books [90], [140] and [194].

6.6.8 *The Inverse Sturm–Liouville Problem for the Spectral Matrix on the Whole Axis*

In this section we consider the inverse problem for the spectral matrix for Sturm–Liouville equations on the real axis with scalar potential $q(x)$ or operator potential $Q(x)$. When the boundary conditions at both ends $\pm\infty$ are imposed, we assume that the conditions are separated in the scalar case (Theorems 6.31 and 6.32 below) as well as in the case of operator equations (Theorems 6.33 and 6.34 below).

Theorem 6.31 **(Rofe-Beketov [771])** *Symmetric real non-decreasing matrix-function* $R(\lambda) = (\rho_{jk}(\lambda))_{j,k=1,2}$ *is a spectral matrix function of the equation (6.1) on the real axis* $-\infty < x < +\infty$ *with arbitrary real-valued potential* $q(x) \in C^{r_\pm}(\mathbb{R}_\pm)$ *if and only if the following two conditions hold:*

(1) *Stieltjes transform of the matrix* $R(\lambda)$ *has constant determinant equal to* $-\frac{1}{4}$, *i.e. the functions*

$$M_{1k}(z) = \int_{-\infty}^{\infty} \frac{d\rho_{1k}(\lambda)}{\lambda - z}, \qquad k = 1, 2,$$

$$M_{22}(z) = \int_{-\infty}^{\infty} \left(\frac{1}{\lambda - z} - \frac{\lambda}{1 + \lambda^2} \right) d\rho_{22}(\lambda) + a,$$

where $M_{22}(z) = \frac{i}{2}\sqrt{z} + o(1)$ *as* $z \to +i\infty$, *exist and*

$$\det(M_{jk}(z))_{j,k=1,2} \equiv -\frac{1}{4}, \qquad (Im\ z \neq 0).$$

This condition guarantees the existence of two non-decreasing functions $\rho_1(\lambda)$, $\rho_2(\lambda)$ *that are uniquely determined by the Stieltjes inversion formula:*

$$\rho_k(\mu) - \rho_k(\nu) = \frac{1}{\pi} \lim_{\varepsilon \downarrow 0} \int_{\nu}^{\mu} Im \frac{M_{11}(\lambda + i\varepsilon)\, d\lambda}{\frac{1}{2} - (-1)^k M_{1k}(\lambda + i\varepsilon)}.$$

(2) *The functions* $\rho_1(\lambda)$, $\rho_2(\lambda)$ *must be spectral functions of some Sturm–Liouville problems on the semi-axes* $\mathbb{R}_-$ *and* $\mathbb{R}_+$ *respectively, with the condition* $y'(0) = 0$ *and with potentials of the class* $C^{r_\pm}(\mathbb{R}_\pm)$. *From the spectral functions* $\rho_1(\lambda)$ *and* $\rho_2(\lambda)$ *the potential* $q(x)$ *is constructed on* $\mathbb{R}_-$ *and* $\mathbb{R}_+$ *respectively.*

Theorem 6.32 **([778])** *The potential* $q(x)$ *is determined only by diagonal elements of the spectral matrix up to mirror symmetry, i.e. up to the replacement of* $q(x)$ *by* $q(-x)$.

With regard to problems of eigenfunction expansion for Sturm–Liouville equations on the whole line, we will mention the paper [628].

We now consider the equation (1.51) in a separable Hilbert space $\mathbf{H}$ with operator potential $Q(x) = Q^*(x) \in B(\mathbf{H})$ on the real axis.

Theorem 6.33 **(Khrabustovskiy [552])** *Non-decreasing operator matrix-function* $R(\lambda) = (\rho_{jk}(\lambda))_{j,k=1,2}$, $\rho_{jk}(\lambda) \in B(\mathbf{H})$, *is a spectral matrix of the equation (1.51) on the real axis* $-\infty < x < +\infty$ *with operator potential* $Q(x) = Q^*(x) \in C^{r_\pm}(\mathbb{R}_\pm)$ *if and only if the following two conditions hold:*

(1) *Let for Im* $z > 0$ *Stieltjes transforms of operator blocks of the spectral matrix be defined by the formulas (6.264) and (6.265) below:*

$$M_{11}(z) = \int_{-\infty}^{\infty} \frac{d\rho_{11}(\lambda)}{\lambda - z},$$
$$M_{jk}(z) = a_{jk} + \int_{-\infty}^{\infty} \left(\frac{1}{\lambda - z} - \frac{\lambda}{1+\lambda^2} \right) d\rho_{jk}(\lambda), \qquad j+k > 2, \tag{6.264}$$

where the integrals converge weakly, and the operator constants $a_{jk} = a_{kj}^* \in B(\mathbf{H})$ *are determined by the following asymptotics:*

$$\|M_{jk}(z)\| \to 0 \quad (j+k < 4),$$
$$\left\| M_{22}(z) - \frac{i\sqrt{z}}{2} I \right\| \to 0, \quad z \to +i\infty. \tag{6.265}$$

These Stieltjes transforms must satisfy at least one of the following identities:

$$M_{11}M_{22} - M_{12}^2 = -\frac{1}{4}I, \qquad M_{22}M_{11} - M_{21}^2 = -\frac{1}{4}I, \tag{6.266}$$

and, also, at least one of the following identities:

$$M_{12}M_{11} = M_{11}M_{21}, \qquad M_{21}M_{22} = M_{22}M_{12}. \tag{6.267}$$

(2) *To Nevanlinna-type analytic operator–functions, which exist for Im* $z \neq 0$ *by the condition (1),*

$$m_{\pm}(z) = \mp M_{11} \left(M_{21} \mp \frac{1}{2} I \right)^{-1} = \mp \left(M_{12} \mp \frac{1}{2} I \right)^{-1} M_{11}$$
$$= \mp \left(M_{12} \pm \frac{1}{2} I \right) M_{22}^{-1} = \mp M_{22}^{-1} \left(M_{21} \pm \frac{1}{2} I \right) = m^*(\bar{z}).$$

we associate, using operator Stieltjes inversion formula, non-decreasing operator–functions $\rho_{\pm}(\lambda)$.
The operator–functions $\rho_{\pm}(\lambda)$ *must be the spectral functions of some Sturm–Liouville problems on semi-axes* $\mathbb{R}_{\pm}$ *with the condition* $y'(0) = 0$ *and with potentials of the class* $C^{r_{\pm}}(\mathbb{R}_{\pm})$.
From the spectral functions $\rho_{\pm}(\lambda)$ *the operator potential* $Q(x)$ *is constructed on* $\mathbb{R}_{\pm}$ *according to a well-known scheme, as it was done for the operator problem in [766].*

Remark 6.11 In [552] it was shown that from the condition (1) of Theorem 6.33 it follows that for Im $z \neq 0$ all four identities (6.266) and (6.267) are satisfied, and the operator–functions

$$M_{jj}^{-1}(z) \in B(\mathbf{H}), \qquad \left(M_{jk}(z) \pm \frac{1}{2}I\right)^{-1} \in B(\mathbf{H}), \quad j \neq k.$$

Theorem 6.34 (Khrabustovskiy [552])

(1) If $R(\lambda) = (\rho_{jk}(\lambda))_{j,k=1,2}$ is a spectral matrix of the equation (1.51) with operator potential $Q(x) = Q^*(x)$ on the real axis, then $((-1)^{j+k}\rho_{jk}(\lambda))_{j,k=1,2}$ is a spectral matrix of the equation (1.51) with potential $Q(-x)$.

(2) The potential $Q(x) = Q^*(x)$ in the equation (1.51) is even, i.e. $Q(x) = Q(-x)$, if and only if the equation (1.51) has block-diagonal spectral matrix (i.e. with blocks $\rho_{12} = \rho_{21} = 0$).

Remark 6.12 In the case $\dim H < \infty$, for the equation (1.51) with matrix potential $Q(x) = Q^*(x)$, the identities of the form (6.266) and (6.267) for the blocks of the Weyl–Titchmarsh matrix are contained in the interesting work of F. Gesztesy and L. Sakhnovich [458] in the case of limit points at $\pm\infty$. Using those identities, a class of Schrödinger operators with $n \times n$ matrix potential and with a given finite-zone spectrum having maximal possible multiplicity of $2n$ was constructed in [458].

* * *

In connection with [458], it is also interesting to note the work [343] by S. Clark, F. Gesztesy, H. Holden and B. M. Levitan, where a Borg-type uniqueness theorem for matrix-valued Schrödinger operators is proven. Namely, assuming a reflectionless potential matrix and its spectrum $[0, \infty)$, the authors of [343] derive the triviality of the potential matrix. As a by-product of their method, they obtained an extension of Borg's classical result (see [296]) from the class of periodic scalar potentials to the class of reflectionless $(n \times n)$-matrix valued periodic potentials having uniform spectral multiplicity $2n$. If, in the latter case, the spectrum coincides with $[a, \infty)$ for some $a \in \mathbb{R}$, then the potential $Q(x) \equiv aI_n$ for a. e. $x \in \mathbb{R}$.

As a continuation and generalization of [343], we note the paper [255] by E. D. Belokolos, F. Gesztesy, K. A. Makarov and L. A. Sakhnovich. In this paper, the authors prove a generalization in the matrix-valued context of the well-known theorems by Borg and Hochstadt for periodic self-adjoint

Schrödinger operators without a spectral gap and, respectively, one gap in their spectrum. This extension of theorems of Borg and Hochstadt replaces the periodicity condition of the potential by the more general property of being reflectionless (the resulting potentials then automatically turn out to be periodic, and [255] recovers Despres' matrix version of Borg's result). In addition, the authors of [255] assume that the spectra have uniform maximum multiplicity (a condition automatically fulfilled in the scalar context considered by Borg and Hochstadt). Moreover, in [255] the authors establish the connection with the stationary matrix KdV-hierarchy. The methods employed in [255] rely on matrix-valued Nevanlinna–Herglotz functions, Weyl–Titchmarsh theory, pencils of matrices, and basic inverse spectral theory for matrix-valued Schrödinger operators.

Note also a recent work [466], where a general Borg-type result for reflectionless unitary CMV (Cantero, Moral, and Velázquez [335]) operators U associated with orthogonal polynomials on the unit circle is proven. The spectrum of U is assumed to be a connected arc on the unit circle. This extends a recent result of B. Simon [181, Sec. 4.5 and 10.5] in connection with a periodic CMV operator whose spectrum is the whole unit circle.

Other versions of direct and inverse problems of spectral analysis for five-diagonal semi-infinite symmetric matrices were solved in the works by M. A. Kudryavtsev [606], [607]. The inverse problem on the axis for the spectral matrix for a difference equation was solved by Yu. M. Berezanskiy [24] by his "doubling" method. The difference analogue of Theorem 6.31 and a useful addition to the "doubling" method from [24] is the following theorem.

Theorem 6.35 **(G. Sh. Guseinov [502])** *The symmetric real-valued non-decreasing matrix-function $R(\lambda) = (\rho_{j,k}(\lambda))_{j,k=1,2}$ is a spectral matrix of an equation of the form*

$$a_{n-1}u_{n-1} + b_n u_n + a_n u_{n+1} = \lambda u_n, \qquad n = 0, \pm 1, \pm 2, \dots,$$

where $a_n > 0$, $Im\, b_n = 0$, if and only if the following three conditions are satisfied:

(1)

$$\int_{-\infty}^{\infty} d\rho_{j,k}(\lambda) = \delta_{j,k},$$

and for $\rho_{j,j}$ there exist all moments on the axis;

(2) there exists a number $a > 0$ such that

$$M_{1,1}(z)M_{2,2}(z) - (M_{1,2}(z))^2 = \frac{-1}{4a^2}, \qquad Im\ z \neq 0,$$

where

$$M_{j,j}(z) = \int_{-\infty}^{\infty} \frac{d\rho_{j,j}(\lambda)}{\lambda - z},$$
$$M_{1,2}(z) = \frac{-1}{2a} + \int_{-\infty}^{\infty} \frac{d\rho_{1,2}(\lambda)}{\lambda - z};$$

(3) the functions

$$\frac{M_{j,j}(z)}{1 - 2aM_{1,2}(z)}$$

are not rational functions of the variable z.

This result was also obtained by N. V. Zhernakov [937] in a similar form. Additionally, using the paper of L. V. Perkolab [739] (the discrete analogue of Marchenko–Ostrovskiy Theorem), N. V. Zhernakov obtained the necessary and sufficient conditions for a non-decreasing 2×2 matrix function to be the spectral matrix of a periodic difference operator of the considered type on the axis. Using the inverse spectral problem theory, Yu. M. Berezanskiy and N. V. Zhernakov [269] found a new class of the initial data for which the system of equations for the Toda lattice which is infinite on both sides can be integrated explicitly—namely, when this lattice corresponds to an L, A-pair with a difference operator $L(t)$, which is compact for $t = 0$ and, as shown, for $t > 0$ as well (this operator is similar to the one considered in Theorem 6.35 with coefficients $a_n(t)$ and $b_n(t)$, where $a_n(0) \to 0$ and $b_n(0) \to 0$ as $|n| \to \infty$). We also mention that M. A. Kudryavtsev [608] used a new type of inverse spectral problem to solve the Cauchy problem for Toda lattice with non-stabilizing initial data on the axis in the case when the corresponding L-operator for $t = 0$ has an interval of absolutely continuous spectrum of multiplicity two. For a semi-infinite Toda lattice with non-stabilizing Cauchy data, the problem was solved by Yu. M. Berezanskiy [265], [266] and M. Kac and P. van Moerbeke [527]. Note also a recent paper in this area by S. N. Bakh and A. A. Pankov [247].

6.6.9 *Three Open Questions*

Open Question 6.1 ([778]) Is it possible to replace the requirement (2) of Theorem 6.31 by requiring sufficient smoothness of $f(x,t)$ and $f(0,0)=0$?

Here,

$$f(x,t) := \int_{-\infty}^{\infty} \sum_{j,k=1}^{2} c_j(x,\lambda)c_k(t,\lambda)\, dS_{j,k}(\lambda), \tag{6.268}$$

where 2×2 matrix

$$S(\lambda) := \begin{cases} R(\lambda) & \text{if } \lambda < 0, \\ R(\lambda) - \pi^{-1}\text{diag}\ \{\lambda^{1/2},\ \frac{1}{3}\lambda^{3/2}\} & \text{if } \lambda > 0, \end{cases} \tag{6.269}$$

and

$$\begin{aligned} c_1(x,\lambda) &:= \cos(x\lambda^{1/2}), \\ c_2(x,\lambda) &:= \lambda^{-1/2}\sin(x\lambda^{1/2}). \end{aligned} \tag{6.270}$$

These conditions are simpler than (2).

Here, it is useful to keep in mind that the function $f(x,t)$ is the kernel of Gel'fand–Levitan type integral equation of the inverse problem on the whole axis $-\infty < x < \infty$ by the spectral matrix $R(\lambda)$:

$$K(x,t)\text{sgn}\ x + f(x,t) + \int_{-x}^{x} K(x,s)f(s,t)ds = 0, \qquad |t| \le |x|. \tag{6.271}$$

Here, $K(x,t)$ is the kernel of the transmutation operator:

$$\phi_j(x,\lambda) = c_j(x,\lambda) + \int_{-x}^{x} K(x,t)c_j(t,\lambda)\, dt, \qquad j = 1,2, \tag{6.272}$$

where $\phi_j(x,\lambda)$ are the solutions of (6.1) such that

$$\begin{aligned} \phi_1(0,\lambda) &= \phi_2'(0,\lambda) = 1, \\ \phi_1'(0,\lambda) &= \phi_2(0,\lambda) = 0. \end{aligned} \tag{6.273}$$

The inverse problem by the spectral matrix on the whole axis was first considered by A. Sh. Blokh [286]. He derived the equation (6.271); however, A. Sh. Blokh did not obtain a complete solution of the considered problem because he did not find specific properties of the spectral matrix that the spectral function of the Sturm–Liouville problem on the half-axis does not possess (see condition (1) of Theorem 6.31). In Theorem 6.31, the solution of the inverse problem on the whole axis for 2×2 spectral matrix is reduced to solutions of two inverse problems on the half-axis (left and right) by the corresponding spectral functions $\rho_j(\lambda)$ (here, $j = 1, 2$ respectively).

However, the solution of the inverse problem on the half-axis, including necessary and sufficient conditions for the spectral function and uniqueness of the solution of the corresponding inverse problem, is well known from the works by I. M. Gel'fand and B. M. Levitan [449], M. G. Krein [589], V. A. Marchenko [653]; see also the books [112], [121] and [145].

The Question 6.1 may be reformulated in the following way:

Open Question 6.1$^{\mathbf{bis}}$ Can the solution of the inverse Sturm–Liouville problem on the axis under the conditions formulated in the Open Question 6.1 be obtained from the equation (6.271) and the formula (6.275) without reduction to the problems on two half-axes? What are the necessary and sufficient conditions here?

In this direction, we can formulate just the following simple theorem, without reduction to the problems on two half-lines, but with another implicit condition (6.274).

Theorem 6.36 **([771], [778], [789])** *The inverse Sturm–Liouville problem by the spectral matrix on the axis is solvable if and only if the solution of (6.271) satisfies*

$$K(x, -x) = 0, \tag{6.274}$$

and, in this case,

$$q(x) = 2\frac{dK(x, x)}{dx}. \tag{6.275}$$

Open Question 6.2 ([789]) Consider the inverse spectral problem for the equation

$$-y'' + q(x)y = \lambda y \tag{6.276}$$

with operator-valued coefficients on the half-axis $0 \leq x < \infty$ and the boundary condition

$$y'(0) - Ay(0) = 0. \tag{6.277}$$

Here, $A = A^* \in B(H)$, $q^*(x) = q(x) \in B(H)$, and H is a separable Hilbert space. As shown in [121], [660] and [766], there exists a non-decreasing operator-valued spectral function $\rho(\lambda) \in B(H)$,

$$\rho^*(\lambda) = \rho(\lambda) \geq 0, \qquad -\infty < \lambda < \infty, \tag{6.278}$$

generating the eigenfunction expansion formulas and Parseval's equality for the boundary problem (6.276), (6.277) on the non-negative x-axis.

Let

$$f(x,t)$$
$$= \int_{-\infty}^{\infty} \cos(x\lambda^{1/2}) \cdot \cos(t\lambda^{1/2})\, d[\rho(\lambda) - 2\pi^{-1}(\max\{\lambda, 0\})^{1/2}]. \tag{6.279}$$

Then the inverse problem can be solved (see [121], [660] and [766]) with the help of operator-valued version of the Gel'fand–Levitan equation:

$$K(x,t) + f(x,t) + \int_0^x K(x,s) f(s,t)\, ds = 0. \tag{6.280}$$

Here, -1 does not belong to the spectrum of the integral operator (6.280) with the operator-valued kernel $f(s,t)$ (at every fixed $x > 0$) acting in the space $L_2(0,x;H)$ of(row)vector-functions, as shown in considering the direct spectral problem. In the scalar case, this property of the spectrum of the Gel'fand–Levitan equation (6.280), which is necessary for solvability of the inverse problem, is guaranteed (by virtue of Marchenko's Tauberian theorem [654]) for arbitrary non-decreasing function $\rho(\lambda)$ by the existence and sufficient smoothness of $f(x,t)$.

In the scalar case, the integral operator (6.280) is compact at each fixed $x > 0$, and, hence, it suffices to prove that -1 is not an eigenvalue of (6.280).

Now, the following question arises: what conditions on arbitrary non-negative operator-valued measure $\rho(\Delta)$ guarantee the above-mentioned property of the spectrum of operator-valued version of Gel'fand–Levitan equation (6.280) (i.e. the property that -1 does not belong to the spectrum of (6.280)), which, in turn, guarantee the solvability of the inverse Sturm–Liouville problem for a given operator-valued $\rho(\lambda)$? (In the case $\dim H = \infty$ the Gel'fand–Levitan operator (6.280) is not compact, and, therefore, it is not sufficient to establish that -1 is not an eigenvalue of (6.280)!)

Here, we need to explain that if $K(x,t)$ is found from the equation (6.280), then we can obtain $q(x)$ and A from the equations (6.276) and (6.277) by the following formulas:

$$q(x) = 2\frac{dK(x,x)}{dx}, \qquad x \geq 0, \quad A = K(0,0). \tag{6.281}$$

The function $K(x,t)$ is the kernel of the transmutation operator. If $y = \omega(x,\lambda)$ is a solution of the problem (6.276), (6.277) such that

$$\omega(0,\lambda) = 1, \qquad \omega'(0,\lambda) = A, \tag{6.282}$$

then

$$\omega(x,\lambda) = \cos(x\lambda^{1/2}) + \int_0^x K(x,t)\cos(t\lambda^{1/2})\,dt, \quad x{\geq}0. \tag{6.283}$$

Open Question 6.3 This question is a follow up to Question 6.2 and might be even simpler than Question 6.2. Let the operator-valued non-decreasing function $\rho(\lambda)$ be a spectral function of some Sturm–Liouville problem (6.276), (6.277) with operator-valued coefficients on the half-axis. Is the truncated function

$$\rho_N(\lambda) := \begin{cases} \rho(\lambda) & \text{if } \lambda > N, \\ 0 & \text{if } \lambda < N \end{cases} \tag{6.284}$$

a spectral function of some Sturm–Liouville problem of type (6.276), (6.277) on the half-axis?

We ask the same question for another truncated function

$$\rho_N(\lambda) := \begin{cases} \rho(\lambda) - \rho(N) & \text{if } \lambda > N, \\ 0 & \text{if } \lambda < N. \end{cases} \tag{6.285}$$

(If $\dim H < \infty$, the answer is affirmative for both (6.284) and (6.285).)

* * *

In connection with the inverse spectral problem on the half-axis for Sturm–Liouville equation with locally bounded self-adjoint operator-valued potential (6.276) and $A = A^*$ in (6.277) by a non-decreasing operator-valued spectral function $\rho(\lambda){\in}B(H)$, touched upon in Open Questions 6.2 and 6.3, note that the works [121], [660] and [766] provide a solution of the inverse spectral problem for the case of non-self-adjoint locally bounded operator-coefficients $q(x)$ and A in (6.276), (6.277). In this case, instead of the spectral operator-function (6.278), we have a generalized spectral matrix R whose entries are Marchenko-type generalized functions (distributions) on the topological space of analytic entire test functions of exponential type. Using the generalized spectral matrix R instead of $\rho(\lambda)$, one can construct the function $f(x,t)$ by a formula that generalizes the construction (6.279). Then the equation (6.280), under the corresponding necessary and sufficient conditions, permits the solution of the inverse problem by the formulas (6.281).

By the way, here is one more open question.

Open Question 6.4 For the non-self-adjoint equation of type (6.276) on the whole axis, what are the analogues of the identities (6.266)?

Generalized 2×2 spectral matrix for the scalar equation was constructed in [121, Chapter 2, Sec. 2, Problem 4], but our question is still open in the scalar case. Naturally, this question is simplified for complex-valued potentials of a special class, for which the eigenfunction expansion has a more explicit form. For example, potentials exponentially small at infinity (see [145]) or with discrete spectrum (see [633]) have the above mentioned property.

For a non-self-adjoint Schrödinger equation in the half-space, generalized spectral matrix was constructed in [584]. Expansion by solutions of the scattering theory problem for a non-self-adjoint Schrödinger operator with potential exponentially decreasing at infinity was constructed in [440], and an approach to spectral analysis of a dissipative Schrödinger operator, based on its dilation, was developed in [737].

In many cases, consideration of scalar partial differential equations can be reduced to consideration of the equation (6.276) on the (half-)axis with unbounded operator-valued coefficient $q(x)$; see [73], [74], [95], [492] and the Bibliographical Comments 6.6.7 and A.4.1 to Chapter 6 and Appendix of the present book. Using this approach, in [682]–[685], generalized spectral matrices of Marchenko type were constructed and uniqueness theorems for inverse problems by generalized spectral matrices were proven for non-self-adjoint abstract hyperbolic and elliptic differential operators on the half-axis of the type

$$L[u] = -u'' \mp Au + c(x)u,$$

where $A = A^* \geq \epsilon > 0$ is an unbounded self-adjoint positive definite constant operator, $c(x) \in B(H)$ is a non-self-adjoint operator-function, $u'(0) = Bu(0)$, and B is a bounded operator. The theory of rigged Hilbert spaces (see Bibliographical Comments A.4.1) was used in these problems.

In this regard, note also the recent work [888].

For uniqueness theorems for the inverse problem by the spectral kernel in the case of a scalar self-adjoint Schrödinger operator in $\mathbb{R}^n$, see [264] and [455]; for the case of an abstract hyperbolic symmetric operator, see [226].

In conclusion, note the books [123], [147], [161], and [163].

Appendix A

Self-Adjoint Extensions of Differential Operators on a Finite Interval in Spaces of Vector-Functions

Our goal is to use boundary conditions to describe all self-adjoint extensions of a symmetric differential operator, which, unlike in the scalar case, has infinite deficiency indices and is, generally speaking, not semi-bounded below. We show that if we start with an arbitrary vector-function $y(t)$ for which the differential expression $l[y]$ (2.1) makes sense, and if in a certain Hilbert space (called the space of boundary values) we construct two vectors $y^\wedge$ and $y^\vee$ expressed only in terms of the values of $y(t)$ and the first $r-1$ derivatives of $y(t)$ at the endpoints of $[0, b]$, where r denotes the order of the differential expression $l[y]$, then any self-adjoint boundary condition for the expression $l[y]$ can be written as

$$\cos A y^\vee - \sin A y^\wedge = 0, \tag{A.1}$$

where A is a self-adjoint operator. Conversely, for any self-adjoint operator A, the condition (A.1) gives rise to a self-adjoint boundary-value problem for the equation $l[y] = \lambda y$. In this case it may happen that none of the operators $\cos A$, $\sin A$ is invertible, and then, unlike in the case when A is a scalar, the equation (A.1) cannot be solved uniquely for at least one of the vectors $y^\wedge$, $y^\vee$.

An important class of self-adjoint boundary conditions are Sturm-type conditions, i.e. separated boundary conditions. They are specified by two equations of the type (A.1):

$$\cos A y_0^\vee - \sin A y_0^\wedge = 0,$$
$$\cos B y_b^\vee + \sin B y_b^\wedge = 0,$$

where vector pairs $y_0^\vee$, $y_0^\wedge$ or $y_b^\vee$, $y_b^\wedge$ are determined by the values of $y(t)$ and its derivatives only at $t = 0$ or $t = b$ respectively. The operators A and B are self-adjoint. If an expression $l[y]$ of even order always has separated self-adjoint boundary conditions, then an expression $l[y]$ of odd-order does not always have separated self-adjoint boundary conditions. Later, we shall give necessary and sufficient condition for an odd-order expression $l[y]$ to have separated self-adjoint boundary conditions.

We will use the method which is based on the notion of a Hermitian binary relation θ in an arbitrary Hilbert space and on the theorem about the general form of Hermitian relations that we prove in Sec. A.1. By this theorem, every Hermitian relation is equivalent to (A.1). On the other hand, using Lagrange's identity, we prove that specifying a self-adjoint extension for the expression $l[y]$ is equivalent to specifying a Hermitian relation. This enables us to obtain the general form of self-adjoint extensions for expressions $l[y]$ using boundary conditions of type (A.1). Unlike most of the existing methods for the description of self-adjoint extensions of symmetric operators, the method used in this book is not based on the well-known Neumann's theorem and other similar theorems about the construction of the domain of the adjoint of a symmetric operator.

In Sec. A.2, we demonstrate our method in the case of the second order equations and there we present all arguments in detail. In Sec. A.3, this enables us to focus our attention on the specifics of equations of arbitrary order without the need to repeat in a more complicated form the arguments that do not principally differ from those given in the simpler case of Sec. A.2.

A.1 Hermitian Relations

A.1.1 *Canonical Form of Hermitian Relations*

Let H be an arbitrary Hilbert space with the inner product $(\cdot,\cdot)$, and let M be an arbitrary subset in $H \oplus H$. We say that an ordered pair of elements $x, x' \in H$ is in a binary relation $\theta = \theta_M$ and write $x\theta x'$ if $\{x, x'\} \in H \oplus H$ belongs to M. The set of first coordinates and the set of second coordinates of elements $\{x, x'\} \in M$ are respectively called the domain D_θ and the range R_θ of the relation θ. The set M is the graph G_θ of the relation θ. Clearly, to different sets $M \subset H \oplus H$ correspond different relations θ_M. We now give the following

Definition A.1 Binary relation θ in a Hilbert space H is said to be a Hermitian relation if the following two properties hold:

(1) if $x\theta x'$ and $y\theta y'$, then

$$(x', y) - (x, y') = 0; \tag{A.2}$$

(2) if x, $x' \in H$ and the equation (A.2) holds for all pairs y, $y' \in H$ such that $y\theta y'$, then $x\theta x'$.

As we shall see later, the graph G_θ of a Hermitian relation θ in H is a Lagrangian plane in $H \oplus H$ with respect to the skew-Hermitian inner product $[\cdot,\cdot]$ in $H \oplus H$, where $[\{x, x'\}, \{y, y'\}] = (x', y) - (x, y')$ as on the left-hand side of (A.2) (i.e. G_θ is a Lagrangian subspace in $H \oplus H$ of dimension $\dim H$ (see, for example, [10, Sec. 41G] and [83]), and every Lagrangian plane in $H \oplus H$ is the graph of some Hermitian relation θ in H).

There is also another terminology: in the language of spaces with indefinite metric, the graph of a Hermitian relation is a hyper-maximal neutral lineal in $H \oplus H$ (see, for example [14, Chapter 1, Sec. 4]), and conversely.

Example A.1 Here we give an example of a Hermitian relation θ in H determined by an arbitrary bounded (or unbounded) self-adjoint operator B in H:

$$x\theta x' \Leftrightarrow Bx = x'. \tag{A.3}$$

This is an important example.

Later we shall derive the general form of Hermitian relations. But first, we note those properties of Hermitian relations that follow directly from the definition:

(3) *Hermitian relation is linear, i.e. if $x\theta x'$ and $y\theta y'$, then*

$$(\alpha x + \beta y)\theta(\alpha x' + \beta y'),$$

where α, and β arbitrary complex numbers.

Proof. Indeed, denote $z = \alpha x + \beta y$ and $z' = \alpha x' + \beta y'$. Then for each u, $u' \in H$ such that $u\theta u'$, we have by virtue of (A.2):

$$(z', u) - (z, u') = \alpha[(x', u) - (x, u')] + \beta[(y', u) - (y, u')] = 0,$$

and, therefore, by property (2) we have $z\theta z'$. □

(4) *If θ and θ_1 are Hermitian relations and $\theta \subset \theta_1$ (i.e. $x\theta x'$ implies $x\theta_1 x'$), then $\theta = \theta_1$. In other words, Hermitian relations do not have non-trivial Hermitian extensions.*

Proof. Indeed, let $x\theta_1 x'$. Then for all $y\theta y'$ we also have $y\theta_1 y'$, and, hence, $(x', y) - (x, y') = 0$. Consequently, by property (2), we have $x\theta x'$, and, hence, $\theta_1 \subset \theta$. This proves that $\theta = \theta_1$. □

By analogy with the Cayley transform U_B of a self-adjoint operator B, we define the Cayley transform U_θ of a Hermitian relation θ as follows: to each pair of vectors x, $x' \in H$, $x\theta x'$, we associate the vectors

$$x^+ = x' + ix, \qquad x^- = x' - ix,$$

and we set

$$U_\theta x^+ = x^-.$$

Thus the domain D_U of the operator U_θ is the set of vectors x^+, and the range $\mathrm{R}_U = \{x^-\}$. The operator U_θ is well-defined since every pair $x\theta x'$ can be recovered from x^+ in a unique way. Indeed, if $x\theta x'$ and $y\theta y'$, and $x^+ = y^+$, i.e. $x' + ix = y' + iy$, then by setting $\nu = x - y$ and $\nu' = x' - y'$, we have

$$\nu\theta\nu', \quad \nu' = -i\nu. \tag{A.4}$$

From (A.4) it follows that $(\nu', \nu) - (\nu, \nu') = 0$, i.e. $-2i(\nu, \nu) = 0$. Thus

$$\nu' = \nu = 0, \tag{A.5}$$

and, hence, $x = y$, $x' = y'$.

Lemma A.1 *Cayley transform U_θ of any Hermitian relation θ is a unitary operator.*

Proof. Let $U = U_\theta$. Since $(x', x) - (x, x') = 0$ for $x\theta x'$, we have

$$|Ux^+|^2 = |x^-|^2 = (x' - ix)(x' - ix) = |x'|^2 + |x|^2 + i\{(x', x) - (x, x')\}$$
$$= |x'|^2 + |x|^2.$$

Similarly, we obtain

$$|x^+|^2 = |x'|^2 + |x|^2, \tag{A.6}$$

i.e. $|Ux^+| = |x^+|$, and the operator U is isometric.

We will now show that $\mathrm{D}_U = \mathrm{R}_U = H$. We will first consider D_U.

Let $\nu \perp \mathrm{D}_U$, i.e. $(\nu, x^+) = 0$ for all x^+. Then we get $(\nu, x') + (\nu, ix) = 0$, and putting $\nu' = i\nu$, we obtain $(\nu', x) - (\nu, x') = 0$ for all $x\theta x'$. By property (2) of Definition A.1, it follows that $\nu\theta\nu'$. Thus the pair ν, ν'

satisfies[44] (A.4), and, hence, (A.5). Therefore, D_U is dense in H. Let $x_n^+ \to \tilde{x}$, where $x_n^+ \in \mathrm{D}_U$. We will show that $\tilde{x} \in \mathrm{D}_U$. We have $x_n^+ = x_n' + ix_n$, where $x_n \theta x_n'$. By (A.6) we have

$$|x_m' - x_n'|^2 + |x_m - x_n|^2 = |x_m^+ - x_n^+|^2 \to 0, \qquad \text{as } m, n \to \infty,$$

and, hence, the sequences x_n', x_n converge: $x_n' \to x'$ and $x_n \to x$. This shows that $\tilde{x} = x' + ix$. Since $(x_n', y) - (x_n, y') = 0$ for all $y\theta y'$, it follows that $(x', y) - (x, y') = 0$ for all $y\theta y'$. Therefore $x\theta x'$, and, hence, $\tilde{x} \in \mathrm{D}_U$. This shows that $\mathrm{D}_U = \overline{\mathrm{D}_U} = H$. The equality $\mathrm{R}_U = H$ can be proven similarly. This concludes the proof of the Lemma. □

Theorem A.1 *Let A be a self-adjoint operator and U a unitary operator in H. Then the relation determined by any of the two equations*

$$\cos A \cdot x' - \sin A \cdot x = 0, \tag{A.7}$$

$$(U - I)x' + i(U + I)x = 0 \tag{A.8}$$

is Hermitian.

Conversely, every Hermitian relation $x\theta x'$ can be represented via (A.7) and (A.8) which we call canonical forms, and the operators $-e^{2iA}$ and U in (A.7) and (A.8) respectively are determined uniquely by the relation θ. The operators $-e^{2iA}$ and U in (A.7) and (A.8) are Cayley transforms U_θ of the relation θ.

(Clearly, the operator A in (A.7) can always be chosen so that $|A| \le \pi/2$ and $-\pi/2$ is not an eigenvalue of A. With this normalization, the operator A is determined uniquely by the relation θ.)

The relation θ is generated by a self-adjoint operator B in the sense of (A.3) if and only if 1 is not an eigenvalue for U_θ. In this case, $U_\theta = U_B = (B - iI)(B + iI)^{-1}$.

We first prove the following Lemma:

Lemma A.2 *Vectors x and $x' \in H$ satisfy (A.7) if and only if they can be represented as*

$$x' = \sin A \cdot \nu, \quad x = \cos A \cdot \nu, \quad \nu \in H. \tag{A.9}$$

By (A.9), vector $\nu \in H$ is determined uniquely as

$$\nu = \sin A \cdot x' + \cos A \cdot x. \tag{A.10}$$

[44] up to its sign

(The Lemma also holds in the case when x, x' and ν are operators in H.)

Proof of the Lemma. If ν is defined by (A.10), then by (A.7) we have $\cos A \cdot \nu = \sin A \cdot \cos A \cdot x' + \cos^2 A \cdot x = \sin^2 A \cdot x + \cos^2 A \cdot x = x$, and, similarly, we obtain $\sin A \cdot \nu = x'$. Now the lemma obviously follows. □

Proof of the Theorem. Suppose that the relation θ is given by (A.7). We will show that θ is Hermitian. Property (1) follows immediately from the representation (A.9) for $x\theta x'$ and $y\theta y'$. We will now check property (2) of Definition A.1. Let $x, x' \in H$ be a pair such that (A.2) holds for all $y\theta y'$. Using the representation (A.9) for y, y', we get

$$(x', \cos A \cdot \nu) - (x, \sin A \cdot \nu) = 0,$$

i.e. $(\cos A \cdot x' - \sin A \cdot x, \nu) = 0$ for all $\nu \in H$. Thus $x\theta x'$, and property (2) of Definition A.1 is satisfied. This shows that θ is Hermitian. We can pass from (A.8) to (A.7) by putting $U = -e^{2iA}$.

Let us find the Cayley transform U_θ of the relation θ in (A.8). From (A.8) we get $U(x' + ix) = x' - ix$ for $x\theta x'$. Thus $U_\theta = U$.

Now suppose that θ is a Hermitian relation. Then for all $x\theta x'$ we have $U_\theta(x' + ix) = x' - ix$. Thus $(U_\theta - I)x' + i(U_\theta + I)x = 0$, where U_θ is the Cayley transform of the relation θ. By the first part of the proof, it follows that the last equation determines a Hermitian relation θ_1. Clearly, $\theta_1 \supset \theta$, and by property (4) of Hermitian relations, it follows that $\theta_1 = \theta$. This concludes the proof of the Theorem. □

A.1.2 *Various Forms of Hermitian Relations. Dissipative and Sectorial Relations*

Every Hermitian relation θ can be represented in canonical form (A.7). However, there are infinitely many equivalent other representations of a relation θ, and these representations can be reduced to (A.7).

We first note the following form of writing a Hermitian relation (this is also considered to be a canonical form; see, for example, [901]). Denote by $H^\perp$ the null-space of the operator $\cos A$. Then there exists a (possibly unbounded) self-adjoint operator $A_1 = \tan A$ in $H_1 = H \ominus H^\perp$. Now the equation (A.7) is equivalent to

$$x' = A_1 x + x^\perp, \quad x \in H_1, \quad x^\perp \in H^\perp. \tag{A.11}$$

A broad class of equations equivalent to (A.7) can be obtained by multiplying (A.7) on the left by an arbitrary injective bounded linear operator K.

Theorem A.2 *Assume that B and C are arbitrary bounded linear operators in H. The relation θ determined by*

$$x\theta x' \Leftrightarrow Cx' - Bx = 0, \tag{A.12}$$

is Hermitian if and only if the following two conditions are satisfied:

(1) The operators $B \pm iC$ are injective.
(2) The operator

$$U = (B + iC)^{-1}(B - iC) \tag{A.13}$$

is unitary (and, in particular, $R_{B-iC} \subseteq R_{B+iC}$).

Under these conditions, U is the Cayley transform of the relation θ, and using U, θ can be represented via (A.8).

Proof. Assume that the relation θ in (A.12) is Hermitian. Then by Theorem A.1 there exists a self-adjoint operator A such that (A.12) is equivalent to (A.7), and for all $\nu \in H$ we have

$$(\cos A \cdot \nu)\theta(\sin A \cdot \nu).$$

By (A.12) we have $(C \sin A - B \cos A)\nu = 0$ for all $\nu \in H$, i.e.

$$C \sin A - B \cos A = 0.$$

By Lemma A.2 there exists a bounded linear operator K such that

$$C = K \cos A, \quad B = K \sin A. \tag{A.14}$$

The operator K is injective. Indeed, if $h \neq 0$ and $Kh = 0$, then (A.12) is satisfied by any pair of vectors x, $x' \in H$ such that

$$\cos A \cdot x' - \sin A \cdot x = h.$$

Here we can take, for example, $x' = \cos A \cdot h$, $x = -\sin A \cdot h$. But this contradicts (A.7), and, thus, K is injective. Now the necessity part of the theorem follows immediately since

$$B \pm iC = \pm iKe^{\mp iA}, \qquad U = (B + iC)^{-1}(B - iC) = -e^{2iA}.$$

We now prove the sufficiency part. Assume that the operator U from (A.13) is defined on all of H. Then the equation $Cx' - Bx = 0$ is equivalent to

$$(U - I)x' + i(U + I)x = 0. \tag{A.15}$$

Indeed, substituting (A.13) into (A.15) and multiplying on the left by the bounded injective operator $B + iC$, we prove the equivalence of (A.15) and (A.12). Since the operator U is unitary by assumption, it follows that the relation given by (A.15) is Hermitian, and so is the relation given by (A.12). This concludes the proof of the Theorem. □

Corollary A.1 *Assume that* $\dim H < \infty$. *The relation* θ *in (A.12) is Hermitian if and only if the following two conditions are satisfied:*

(1) non-singularity of (A.12), i.e.

$$\det(BB^* + CC^*) \neq 0; \tag{A.16}$$

(2) self-adjointness of (A.12), i.e.

$$BC^* = CB^*. \tag{A.17}$$

(Compare (A.16) and (A.17) with the general form of self-adjoint boundary conditions for the scalar equation of order $2n$ in [586], [6], [145] and also [599].) Note that the injectivity of the operator $BB^* + CC^*$, which holds in the case $\dim H < \infty$ by (A.16), does not necessarily hold for the Hermitian relation θ from (A.12) in the case $\dim H = \infty$. We can see this from (A.14) by taking

$$K = \begin{pmatrix} 0 & & & \\ 1 & 0 & & \\ & 1 & 0 & \\ & & \ddots & \ddots \end{pmatrix}.$$

We can also consider relations θ determined by (A.12) with unbounded operators B and C. In this case, it is natural to require that the closure $\theta^\sim$ of θ is a Hermitian relation. If M is the set in $H \oplus H$ corresponding to θ, then $\widetilde{M} = \overline{M}$ is the set in $H \oplus H$ corresponding[45] to $\theta^\sim$. We give the simplest two examples. 1) $x' - Bx = 0$, where $B = B^*$ is an unbounded operator. Here the relation $x\theta x'$ is Hermitian. 2) $Bx' - Bx = 0$, where B is an injective, densely defined operator in H with $D_B \neq H$. In this case,

[45] i.e. the closure of the graph $G_\theta = M$ of θ corresponds to the closure of θ.

the relation $x\theta x'$ is not Hermitian. However, the closure $\theta^{\sim}$ is Hermitian; indeed, $x\theta^{\sim}x'$ is equivalent to $x = x'$. (We also say that the relation θ is essentially Hermitian.) In general, a relation θ_1 is an extension of a relation θ if $G_\theta \subset G_{\theta_1}$. If, in addition, $G_\theta \neq G_{\theta_1}$, we say that θ_1 is a proper[46] extension of θ. We will investigate the case of equation (A.12) with unbounded operators B and C and the corresponding relation θ later in Theorem A.5.

We now consider the notion of numerical range of a relation and some related topics.

Define the value of a relation θ on an element x as the set $\theta(x) := \{x' \in H\colon x\theta x'\}$. The operator $T\colon \mathrm{D}_\theta \to H$ is well-defined by the condition $x\theta x' \Leftrightarrow x' = Tx$, with $\mathrm{D}_T = \mathrm{D}_\theta$, if and only if the following condition holds: $\theta(0) = \{0\}$.

Definition A.2 The numerical range of a linear relation θ in H is the set

$$W(\theta) = \{(x', x) \in \mathbb{C}\colon x\theta x', |x| = 1\}.$$

By the Hausdorff–Stone Theorem, the numerical range of an arbitrary linear operator T in H is a convex set in $\mathbb{C}$ (closed if $\dim H < \infty$). As it is well known and easily verified, the spectrum $\sigma(T)$ of an arbitrary linear operator T is contained in the closure of the numerical range $W(T)^{\sim}$, and by the convexity of $W(T)$, the closed convex hull of $\sigma(T)$ is also contained in $W(T)^{\sim}$[47]:

$$(\mathrm{conv}\ \sigma(T))^{\sim} \subseteq W(T)^{\sim}. \tag{A.18}$$

We have the following analogue of the Hausdorff–Stone Theorem for linear relations in H:

Proposition A.1 *The numerical range $W(\theta)$ of a linear relation θ in H is a convex set in the complex plane. (If $\dim D_\theta < \infty$, then $W(\theta)$ is closed and bounded or $W(\theta) = \mathbb{C}$.) In particular, for Hermitian relations, we have $W(\theta) \subset \mathbb{R}$.*

---

[46] The term "proper extension" is sometimes used in a somewhat different sense; see [73].

[47] It is interesting how the form of the factorization of holomorphic operator-valued function is changed in the A. S. Markus–V. I. Matsaev Theorem [661] if in the conditions of this Theorem the numerical range $W(A(\zeta))$ of the operator-valued function $A(\zeta)$ on the circle $|\zeta| = 1$ is replaced by the set conv $\sigma(A(\zeta))$; see A. M. Gomilko [485]. (In these Theorems, it is required that 0 does not belong to the above-mentioned sets.)

(For the proof, which is based on Lemma A.3 below, see [788].)

Lemma A.3 *If* $0\theta h$ *and* $h \notin D_\theta^\perp = H \ominus \overline{D_\theta}$, *then* $W(\theta) = \mathbb{C}$. *If* $W(\theta) \neq \mathbb{C}$, *then* $\theta(0) \perp D_\theta$.

Proof. Let $0\theta h$ and $h \notin \mathrm{D}_\theta^\perp$. Then there exists $x \in \mathrm{D}_\theta$, $|x| = 1$, such that $(h, x) \neq 0$. But then $x\theta x'$ for some $x' \in H$, and, hence, $x\theta(x' + \zeta h)$ for all $\zeta \in \mathbb{C}$ and $W(\theta) \supset \{(x' + \zeta h, x), \forall \zeta \in \mathbb{C}\} = \mathbb{C}$. The Lemma is proven. □

Definition A.3 A relation θ is said to be dissipative [489] (see also [73] and [74]) if $W(\theta) \subset \mathbb{C}^+ = \{\zeta \in \mathbb{C}: \mathrm{Im}\, \zeta \geq 0\}$, accumulative if $W(\theta) \subset \mathbb{C}^- = \{\zeta \in \mathbb{C}: \mathrm{Im}\, \zeta \leq 0\}$, accretive if $W(\theta) \subset \{\zeta \in \mathbb{C}: \mathrm{Re}\, \zeta \geq 0\}$, symmetric if (A.2) holds, i.e. for $W(\theta) \subset \mathbb{R}$. If a dissipative relation does not have a proper dissipative extension, then it is said to be maximal dissipative (m-dissipative). In a similar manner, one defines an m-accretive relation, *etc.*

Theorem A.3 **([489]; see also [73, Chapter 3, Theorem 1.4])** *If* $U \in B(H)$ *is a contraction (i.e.* $\|U\| \leq 1$*), then the equation (A.8) defines an* m*-dissipative relation, and each* m*-dissipative relation can be represented in this form. The* m*-dissipative relation (A.8) is maximally symmetric if and only if the operator* U *is isometric.*

Theorem A.4 **([744], [188, Chapter 4, Theorem 4.1])**[48] *Every dissipative (or accumulative) extension of a symmetric operator* A *is contained in* A^**. A closed operator* A *is maximal dissipative if and only if* A^* *is maximal accumulative.*

Definition A.4 ([**782**]) By analogy with operators in [87], a relation θ is said to be sectorial if $W(\theta)$ is contained in a sector of the complex ζ-plane:

$$|\arg\,(\zeta - \gamma)| \leq \phi < \pi/2, \quad \gamma \in \mathbb{R}, \tag{A.19}$$

and m-sectorial if it does not have proper extensions with numerical range in the half-plane $\mathrm{Re}\, \zeta \geq \alpha$ for any $\alpha \in \mathbb{R}$. Here γ and ϕ are called the vertex and the semi-angle of the sectorial relation (they are not defined uniquely).

Definition A.5 Let $K \subset \mathbb{C}$, $K \neq \mathbb{C}$ be a closed set. Operator $T: \mathrm{D}_T \to \overline{\mathrm{D}_T}$ is said to be K-maximal in $\overline{\mathrm{D}_T}$ if $W(T) \subset K$ and T does not have proper extensions T_1 in $\overline{\mathrm{D}_T}$ such that $W(T_1) \subset K$. Analogously, linear relation θ is said to be K-maximal if $W(\theta) \subset K$ and θ does not have proper extensions θ_1 such that $W(\theta_1) \subset K \neq \mathbb{C}$.

[48] For the proof of the first part of this theorem, see also [73, Chapter 3, Theorem 1.3].

Example A.2 The operator $\Lambda y = -iy'$, $y(0) = 0$, $0 \le t < \infty$ is K-maximal in $L_2(0,\infty)$ for $K = \mathbb{R}$ and $K = \mathbb{C}^-$. However, Λ is not K-maximal for $K = \mathbb{C}^+$ since Λ has a proper dissipative extension $\Lambda_1 = \Lambda^*$ in $L_2(0,\infty)$: $\Lambda \subset \Lambda_1 \neq \Lambda$, $\Lambda_1 y = -iy'$, $\mathrm{Im}(\Lambda_1 y, y) = |y(0)|^2/2 \ge 0$.

Proposition A.2 *In order that a linear relation θ in H be K-maximal, $W(\theta) \subset K = K^\sim \neq \mathbb{C}$, it is necessary and sufficient that there exists an operator $T\colon D_T = D_\theta \to \bar{D}_\theta$ K-maximal in $\bar{D}_\theta$ such that*

$$x\theta x' \Leftrightarrow x' - Tx \in D_\theta^\perp, \quad R_T \subset \bar{D}_\theta. \tag{A.20}$$

By virtue of (A.20), the correspondence $\theta \sim T$ is one-to-one[49]*. The operator T and the relation θ are closed and $W(T) = W(\theta)$.*

Proposition A.2 generalizes the representation of Hermitian relations in the form (A.11) and m-sectorial relations given by Theorem 4 from [782]: A relation ρ is m-sectorial if and only if $x\rho y \Leftrightarrow y = Tx + x^\perp$, $x \in \mathrm{D}_T$, where $T\colon H_1 \to H_1 = \overline{\mathrm{D}_T} \subseteq H$ is an m-sectorial operator and $x^\perp \in H_1^\perp$.

Lemma A.4 *Assume that θ is K-maximal for some $K \supset W(\theta)$. Then $y \perp D_\theta$ implies $0\theta y$, i.e. $D_\theta^\perp \subset \theta(0)$.*

Proof. Suppose that there exists $y \perp \mathrm{D}_\theta$ such that $0\theta y$ does not hold. Consider the relation θ_1 given by

$$x\theta_1 x' \Leftrightarrow \exists \zeta \in \mathbb{C}\colon x\theta(x' - \zeta y). \tag{A.21}$$

Clearly, θ_1 is a proper extension of θ. By (A.21) it follows that $W(\theta_1) = W(\theta) \subset K$ since $(x', x) = (x' - \zeta y, x)$, due to $y \perp \mathrm{D}_\theta$, $x \in \mathrm{D}_\theta$. But this contradicts the K-maximality of θ, and the Lemma is proven. □

Proof of Proposition A.2. By Lemmas A.3 and A.4, for a K-maximal relation θ in H, we have

$$0\theta y \Leftrightarrow y \perp \mathrm{D}_\theta; \quad \theta(0) = \mathrm{D}_\theta^\perp. \tag{A.22}$$

We denote by P the orthogonal projection from H onto $\bar{\mathrm{D}}_\theta$. By the condition

$$x\theta x' \Leftrightarrow Tx = Px', \tag{A.23}$$

by virtue of (A.22), there exists an operator $T\colon \mathrm{D}_T = \mathrm{D}_\theta \to \bar{\mathrm{D}}_\theta \subset H$ satisfying (A.20), and, therefore, we also have $W(T) = W(\theta)$. If T had a

[49]If θ in (A.20) is K-maximal, then $R_\lambda(\theta) = (T - \lambda)^{-1} P_\theta$, where P_θ is the orthogonal projection onto $\bar{\mathrm{D}}_\theta$.

proper extension T_1 in $\bar{D}_\theta$ for which $W(\theta_1) \subset K$, then this extension, by the formula (A.20), would generate a proper extension θ_1 of the relation θ for which we would have $W(T_1) \subset K$. But this is not allowed by the conditions of the Proposition. Thus, the operator T is K-maximal in $H_1 = \bar{D}_\theta \subset H$. Since T is defined on a set that is dense in H_1 and $W(T) \neq \mathbb{C}$, it follows that T is closable by Theorem V.3.4 of [87]. Therefore, T is closed since, otherwise, its closure $T^{\sim}$ would be its proper extension in H_1 with $W(T^{\sim}) \subset W(T)^{\sim} \subset K^{\sim} = K$. In a similar way, also the K-maximal relation θ is closed since $W(\theta)^{\sim} \subset K \neq \mathbb{C}$. The equality $W(T) = W(\theta)$ is obvious due to (A.20). The necessity of the conditions of the Proposition is proven. We now prove the sufficiency, i.e. the K-maximality of the relation θ in H generated according to the formula (A.20) by the operator T which is K-maximal in $\overline{D_T}$. We have $W(\theta) = W(T) \subset K$. We assume that there exists a proper extension $\theta_1 \supset \theta$ with $W(\theta_1) \subset K$ and that $x_1\theta_1 x'_1$, but not $x_1\theta x'_1$. Then $\bar{D}_{\theta_1} = \bar{D}_\theta$ since, otherwise, we would not have $\bar{D}_{\theta_1} \perp \theta_1(0) \supset \theta(0)$, and by Lemma A.3 we should have $W(\theta_1) = \mathbb{C}$, contradicting the assumption. Therefore, $(x'_1, x_1) = (Px'_1, x_1) \in K$, where P is the orthogonal projection onto $\bar{D}_\theta$. Moreover, $x_1 \notin D_T = D_\theta$ since, otherwise, we would have $x_1\theta(Tx_1)$ and $Px'_1 \neq Tx_1$ (because $\{x_1, x'_1\} \notin G_\theta$), and from here, $x_1\theta_1(Tx_1)$, $x_1\theta_1(Px'_1)$ and $0\theta_1(Tx_1 - Px'_1)$, $0 \neq Tx_1 - Px'_1 \in \bar{D}_{\theta_1}$, which by Lemma A.3, would lead to $W(\theta_1) = \mathbb{C}$, contradicting the assumption. Consequently, setting $T_1 \supset T$, $T_1 x_1 = Px'_1$, we would obtain a proper extension T_1 of the operator T in $\overline{D_T}$ and $W(T_1) \subset K$, contradicting the K-maximality of T in D_θ. The Proposition is proven. □

Corollary A.2 *([782], [788]) The relation θ is m-sectorial in H if and only if there exists an operator $T\colon D_T = D_\theta \to \bar{D}_\theta$ m-sectorial in $\bar{D}_\theta$ such that $W(T) = W(\theta)$ and (A.20) holds.*

Based on the previous considerations, we extend the concept of the Friedrichs extension of sectorial densely defined operators (see [87]) to sectorial relations, and, in particular, to sectorial operators with a domain of definition that is not dense in H. We mention that symmetric semi-bounded relations are special cases of sectorial relations. Regarding their Friedrichs extensions, see [345], [782], [788], and the earlier investigations of A. V. Shtraus [855], [856] based on M. G. Krein's theory of the extension of semi-bounded operators [586] (see also [6]).

As shown in Lemma A.3, if $W(\theta) \neq \mathbb{C}$, then $\theta(0) \perp D_\theta$. Therefore, every linear relation θ with $W(\theta) \neq \mathbb{C}$ generates by the formula (A.23) an operator T, which, in turn, by the formula

$$x\theta_1 x' \Leftrightarrow x' - Tx \in \mathrm{D}_\theta^\perp, \quad \mathrm{R}_T \subset \bar{\mathrm{D}}_\theta, \tag{A.24}$$

generates a relation $\theta_1 \supset \theta$, $\mathrm{D}_{\theta_1} = \mathrm{D}_\theta$ and $\theta_1(0) = \mathrm{D}_\theta^\perp \supset \theta(0)$, $W(\theta_1) = W(\theta) = W(T)$. If θ is a sectorial relation in H, then the operator T is sectorial and densely defined in $H_1 = \bar{\mathrm{D}}_\theta$, and, therefore, admits in H_1 an m-sectorial Friedrichs extension T^{F_1} (see [87, Chapter VI, Sec. 2]), which is constructed by means of the closure in H_1 of the quadratic form (Tx, x), $x \in \mathrm{D}_\theta$, and, therefore, $W(T^{F_1}) \subset W(T)^\sim = W(\theta)^\sim$, $\mathrm{R}_{T^{F_1}} \subset H_1 = \bar{\mathrm{D}}_\theta$.

By the formula (A.20), with T^{F_1} instead of T, we define an m-sectorial relation $\theta^F \supset \theta$, $W(\theta^F) \subset W(\theta)^\sim$. The relation θ^F is called the *Friedrichs extension* (in H) of the *sectorial relation* θ.

To each sectorial relation θ in H there corresponds a quadratic closable sectorial form

$$a_\theta(x) = (x', x), \qquad x\theta x'. \tag{A.25}$$

This form does not depend on the choice of the vector x' since $\theta(0) \perp \mathrm{D}_\theta$, and we have

$$a_\theta(x) = (Tx, x), \qquad x \in \mathrm{D}_\theta. \tag{A.26}$$

Proposition A.3 *([788]) Among all possible m-sectorial extensions of the sectorial relation θ in H, there exists a unique one having the domain of definition which is contained in the domain of definition of the closure of the quadratic form $a_\theta(x)$. This extension is also the unique extension for which the closure of its quadratic form has the smallest domain of definition, i.e. contained in the domain of definition of the closure of the form corresponding to any other m-sectorial extension of the relation θ. This relation is the above constructed Friedrichs extension θ^F of the relation θ.*

The proof of Proposition A.3 follows from previous considerations and from Theorems VI.2.10 and VI.2.11 of [87].

Corollary A.3 *If $T\colon D_T \to H$ is a sectorial operator whose domain of definition is not dense in H (the inclusion $R_T \subset \overline{D_T}$ is not assumed), then its Friedrichs extension in H is an m-sectorial relation T^F generated by the formula $xT^Fx' \Leftrightarrow x' - (PT)^{F_1}x \in D_T^\perp$, where P denotes the orthogonal projection onto D_T, F_1 denotes the Friedrichs extension of the operator in the subspace $H_1 = \overline{D_T}$, and F denotes the Friedrichs extension in H. For the resolvent of the Friedrichs extension T^F for $\lambda \in \mathbb{C}\backslash W(T)^\sim$, we have* *the formula*

$$R_\lambda(T^F) = R_\lambda^{(1)}((PT)^{F_1})P, \tag{A.27}$$

where $R_\lambda^{(1)}(A)$ is the resolvent of the operator A in H_1, which is a bounded operator for $\lambda \notin W(A)^\sim$ (T^F reduces to an operator only for $H_1 = H$).

The boundedness of the resolvent follows here from Theorem 42 in [70, Chapter 1, Sec. 4, no. 21].

The formula (A.27) is a generalization for sectorial operators of the Ando–Nishio–Shtraus formula (3.44) that was established for semi-bounded operators.

Remark A.1 Using the known fact (see, for example, [188]) on the existence of m-dissipative extensions of a dissipative operator, Proposition A.2 and the formula (A.24), it is easy to show the existence of m-dissipative extensions for any dissipative relation (which is also known).

We now return to investigation of the question when the relation θ generated by the condition $x\theta x' \Leftrightarrow Cx' - Bx = 0$, with unbounded operators B and C, is Hermitian.

Together with the relation θ in (A.12), we consider its restriction

$$\theta_1 \subset \theta\colon x\theta_1 x' \Leftrightarrow C_1 x' - B_1 x = 0, \tag{A.28}$$

where

$$\begin{aligned} B_1 \subset B, \quad \mathrm{D}_{B_1} &= \mathrm{D}_B \cap (\mathrm{D}_C \oplus \mathrm{D}_{\overline{C}}^{\perp}), \\ C_1 \subset C, \quad \mathrm{D}_{C_1} &= \mathrm{D}_C \cap (\mathrm{D}_B \oplus \mathrm{D}_{\overline{B}}^{\perp}). \end{aligned} \tag{A.29}$$

Theorem A.5 *Assume that B and C in (A.12) are linear operators (possibly unbounded, non-closed and with not necessarily dense domains D_B and D_C) mapping H into an arbitrary linear space E*[50].

The conditions (1)–(4) below are necessary for the relation θ defined by (A.12) to be Hermitian, and are sufficient for the closure $\theta_1^\sim$ of the relation $\theta_1 \subset \theta$ defined by (A.28) and (A.29) to be Hermitian (clearly, $\theta_1^\sim \subset \theta^\sim$).

(1) $D_{\overline{B}}^{\perp} \subset \operatorname{Ker} C$, $D_{\overline{C}}^{\perp} \subset \operatorname{Ker} B$.
(2) $\overline{D_B \cap D_C} = \overline{D_B} \cap \overline{D_C}$.
(3) Operators $B \pm iC$ are injective.

[50] We assume that E is such that the notions "closed" and "densely defined" make sense. For example, E may be a topological vector space.

(4) The closure $U_1^{\sim}$ of the operator U_1, where

$$U_1 = (B + iC)^{-1}(B - iC), \tag{A.30}$$

in the subspace $H_1 = \overline{D_B \cap D_C}$ exists and is unitary.

Under these conditions, the operator

$$U = U_1^{\sim} \oplus I_{D_B^{\perp}} \oplus \left(-I_{D_C^{\perp}}\right) \tag{A.31}$$

is unitary. Moreover, U is the Cayley transform of the relation $\theta_1^{\sim}$ which can be represented with the help of U from (A.31) via (A.8).

Proof. *Necessity.* Assume that θ in (A.12) is Hermitian. Then θ is $\mathbb{R}$-maximal, and by Lemma A.4 we have $\mathrm{D}_\theta^{\perp} \subset \theta(0)$. This shows that $\mathrm{D}_B^{\perp} \subset \mathrm{D}_\theta^{\perp} \subset \theta(0) = \ker C$. Since θ in (A.12) is Hermitian, its inverse relation $Bx' - Cx = 0$ is also Hermitian. This shows that $\mathrm{D}_C^{\perp} \subset \ker B$, and property (1) is proven.

We now prove property (2). By (A.7), where $A = A^*$ and $\|A\| \leq \pi/2$, we can represent the Hermitian relation θ in parametric form: $x = \cos A \cdot h$, $x' = \sin A \cdot h$, $\forall h \in H$, and we see that $\mathrm{D}_B \supset \mathrm{R}_{\cos A}$ and $\mathrm{D}_C \supset \mathrm{R}_{\sin A}$.

Let $E(t)$ be the resolution of identity of the operator A, and let

$$H_c = E([-\pi/4, \pi/4])H,$$

and

$$H_s = E([-\pi/2, -\pi/4) \cup (\pi/4, \pi/2])H.$$

It is easy to see that

$$H_c \subset \mathrm{R}_{\cos A} \subset \mathrm{D}_B; \quad H_s \subset \mathrm{R}_{\sin A} \subset \mathrm{D}_C, \tag{A.32}$$

and

$$H_c \oplus H_s = H; \quad H_s = H_c^{\perp}. \tag{A.33}$$

Indeed, if, for example, $h \in H_c$, then $h = \cos A \cdot g$, where

$$g = \int_{-\pi/4-0}^{\pi/4+0} \frac{1}{\cos t}\, dE(t)\, h,$$

and, hence, $h \in \mathrm{R}_{\cos A}$, so $H_c \subset \mathrm{R}_{\cos A}$.

Let P_c and $P_s = P_c^\perp$ be the orthogonal projections onto H_c and H_s respectively. By (A.32) we have

$$\mathrm{D}_B = H_c \oplus P_s\mathrm{D}_B = H_c \oplus L_s, \quad \mathrm{D}_C = H_s \oplus P_c\mathrm{D}_C = H_s \oplus L_c, \tag{A.34}$$

where H_c and H_s are closed subspaces as in (A.33), and $L_s \subset H_s$ and $L_c \subset H_C$ are some linear manifolds. From (A.34) it follows that $\mathrm{D}_B \cap \mathrm{D}_C = L_s \oplus L_c$, $\overline{\mathrm{D}_B \cap \mathrm{D}_C} = \overline{L_s \oplus L_c} = \overline{L_s} \oplus \overline{L_c}$.

On the other hand, we have

$$\overline{\mathrm{D}_B} = \overline{H_c \oplus L_s} = H_c \oplus \overline{L_s}, \quad \overline{\mathrm{D}_C} = \overline{H_s \oplus L_c} = H_s \oplus \overline{L_c};$$

hence,

$$\overline{\mathrm{D}_B} \cap \overline{\mathrm{D}_C} = \overline{L_s} \oplus \overline{L_c}.$$

Therefore,

$$\overline{\mathrm{D}_B \cap \mathrm{D}_C} = \overline{\mathrm{D}_B} \cap \overline{\mathrm{D}_C},$$

and property (2) is proven.

We now prove property (3). Suppose that $(B + iC)x = 0$. Then $Cx - B(ix) = 0$, i.e. $(ix)\theta x$, and thus $x = 0$ since $W(\theta) \subset \mathbb{R}$. Therefore, the operator $B + iC$ is injective. The injectivity of $(B - iC)$ is proven similarly.

We now prove property (4). Since the operators $(B \pm iC)$ are injective, the operator $U_1 = (B + iC)^{-1}(B - iC)$ exists and is injective. Clearly, $\mathrm{D}_{U_1} \subset \mathrm{D}_B \cap \mathrm{D}_C$ and $\mathrm{R}_{U_1} \subset \mathrm{D}_B \cap \mathrm{D}_C$.

We will show that here we have not only inclusions but also equalities. We will also show that U_1 is an isometry. Letting $f = U_1 h$, where $h \in \mathrm{D}_B \cap \mathrm{D}_C$, we have

$$(B + iC)f = (B - iC)h; \tag{A.35}$$

hence,

$$C(i(f + h)) - B(h - f) = 0, \quad \text{i.e. } (h - f)\theta(i(h + f)). \tag{A.36}$$

Substituting $U = U_\theta$ and

$$x = h - f, \quad x' = i(h + f) \tag{A.37}$$

into (A.8), we obtain $f = U_\theta h$, i.e. $U_1 \subset U_\theta|_{(\mathrm{D}_B \cap \mathrm{D}_C)}$. Therefore, U_1 is an isometry, and it is closable with closure denoted by $U_1^\sim$. To prove that $U_1^\sim$

is unitary in the subspace $H_1 = \overline{D_B \cap D_C}$, it is enough to show that

$$U_\theta(D_B \cap D_C) = D_B \cap D_C. \tag{A.38}$$

Then for $h \in D_B \cap D_C$, we have $U_\theta h = f \in D_B \cap D_C$, and there exists a pair $x, x' \in H$, $x\theta x'$ such that (A.37) and hence (A.35) holds, i.e. $U_\theta|_{(D_B \cap D_C)} \subset U_1$. Thus $U_1 = U_\theta|_{(D_B \cap D_C)}$, and, hence, $U_1^\sim = U_\theta|_{H_1}$, $D_{U_1^\sim} = R_{U_1^\sim} = H_1 = \overline{D_B \cap D_C}$, i.e. $U_1^\sim$ is a unitary operator in H_1.

It remains to show (A.38). From (A.7) and (A.8) it follows that $U_\theta = -e^{2iA}$. Therefore, for all $h \in H$, we have

$$(U_\theta - I)h = -2\cos A \cdot e^{iA} h \in R_{\cos A} \subset D_B,$$

and

$$(U_\theta + I)h = -2i \sin A \cdot e^{iA} h \in R_{\sin A} \subset D_C.$$

Thus if $h \in D_B$, then $U_\theta h = (U_\theta - I)h + h \in D_B$, i.e. $U_\theta D_B \subset D_B$. Similarly, $U_\theta^{-1} D_B \subset D_B$, and hence $U_\theta D_B = D_B$. In the same way we get $U_\theta D_C = D_C$. Hence,

$$U_\theta(D_B \cap D_C) = (U_\theta D_B) \cap (U_\theta D_C) = D_B \cap D_C,$$

and the necessity part of the theorem is proven.

Sufficiency. Let

$$U' = U_1 \oplus I_{D_B^\perp} \oplus \left(-I_{D_C^\perp}\right), \tag{A.39}$$

where U_1 is defined by formula (A.30). Consider the linear relation $x\rho x'$ given by

$$x\rho x' \Leftrightarrow (U' - I)x' + i(U' + I)x = 0. \tag{A.40}$$

We have

$$U' - I = (U_1 - I)_{D_B \cap D_C} \oplus 0_{D_B^\perp} \oplus \left(-2I_{D_C^\perp}\right);$$

and

$$U' + I = (U_1 + I)_{D_B \cap D_C} \oplus 2I_{D_B^\perp} \oplus 0_{D_C^\perp}.$$

Thus if $x\rho x'$, then

$$P_{D_B^\perp} x = 0, \quad P_{D_C^\perp} x' = 0, \quad P_{H_1} x \in D_B \cap D_C, \quad P_{H_1} x' \in D_B \cap D_C,$$

and

$$(U'-I)x' \in \mathrm{D}_B \cap \mathrm{D}_C, \quad (U'+I)x \in \mathrm{D}_B \cap \mathrm{D}_C.$$

Multiplying (A.40) by the injective operator $B+iC$ with the domain $\mathrm{D}_B \cap \mathrm{D}_C$, we obtain an equivalent equation

$$\left(\{(B-iC)-(B+iC)\}_{\mathrm{D}_B\cap\mathrm{D}_C} \oplus 0_{\mathrm{D}_B^\perp}\right) x' \\ +i\left(\{(B-iC)+(B+iC)\}_{\mathrm{D}_B\cap\mathrm{D}_C} \oplus 0_{\mathrm{D}_C^\perp}\right) x = 0,$$

which also determines the relation ρ in H. This equation can be written as

$$x\rho x' \Leftrightarrow (C|_{(\mathrm{D}_B\cap\mathrm{D}_C)} \oplus 0_{\mathrm{D}_B^\perp})x' - (B|_{(\mathrm{D}_B\cap\mathrm{D}_C)} \oplus 0_{\mathrm{D}_C^\perp})x = 0.$$

By property (1) we have $\mathrm{D}_B^\perp \subset \ker C$, $\mathrm{D}_C^\perp \subset \ker B$. Hence (see also (A.29)) we have

$$C|_{(\mathrm{D}_B\cap\mathrm{D}_C)} \oplus 0_{\mathrm{D}_B^\perp} = C|_{(\mathrm{D}_B\cap\mathrm{D}_C\oplus\mathrm{D}_B^\perp)} = C|_{\{\mathrm{D}_C\cap(\mathrm{D}_B\oplus\mathrm{D}_B^\perp)\}} = C_1$$

and

$$B|_{(\mathrm{D}_B\cap\mathrm{D}_C)} \oplus 0_{\mathrm{D}_C^\perp} = B_1.$$

Thus $\rho = \theta_1 \subset \theta$, $\rho^\sim = \theta_1^\sim \subset \theta^\sim$.

Note that ρ in (A.40) can be written in the parametric form

$$x' = -i(U'+I)h, \quad x = (U'-I)h, \quad \forall h \in \mathrm{D}_B \cap \mathrm{D}_C \oplus \mathrm{D}_B^\perp \oplus \mathrm{D}_C^\perp = \mathrm{D}_{U'},$$

where $h = (ix'-x)/2$ and the operator U' is bounded. Therefore, we can obtain the closure $\rho^\sim$ by passing from U' to $U'^\sim$ in the parametric form of the relation ρ, and, hence, in (A.40):

$$x\rho^\sim x' \Leftrightarrow (U'^\sim - I)x' + i(U'^\sim + I)x = 0.$$

It is easy to see that the operator $U'^\sim$ in (A.39) is unitary in H because, by assumption, $U_1^\sim$ is unitary in

$$H_1 = \overline{\mathrm{D}_B \cap \mathrm{D}_C} = \overline{\mathrm{D}_B} \cap \overline{\mathrm{D}_C}$$

and

$$\overline{\mathrm{D}_B} \cap \overline{\mathrm{D}_C} \oplus \mathrm{D}_B^\perp \oplus \mathrm{D}_C^\perp = H.$$

Thus the relation $\rho^\sim$, i.e. $\theta_1^\sim$ is Hermitian, and its Cayley transform is

$$U'^\sim = U_1^\sim \oplus I_{\mathrm{D}_B^\perp} \oplus \left(-I_{\mathrm{D}_C^\perp}\right) = U.$$

This proves all properties of the operator U in (A.31) and the sufficiency part of the theorem. Theorem A.5 is proven. □

In conclusion of this section, we give an example showing that under the conditions (1)–(4) of Theorem A.5, it can happen that the closure $\theta^{\sim}$ of the relation θ in (A.12) is not Hermitian but $\theta^{\sim} \supset \theta_1^{\sim}$, where θ_1 is as in (A.28) and (A.29). Indeed, let B be a densely defined invertible operator in H, $\mathrm{D}_B \neq H$. Let $C \supset B$ be its proper extension such that D_C is the linear span of D_B and $h \notin \mathrm{D}_B$, and let $Ch = 0$. Then all conditions of Theorem A.5 are satisfied. But $0\theta h$ by (A.12) and, by assumption, $\mathrm{D}_\theta = \mathrm{D}_B$ is dense in H. From Lemma A.3 we have $W(\theta) = \mathbb{C}$ so that θ is not even a symmetric relation. At the same time, $x\theta_1^{\sim}x' \Leftrightarrow x' = x$ is a Hermitian relation.

A.2 Self-Adjoint Boundary Conditions for Infinite Systems of Second Order Differential Equations

Consider the differential equation

$$l[y] := -(p(t)y')' + \frac{i}{2}[r^*(t)y' + (r(t)y)'] + q(t)y = \lambda y, \tag{A.41}$$

whose coefficients $p(t)$, $q(t)$, $r(t)$, $t \in [0, b]$ are bounded linear operators in a separable Hilbert space H,

$$p^*(t) = p(t), \quad q^*(t) = q(t),$$

and $p(t)$, $q(t)$, $r(t)$, $p'(t)$, $r'(t)$ are continuous in t (in the strong sense), and for all $t \in [0, b]$, $p(t)$ has a bounded inverse $p^{-1}(t)$. (The requirement on the continuity of coefficients can be weakened.)

The solutions of the equation (A.41) are operator-valued functions or vector-valued functions $y(t)$ (with values in H).

To each pair of vectors y_0, $y_1 \in H$ corresponds a solution $y(t, \lambda)$ of (A.41) with the Cauchy data y_0 and y_1: $y(0, \lambda) = y_0$, $y'(0, \lambda) = y_1$. The solution $y(t, \lambda)$ is given by the formula

$$y(t, \lambda) = Y_0(t, \lambda)y_0 + Y_1(t, \lambda)y_1, \tag{A.42}$$

where $Y_0(t, \lambda)$, $Y_1(t, \lambda)$ are operator solutions of (A.41) with the initial data

$$Y_0(0, \lambda) = I, \quad Y_0'(0, \lambda) = 0,$$

$$Y_1(0, \lambda) = 0, \quad Y_1'(0, \lambda) = I.$$

We denote by $H(0,b)$ the Hilbert space formed by vector-valued functions $y(t)$, $t \in [0,b]$ such that $\langle y, y\rangle < +\infty$, where

$$\langle x, y\rangle = \int_0^b (x(t), y(t))_H \, dt.$$

Lemma A.5 *The correspondence between pairs $\{y_0, y_1\} \in H \oplus H$ and solutions $y(t,\lambda) \in H(0,b)$ of (A.41) defined by the formula (A.42) is bijective and bi-continuous (i.e. it is a homeomorphism). Thus the set of solutions of $l[y] = \lambda y$ is a closed subspace of $H(0,b)$.*

Proof. We will prove that

$$\langle y, y\rangle = \int_0^b |y(t,\lambda)|^2 \, dt \geq C(\lambda)\left(|y_0|^2 + |y_1|^2\right), \quad C(\lambda) > 0. \qquad \text{(A.43)}$$

Since

$$\max_{0\leq t\leq b}\left\{\left|Y_0''(t,\lambda)\right| + \left|Y_1''(t,\lambda)\right|\right\} < \infty,$$

we have

$$y(t,\lambda) = y_0 \cos t + y_1 \cos t + t^2 \cdot \left(|y_0|^2 + |y_1|^2\right)^{1/2} \cdot \vec{O}(1),$$

where $\vec{O}(1)$ is a vector quantity with bounded norm (depending, obviously, on λ). Hence, for $\gamma = \gamma(y_0, y_1)$, we obtain

$$|y(t,\lambda)| \geq \left(|y_0|^2 + |y_1|^2\right)^{1/2} \{\sin(t-\gamma) + O(t^2)\},$$

and (A.43) follows. Now continuous dependence of the initial data on $y(t,\lambda)$ follows from (A.43), and the remaining assertions of the Lemma are easily checked. □

Let $\mathrm{D} \subset H(0,b)$ be the set of vector-valued functions $y(t)$ such that $p(t)y'(t)$ is absolutely continuous[51] (in the strong sense) and $l[y] \in H(0,b)$. The expression $l[y]$ in (A.41) gives rise to a differential operator L in $H(0,b)$ with domain D: $Ly = l[y]$, $y \in \mathrm{D}$. The operator L is called the maximal operator corresponding to the differential expression $l[y]$ in $H(0,b)$.

By L_0 we denote the restriction of L on the linear manifold

$$\mathrm{D}_0 = \{y(t) \in \mathrm{D}\colon y(0) = y'(0) = y(b) = y'(b) = 0\}.$$

[51] As in [73] and [74], we say that a vector valued function $u(t)$ is absolutely continuous if it can be represented in the form $u(t) = u(0) + \int_0^t \nu(s)\, ds$, where $\nu(s)$ is a Bochner integrable vector-valued function. In this case $u'(t)$ exists almost everywhere on $[0,b]$ and $u'(t) = \nu(t)$.

Clearly, the operator L_0 is densely defined in $H(0,b)$ and is symmetric due to Lagrange's formula

$$\langle l[x], y\rangle = \{-(p(t)x'(t), y(t)) + (p(t)x(t), y'(t)) + i([\mathrm{Re}\ r(t)]x(t), y(t))\}_0^b + \langle x, l[y]\rangle. \tag{A.44}$$

We call L_0 the minimal operator corresponding to the differential expression $l[y]$ in $H(0,b)$. (Sometimes, especially in the case of differential operators with unbounded operator-valued coefficients, D_0 is restricted by an additional condition—each vector-valued function $y(t) \in \mathrm{D}_0$ has values belonging to some finite-dimensional subspace of H at all $t \in (0,b)$. This subspace depends on the function $y(t)$; see, for example, [896] and [913].)

Lemma A.6 *The adjoint of the minimal operator L_0 coincides with the maximal operator L corresponding to the differential expression $l[y]$ in $H(0,b)$, i.e. the following equality holds:*

$$L_0^* = L. \tag{A.45}$$

Proof. Using Lemma A.5, the proof is analogous to the one in the scalar case; see [6], [145] and [586]. By (A.44) we have $L \subset L_0^*$, so it remains to show $L_0^* \subset L$. Let $x(t) \in \mathrm{D}_{L_0^*}$ and let $y(t)$ be a solution of $l[y] = L_0^* x$. Then $\langle L_0 g, x - y\rangle = 0$ for all $g(t) \in \mathrm{D}_0$. It follows that $u(t) = x - y$ is a solution of $l[u] = 0$ since by Lemma A.5, the space N_0 of solutions of $l[y] = 0$ is a subspace of $H(0,b)$, and, as we will show below,

$$H(0,b) \ominus N_0 = \mathrm{R}_{L_0} := L_0 \mathrm{D}_0. \tag{A.46}$$

Indeed, let $h(t) \in H(0,b) \ominus N_0$ and let $g(t)$ be the solution of $l(g) = h$ with the initial data $g(b) = g'(b) = 0$. Then for all $u \in N_0$, from (A.44) we get

$$0 = \langle l[g], u\rangle = (p(0)g'(0), u(0)) - (p(0)g(0), u'(0)) - i([\mathrm{Re}\ r(0)]g(0), u(0)).$$

Since $u(0)$, $u'(0) \in H$ are arbitrary, and since $p(0)$ is invertible with the inverse $p^{-1}(0)$, we get $g(0) = g'(0) = 0$. Hence $g(t) \in \mathrm{D}_0$, and (A.46) is proven. Therefore $x = y + u$, where $y \in \mathrm{D}$, $u \in N_0$, i.e. $x \in \mathrm{D}$ and $L_0^* \subset L$. This concludes the proof of the Lemma. □

The result obtained in (A.45) shows that any self-adjoint extension $\widetilde{L}$ of the operator L_0 is determined on its domain $\widetilde{\mathrm{D}}$, $\mathrm{D}_0 \subset \widetilde{\mathrm{D}} \subset \mathrm{D}$, by the differential expression l in (A.41):

$$\widetilde{L}y = l[y], \quad y \in \widetilde{\mathrm{D}}. \tag{A.47}$$

By (A.44), a vector function $x(t) \in \mathrm{D}$ belongs to $\widetilde{\mathrm{D}}$ if and only if for all $y \in \widetilde{\mathrm{D}}$,

$$\{-(p(t)x'(t), y(t)) + (p(t)x(t), y'(t)) + i([\mathrm{Re}\ r(t)]x(t), y(t))\}_0^b = 0, \quad \text{(A.48)}$$

and any linear manifold $\widetilde{\mathrm{D}} \subset \mathrm{D}$ satisfying this property is a domain of some self-adjoint extension $\widetilde{L}$ of L_0.

We denote $H^2 = H \oplus H$ and associate to each element $\nu(t) \in \mathrm{D}$ a pair of vectors $\nu^\wedge$, $\nu^\vee \in H^2$:

$$\nu^\wedge = \{\nu(0), \nu(b)\},$$

$$\nu^\vee = \{p(0)\nu'(0) - \frac{i}{2}r(0)\nu(0); -p(b)\nu'(b) + \frac{i}{2}r(b)\nu(b)\}. \quad \text{(A.49)}$$

Since $p^{-1}(0)$ and $p^{-1}(b)$ are defined on all of H, then any pair of vectors $\nu^\wedge$, $\nu^\vee \in H^2$ can be represented as in (A.49), where $\nu(t) \in \mathrm{D}$. By (A.48), only the values of $\nu^\wedge$ and $\nu^\vee$ determine whether $\nu(t) \in \widetilde{\mathrm{D}}$ (the validity of the last inclusion is independent of the choice of $\nu(t) \in \mathrm{D}$).

Therefore, every self-adjoint extension $\widetilde{L}$ of L_0 defines in H^2 a binary relation θ, as follows: $\nu^\wedge \theta \nu^\vee$ if and only if the elements $\nu(t) \in \widetilde{\mathrm{D}}$ correspond to the pair $\nu^\wedge$, $\nu^\vee$. By (A.48) it follows that if $x^\wedge \theta x^\vee$ and $y^\wedge \theta y^\vee$, then

$$(x^\vee, y^\wedge)_2 - (x^\wedge, y^\vee)_2 = 0, \quad \text{(A.50)}$$

where $(\cdot,\cdot)_2$ denotes the inner product in H^2. By (A.48) it also follows that if $x^\wedge$, $x^\vee$ is a pair such that (A.50) holds for all $y^\wedge$, $y^\vee \in H^2$ such that $y^\wedge \theta y^\vee$, then $x^\wedge \theta x^\vee$. Thus the relation θ is Hermitian; see Sec. A.1. In other words, the condition (A.48) establishes a one-to-one correspondence between all self-adjoint extensions $\widetilde{L}$ of the operator L_0 and all Hermitian relations θ in $H^2 = H \oplus H$. Recalling Theorem A.1 on the general form of Hermitian relations, we see that we have proven the following:

Theorem A.6 *Every self-adjoint extension $\widetilde{L}$ of the operator L_0 determined in $H(0, b)$ by the differential expression $l[y]$ in (A.41) is defined by (A.47) and by any of the two types of boundary conditions:*

$$\cos \widehat{A} \cdot y^\vee - \sin \widehat{A} \cdot y^\wedge = 0, \quad \text{(A.51)}$$

or

$$(\widehat{U} - \widehat{I}) \cdot y^\vee + i(\widehat{U} + \widehat{I}) \cdot y^\wedge = 0, \quad \text{(A.52)}$$

where $\widehat{A}$ is a self-adjoint operator, $\widehat{U}$ is a unitary operator in $H^2 = H \oplus H$, and vectors $y^\wedge$, $y^\vee \in H^2$ are determined from $y(t)$ via (A.49). Conversely, any of the two boundary conditions (A.51) or (A.52) defines a self-adjoint extension of L_0. There is a one-to-one correspondence between unitary operators $\widehat{U}$ or $-\exp\{2i\widehat{A}\}$ and self-adjoint extensions $\widetilde{L}$.

Remark A.2 Any of the conditions (A.51), (A.52) is equivalent to specifying a "Lagrangian plane," and this plane is a hyper-maximal neutral lineal in $H \oplus H$. The requirement that the vector $\{y^\wedge, y^\vee\}$ belongs to this "Lagrangian plane" is equivalent to specifying a self-adjoint extension $\widetilde{L}$.

Remark A.3 Determined by Theorem A.4, the general form of self-adjoint boundary conditions for the equation $l[y] = \lambda y$ (A.41) in the space $H(0,b) = L_2(H;\ (0,b);\ dt)$ is valid without any changes also for equation $l_W[y] := W^{-1}l[y] = \lambda y$ in the space $H_W(0,b) := L_2(H;\ (0,b);\ W(t)dt)$, because the formulas (A.44)–(A.50) remain valid under simultaneous substitution $l_W[y]$ instead of $l[y]$ and scalar product $\langle\cdot,\cdot\rangle_W$ in $H_W(0,b)$ instead of the scalar product $\langle\cdot,\cdot\rangle$ in $H(0,b)$. This is obvious because $\langle l[x], y\rangle = \langle l_W[x], y\rangle_W := \langle WW^{-1}l[x], y\rangle$ and because, at $b < \infty$, $l[x] \in H(0,b)$ and $l_W[x] \in H_W(0,b)$ only simultaneously.

Example A.3 Separated boundary conditions. (Sturm-type conditions). Their general form is

$$\begin{cases} \cos A \cdot p(0)y'(0) - \left(\sin A + \frac{i}{2}\cos A \cdot r(0)\right) y(0) = 0, \\ \cos B \cdot p(b)y'(b) + \left(\sin B - \frac{i}{2}\cos B \cdot r(b)\right) y(b) = 0, \end{cases} \tag{A.53}$$

where A, B are arbitrary self-adjoint operators in H. In (A.51) and (A.52) the corresponding operators $\widehat{A}$ and $\widehat{U}$ are given by

$$\widehat{A} = \begin{pmatrix} A & 0 \\ 0 & B \end{pmatrix}, \qquad \widehat{U} = -\begin{pmatrix} \exp 2iA & 0 \\ 0 & \exp 2iB \end{pmatrix}.$$

If, in particular, there exits $(\cos A)^{-1}$, then the first condition in (A.53) becomes

$$p(0)y'(0) - \left(A_1 + \frac{i}{2}r(0)\right) y(0) = 0, \tag{A.54}$$

where A_1 is an arbitrary self-adjoint (possibly unbounded) operator in H. Conditions of type (A.53) can also be obtained directly from (A.48) by Theorem A.1.

Example A.4 Generalized periodic conditions have the following form:

$$\begin{aligned} & y(b) = Uy(0), \\ & p(b)y'(b) - \frac{i}{2}r(b)y(b) = U\left[p(0)y'(0) - \frac{i}{2}r(0)y(0)\right], \end{aligned} \tag{A.55}$$

where U is an arbitrary unitary operator in H. In (A.52) and (A.51) the corresponding operators $\widehat{U}$ and $\widehat{A}$ are given by

$$\widehat{U} = -\begin{pmatrix} 0 & U^* \\ U & 0 \end{pmatrix}, \qquad \widehat{A} = \frac{\pi}{4}\begin{pmatrix} I & -U^* \\ -U & I \end{pmatrix}.$$

Example A.5 (Mixed Sturm-periodic conditions.) For simplicity of notation, we let $p(t) \equiv I$, $r(t) \equiv 0$. Let H_0, H_b be subspaces of H with $\dim H_0 = \dim H_b \leq \infty$. Let P_0, P_b be the orthogonal projections onto H_0 and H_b respectively, and let $P_0^\perp$ and $P_b^\perp$ be the orthogonal projections onto $H_0^\perp = H \ominus H_0$ and $H_b^\perp = H \ominus H_b$. If H is infinite-dimensional, it can happen that $\dim H_0^\perp \neq \dim H_b^\perp$. Let V be an isometry, $VH_0 = H_b$, and let A and B be arbitrary self-adjoint operators in $H_0^\perp$ and $H_b^\perp$ respectively. Now we can write mixed Sturm-periodic conditions as follows:

$$\begin{aligned} & P_b y'(b) = VP_0 y'(0), \quad P_b y(b) = VP_0 y(0), \\ & \cos A \cdot P_0^\perp y'(0) - \sin A \cdot P_0^\perp y(0) = 0, \\ & \cos B \cdot P_b^\perp y'(b) + \sin B \cdot P_b^\perp y(b) = 0. \end{aligned} \tag{A.56}$$

In the formulas (A.52) and (A.51), the corresponding operators $\widehat{U}$ and $\widehat{A}$ in the space $H \oplus H = H_0^\perp \oplus H_0 \oplus H_b \oplus H_b^\perp$ are given by

$$\widehat{U} = -\begin{pmatrix} \exp 2iA & & & \\ & 0 & V^* & \\ & V & 0 & \\ & & & \exp 2iB \end{pmatrix},$$

$$\widehat{A} = \frac{\pi}{4}\begin{pmatrix} \frac{4}{\pi}A & & & \\ & I_0 & -V^* & \\ & -V & I_b & \\ & & & \frac{4}{\pi}B \end{pmatrix},$$

where all blank elements in these matrices are zeros, and I_0, I_b are the identity operators in H_0 and H_b.

A.3 Self-Adjoint Extensions of Differential Operators of Arbitrary Order with Operator Coefficients

We will separately consider the cases of even-order and odd-order differential expressions defined on the space of sufficiently smooth vector-valued functions $y(t) \in H(0, b)$.

Remark A.4 Here, as in Sec. A.2 (see Remark A.3 there), we put $W(t) \equiv I$, and all results in Sec. A.3 are valid for the general case. In particular, self-adjoint boundary conditions obtained below have the same form—invariant with respect to simultaneous substitutions $l_W[y]$ instead of $l[y]$ and $H_W(0, b)$ instead of $H(0, b)$, where $W^*(t) = W(t) >> 0$ and $b < \infty$.

A.3.1 *The Expressions of Order 2n*

Let

$$l[y] = \sum_{k=1}^{n} (-1)^k \left\{ \left(p_k y^{(k)}\right)^{(k)} - \frac{i}{2}\left[\left(q_k^* y^{(k)}\right)^{(k-1)} + \left(q_k y^{(k-1)}\right)^{(k)}\right]\right\} + p_0 y, \tag{A.57}$$

where the coefficients $p_k(t)$ are self-adjoint operators,

$$p_k(t) = p_k^*(t), \quad (k = 0, 1, \ldots, n),$$

and all coefficients $p_k(t)$, $q_k(t)$ continuously depend on t together with all of its derivatives of order up to and including k, and $p_n^{-1}(t)$ exists and is bounded for $t \in [0, b]$. (These requirements can be weakened by viewing l as a quasi-differential expression.)

We will need the expressions for quasi-derivatives $y^{[k]}$ corresponding to l. We define the quasi-derivatives as follows[52]:

$$y^{[k]} = y^{(k)}, \quad (k = 0, 1, \ldots, n-1; y^{(0)} = y)$$

$$y^{[n]} = p_n y^{(n)} - \frac{i}{2} q_n y^{(n-1)}$$

[52]For $q_1 = q_2 = \cdots = q_n \equiv 0$, our definition coincides with the one used in [6], [145] and [586].

$$y^{[2n-k]} = -\frac{d}{dt}y^{[2n-k-1]} + p_k y^{(k)} + \frac{i}{2}\left[q_{k+1}^* y^{(k+1)} - q_k y^{(k-1)}\right],$$

$$(k = 0, 1, \ldots, n-1; q_0 \equiv 0) \tag{A.58}$$

With these notations, we have

$$l[y] = y^{[2n]}. \tag{A.59}$$

It is easily checked that

$$\begin{aligned}(l[x], y)_H = (x^{[2n]}, y) = &-\frac{d}{dt}\sum_{k=1}^{n}\left(x^{[2n-k]}, y^{(k-1)}\right) \\ &+ \sum_{k=0}^{n}\left(p_k x^{(k)}, y^{(k)}\right) + \frac{i}{2}\sum_{k=1}^{n}\left[\left(q_k^* x^{(k)}, y^{(k-1)}\right)\right. \\ &\left. - \left(q_k x^{(k-1)}, y^{(k)}\right)\right].\end{aligned} \tag{A.60}$$

Subtracting the analogous expression for $(x, l[y])_H$ and integrating, we obtain the Lagrange's identity:

$$\begin{aligned}&\langle l[x], y\rangle - \langle x, l[y]\rangle \\ &= \sum_{k=1}^{n}\left[\left(x^{(k-1)}, y^{[2n-k]}\right) - \left(x^{[2n-k]}, y^{(k-1)}\right)\right]_{t=0}^{t=b}.\end{aligned} \tag{A.61}$$

We will rewrite the right hand side of (A.61). To this end, to each $2n-1$ times continuous quasi-differentiable vector-valued function $u(t) \in H(0,b)$ we associate a pair of vectors $u^\wedge$, $u^\vee \in H^{2n} = \underbrace{H \oplus H \oplus \cdots \oplus H}_{(2n)}$ by the following formulas:

$$\begin{aligned}u^\wedge &= \text{col}\,\{u(0), u'(0), \ldots, u^{(n-1)}(0), u(b), u'(b), \ldots, u^{(n-1)}(b)\} \\ u^\vee &= \text{col}\,\{u^{[2n-1]}(0), u^{[2n-2]}(0), \ldots, u^{[n]}(0), -u^{[2n-1]}(b), \\ &\quad -u^{[2n-2]}(b), \ldots, -u^{[n]}(b)\}.\end{aligned} \tag{A.62}$$

Using these notations, we can write (A.61) as follows:

$$\langle l[x], y\rangle - \langle x, l[y]\rangle = (x^\vee, y^\wedge)_{2n} - (x^\wedge, y^\vee)_{2n}, \tag{A.63}$$

where $(\cdot,\cdot)_{2n}$ denotes the inner product in H^{2n}.

We now define the operators L_0 (minimal) and L (maximal) corresponding to the expression $l[y]$ in (A.57) with the domains D_0 and D respectively. Here D is the set of vector-valued functions $y(t) \in H(0,b)$ with absolutely continuous quasi-derivatives of order up to and including $2n-1$ and such that $l[y] \in H(0,b)$, and D_0 is the set of $y(t) \in \mathrm{D}$ such that $y(0) = y^{[1]}(0) = \cdots = y^{[2n-1]}(0) = 0$ and $y(b) = y^{[1]}(b) = \cdots = y^{[2n-1]}(b) = 0$. As in Sec. A.2, we can prove the following analogue of Lemma A.6:

Lemma A.7 *The adjoint of the minimal operator L_0 coincides with the maximal operator L corresponding to the differential expression $l[y]$ in $H(0,b)$, i.e. the following equality holds: $L_0^* = L$.*

Here we only need to check that the set of solutions of the homogeneous equation $l[y] = 0$ is a closed subspace of $H(0,b)$. We will do this in Lemma A.10 at the end of this section. Then the proof of Lemma A.7 follows the well-known scheme as in the scalar case; see [6], [145] and [586].

Thus by (A.63), every self-adjoint extension $\widetilde{L}$ of L_0 is generated by the expression $l[y]$ on a domain $\widetilde{\mathrm{D}}$, $\mathrm{D}_0 \subset \widetilde{\mathrm{D}} \subset \mathrm{D}$, where $\widetilde{\mathrm{D}}$ satisfies the following property: $x(t) \in \mathrm{D}$ belongs to $\widetilde{\mathrm{D}}$ if and only if

$$(x^\vee, y^\wedge)_{2n} - (x^\wedge, y^\vee)_{2n} = 0 \qquad \text{for all } y(t) \in \widetilde{\mathrm{D}}. \tag{A.64}$$

Here the vectors $x^\wedge$, $x^\vee$, $y^\wedge$, $y^\vee \in H^{2n}$ are determined from $x(t)$, $y(t)$ via (A.62). Conversely, every linear manifold $\widetilde{\mathrm{D}} \subset \mathrm{D}$ satisfying the above property gives rise to a self-adjoint extension $\widetilde{L}$ of L_0. As in Sec. A.2, we have the following:

Theorem A.7 *Every self-adjoint extension $\widetilde{L}$ of the operator L_0 gives rise to a Hermitian relation θ in H^{2n}, and conversely. This correspondence between self-adjoint extensions of L_0 and Hermitian relations in H^{2n} is one-to-one. In other words, every self-adjoint extension $\widetilde{L}$ of L_0 is determined by any of the following two types of boundary conditions:*

$$\cos\widehat{A} \cdot y^\vee - \sin\widehat{A} \cdot y^\wedge = 0, \tag{A.65}$$

or

$$(\widehat{U} - \widehat{I}) \cdot y^\vee + i(\widehat{U} + \widehat{I}) \cdot y^\wedge = 0, \tag{A.66}$$

where $\widehat{A}$ is a self-adjoint operator, $\widehat{U}$ is a unitary operator in H^{2n}, and vectors $y^\wedge$, $y^\vee \in H^{2n}$ are determined from $y(t)$ via (A.62). Conversely, any of the two boundary conditions (A.65), (A.66) defines a self-adjoint extension $\widetilde{L}$ of L_0. (Here, as in Theorem A.6, the correspondence between unitary operators $\widehat{U}$ or $-\exp\{2i\widehat{A}\}$ and self-adjoint extensions $\widetilde{L}$ is one-to-one.)

Example A.6 Separated boundary conditions for $t = 0$ (similarly for $t = b$, cf. (A.53)) are written in the following form:

$$\cos A \cdot y^{\vee}(0) - \sin A \cdot y^{\wedge}(0) = 0, \tag{A.67}$$

where A is an arbitrary self-adjoint operator in H^n, and

$$\begin{aligned} y^{\wedge}(0) &= \text{col }\{y(0), y'(0), \ldots, y^{(n-1)}(0)\} \in H^n, \\ y^{\vee}(0) &= \text{col }\{y^{[2n-1]}(0), y^{[2n-2]}(0), \ldots, y^{[n]}(0)\} \in H^n. \end{aligned} \tag{A.68}$$

In particular, if $A = \text{diag }\{\alpha_1, \ldots, \alpha_n\}$ is block-diagonal, we have:

$$(\cos\alpha_k) \cdot y^{[2n-k]}(0) - (\sin\alpha_k) \cdot y^{(k-1)}(0) = 0, \qquad k = 1, \ldots, n. \tag{A.69}$$

Example A.7 Generalized periodic conditions:

$$y^{\vee}(b) = Uy^{\vee}(0), \qquad y^{\wedge}(b) = Uy^{\wedge}(0), \tag{A.70}$$

where U is a unitary operator in H^n.

In particular, if $U = \text{diag }\{U_1, \ldots, U_n\}$, where U_k are unitary operators in H, we have:

$$y^{[2n-k]}(b) = U_k y^{[2n-k]}(0), \quad y^{(k-1)}(b) = U_k y^{(k-1)}(0), \tag{A.71}$$

where $k = 1, \ldots, n$.

A.3.2 *The Expressions of Order $2n - 1$*

Let

$$\begin{aligned} l[y] = \sum_{k=0}^{n-1} (-1)^k \Big\{ \frac{i}{2} \Big[\left(q_{k+1} y^{(k)}\right)^{(k+1)} + \left(q_{k+1}^* y^{(k+1)}\right)^{(k)} \Big] \\ + \left(p_k y^{(k)}\right)^{(k)} \Big\}, \end{aligned} \tag{A.72}$$

where the coefficients $p_k(t)$ are self-adjoint operators, and all $p_k(t)$, $q_k(t)$ continuously depend on t together with all of its derivatives of order up to and including k, and $[\text{Re }q_n(t)]^{-1}$ exists and is bounded for $t \in [0, b]$.

We define the expressions for quasi-derivatives $y^{[k]}$ corresponding to l as follows:

$$y^{[k]} = y^{(k)}, \quad (k = 0, 1, \ldots, n-2); \quad y^{[n-1]} = -\frac{i}{2} q_n y^{(n-1)},$$

$$y^{[2n-k-1]} = -\frac{d}{dt}y^{[2n-k-2]} + p_k y^{(k)} + \frac{i}{2}\left[q_{k+1}^* y^{(k+1)}\right.$$
$$\left. - q_k y^{(k-1)}\right], \quad (k = 0, 1, \ldots, n-1; q_0 \equiv 0). \tag{A.73}$$

With these notations, we have

$$l[y] = y^{[2n-1]}. \tag{A.74}$$

Instead of (A.60), we now have

$$(l[x], y)_H = (x^{[2n-1]}, y) = -\frac{d}{dt}\sum_{k=0}^{n-1}\left(x^{[2n-k-2]}, y^{(k)}\right)$$
$$+ \sum_{k=0}^{n-1}\left(p_k x^{(k)}, y^{(k)}\right) + \frac{i}{2}\sum_{k=0}^{n-1}\left[\left(q_{k+1}^* x^{(k+1)}, y^{(k)}\right)\right.$$
$$\left. - \left(q_{k+1} x^{(k)}, y^{(k+1)}\right)\right],$$

and Lagrange's identity becomes

$$\langle l[x], y\rangle - \langle x, l[y]\rangle$$
$$= \sum_{k=0}^{n-1}\left[\left(x^{(k)}, y^{[2n-k-2]}\right) - \left(x^{[2n-k-2]}, y^{(k)}\right)\right]_{t=0}^{t=b}$$
$$= i\left([\text{Re } q_n(t)]\, x^{(n-1)}, y^{(n-1)}\right)\Big|_0^b \tag{A.75}$$
$$+ \sum_{k=0}^{n-2}\left[\left(x^{(k)}, y^{[2n-k-2]}\right) - \left(x^{[2n-k-2]}, y^{(k)}\right)\right]_{t=0}^{t=b}.$$

The right hand side of the last equality in (A.75) contains a term with derivatives $x^{(n-1)}$ and $y^{(n-1)}$ of the same order, so it is more difficult to pass to the form (A.63) of Lagrange's identity. To do this, we argue as follows. Since the operator Re $q_n(t)$ is self-adjoint and has bounded inverse, then

$$H = H_t^+ \oplus H_t^-,$$

where H_t^+, H_t^- are invariant subspaces of Re $q_n(t)$ such that Re $q_n(t)$ is a positive operator (i.e. it has a positive greatest lower bound) on H_t^+ and $-$Re $q_n(t)$ is a positive operator on H_t^-. We denote by P_t^+, P_t^- the orthogonal projections onto H_t^+, H_t^- respectively and let

$$q_+(t) = \frac{1}{2}[\text{Re } q_n(t)]P_t^+, \quad q_-(t) = -\frac{1}{2}[\text{Re } q_n(t)]P_t^-. \tag{A.76}$$

Clearly, the operators $q_+(t)$, $q_-(t)$ are nonnegative and

$$\frac{1}{2}\text{Re } q_n(t) = q_+(t) - q_-(t).$$

We also let

$$Q_t^+ = q_+^{\frac{1}{2}}(t) + q_-^{\frac{1}{2}}(t), \quad Q_t^- = q_+^{\frac{1}{2}}(t) - q_-^{\frac{1}{2}}(t), \tag{A.77}$$

where $q_\pm^{\frac{1}{2}}(t) \geq 0$. The operators $Q_t^\pm$ are self-adjoint, they commute with each other, and have bounded inverses. It is easily checked that

$$Q_t^+ Q_t^- = Q_t^- Q_t^+ = \frac{1}{2}\text{Re } q_n(t), \tag{A.78}$$

and also

$$Q_t^- \left(Q_t^+\right)^{-1} = Q_t^+ \left(Q_t^-\right)^{-1} = P_t^+ - P_t^-. \tag{A.79}$$

(The last formula will only be needed in the proof of Theorem A.9.)

Finally, note that

$$\dim H_b^+ = \dim H_0^+ = H_t^+, \dim H_b^- = \dim H_0^- = \dim H_t^-, \quad t \in [0, b]. \tag{A.80}$$

The equality of dimensions (finite or infinite) follows since Re $q_n(t)$ continuously depends on $t \in [0, b]$, and $[\text{Re } q_n(t)]^{-1}$ is bounded for all $t \in [0, b]$, i.e. 0 does not belong to the spectrum of Re $q_n(t)$ for any $t \in [0, b]$. In this case, the orthogonal projections $P_t^- = E_{0t}$, $P_t^+ = I - E_{0t}$, where $E_{\lambda t}$ is the resolution of identity of the operator Re $q_n(t)$, continuously (uniformly) depend on t[53], and, hence, (A.80) holds.

By (A.80) there exist infinitely many unitary operators mapping H_b^- to H_0^- and H_b^+ to H_0^+. We arbitrarily fix one such operator U_q. (In particular, if $\dim H_0^+ = 0$, or $\dim H_0^- = 0$, or $H_0^+ = H_b^+$, we may take $U_q = I$.) Clearly,

$$P_0^- U_q = U_q P_b^-, \quad P_0^+ U_q = U_q P_b^+. \tag{A.81}$$

[53] See [169, Chapter IX, Sec. 135].

We will show that the following holds:

$$Q_0^+ U_q Q_b^- = Q_0^- U_q Q_b^+. \tag{A.82}$$

To do this, note that $Q_t^{\pm} P^+(t) = q_+^{\frac{1}{2}}(t)$, $Q_t^{\pm} P^-(t) = \pm q_-^{\frac{1}{2}}(t)$, and, hence, using (A.81), we obtain

$$Q_0^+ U_q Q_b^- P_b^+ = Q_0^+ P_0^+ U_q q_+^{\frac{1}{2}}(b) = q_+^{\frac{1}{2}}(0) U_q q_+^{\frac{1}{2}}(b),$$

$$Q_0^- U_q Q_b^+ P_b^+ = Q_0^- P_0^+ U_q q_+^{\frac{1}{2}}(b) = q_+^{\frac{1}{2}}(0) U_q q_+^{\frac{1}{2}}(b),$$

i.e.

$$Q_0^+ U_q Q_b^- P_b^+ = Q_0^- U_q Q_b^+ P_b^+.$$

Similarly,

$$Q_0^+ U_q Q_b^- P_b^- = Q_0^- U_q Q_b^+ P_b^-.$$

Since $P_b^+ + P_b^- = I$, we obtain (A.82).

Now note that if f_0, $f_b \in H$ are arbitrary vectors, and if we associate f, $f' \in H$ to f_0, $f_b \in H$ by formulas

$$f = Q_0^+ f_0 + U_q Q_b^+ f_b, \quad f' = -iQ_0^- f_0 + iU_q Q_b^- f_b, \tag{A.83}$$

and if we associate g, $g' \in H$ to g_0, $g_b \in H$ by the same formulas, then using (A.83), (A.82) and (A.78), we get

$$(f', g) - (f, g') = i\left\{([\mathrm{Re}\ q_n(b)]\, f_b, g_b) - ([\mathrm{Re}\ q_n(0)]\, f_0, g_0)\right\}. \tag{A.84}$$

It remains to associate to each $(2n-2)$ times continuous (quasi)-differentiable vector-valued function $\nu(t) \in H(0, b)$ a pair of vectors $\nu^\wedge$, $\nu^\vee \in H^{2n-1} = H \oplus H^{2n-2}$ by the following formulas:

$$\begin{aligned} \nu^\wedge = \mathrm{col}\ \{&Q_0^+ \nu_0^{(n-1)} + U_q Q_b^+ \nu_b^{(n-1)},\\ &\nu_0, \nu_0', \ldots, \nu_0^{(n-2)}, \nu_b, \nu_b', \ldots, \nu_b^{(n-2)}\} \end{aligned}$$

$$\begin{aligned} \nu^\vee = \mathrm{col}\ \{&iU_q Q_b^- \nu_b^{(n-1)} - iQ_0^- \nu_0^{(n-1)},\\ &\nu_0^{[2n-2]}, \nu_0^{[2n-3]} \ldots, \nu_0^{[n]}, -\nu_b^{[2n-2]}, -\nu_b^{[2n-3]}, \ldots, -\nu_b^{[n]}\}, \end{aligned} \tag{A.85}$$

where $\nu_0 = \nu(0)$, $\nu_b = \nu(b)$ *etc.*

Lagrange's identity (A.75) can now be written as

$$\langle l[x], y\rangle - \langle x, l[y]\rangle = (x^\vee, y^\wedge)_{2n-1} - (x^\wedge, y^\vee)_{2n-1}. \tag{A.86}$$

Here $(\cdot,\cdot)_{2n-1}$ denotes the inner product in H^{2n-1}, and the vectors $x^\wedge$, $y^\wedge$, $x^\vee$, $y^\vee$ are given by (A.85).

By L_0 we denote the minimal operator corresponding to the expression $l[y]$ with the domain consisting of vector-valued functions y such[54] that $y(0) = y^{[1]}(0) = \cdots = y^{[2n-2]}(0) = 0$ and $y(b) = y^{[1]}(b) = \cdots = y^{[2n-2]}(b) = 0$. Arguing as before, we have the following:

Theorem A.8 *Theorem A.7 holds for self-adjoint extensions $\widetilde{L}$ of the operator L_0 corresponding to the $(2n-1)$-order differential expression $l[y]$ in (A.72) with the following notational replacements: H^{2n} replaced by H^{2n-1}, the formulas (A.62) replaced by the formulas (A.85) which now define vectors $y^\wedge$, $y^\vee \in H^{2n-1}$ in the boundary conditions (A.65), (A.66).*

For the expression $l[y]$ of even order, there always exist separated self-adjoint boundary conditions. However, this is not always the case for the expression $l[y]$ of odd order.

Theorem A.9 *Assume that $l[y]$ has order $2n-1$. Assume that* $\dim H < \infty$ *(n is arbitrary) or* $\dim H = \infty$ *but $n = 1$ (i.e. $l[y]$ is of order 1).*

There exist separated self-adjoint boundary conditions for the expression $l[y]$ if and only if the following hold:

$$\dim H_0^+ = \dim H_0^- (\le \infty), \dim H_b^+ = \dim H_b^- \tag{A.87}$$

(by (A.80), the equalities in (A.87) imply each other).

If (A.87) holds, then all such boundary conditions (for example, for $t = 0$) can be written, as any Hermitian relation, in each of the following forms:

$$\cos A \cdot y^\vee(0) - \sin A \cdot y^\wedge(0) = 0 \tag{A.88}$$

or

$$(U - I)y^\vee(0) + i(U + I)y^\wedge(0) = 0, \tag{A.89}$$

[54] Of course, we mean that y has $2n-2$ absolutely continuous derivatives and $l[y] \in H(0, b)$. On this set of functions (without any boundary conditions) the expression $l[y]$ generates the maximal operator L, and we have $L = L_0^*$.

where A is self-adjoint and U is a unitary operator in $\mathbf{H}^{\sim} = H_0^+ \oplus H^{n-1}$, and vectors $y^{\wedge}(0)$, $y^{\vee}(0) \in \mathbf{H}^{\sim}$ are defined from $y(t)$ via

$$y^{\wedge}(0) = col\,\{(q_+^{\frac{1}{2}}(0) + V_0 q_-^{\frac{1}{2}}(0))y_0^{(n-1)}, y_0, y_0', \ldots, y_0^{(n-2)}\}, \qquad \text{(A.90)}$$
$$y^{\vee}(0) = col\,\{i(V_0 q_-^{\frac{1}{2}}(0) - q_+^{\frac{1}{2}}(0))y_0^{(n-1)}, y_0^{[2n-2]}, y_0^{[2n-3]}, \ldots, y_0^{[n]}\},$$

where V_0 is an arbitrarily fixed (for all boundary conditions with $t = 0$) isometric operator mapping H_0^- onto H_0^+: $V_0 H_0^- = H_0^+$.

If $\dim H = \infty$ and $n \geq 2$ (i.e. the expression l is of order at least 3), then separated self-adjoint boundary conditions for l always exist, and for $t = 0$ have the same general form (A.88), where $y^{\wedge}(0)$, $y^{\vee}(0) \in \mathbf{H}^{\sim}$ are defined by

$$y^{\wedge}(0) = col\,\{y_0^+ + V_1 y_0^-, y_0, y_0', \ldots, y_0^{(n-3)}\}, \qquad \text{(A.91)}$$
$$y^{\vee}(0) = col\,\{iV_1 y_0^- - iy_0^+, y_0^{[2n-2]}, y_0^{[2n-3]}, \ldots, y_0^{[n+1]}\},$$

where

$$y_0^+ = \left\{q_+^{\frac{1}{2}}(0)y_0^{(n-1)}, \frac{1}{2}\left(y_0^{[n]} - iy_0^{(n-2)}\right)\right\} \in H_0^+ \oplus H,$$

$$y_0^- = \left\{q_-^{\frac{1}{2}}(0)y_0^{(n-1)}, \frac{1}{2}\left(y_0^{[n]} + iy_0^{(n-2)}\right)\right\} \in H_0^- \oplus H, \qquad \text{(A.92)}$$

and V_1 is an arbitrarily fixed isometric operator mapping $H_0^- \oplus H$ onto $H_0^+ \oplus H$:

$$V_1\{H_0^- \oplus H\} = H_0^+ \oplus H. \qquad \text{(A.93)}$$

If $\dim H = \infty$, $n \geq 2$ and (A.87) holds, then instead of (A.91), we can use formulas (A.90).

Proof. We will prove that the condition (A.87) is necessary. Assume that there exist separated self-adjoint boundary conditions for the expression l in (A.72). These boundary conditions give rise to a self-adjoint extension of the operator L_0, and, hence, we can associate to the given boundary conditions a Hermitian relation θ in H^{2n-1} with the following property: if a vector-valued function $\nu(t)$ satisfies those boundary conditions, then $\nu^{\wedge}\theta\nu^{\vee}$ and conversely. The vectors $\nu^{\wedge}$, $\nu^{\vee} \in H^{2n-1}$ are defined by the formulas (A.85). Since the boundary conditions are separated, together with $\nu^\wedge\theta\nu^\vee$ we also have $\nu_0^\wedge\theta\nu_0^\vee$, where the vectors $\nu_0^\wedge$, $\nu_0^\vee \in H^{2n-1}$ are obtained by putting $\nu_b = \nu_b' = \cdots = \nu_b^{[2n-2]} = 0$[55], i.e.

$$\nu_0^\wedge = \{\nu^\sim, \nu, 0\} \in H^{2n-1}, \quad \nu_0^\vee = \{(\nu^\sim)', \nu', 0\} \in H^{2n-1}, \tag{A.94}$$

where

$$\begin{gathered} \nu^\sim = Q_0^+\nu_0^{(n-1)} \in H, \nu = \{\nu_0, \nu'_0, \dots \nu_0^{(n-2)}\} \in H^{n-1}, \qquad \text{(A.95)} \\ (\nu^\sim)' = -iQ_0^-\nu_0^{(n-1)} \in H, \quad \nu' = \{\nu_0^{[2n-2]}, \nu_0^{[2n-3]}, \dots \nu_0^{[n]}\} \in H^{n-1}, \end{gathered}$$

and the element 0 in braces denotes the zero-vector in H^{2n-1}. The relation $\nu_0^\wedge\theta\nu_0^\vee$ means that the function $\nu(t)$ satisfies the given boundary conditions for $t = 0$.

From (A.95) we obtain following connection between the vectors $\nu^\sim$, $(\nu^\sim)' \in H$: $(\nu^\sim)' = -iQ_0^-(Q_0^+)^{-1}\nu^\sim$. Using (A.79) we get

$$(\nu^\sim)' = i(P_0^- - P_0^+)\nu^\sim. \tag{A.96}$$

Let $u(t)$ be a vector-valued function, and let $u_0^\wedge$, $u_0^\vee \in H^{2n-1}$ be the vectors defined from $u(t)$ via (A.94) and (A.95). Since the given separated boundary conditions are self-adjoint, the following holds: if

$$\begin{gathered} (u_0^\vee, \nu_0^\wedge)_{2n-1} - (u_0^\wedge, \nu_0^\vee)_{2n-1} \equiv 2i((P_0^- - P_0^+)u^\sim, \nu^\sim) \\ +(u', \nu)_{n-1} - (u, \nu')_{n-1} = 0, \qquad \text{(A.97)} \end{gathered}$$

for all $\nu(t)$ satisfying the boundary conditions $\nu_0^\wedge\theta\nu_0^\vee$ at $t = 0$, then $u(t)$ satisfies the same boundary conditions for $t = 0$, i.e. $u_0^\wedge\theta u_0^\vee$.

By M we will denote the set of vectors $X = \{x^\sim, x_1, x_2\} \in H^{2n-1}$, where $x^\sim \in H$, x_1, $x_2 \in H^{n-1}$, such that if

$$X^\wedge = \{x^\sim, x_1 + ix_2, 0\} \in H^{2n-1},$$

[55]This property may be taken as a basis for the definition of separated boundary conditions as it is done for partial differential operators (see [70, p. 85]). This leads to the following assertion:

Proposition A.4 *Hermitian relation θ in H^{2n-1} corresponds to separated boundary conditions in accordance with (A.85) if and only if $\{w, \nu, u\}\theta\{w', \nu', u'\}$ implies*

$$\left\{\frac{1}{2}\left[w + i(P_0^+ - P_0^-)w'\right], \nu, 0\right\}\theta\left\{\frac{1}{2}\left[w' - i(P_0^+ - P_0^-)w\right], \nu', 0\right\},$$

where w, $w' \in H$, and ν, ν', u, $u' \in H^{n-1}$. Hermitian relation θ in H^{2n} corresponds to separated boundary conditions in accordance with (A.62) if and only if $\{v, u\}\theta\{v', u'\}$ implies $\{v, 0\}\theta\{v', 0\}$, where v, v', u, $u' \in H^n$.

$$X^{\vee} = \{i(P_0^- - P_0^+)x^{\sim}, ix_1 + x_2, 0\} \in H^{2n-1}, \tag{A.98}$$

then $X^{\wedge}\theta X^{\vee}$. From the properties of Hermitian relations it follows that M is a closed linear set in H^{2n-1} (i.e. a subspace of H^{2n-1}). It also follows that if $X \in M$ and $Y = \{y^{\sim}, y_1, y_2\} \in M$, then

$$\begin{aligned}(X^{\vee}, Y^{\wedge})_{2n-1} - (X^{\wedge}, Y^{\vee})_{2n-1} \equiv 2i[((P_0^- - P_0^+)x^{\sim}, y^{\sim})\\ + (x_1, y_1)_{n-1} - (x_2, y_2)_{n-1}] = 0.\end{aligned} \tag{A.99}$$

The last equality can be written as

$$(TX, Y)_{2n-1} = 0, \tag{A.100}$$

where T is the operator in H^{2n-1} given by the matrix

$$T = \begin{pmatrix} P_0^- - P_0^+ & 0 & 0 \\ 0 & I_{n-1} & 0 \\ 0 & 0 & -I_{n-1} \end{pmatrix}, \tag{A.101}$$

where I_{n-1} is the identity operator in H^{n-1}.

The property of self-adjoint separated boundary conditions and the corresponding Hermitian relations θ formulated in connection with (A.97) can now be interpreted as follows: if $X \in H^{2n-1}$ and (A.99) (or, which is the same, (A.100)) holds for all $Y \in M$, then $X \in M$. In other words, the operator T maps M onto its orthogonal complement in H^{2n-1}:

$$H^{2n-1} = M \oplus TM. \tag{A.102}$$

Since the operator T in (A.101) is unitary, we have

$$\dim M = \dim TM. \tag{A.103}$$

On the other hand, $T = P_1 - P_2$, where

$$P_1 = \begin{pmatrix} P_0^- & & O \\ & I_{n-1} & \\ O & & 0 \end{pmatrix}, \quad P_2 = \begin{pmatrix} P_0^+ & & O \\ & 0 & \\ O & & I_{n-1} \end{pmatrix}$$

are the orthogonal projections in H^{2n-1} whose images are mutually orthogonal subspaces H_1 and H_2 respectively:

$$H_1 = H_0^- \oplus H^{n-1}, \quad H_2 = H_0^+ \oplus H^{n-1}, \tag{A.104}$$

whereby

$$H^{2n-1} = H_1 \oplus H_2. \tag{A.105}$$

Thus if $X \in M$, and, hence, $(TX, X)_{2n-1} = 0$, then

$$((P_1 - P_2)X, X)_{2n-1} = 0,$$

or

$$|P_1 X| = |P_2 X|, \quad X \in M. \tag{A.106}$$

It follows that

$$\dim M \leq \min\{\dim H_1, \dim H_2\}. \tag{A.107}$$

Indeed, suppose that, for example, $\dim H_2 \geq \dim H_1 = m$. Then for any $m+1$ vectors $f_k \in M$ $(k = 1, \ldots, m+1)$, there exists a nontrivial linear combination $f = \sum_{k=1}^{m+1} a_k f_k$ such that $P_1 f = 0$ because the vectors $P_1 f_k$ $(k = 1, \ldots, m+1)$ are linearly dependent. But then from (A.106) we have $P_2 f = 0$, and, hence, $f = 0$ because $P_1 + P_2 = I_{2n-1}$. Thus $\dim M \leq m$[56].

Comparing (A.105), (A.107), (A.103) and (A.102), we see that $\dim M \leq m$ is possible only when

$$\dim H_1 = \dim H_2 = \dim M.$$

Now from (A.104), it follows that

$$\dim\{H_0^- \oplus H^{n-1}\} = \dim\{H_0^+ \oplus H^{n-1}\}. \tag{A.108}$$

The equation (A.108) is possible only in the following cases: 1) $\dim H_0^- = \dim H_0^+$, $\dim H$ is arbitrary (more precisely, even or infinite), or 2) $\dim H^{n-1} = \infty$, i.e. $\dim H = \infty$ and $n \geq 2$ (since in the case $n = 1$, $\dim H^{n-1} = 0$). This concludes the proof of the necessity part of the Theorem. We leave it to the reader to check the remaining assertions of the Theorem. □

Corollary A.4 *If $\dim H < \infty$ is odd, then the differential expression $l[y]$ of odd order does not have any self-adjoint separated boundary conditions.*

Example A.8 *General form of self-adjoint boundary conditions for first order operators, and, in particular, for canonical systems.*

[56] Here $m < \infty$. For $m = \infty$, (A.107) is trivial. (H is separable!)

Consider the system

$$J\frac{d}{dt}y + p(t)y = \lambda W(t)y, \qquad 0 \le t \le b, \tag{A.109}$$

where

$$J^* = -J, \qquad J^2 = -I, \qquad J = i(P^+ - P^-), \tag{A.110}$$

and $P^\pm$ are, respectively, orthogonal projections onto the positive and negative subspaces $H^\pm$ of the operator J/i.

In accordance with Theorem A.8, by formulas (A.85) for $n = 1$, we construct the expressions $y^\wedge$ and $y^\vee$. We have $J = const$; hence, $H_0^+ = H_b^+$, $H_0^- = H_b^-$, and, thus, we can take $U_q = I$ in (A.85). Then by (A.76) and (A.77) we have

$$q_+(t) = \frac{1}{2i}JP^+ = \frac{1}{2}P^+, \qquad q_-(t) = -\frac{1}{2i}JP_- = \frac{1}{2}P_-,$$
$$Q_t^+ = \frac{1}{\sqrt{2}}(P^+ + P^-) = \frac{1}{\sqrt{2}}I, \qquad Q_t^- = \frac{1}{\sqrt{2}}(P^+ - P^-) = \frac{1}{i\sqrt{2}}J.$$

Thus, (A.85) gives[57]

$$y^\wedge = \frac{1}{\sqrt{2}}\{y(0) + y(b)\}, \qquad y^\vee = \frac{1}{\sqrt{2}}J\{y(b) - y(0)\}, \tag{A.111}$$

and by formulas (A.65), (A.66) we get

$$(\cos A) \cdot J \cdot \{y(b) - y(0)\} - (\sin A) \cdot \{y(b) + y(0)\} = 0 \tag{A.112}$$

or

$$(U - I) \cdot J \cdot \{y(b) - y(0)\} + i(U + I) \cdot \{y(b) + y(0)\} = 0. \tag{A.113}$$

Example A.9 *Self-adjoint separated boundary conditions for the system (A.109).*

Assume additionally that

$$\dim H^+ = \dim H^- \le \infty. \tag{A.114}$$

With the help of formula (A.89), separated self-adjoint boundary conditions can be written in the following form:

[57] It is interesting to note that in a qualitatively different and more complicated case when the equation contains an unbounded operator coefficient, general self-adjoint boundary conditions retain the same form, as shown by V. I. Gorbachuk and M. L. Gorbachuk in [495]. Note also a later work [672].

$$P^+y(0) = V_0 P^- y(0), \qquad P^+y(b) = V_b P^- y(b), \tag{A.115}$$

where V_0, V_b are arbitrary isometric operators mapping H^- onto H^+.

This form of writing boundary condition is convenient if J is represented as a diagonal or block-diagonal matrix. However, J is often written as

$$J = \begin{pmatrix} 0 & -I \\ I & 0 \end{pmatrix} = \left(\begin{array}{ccc|ccc} & & & -1 & & \\ & 0 & & & -1 & \\ & & & & & \ddots \\ \hline 1 & & & & & \\ & 1 & & & 0 & \\ & & \ddots & & & \end{array}\right). \tag{A.116}$$

In this case, in accordance with (A.116), it is more convenient to put

$$H = H' \oplus H', \qquad y(t) = y_1(t) \oplus y_2(t), \tag{A.117}$$

where $y_1(t), y_2(t) \in H'$, instead of $H = H^+ \oplus H^-$, as in (A.114) and (A.115).

This brings us to writing self-adjoint separated boundary conditions in the following form:

$$\begin{aligned} \cos A \cdot y_1(0) - \sin A \cdot y_2(0) &= 0, \\ \cos B \cdot y_1(b) + \sin B \cdot y_2(b) &= 0, \end{aligned} \tag{A.118}$$

where A and B are arbitrary self-adjoint operators in H', and, in accordance with (A.116) and (A.117),

$$y(0) = y_1(0) \oplus y_2(0), \qquad y(b) = y_1(b) \oplus y_2(b) \in H' \oplus H'. \tag{A.119}$$

The conditions (A.118) are of the type (A.88), though of a somewhat modified form.

Example A.10 *A system with factorized leading coefficient:*

$$l[y] := q^*(t)J(q(t)y)' + p(t)y = \lambda W(t)y. \tag{A.120}$$

Here, $q^{-1} \in B(H)$ is not assumed to be self-adjoint, $p(t) = p^*(t)$ and J is as in Example A.8 (see (A.110)).

By analogy with Example A.9, separated self-adjoint boundary conditions can be represented in the form

$$P^+q(0)y(0) = V_0 P^- q(0)y(0),$$
$$P^+q(b)y(b) = V_b P^- q(b)y(b), \tag{A.121}$$

corresponding to (A.115) with the same notations, and, also, in the form

$$\cos A \cdot (q(0)y(0))_1 - \sin A \cdot (q(0)y(0))_2 = 0,$$
$$\cos B \cdot (q(b)y(b))_1 + \sin B \cdot (q(b)y(b))_2 = 0, \tag{A.122}$$

with the notations as in (A.117)–(A.119). (The system (A.120) can be reduced to a canonical one. If, in particular, $p(t) \equiv 0$, then it suffices to introduce a new unknown $z(t) = q(t)y(t)$, and we get the following canonical system for $z(t)$:

$$J\frac{dz}{dt} = \lambda W_q(t)z, \qquad \text{with} \qquad W_q(t) = (q^*)^{-1}Wq^{-1}(t).) \tag{A.123}$$

Example A.11 (General first order system.)

$$l[y] := \frac{i}{2}\{(q(t)y)' + q(t)y'\} + p(t)y = \lambda W(t)y. \tag{A.124}$$

We assume that $q^*(t) = q(t)$ (general case is reduced to this case by decomposition $q(t) = \text{Re}\, q(t) + i\text{Im}\, q(t)$), $\dim H_t^+ = \dim H_t^-$, where $H_t^\pm$ are reducing positive and negative subspaces for $q(t)$ respectively. Using the conditions (A.89) we get the general form of separated boundary conditions in this case:

$$q_+^{1/2}(0)y(0) = V_0 q_-^{1/2}(0)y(0),$$
$$q_+^{1/2}(b)y(b) = V_0 q_-^{1/2}(b)y(b), \tag{A.125}$$

where V_0, V_b are arbitrary isometric operators mapping H_0^- onto H_0^+ and H_b^- onto H_b^+.

Fixing the choice of V_0, the conditions (A.125) can be written in the form corresponding to (A.88), e.g. for $t = 0$:

$$i\cos A_0 \cdot (-q_+^{1/2}(0) + V_0 q_-^{1/2}(0))y(0)$$
$$- \sin A_0 \cdot (q_+^{1/2}(0) + V_0 q_-^{1/2}(0))y(0) = 0, \tag{A.126}$$

where A_0 is an arbitrary self-adjoint operator in H_0^+. For $t = b$ we obtain an analogous formula.

Example A.12 For an arbitrary odd order expression of the form (A.72), under the condition (A.87), a broad class of separated boundary conditions is given by the formulas

$$\begin{aligned}
&q_+^{1/2}(0)y^{(n-1)}(0) = V_0 q_-^{1/2}(0)y^{(n-1)}(0),\\
&\cos A \cdot y^\vee(0) - \sin A \cdot y^\wedge(0) = 0,\\
&q_+^{1/2}(b)y^{(n-1)}(b) = V_b q_-^{1/2}(b)y^{(n-1)}(b),\\
&\cos B \cdot y^\vee(b) + \sin B \cdot y^\wedge(b) = 0,
\end{aligned} \tag{A.127}$$

where $q_\pm(t)$ are defined in (A.76); A, B are arbitrary self adjoint operators in H^{n-1}; V_0, V_b are the same as in (A.125);

$$\begin{aligned}
y^\wedge(t) &= \text{col } \{y(t), y'(t), \dots, y^{(n-2)}(t)\},\\
y^\vee(t) &= \text{col } \{y^{[2n-2]}(t), y^{[2n-3]}(t), \dots, y^{[n]}(t)\}.
\end{aligned} \tag{A.128}$$

Example A.13 (The case when the condition (A.87) is not satisfied.)

Let $\dim H = \infty$, and consider the simplest third order equation

$$l[y] := -iy''' = \lambda y. \tag{A.129}$$

Here $n = 2$, and, in accordance with (A.72), (A.76), (A.92), (A.93) and (A.91), we have $q_2(t) \equiv I$, $q_+(t) \equiv \frac{1}{2}I$, $q_-(t) \equiv 0$,

$$\begin{aligned}
y_0^+ &= \frac{1}{\sqrt{2}}y'(0) \oplus \frac{i}{2}(y''(0) - y(0)) \in H^2, \quad (H_0^+ = H),\\
y_0^- &= \frac{i}{2}(y''(0) + y(0)) \in H, \quad (H_0^- = \{0\}),
\end{aligned} \tag{A.130}$$

$$y_0^\wedge = y_0^+ + Vy_0^- \in H^2, \qquad y_0^\vee = iVy_0^- - iy_0^+ \in H^2, \tag{A.131}$$

where V is an arbitrarily fixed operator isometrically mapping H onto H^2.

For example, if $\{e_k\}_{k=1}^\infty$ is a given orthonormal basis in H, then we choose a basis $\{e_k, 0\}_{k=1}^\infty \cup \{0, e_j\}_{j=1}^\infty$ in H^2 and set

$$Ve_{2k} = \{e_k, 0\}, \qquad Ve_{2k-1} = \{0, e_k\}, \qquad k = 1, 2, \dots \tag{A.132}$$

Then, using (A.131) and (A.132), separated self-adjoint boundary conditions can be written for $t = 0$ (and, analogously, for $t = b$) in any of the forms (A.88), (A.89).

Example A.14 Generalized periodic conditions for the expression l of order $2n-1$:

$$y^{[2n-k-2]}(b) = U_k y^{[2n-k-2]}(0), \quad y^{(k)}(b) = U_k y^{(k)}(0), \tag{A.133}$$

$$q_{\pm}^{\frac{1}{2}}(b) y^{(n-1)}(b) = V^{\pm} q_{\pm}^{\frac{1}{2}}(0) y^{(n-1)}(0), \tag{A.134}$$

where $k = 0, 1, \ldots, n-2$, U_k are unitary operators in H, and $V^{\pm}$ are isometric operators mapping $H_0^{\pm}$ onto $H_b^{\pm}$. (Here we always take either only $+$ or only $-$.)

More general form of generalized periodic conditions can be obtained if, instead of the formula (A.134), we write the formula (A.70) with $y^{\wedge}(t)$, $y^{\vee}(t)$ given by (A.128) at $t = 0$ and $t = b$ respectively, where U is a unitary operator in H^{n-1}.

A.3.3 *Continuous Dependence of Initial Data on Solutions of Differential Equations*

Before we prove the main result of this subsection—Lemma A.10[58], we state the following simple Lemma:

Lemma A.8 *Let E_m be a finite-dimensional Hilbert space. Let*

$$A(x) = \sum_{j,k=1}^{m} a_{jk} x_j \bar{x}_k, \quad a_{jk} = \bar{a}_{kj},$$

where $x = (x_1, \ldots, x_m) \in E_m$, be a Hermitian quadratic form on E_m with a lower bound μ, i.e. $A(x) \geq \mu |x|^2$. Let $\widehat{A}(y^{\wedge})$ be the quadratic form in the Hilbert space $H^m = \underbrace{H \oplus \cdots \oplus H}_{m}$ defined by

$$\widehat{A}(y^{\wedge}) = \sum_{j,k=1}^{m} a_{jk}(y_j, y_k),$$

where $y^{\wedge} = col\,\{y_1, \ldots, y_m\} \in H^m$, $y_k \in H$, $(k = 1, \ldots, m)$.

[58] This lemma was used in the proof of Lemma A.7.

Then $\widehat{A}$ has the same lower bound μ,

$$\widehat{A}(y^\wedge) \geq \mu |y^\wedge|^2 = \mu \sum_{k=1}^{m} (y_k, y_k).$$

This Lemma is an obvious consequence of the following statement which is readily checked:

Lemma A.9 *Let E_m be a finite-dimensional Hilbert space. Let $A = \|a_{jk}\|_{j,k=1}^m$, $a_{jk} = \bar{a}_{jk}$ be a Hermitian operator on E_m with eigenvalues $\lambda_1 \leq \lambda_2 \leq \ldots \leq \lambda_m$. Then the operator $\widehat{A}$ in H^m*

$$\widehat{A} = \|a_{jk} \cdot I\|_{j,k=1}^m ,$$

where I is the identity operator in H, is self-adjoint. Moreover, the spectrum of $\widehat{A}$ consists of the same eigenvalues λ_r, $(r = 1, \ldots, m)$, and the multiplicity of each eigenvalue λ_r now increases $\dim H \leq \infty$ times. If $h = (h_1, h_2, \ldots, h_m) \in E_m$ is an eigenvector corresponding to the eigenvalue λ_r of A, then to the same λ_r corresponds the eigenspace of $\widehat{A}$ in H^m consisting of all vectors of the form

$$h^\wedge = \{h_1 x, h_2 x, \ldots, h_m x\} \in H^m, \quad x \in H.$$

Lemma A.10 *Linear space of vector-valued solutions of $y = y(t)$ of the m-th order homogeneous equation*

$$l[y] \equiv y^{(m)} + g_1(t) y^{(m-1)} + \cdots + g_m(t) y = 0, \tag{A.135}$$

with arbitrary operator continuous coefficients $g_k(t)$ is a subspace of $H(0,b)$ (hence, it is closed in $H(0,b)$). This follows from the one-to-one correspondence and bicontinuous dependence (homeomorphism) between solutions $y(t) \in H(0,b)$ and their Cauchy data $y^\wedge$:

$$y^\wedge = \{y_0, y_1, \ldots, y_{m-1}\} \in H^m, \quad y_k = y^{(k)}(0).$$

Proof. The proof of the continuous dependence of solutions on the initial data follows a well-known scheme, and we will omit it here. We will prove that the dependence of the initial data $y^\wedge$ on solutions $y(t)$ of (A.135) is continuous. By $Y_k(t)$ we denote the operator solutions of (A.135) with the following initial conditions:

$$Y_k^{(j)}(0) = \delta_{jk} \cdot I, \quad (j, k = 0, 1, \ldots, m-1).$$

Then any vector-valued solution $y(t)$, $y^{(k)}(0) = y_k$ can be written as

$$y(t) = \sum_{k=0}^{m-1} Y_k(t) y_k.$$

Since

$$Y_k(t) = \frac{1}{k!} t^k \cdot I + O(t^m), \quad (k = 0, 1, \ldots, m-1),$$

where the operator quantity $O(t^m)$ satisfies the uniform estimate

$$|O(t^m)| \leq Ct^m,$$

we obtain

$$y(t) = \sum_{k=0}^{m-1} y_k \cdot \frac{t^k}{k!} + |y^\wedge| \cdot t^m \cdot \vec{O}(1).$$

Therefore, for $0 < \epsilon < b$, we have by the triangle inequality

$$\left(\int_0^b |y(t)|^2 \, dt \right)^{\frac{1}{2}} > \left(\int_0^\epsilon |y(t)|^2 \, dt \right)^{\frac{1}{2}}$$

$$\geq \left(\int_0^\epsilon \left| \sum_{k=0}^{m-1} y_k \cdot \frac{t^k}{k!} \right|^2 dt \right)^{\frac{1}{2}} - |y^\wedge| \cdot O\left(\epsilon^{m+\frac{1}{2}}\right). \tag{A.136}$$

Consider the square of the first term on the right hand side of (A.136):

$$\widehat{A}_\epsilon(y^\wedge) \equiv \int_0^\epsilon \left| \sum_{k=0}^{m-1} y_k \cdot \frac{t^k}{k!} \right|^2 dt = \sum_{j,k=0}^{m-1} \frac{\epsilon^{j+k+1}}{j!k!(j+k+1)} (y_j, y_k).$$

By Lemma A.8, we have

$$\widehat{A}_\epsilon(y^\wedge) \geq \mu(\epsilon) \left| y^\wedge \right|^2,$$

where $\mu(\epsilon)$ is the greatest lower bound of the quadratic form $A_\epsilon(x)$,

$$A_\epsilon(x) = \sum_{j,k=0}^{m-1} \frac{\epsilon^{j+k+1}}{j!k!(j+k+1)} x_j \bar{x}_k = \epsilon A_1(x_\epsilon),$$

where

$$x = (x_0, x_1, \ldots, x_{m-1}) \in E_m, \quad x_\epsilon = (x_0, \epsilon x_1, \ldots, \epsilon^{m-1} x_{m-1}) \in E_m.$$

Clearly, $\mu(\epsilon) > 0$. We will now estimate $\mu(\epsilon)$ from below more precisely. For $0 < \epsilon < 1$, we have

$$\mu(\epsilon) = \inf_{x \in E_m} \frac{A_\epsilon(x)}{|x|^2} \geq \epsilon \cdot \inf_{x_\epsilon \in E_m} \frac{A_1(x_\epsilon)}{|x_\epsilon|^2} \cdot \inf_{x \in E_m} \frac{|x_\epsilon|^2}{|x|^2} = \epsilon^{2m-1} \cdot \mu(1).$$

Therefore,

$$\widehat{A}_\epsilon(y^\wedge) \geq \mu(1) \cdot \epsilon^{2m-1} \cdot |y^\wedge|^2 .$$

Hence, for sufficiently small ϵ, from (A.136) we get

$$\left(\int_0^b |y(t)|^2 \, dt \right)^{\frac{1}{2}} \geq |y^\wedge| \, \epsilon^{m-\frac{1}{2}} \left[\mu^{\frac{1}{2}}(1) - O(\epsilon) \right] > C|y^\wedge|,$$

where $C > 0$ does not depend on $y^\wedge$. This shows the continuity of dependence of $y^\wedge \in H^m$ on the corresponding solution $y(t) \in H(0,b)$. The Lemma is proven. □

A.4 Bibliographical Comments

This chapter is based on the works [772], [773], [777] and [788]. In the case of a scalar quasi-differentiable expression of even order with real coefficients, a complete description of all self-adjoint extensions on a bounded interval is contained as one of the results in a fundamental work of M. G. Krein [586]. However, this description is given in a form different from the one in the present book and with abundant number of parameters satisfying a system of non-linear constraints. This description is based on the general extension theory of abstract symmetric operators in Hilbert space, due to J. von Neumann. Algebraic investigation of linear boundary value problems for scalar differential equations due to M. Bôcher and others (see [29]) is contained in the book [45] (see also [501]).

Boundary conditions of the form (A.67) for scalar equations of even order appear as self-adjoint conditions in the work [587] of M. G. Krein. For the case of Schrödinger equation with operator potential, separated self-adjoint conditions of the form $y'(0) - Ay(0) = 0$ were used in [766, Chapter III, sec. 7], where eigenfunction expansion of the Schrödinger equation was constructed. Separated boundary conditions of the general form (A.67) were used as self-adjoint conditions in [768]. The general form of separated boundary conditions for the Schrödinger equation was established by M. L. Gorbachuk [486] by a method different from the one in the present book and in a somewhat more complicated form than in (A.67). Several boundary value problems of special type were thoroughly studied in the work of S. G. Krein and G. I. Laptev [599] and in the monograph [97] by S. G. Krein.

An approach to the description of self-adjoint extensions with the help of Hermitian binary relations, explained in the present chapter and employed in the present book, has something in common with the method of Calkin [334] (see also [58] and [854], but all these references are related to extensions of operators with finite deficiency indices).

The well-known general form of self-adjoint boundary conditions for elliptic differential equations, established by M. I. Vishik [901], can also be explained from the viewpoint of Hermitian relations and can be represented in the form (A.65).

The apparatus of the theory of binary relations (arbitrary non-Hermitian) in finite-dimensional spaces was systematically developed and used in the work of I. M. Gel'fand and V. A. Ponomarev [451] in conjunction with the questions studied there. The theory of binary relations in Hilbert or Banach spaces was later considered in the works of many mathematicians, such as C. Bennewitz [259]; R. C. Brown [313]; V .M. Bruk [319], [320]; E. A. Coddington [45]; E. A. Coddington and H. S. V. de Snoo [345]; A. Dijksma and H. S. V. de Snoo [364]; I. S. Kats [535]; A. N. Kochubei [556], [557]; H. Langer and B. Textorius [615]; S. J. Lee [618] and A. Pleijel [748].

The notion of a self-adjoint binary relation (equivalent to the notion of a Hermitian relation), as an object of research, but unlike in our case, not as an instrument of the description of self-adjoint extensions, first appeared in the work of R. Arens [229] in a somewhat different form. Unlike us, Arens [229] postulates the linearity of a self-adjoint binary relation, while we deduce the properties (3) (linearity) and (4) below the equation (A.3) only from the properties (1) and (2) in Definition A.1. The term "Hermitian relation" was introduced in the works [772], [773] and in the talk of F. S. Rofe-Beketov [777] at the Summer Mathematical School in Baku–Zagul'ba in 1968, where a complete description of self-adjoint extensions of differential operators with continuous operator coefficients in a space of vector-functions (Hilbert-space-valued on a bounded interval) was given with the help of Hermitian relations, and, most importantly, a new and efficient method in the theory of self-adjoint extensions (including the case of operators with infinite deficiency indices) was shown in action. (This approach can be considered as constructive theory of extensions.) We note that this method often allows us to get self-adjoint extensions bypassing the investigation of deficiency subspaces, and it allows us to give the description of extensions directly in terms of boundary conditions (or abstract boundary conditions; see the monograph by N. Dunford and J. T. Schwartz [58] for the case of finite deficiency indices and the works by V. M. Bruk [318] and A. N. Kochubei [556] for the general case). Besides, the method of Hermitian relations gives the description of self-adjoint extensions in terms of free parameters only, unlike [586] and the first two editions of [6] (1950, 1966).

A.4.1 *M. L. Gorbachuk Theory of Generalized Boundary Values. Abstract Boundary Conditions*

In connection with the method of rigged spaces[59] due to I. M. Gel'fand–A. G. Kostyuchenko [448] and Yu. M. Berezanskiy [24], [25], [263] (used in the form of [24]), the method of binary relations, turned out, starting with the works of M. L. Gorbachuk [487], to be effective in applications to equations with unbounded operator coefficients and also in the theory of extensions of abstract operators. (See the monographs [73] and [74] by V. I. Gorbachuk and M. L. Gorbachuk, and the monograph [119] by V. E. Lyantse and O. G. Storozh. For extensions of normal subspaces, see [45]; for binary relations (or so-called multivalued operators) in Banach spaces, see [49] and [65]. The method of Hermitian relations also appeared in the third edition of the book by N. I. Akhiezer and I. M. Glazman [6]. Note also the review [496] with a very rich content.)

We will briefly explain the method of M. L. Gorbachuk [487] that lies at the foundation of his theory of generalized boundary values (see also [73], [74] and [496]). Let

$$l[y] = -y''(t) + qy(t), \qquad -\infty < a \leq t \leq b < \infty, \tag{A.137}$$

where $q \geq 0$ is an unbounded self-adjoint operator (constant for simplicity).

[59]Here we consider Hilbert space H with the Hilbert rigging H^+ and H^-, where the Hilbert space H^+ is dense in H with respect to the inner product of H, the space H is dense in the Hilbert space H^- with respect to the inner product in H^-, and H^+ and H^- are dual to each other with respect to the inner product in H. The inner product in H can be extended up to a continuous bilinear (sesquilinear) form on $H^- \times H^+$. Hilbert riggings are most conveniently and most frequently used in applications. There are other types of riggings known, such as, for example, topological riggings. When H is a function space, the space H^- of generalized elements is called the space of distributions or the space of generalized functions. In this way, we can obtain Sobolev spaces and Schwartz distributions.

The unboundedness of q essentially changes the situation. Here, for the expression $l[y]$, the minimal and the maximal operators L_m and L_M respectively are constructed, and $L_m^* = L_M$; however, in the case of an unbounded q, functions belonging to the domain $\mathrm{D}(L_M)$ of the maximal operator L_M are sufficiently smooth only inside the interval (a, b). Boundary values of elements of $\mathrm{D}(L_M)$ and boundary values of their derivatives may not exist in H, but they exist in a larger than H space H^- of generalized elements (or generalized vectors), so that the space H^- and its dual H^+ relative to H form a Hilbert rigging of the space H. Therefore, in this case, we cannot write the Lagrange–Green formulas in the form (A.44), (A.48), and we cannot write the self-adjoint extensions $L^{\sim}$ of the minimal operator L_m in the form (A.51)–(A.53). Likewise, we cannot write the formula (A.50) using (A.49) for $y^{\wedge}$ and $y^{\vee}$.

However, M. L. Gorbachuk [487] found linear combinations of generalized boundary values of vector-functions $y(t) \in \mathrm{D}(L_M) = \mathrm{D}(L_m^*)$ and their derivatives belonging to H, allowing the construction of expressions for $y^{\wedge}$ and $y^{\vee}$ instead of (A.49) so that the formulas (A.50)–(A.52) are valid again for the new $y^{\wedge}$ and $y^{\vee}$. The corresponding construction goes as follows. Denote by Q the extension in a special way to the space H^- of the operator q defined in H. Suppose, for simplicity, that the operator q is strongly positive definite in H: $q >> 0$. Then, for the expression (A.137), one can construct $y^{\wedge}$ and $y^{\vee}$ in $H \oplus H$ according to the following formulas (see [487]):

$$y^{\wedge} = \{Q^{-1/4}y(a),\, Q^{-1/4}y(b)\}, \tag{A.138}$$

$$y^{\vee} = \{Q^{1/4}(y'(a) + Q^{1/2}y(a)),\, Q^{1/4}(-y'(b) + Q^{1/2}y(b))\}. \tag{A.139}$$

The papers [489] and [490] by M. L. Gorbachuk, A. N. Kochubei and M. A. Rybak describe a representation for dissipative linear relations and, correspondingly, maximal dissipative and self-adjoint extensions for minimal differential operators generated by the two-term $2n$-th order differential expression $l_{2n}[\mathrm{y}]$ of the type

$$l_{2n}[y] = (-1)^n y^{(2n)}(t) + qy(t).$$

Analogous constructions were later offered for differential operators with variable unbounded operator coefficients; see [314] and [894].

The case of unbounded $q < 0$ was investigated in [897].

More complete and detailed account of this theory can be found in the monographs [73] and [74] by V. I. Gorbachuk and M. L. Gorbachuk, where similar problems for many types of differential operators are considered and extensive bibliography is contained.

As it turned out, the method of constructing and classifying extensions, developed in the works [487], [493], [494], [772] and [773] for operator-differential equations, has a universal nature. Independently from each other, A. N. Kochubei [556] and V. M. Bruk [318] extended this method to arbitrary symmetric operators. Let A be a closed densely defined symmetric operator with deficiency indices (n, n) (n can be finite or infinite) in a Hilbert space H. The triplet $(\mathbf{H}, \Gamma_1, \Gamma_2)$, where $\mathbf{H}$ is some Hilbert space and Γ_j, $j = 1, 2$, are linear mappings from $\mathrm{D}(A^*)$ into $\mathbf{H}$, is called a **boundary values space** of the operator A, if

(1) an abstract Green–Lagrange formula holds for every $f, g \in \mathrm{D}(A^*)$, i.e.

$$(A^*f, g) - (f, A^*g) = (\Gamma_1 f, \Gamma_2 g)_{\mathbf{H}} - (\Gamma_2 f, \Gamma_1 g)_{\mathbf{H}},$$

and

(2) for arbitrary $f^\vee, f^\wedge \in \mathbf{H}$, there exists $f \in \mathrm{D}(A^*)$ such that

$$\Gamma_1 f = f^\vee, \quad \Gamma_2 f = f^\wedge. \tag{A.140}$$

From this definition it follows that (see [318])

$$f \in \mathrm{D}(A) \Leftrightarrow \Gamma_1 f = \Gamma_2 f = 0.$$

Theorem A.10 **(A. N. Kochubei [556] and V. M. Bruk [318])** *For the above considered operator A with deficiency indices (n, n), there always exists a boundary values space with* $\dim \mathbf{H} = n (\leq \infty)$.

More specifically, we may choose as the space $\mathbf{H}$ the deficiency subspace N_{-i} of the operator A, and

$$\Gamma_1 = P_{-i} + UP_i, \qquad \Gamma_2 = -iP_{-i} + iUP_i,$$

where $P_{\pm i}$ are the projectors of $D(A^*)$ onto $N_{\pm i}$ parallel to $D(A) \dot{+} N_{\mp i}$ respectively, and U is an isometry of N_{+i} onto N_{-i}[60].

For every contraction K in $\mathbf{H}$, the restriction of the operator A^* by the condition

$$(K - I)\Gamma_1 f + i(K + I)\Gamma_2 f = 0$$

[60] In applications, it is convenient to choose the boundary values space corresponding to the considered boundary value problem.

gives rise to maximal dissipative extension $\widetilde{A}$ *of the operator* A, *and* vice versa, *i.e. every maximal dissipative extension* $\widetilde{A}$ *of* A *is generated by (A.140), and the correspondence* $\widetilde{A} \Leftrightarrow K$ *is one to one. To self-adjoint extensions of* A *there correspond unitary operators* K.

For example (see [487]), the minimal operator (A.137) in $H(a, b)$ has the boundary values space $(H \oplus H, \Gamma_1, \Gamma_2)$ in the notations (A.138)–(A.140). For similar boundary values spaces for many differential operators, see [73] and [74]. For differential operators with bounded operator coefficients, many boundary values spaces are contained in Chapter 7 and in Theorem 5.1 of the present book. The references [73], [74], [318], and [556] also contain a description of various classes of extensions by means similar to the above mentioned construction. In the case of a symmetric operator A with different deficiency numbers, a description of maximal dissipative extensions with the help of abstract boundary values was obtained in [873]. For the maximal operator corresponding to the equation

$$y' - Ay = 0$$

with uniformly well-posed Cauchy problem (A is closed linear operator in H), the boundary values space was constructed in [323]. The case $A \geq 0$ was considered before in [73] and [74]. For applications of the theory of abstract boundary values to the theory of Weyl functions, see, for example, [358] and [362].

If a closed symmetric operator A satisfies $A >> 0$, i.e. if A is strongly positive definite: $A \geq \epsilon I$, $\epsilon > 0$, then among boundary values spaces of the operator A we can select a subset of so-called **positive boundary values spaces**. More specifically, let $\widetilde{A}$ be a fixed positive definite self-adjoint extension of A. Then

$$\mathrm{D}(A^*) = \mathrm{D}(\widetilde{A}) \dotplus \mathrm{Ker}\ A^*. \tag{A.141}$$

Following [73], [74] and [496], denote by $\widetilde{P}$ and P_0 the projectors of $\mathrm{D}(A^*)$ onto $\mathrm{D}(\widetilde{A})$ and Ker A^* corresponding to the expansion (A.141). Then (see [558]) a boundary values space $(\mathbf{H}, \Gamma_1, \Gamma_2)$ of the operator A is called a positive boundary values space associated with the expansion (A.141) if for any f, $g \in \mathrm{D}(A^*)$ we have

$$(A^* f, g) = (\widetilde{A}\widetilde{P} f, \widetilde{P} g) + (\Gamma_1 f, \Gamma_2 g)_{\mathbf{H}}.$$

Note that this condition is equivalent to

$$(A^*f, P_0g) = (\Gamma_1 f, \Gamma_2 g)_{\mathbf{H}},$$

or

$$(\Gamma_1 f, \Gamma_2 g)_{\mathbf{H}} = (\Gamma_1 f, \Gamma_2 P_0 g)_{\mathbf{H}} - (\Gamma_2 f, \Gamma_1 P_0 g)_{\mathbf{H}}$$

or

$$(\Gamma_1 f, \Gamma_2 \widetilde{P} g)_{\mathbf{H}} + (\Gamma_2 f, \Gamma_1 P_0 g)_{\mathbf{H}} = 0.$$

It was shown in [558] (see also [73, Chapter 3, **Lemma 2.1**] and [74]) that a positive boundary values space $\mathbf{H}$ can always be constructed for any positive definite symmetric operator A as follows:

$$\mathbf{H} = \text{Ker } A^*, \qquad \Gamma_1 = P\widetilde{A}\widetilde{P}, \qquad \Gamma_2 = P_0,$$

where P is the orthogonal projection onto Ker A^* in the space H.

Theorem A.11 **(A. N. Kochubei [558])** [61] *If an arbitrary fixed positive boundary values space is given, then to every proper solvable extension*[62] *A' of a symmetric operator $A >> 0$ corresponds an abstract boundary condition*

$$\Gamma_2 f = \mathbf{B}\Gamma_1 f,$$

where $\mathbf{B} \in B(\mathbf{H})$ is arbitrary, and vice versa. *This correspondence is bijective.*

Theorem A.12 **(see [73, Chapter 3, Theorem 3.5], [74] and [675])** *If $\mathbf{B}$ in the above mentioned abstract boundary condition is a non-negative operator, then this boundary condition gives us a positive definite self-adjoint extension A' of the operator A, and if, additionally, $\widetilde{A}$ is the Friedrichs extension of A, then every positive definite self-adjoint extension of A can be described by $\Gamma_2 f = \mathbf{B}\Gamma_1 f$ with a non-negative $\mathbf{B} \in B(\mathbf{H})$.*

In a similar way, one can also describe maximal sectorial extensions; see [558] and [675]. The case of nonnegative A is more complicated and was considered in the works [231], [363] and [874]. Some other classes of extensions, such as closed, normally solvable and Fredholm-type extensions

[61] See also [73, Chapter 3, Theorem 2.2] and the reference [74].

[62] that is, an operator A' such that $A \subset A' \subset A^*$ and $(A')^{-1} \in B(H)$ exists in terminology of [73], [74].

were investigated by V. A. Mikhailets (see [674] and [675]) and L. I. Vainerman [895]. In [359] and [360] V. A. Derkach and M. M. Malamud obtained all self-adjoint extensions of a symmetric operator with spectral gaps, such that the self-adjoint extension has the same spectral gaps as the corresponding original operator, with a possible exception of a finite number of eigenvalues in each of the gaps.

We also note the classical works by M. G. Krein [586], M. I. Vishik [901] and M. Sh. Birman [275] devoted to extension theory of positive operators. The review by A. Alonso and B. Simon [222] is devoted to the Birman–Krein–Vishik theory.

Classical approach to the theory of normally solvable extensions and general problems of the theory of boundary value problems are also considered by M. I. Vishik [901] and in the book by A. A. Dezin [54].

* * *

In the text of the Appendix, we have already noted the equivalence of Hermitian relations with Lagrangian planes or hyper-maximal neutral lineals in doubled Hilbert spaces. This approach was also used in a relatively recent work of S. P. Novikov [708], where it was said, "In 1971 I. M. Gel'fand communicated to the author an interesting idea about a connection between self-adjoint extensions of symmetric operators and Lagrangian planes. This was Gel'fand's reaction to the work of the author [705] built on a connection between the Hamiltonian formalism and differential topology." In this regard, we also note the following papers and reviews: [218], [413], [414], [605], [709] and [750].

Very broad literature is devoted to different approaches to the theory of extensions. Relations-subspaces (not necessarily self-adjoint, i.e. Hermitian) are applicable to the description of non-symmetric operators (for example, normal, dissipative or sectorial) or non-self-adjoint extensions of symmetric operators. Here we will mention the works [45], [232], [235], [248], [308], [359]–[357], [489], [490], [505], [557], [646], [647], [650], [651] and [874]–[878]. (The notion of a sectorial relation was introduced in [782]; see also [788].) There is also a well-known approach to the theory of extensions with the help of quadratic forms starting from Friedrichs extensions. The work [307] is devoted to new aspects of this approach.

Regarding Theorem A.2 from Sec. A.1, we will mention that another criterion (conditional) for a linear relation θ in (A.12) to be self-adjoint has been obtained in the works [357] and [358] by V. A. Derkach, S. Hassi, M. M. Malamud and H. S. V. de Snoo.

Theorem A.13 ([357], [358]) *Let the linear relation θ be defined by (A.12) with bounded operators B and C. Assume that (A.17) is satisfied and the operator $BB^* + CC^*$ has a bounded inverse operator. Then the relation θ in (A.12) is Hermitian if and only if the operators*

$$K_1 := B^*B + C^*C, \qquad K_2 := K_1 - B_{21}{K_1}^{-1}B_{21}^*$$

where $B_{21} := C^*B - B^*C$, are strongly positive definite.

Corollary A.5 ([357]) *If $B^*C = C^*B$ then $K_2 = K_1$, and the above conditions simplify accordingly.*

The work [534] by R. M. Kashaev is devoted to self-adjoint extensions of a difference operator.

Regarding Proposition A.1 from Sec. A.1, we will mention that a further generalization of Hausdorff–Stone Theorem was obtained by Ts. Bayasgalan [251] who showed that the numerical range of a linear relation in a space with indefinite metric is convex. For a higher-dimensional generalization of numerical ranges of quadratic operators, see the work [339] by M.-T. Chien, S.-H. Tso and P. Y. Wu.

In 1980 A. V. Kuzhel' introduced the notion of a regular extension of Hermitian (non-densely-defined) operators, which allowed him to investigate different classes of extensions of Hermitian operators without using the method of linear relations. This approach was used in monographs [102] and [103].

The monographs of B. I. Ptashnik [162] and V. P. Burskiy [33] are devoted to methods of investigation of boundary-value problems for general partial differential equations, including the ill-posed boundary-value problems. It is interesting to note that, in this scope of problems, the well-known problem of small denominators occurs.

In various applications or in considering extensions outside the corresponding space, often arise boundary value problems with a spectral parameter not only in the differential equation but also in the boundary conditions[63]. A. V. Shtraus [850]–[853] and J. Walter [909] considered this type of problem with Nevanlinna functions of spectral parameter. In this case, the corresponding boundary problem can be interpreted as an eigenvalue problem for some self-adjoint operator in a wider Hilbert space[64] contain-

[63] See, for example, the books L. Collatz [47], A. N. Krylov [99], F. Roseau [170] and S. P. Timoshenko [190].

[64] Here, the original space H and the wider space are Hilbert spaces. Below we consider the case when the wider space is the space with indefinite metric (Pontryagin space).

ing the space under consideration as a subspace. For more on extensions of operators with emergence from the original space, see [6] and [145]. However, operator-theoretical interpretation of the problem with a polynomial entering of a spectral parameter into the boundary conditions at both endpoints, which was proposed and developed by E. M. Russakovskiy in the papers [813]–[817] requires the theory of operators in a space with indefinite metric (see [14] and [591]) encompassing the starting Hilbert space.

Important results concerning boundary problems with a spectral parameter in boundary conditions belong to A. A. Shkalikov [847]. We also mention the works [256], [257], [271], [404], [644] and [899]. In [404] A. E. Etkin considers an abstract problem with a spectral parameter in boundary conditions and suggests a different approach to this problem, relying on results and ideas of A. V. Shtraus ([854] *etc.*). The recent paper [217] is devoted to eigenfunction expansions for problems with boundary conditions containing a parameter (see also [515]).

A comprehensive review paper by B. S. Pavlov [738] is devoted to a method of construction of explicitly solvable physical models based on the theory of extensions outside the corresponding space. In this review, many important results belonging to representatives of St. Petersburg mathematical school as well as historical comments were given. We also mention the reviews by V. I. Gorbachuk, M. L. Gorbachuk and A. N. Kochubei [496], V. D. Koshmanenko [573], and the monograph [7] by S. Albeverio, F. Gesztesy, R. Hoegh-Kron and H. Holden devoted to solvable models in quantum mechanics. Additionally, we mention more recent works by A. N. Kochubei [559], F. Gesztesy, A. Kiselev and K. A. Makarov [457], and an earlier work by F. A. Berezin and L. D. Faddeev [270] containing a description of self-adjoint extensions for the Schrödinger operator in $\mathbb{R}^3$ with the singularity at the origin (potential of null-radius of action, or Dirac δ-function). The work [533] by Yu. E. Karpeshina and B. S. Pavlov is devoted to analogous problem for biharmonic and polyharmonic operators. In the work [252] by M. Bayramogly, an operator equation with singularity is considered. General perturbation theory of self-adjoint semi-bounded operators by singular bilinear forms is constructed in the works by T. V. Karataeva and V. D. Koshmanenko [532] and M. I. Neyman-zade and A. A. Shkalikov [704]. In the works by Yu. Berezanskiy and J. Brasche [267], J. Brasche [306], T. Kappeler and C. Möhr [531] and C. Möhr [694], singular perturbed semi-bounded operators are considered in the corresponding negative Hilbert spaces. This approach simplifies the investigation of spectral properties.

Problems for equations of elliptic type in domains with complicated structure, for example, in domains with so called fine-grained boundary, in perforated regions *etc.*, play an important role in physics and various applications (radio-physics, rheology, hydromechanics, elasticity theory). Finding solutions to such problems by traditional methods (exact and approximate) leads to insurmountable difficulties. However, in many cases, passage to the limit enables us to replace a complicated boundary of the starting problem by a relatively simple surface on which so called homogenized boundary conditions are defined, leading to a new method for obtaining extensions. Passage to the limit (homogenization) for the problems in perforated regions, i.e. in regions with cuts that are infinitely refined and concentrated (like cheese with finer and finer holes which are simultaneously concentrating), may lead to a change of the form of the considered operator (differential expression). The book [124] by V. A. Marchenko and E. Ya. Khruslov was the world's first monograph devoted to the above described homogenization method. Important results on the homogenization method and its further developments are contained in the well known monographs G. P. Panasenko [18]; A. Bensoussan, J.-L. Lions and G. Papanicolaou [23]; D. Cioranescu and J. Saint Jean Paulin [44]; O. A. Oleinik, G. A. Iosifyan, A. S. Shamaev [150]; E. Sánchez-Palencia [175]; I. V. Skrypnik [183]; V. V. Zhikov, S. M. Kozlov, O. A. Oleinik [207]; in the new monographs [125] and [126] by V. A. Marchenko and E. Ya. Khruslov and many other works.

The question of resolvent convergence of Dirichlet problems for $-\Delta$ in a region with infinitely fine cuts, important for applications to the problem for $-\Delta + q(x)$ in the same region but without any cuts, is considered, among other questions, in the reference [124]. Here we will consider a much simpler and, in a sense, inverse problem; namely, the description of a class of perturbations $A+C$ of the initial operator $A > 0$ that can be obtained by passing to the limit from Friedrichs extensions of a sequence of restrictions A_ξ^F of the operator $A = A^* > 0$.

Theorem A.14 ([**782**], [**784**]) *Let the operator $A = A^* > 0$ have a compact inverse A^{-1}, and let $C \geq 0$ be an arbitrary symmetric and relatively compact operator, i.e. $D(C) \supset D(A^{1/2})$ and $A^{-1/2}CA^{-1/2}$ is compact. Then a sequence (not unique) of linear, generally non-dense in H submanifolds $D_\xi \subset D(A)$, can be constructed such that for $A_\xi = A|_{D_\xi}$ and $\lambda \in C\backslash[0, \infty)$, the corresponding sequence of resolvents converges uniformly,* *i.e.*

$$\|R_\lambda(A_\xi^F) - R_\lambda(A+C)\| \to 0, \quad \textit{as } \xi \to \infty.$$

The submanifolds D_ξ *can be chosen so that* $\dim D(A)/D_\xi < \infty$.

(We recall that $R_\lambda(A_\xi^F) = [(P_\xi A_\xi)^{F_\xi} - \lambda]^{-1} P_\xi$, where P_ξ is the orthogonal projection onto $H_\xi := \overline{\mathrm{D}_\xi} \subseteq H$, and F_ξ denotes the Friedrichs extension in H_ξ; see also the notations in the Bibliographical Comments at the end of Chapter 3.)

Example A.15 If $C = A^\alpha$, then the conditions of Theorem A.14 are satisfied for $\alpha \leq 1/2$. With such special perturbations, the statement of the Theorem remains valid for any $\alpha < 1$.

Example A.16 If $A = -\Delta + |x|^2$, where $x \in \mathbb{R}^n$, then the perturbation $C \geq 0$ in Theorem A.14 can be any bounded operator in $L^2(\mathbb{R}^n)$.

Example A.17 If $A = \Delta^2, x \in \Omega \subset \mathbb{R}^n$, and if Ω is a bounded domain,

$$u|_{\partial\Omega} = \left(\frac{\partial u}{\partial \nu}\right)\Big|_{\partial\Omega} = 0,$$

then in Theorem A.14 we can take $C = (i\nabla + b(x))^2 + C_1$, where $C_1 \geq 0$ is a bounded operator.

A.4.2 *Strong Resolvent Convergence of Hermitian Relations and Operators with Non-Dense Domains and Resolution of the Identity. Stability of Eigenvalues*

Consideration of boundary value problems in $L^2(\Omega)$ under perturbation of a domain $\Omega \subset \mathbb{R}^n$ naturally leads to the question about perturbation of non-densely defined semi-bounded or sectorial operators, forms and relations in a Hilbert space H with variable domain. By Corollary A.2 and the formulas (A.25) and (A.26) there is a one-to-one correspondence among relations θ, forms a_θ and self-adjoint or m-sectorial operators A_θ on the closures of domains of A_θ:

$$\theta \sim A_\theta \sim a_\theta, \qquad \mathrm{D}(A_\theta) = \mathrm{D}(\theta), \qquad \overline{\mathrm{D}(A_\theta)} = \overline{\mathrm{D}(a_\theta)} = H_\theta \subseteq H.$$

The resolvent of such relation θ is the operator

$$R_\lambda(\theta) = (A_\theta - \lambda)^{-1} P_\theta, \qquad (A_\theta \sim \theta), \tag{A.142}$$

where P_θ is the orthogonal projection onto H_θ. In particular, the resolvent of the Friedrichs extension of a sectorial (or a symmetric semi-bounded below) operator is defined by the formula (A.27). Every Hermitian relation θ has resolution of the identity—spectral function

$$E_\lambda(\theta) = E_\lambda(A_\theta)P_\theta, \tag{A.143}$$

where $E_\lambda(A_\theta)$ is resolution of the identity of the operator A_θ in the subspace $H_\theta \subseteq H$.

Hermitian relation θ is semi-bounded simultaneously with A_θ and a_θ.

The inequalities $(\gamma \leq)\, \theta_1 \leq \theta_2$ and $A_{\theta_1} \leq A_{\theta_2}$ are understood in the sense of quadratic forms:

$$a_{\theta_1}[x] \leq a_{\theta_2}[x], \qquad x \in \mathrm{D}(a_{\theta_2}) \subseteq \mathrm{D}(a_{\theta_1}).$$

As in the case of densely defined operators (see [87] and [165]), this inequality is equivalent to

$$R_\lambda(\theta_1) \geq R_\lambda(\theta_2), \qquad \text{for } \lambda < \gamma.$$

(The set $\mathrm{D}(a)$ is the closure of $\mathrm{D}(A)$ in A-metric or, equivalently, in a-metric, where $a \sim A$; i.e. in the metric of the positive quadratic form $a - \alpha$ associated with the operator $A - \alpha$ with $\alpha < \gamma \leq A$.)

Many results about strong resolvent convergence of sequences of densely defined closed sectorial or symmetric semi-bounded forms and operators, such as, for example, Theorems VIII.3.6, VIII.3.11 and VIII.3.13 from the book by Kato [87], hold in the case of non-densely defined forms and associated binary relations (m-sectorial or Hermitian) or non-densely defined operators, as noted in [860] (in the case of semi-bounded forms) and in [782]. The well-known proofs can be carried to the above mentioned cases practically without changes, and we shall not repeat them here.

Nevertheless, for the case of strong resolvent $(s - R)$ convergence of Hermitian relations, we will prove a generalization of the well-known theorem about the convergence of spectral projections of a convergent sequence of self-adjoint (densely defined) operators (see [165, Theorem VIII.24(b)] or [87, Theorem VIII.1.15]). This will establish the corresponding result also for a sequence of non-densely defined operators, which we will call $s - R$ convergent if the sequence of associated binary relations is $s - R$ convergent. The following theorem generalizes Theorem 6 from [782] to the case of arbitrary (not necessarily semi-bounded) Hermitian relations and is due to the same author.

Theorem A.15 *Let θ_ω and θ_n $(n = 1, 2, \ldots)$ be Hermitian relations and let θ_ω be the strong resolvent limit of the sequence $\{\theta_n\}$: $s-R-\lim \theta_n = \theta_\omega$. If $\mu < \nu$ are continuity points of the spectral function $E_\lambda(\theta_\omega)$, then*

$$s - \lim E_{(\mu,\nu)}(\theta_n) = E_{(\mu,\nu)}(\theta_\omega). \tag{A.144}$$

(By virtue of (A.143) the equality (A.144) is equivalent to the following: for all $A_n \sim \theta_n$ and $A_\omega \sim \theta_\omega$, we have

$$s - \lim E_{(\mu,\nu)}(A_n)P_n = E_{(\mu,\nu)}(A_\omega)P_\omega, \tag{A.145}$$

where $A_n^* = A_n$ in $H_n = \overline{D(A_n)}$ and $A_\omega^* = A_\omega$ in $H_\omega = \overline{D(A_\omega)} \subseteq H$.)

Proof. As in the case of a sequence of densely defined self-adjoint operators (see Theorems VIII.24 and VIII.20 in [165]), we will rely on Stone–Weierstrass Theorem about the density of the set of polynomials in $(\lambda+i)^{-1}$ and $(\lambda - i)^{-1}$ in the space $C_\infty(\mathbb{R})$ of continuous functions on $\mathbb{R}$ going to 0 at $\pm\infty$.

Let us take an arbitrary vector $h \in H$ and $\epsilon > 0$. Since μ and ν are continuity points of $E_\lambda(\theta_\omega) = E_\lambda(A_\omega)P_\omega$, there exist functions f, $g \in C_\infty(\mathbb{R})$ such that

$$f(\lambda) \leq \chi_{(\mu,\nu)}(\lambda) \leq \chi_{[\mu,\nu]}(\lambda) \leq g(\lambda),$$

where χ is the characteristic function of the corresponding interval, and

$$\|f(A_\omega)P_\omega h - g(A_\omega)P_\omega h\| \leq \epsilon, \tag{A.146}$$

and, hence,

$$\|f(A_\omega)P_\omega h - E_{(\mu,\nu)}(A_\omega)P_\omega h\| \leq \epsilon. \tag{A.147}$$

Let us approximate $f(\lambda)$ and $g(\lambda)$ uniformly on $\mathbb{R}$ by polynomials $Q_f\left((\lambda + i)^{-1}, (\lambda - i)^{-1}\right)$ and $Q_g\left((\lambda + i)^{-1}, (\lambda - i)^{-1}\right)$ with accuracy up to $\epsilon/\|h\|$.

Then for $A = A_n$, A_ω and $P = P_n$, P_ω we have

$$\left\|f(A)Ph - Q_f\left((A + i)^{-1}, (A - i)^{-1}\right) Ph\right\| \leq \epsilon, \tag{A.148}$$

and analogous inequalities hold for $g(A_n)$ and $g(A_\omega)$.

Since, by hypothesis,

$$R_{\pm i}(\theta_n) = (A_n \mp i)^{-1}P_n \xrightarrow{s} (A_\omega \mp i)^{-1}P_\omega = R_{\pm i}(\theta_\omega),$$

then

$$Q_f\left((A_n+i)^{-1},(A_n-i)^{-1}\right)P_n = Q_f\left((A_n+i)^{-1}P_n,(A_n-i)^{-1}P_n\right)$$
$$\xrightarrow{s} Q_f\left((A_\omega+i)^{-1},(A_\omega-i)^{-1}\right)P_\omega. \tag{A.149}$$

The analogous relation holds for Q_g.

Thus, for sufficiently large n, by (A.148) and (A.149) we have

$$\|f(A_n)P_nh - f(A_\omega)P_\omega h\| \le 3\epsilon, \tag{A.150}$$

and an analogous inequality holds for $g(A_n)$ and $g(A_\omega)$.

Thus, from (A.146), by virtue of (A.149) and (A.150), for large n we get

$$\|f(A_n)P_nh - g(A_n)P_nh\| \le 7\epsilon. \tag{A.151}$$

Since

$$\|E_{(\mu,\nu)}(A_n)P_nh - E_{(\mu,\nu)}(A_\omega)P_\omega h\| \le \|f(A_n)P_nh - E_{(\mu,\nu)}(A_n)P_nh\|$$
$$+\|f(A_\omega)P_\omega h - E_{(\mu,\nu)}(A_\omega)P_\omega h\| + \|f(A_n)P_nh - f(A_\omega)P_\omega h\| \tag{A.152}$$

and

$$\|f(A_n)P_nh - E_{(\mu,\nu)}(A_n)P_nh\| \le \|f(A_n)P_nh - g(A_n)P_nh\|, \tag{A.153}$$

from (A.152), (A.153), (A.151), (A.147) and (A.150) it follows that

$$\|E_{(\mu,\nu)}(\theta_n)h - E_{(\mu,\nu)}(\theta_\omega)h\| = \|E_{(\mu,\nu)}(A_n)P_nh - E_{(\mu,\nu)}(A_\omega)P_\omega h\| \le 11\epsilon,$$

and the Theorem is proven. □

Corollary A.6 **([782, Theorem 6])** *If, under the conditions of Theorem A.15, Hermitian relations θ_n (or operators $A_n \sim \theta_n$) are semi-bounded below uniformly in n, i.e. $\theta_n \ge \gamma$ ($A_n \ge \gamma$) for all $n = 1, 2, \ldots$, then at the continuity points of $E_\lambda(\theta_\omega)$ we have*

$$s-\lim E_\lambda(\theta_n) = E_\lambda(\theta_\omega),$$
$$s-\lim E_\lambda(A_n)P_n = E_\lambda(A_\omega)P_\omega. \tag{A.154}$$

Proof. The proof follows from (A.144) and (A.145) if we put $\nu = \lambda$, $\mu < \nu$, because then

$$E_{(\mu,\nu)}(\theta_n) = E_\lambda(\theta_n) \qquad \text{since } E_\mu(\theta_n) = 0.$$

□

Remark A.5 If we do not require semi-boundedness below, unlike in the case of a given sequence of self-adjoint operators converging in $S-R$ sense to a self-adjoint operator, the assertion (A.154) might not hold. For example, if $H = \mathbb{C}^2$, $A_n = \begin{pmatrix} 1 & 0 \\ 0 & -n \end{pmatrix}$, $P_n = I$, then it is easy to see that A_ω acts in a one dimensional subspace,

$$A_\omega P_\omega = (s - R - \lim A_n) P_\omega = \begin{pmatrix} 1 & 0 \\ 0 & 0 \end{pmatrix},$$

$$s - \lim E_\lambda(A_n) P_n = \begin{cases} \begin{pmatrix} 0 & 0 \\ 0 & 1 \end{pmatrix}, & \lambda < 1, \\ I, & \lambda > 1, \end{cases}$$

while

$$E_\lambda(A_\omega) P_\omega = \begin{cases} 0, & \lambda < 1, \\ \begin{pmatrix} 1 & 0 \\ 0 & 0 \end{pmatrix}, & \lambda > 1. \end{cases}$$

We have already given Definition 3.1 which generalizes the notion of stability in the sense of Kato of eigenvalues for the case of non-densely defined operators. We now extend this notion to the case of a sequence of Hermitian relations.

Definition A.6 *Let θ_n be a sequence of Hermitian relations (or self-adjoint operators in $\overline{D(A_n)} = H_n \subseteq H$) converging in strong resolvent sense to a Hermitian relation θ (or self-adjoint operator A in $\overline{D(A)}$). An isolated eigenvalue λ' of the relation θ (operator A) is said to be stable (with respect to the considered limit, or with respect to the perturbation) if some punctured neighborhood $\Delta \backslash \{\lambda'\}$ belongs to the region of strong convergence of resolvents $R_\lambda(\theta_n)$ (respectively, $R_\lambda(A_n)P_n$) and the spectral projections corresponding to Δ converge uniformly as $n \to \infty$:*

$$\|E_\Delta(\theta_n) - E_{\{\lambda'\}}(\theta)\| \to 0,$$

$$\|E_\Delta(A_n)P_n - E_{\{\lambda'\}}(A)P\| \to 0, \tag{A.155}$$

where P_n is the orthogonal projection onto H_n and P is the orthogonal projection onto $\overline{D(A)}$. In particular, this implies that for sufficiently *large* n

$$\dim E_\Delta(\theta_n) = \dim E_{\{\lambda'\}}(\theta).$$

Lemma A.11 **(Kato [87, Lemma VIII.1.21])** *Assume that* Q_n *is a strongly convergent sequence of orthogonal projections in* H, $Q_n \xrightarrow{s} Q$ *(where, as is well-known,* Q *is also an orthogonal projection)*[65]. *Assume that* $\dim Q_n \leq \dim Q < \infty$ *for sufficiently large* n. *Then the sequence* Q_n *converges in norm, i.e.*

$$\|Q_n - Q\| \to 0$$

and, hence, $\dim Q_n = \dim Q$ *for large* n.

The following Proposition generalizes Theorem VIII.3.15 from [87] about the stability of the discrete part for $\lambda < \beta$ of the spectrum of the $s-R$ limit of a convergent above sequence of densely defined operators (cf. Proposition 3.1).

Proposition A.5 **([782])** *If a sequence of semi-bounded below Hermitian relations* θ_n *(or operators* $A_n \sim \theta_n$*) converges above in the* $s-R$ *sense to a semi-bounded Hermitian relation* θ *(respectively, operator* $A \sim \theta$*) whose lower part of the spectrum for* $\lambda < \beta$ *is discrete, i.e. consists of eigenvalues of finite multiplicity and* $\lambda_1 \leq \lambda_2 \leq \cdots (< \beta)$ *(enumerated counting multiplicity), then those eigenvalues are stable, and*

$$\lambda_j(A_n) \to \lambda_j(A) + 0 \tag{A.156}$$

for $\lambda_j(A) < \beta$.

In the case of a monotone convergent (non-increasing) sequence $\theta_n \sim A_n$, *the convergence of* $\lambda_j(A_n)$ *in (A.156) is monotone:*

$$\lambda_j(A_n) \searrow \lambda_j(A)(< \beta).$$

Proof. By Theorem A.15 and Corollary A.6 the spectral functions $E_\lambda(\theta_n) = E_\lambda(A_n)P_n$ converge strongly to $E_\lambda(\theta) = E_\lambda(A)P$ at the continuity points of $E_\lambda(\theta)$ with respect to λ. From the condition $\theta_n \geq \theta$ ($A_n \geq A$), for $\lambda < \beta$, we have $\dim E_\lambda(\theta_n) = \dim E_\lambda(\theta)$, and this, by Lemma A.11, implies the convergence in norm

$$\|E_\lambda(\theta_n) - E_\lambda(\theta)\| \to 0$$

[65] In [87] this Lemma is formulated for projections in Banach space.

and the stability of eigenvalues $\lambda_j(A) < \beta$. The convergence above (A.156) (monotone when the sequence $\{\theta_n\}$ is monotone) is obvious from min-max properties of eigenvalues. The Proposition A.5 (and, hence, Proposition 3.1) is proven. □

Proposition A.6 **([782, Theorem 7])** *Assume that for a monotone (non-increasing or non-decreasing) sequence of semi-bounded Hermitian relations $\theta_n \geq \gamma$ the difference of resolvents $R_{\zeta_0}(\theta_n) - R_{\zeta_0}(\theta)$ is compact for all n and some $\zeta_0 \in \mathbb{C}\backslash[\gamma,\infty)$, where $\theta = s - R - \lim \theta_n$. Then*

$$\|R_\zeta(\theta_n) - R_\zeta(\theta)\| \to 0, \qquad \textit{as } n \to \infty, \tag{A.157}$$

for all $\zeta \in \mathbb{C}\backslash[\gamma,\infty)$.

Proof. Note that the compactness of the difference of resolvents at $\zeta_0 \in \mathbb{C}\backslash[\gamma,\infty)$ implies the compactness of the difference $R_\zeta(\theta_n) - R_\zeta(\theta)$ for all $\zeta \in \mathbb{C}\backslash[\gamma,\infty)$ by virtue of the identity

$$\begin{aligned} R_\zeta(\theta_n) - R_\zeta(\theta) &= (1 + (\zeta - \zeta_0)R_\zeta(\theta_n))(R_{\zeta_0}(\theta_n) - R_{\zeta_0}(\theta)) \\ &\quad \cdot (1 + (\zeta - \zeta_0)R_\zeta(\theta)) \end{aligned} \tag{A.158}$$

which is known for resolvents of densely defined operators (see the formula (VIII.1.3) in [87]) and can be analogously proven for resolvents of relations using Hilbert's identity.

Thus, in the hypotheses of the Proposition, we may assume that the difference of resolvents is compact at the point $\zeta_0 = \gamma - 1$, and this difference converges to 0 at ζ_0

$$R_{\zeta_0}(\theta_n) - R_{\zeta_0}(\theta) \to 0, \qquad \text{as } n \to \infty, \tag{A.159}$$

strongly and monotonely by Theorems VIII.3.11 and VIII.3.13 from [87] about strong resolvent convergence of monotone sequences of operators, which also hold for sequences of non-densely defined operators (see [860]) and relations (see [782]). Thus, by Theorem VIII.3.5 from [87], in (A.159) we have the convergence in norm, and, hence, by (A.158) we get (A.157). This concludes the proof of the Proposition. □

Unlike in the case of a convergent above sequence of operators or relations, in the case of an increasing sequence of semi-bounded below operators $A_n \nearrow A_\omega$, the inequality

$$\dim E_\lambda(A_n)P_n \leq \dim E_\lambda(A_\omega)P_\omega,$$

which by Lemma A.11 implies the stability of eigenvalues of the limit operator, does not follow directly from variational principles, and the proof of the stability becomes considerably more difficult. The corresponding result (see [782, Theorem 8] and [784, Theorem 5]) is also new for sequences of densely defined operators with densely defined $s-R$-limit.

Theorem A.16 **([782])** *Assume that a sequence of semi-bounded below Hermitian relations θ_n or associated operators $A_n \sim \theta_n$ is non-decreasing:*

$$-\infty < \gamma I \leq A_1 \leq A_2 \leq \cdots \tag{A.160}$$

(the operators A_n are semi-bounded and self-adjoint in $H_n = \overline{D(A_n)} \subseteq H$).

Assume that the spectrum $\sigma(A_1)$ of A_1 is discrete for $\lambda < \beta$, and hence all spectra $\sigma(A_n)$ and $\sigma(A_\omega)$, where $A_\omega = s - R - \lim A_n$, are discrete for $\lambda < \beta$. (The existence of A_ω and its self-adjointness in $H_\omega = \overline{D(A_\omega)} \subseteq H$ were proven by B. Simon [860], [861][66].)

Then for $\lambda < \beta$ the eigenvalues converge monotonely

$$\lambda_j(A_n) \nearrow \lambda_j(A_\omega)(< \beta), \qquad \text{as } n \to \infty,\ j = 1, 2, \ldots, \tag{A.161}$$

and the spectral functions converge in norm for $\beta > \lambda \notin \sigma(A_\omega)$, $j = 1, 2, \ldots$

$$\|E_\lambda(\theta_n) - E_\lambda(\theta_\omega)\| = \|E_\lambda(A_n)P_n - E_\lambda(A_\omega)P_\omega\| \to 0. \tag{A.162}$$

(Here, $\theta_\omega \sim A_\omega$, P_n, P_ω are orthogonal projections onto H_n, H_ω, and the eigenvalues are put in the increasing order and enumerated counting multiplicity.)

Remark A.6 The non-decreasing condition on $\{A_n\}$ cannot be replaced by convergence below. For example, if $A_n = I - Q_n$, where Q_n is the orthogonal projection onto e_n (here, $\{e_n\}_{n=1}^{\infty}$ is an orthonormal basis in H), then

$$A_n \leq A_\omega = s - \lim A_n = s - R - \lim A_n = I;$$

however, $\lim \lambda_1(A_n) = 0 \notin \sigma(A_\omega)$.

A priori assumption on the discreteness of $\sigma(A_1)$ for $\lambda < \beta$ cannot be replaced by the assumption on the discreteness of $\sigma(A_\omega)$ for $\lambda < \beta$, since the latter does not guarantee the discreteness of $\sigma(A_n)$ for $\lambda < \beta$. For example, let

$$A_n = -\Delta + q_n(x), \qquad x \in \mathbb{R}^m,$$

[66] See Remark A.7 below.

$$q_n(x) = \begin{cases} |x|^2, \ |x| \le n, \\ 0, \quad |x| > n. \end{cases}$$

Then $s - R - \lim A_n = A_\omega = -\Delta + |x|^2$, the operator A_ω has purely discrete spectrum, but $\sigma(A_n)$ are absolutely continuous for all n. (Here, for the sequence $\{A_n\}_{n=1}^\infty$, spectral concentration occurs around $\lambda_j(A_\omega)$, $j = 1, 2, \ldots$; see [191], [87] and [165].)

Remark A.7 By a Theorem of B. Simon [860] the operator A_ω is associated with a closed form a_ω, where

$$\mathrm{D}(a_\omega) = \left\{ x \in \bigcap_n \mathrm{D}(a_n) : \sup_n a_n[x] < \infty \right\},$$
$$a_\omega[x, y] = \lim_{n\to\infty} a_n[x, y], \qquad x, y \in \mathrm{D}(a_\omega),$$

and a_n are the forms associated with A_n: $a_n \sim A_n$.

Proof of Theorem A.16. For $\lambda_j := \lambda_j(A_\omega) < \beta$ we have

$$\lambda_j(A_n) \le \lambda_j(A_{n+1}) \le \lambda_j,$$

and, hence, for $\lambda < \beta$ and $E_{\lambda,n} := E_\lambda(A_n)P_n$:

$$\dim E_{\lambda,\omega} \le \dim E_{\lambda,n+1} \le \dim E_{\lambda,n} \le E_{\lambda,1} < \infty.$$

Thus, $\dim E_{\lambda,n} = \dim E_{\lambda,\omega} + m_\lambda < \infty$, where $m_\lambda \ge 0$ and it does not depend on n for $n > n_\lambda$.

Denote

$$\mu_j := \lim_{n\to\infty} \lambda_j(A_n) \le \lambda_j, \qquad j = 1, 2, \ldots$$

Then $\bigcup_{\lambda_j<\beta} \{\lambda_j\} \subseteq \bigcup_{\mu_j<\beta} \{\mu_j\}$. We will show that $m_\lambda = 0$ for

$$\lambda \in (\gamma, \beta) \backslash \cup \{\mu_j\} := \Lambda \quad \text{and} \quad n > n_\lambda,$$

i.e. that

$$\dim E_{\lambda,n} = \dim E_{\lambda,\omega}, \qquad (\lambda \in \Lambda,\ n > n_\lambda).$$

By Lemma A.11, this implies (A.162) since, by Theorem A.15,

$$s - \lim E_{\lambda,n} = E_{\lambda,\omega}, \qquad n \to \infty,\ \beta > \lambda \neq \lambda_j,\ j = 1, 2, \ldots \tag{A.163}$$

Assume the opposite, i.e. let $m_\lambda > 0$ for some $\lambda \in \Lambda$. Then, there exist $\delta > 0$ and $N > 0$ such that

$$(\lambda - \delta, \lambda + \delta) \cap [\sigma(A_n) \cup \sigma(A_\omega)] = \emptyset, \qquad n \geq N. \tag{A.164}$$

For $\lambda < \beta$, let

$$H_{\lambda,n} := (E_{\lambda,n} H) \cap ((I - E_{\lambda,\omega})H) \subset \mathrm{D}(A_n). \tag{A.165}$$

Then

$$\infty > \dim H_{\lambda,n} \geq m_\lambda > 0, \qquad n > n_\gamma,$$

and

$$\forall x_n \in H_{\lambda,n} : \quad a_n[x_n] = (A_n x_n, x_n) \leq (\lambda - \delta)|x_n|^2. \tag{A.166}$$

Let $Q_{\lambda,n}$ be the orthogonal projection onto $H_{\lambda,n}$. Then the operator

$$F_{\lambda,n} := E_{\lambda,n} - Q_{\lambda,n} \tag{A.167}$$

is also an orthogonal projection, and

$$\dim F_{\lambda,n} = \dim E_{\lambda,n} - \dim H_{\lambda,n} \leq \dim E_{\lambda,\omega} \tag{A.168}$$

because

$$\dim H_{\lambda,n} \geq \dim E_{\lambda,n} - \dim E_{\lambda,\omega}.$$

By (A.165), (A.167) and (A.163) we have

$$Q_{\lambda,n} = Q_{\lambda,n}(Q_{\lambda,n} + F_{\lambda,n} - E_{\lambda,\omega}) \xrightarrow{s} 0, \qquad \text{as } n \to \infty. \tag{A.169}$$

Thus,

$$s - \lim F_{\lambda,n} = E_{\lambda,\omega},$$

and by virtue of (A.168), from Lemma A.11, we get

$$\|F_{\lambda,n} - E_{\lambda,\omega}\| \to 0, \qquad \text{as } n \to \infty. \tag{A.170}$$

Since

$$\dim Q_{\lambda,s} \leq \dim E_{\lambda,s} \leq \dim E_{\lambda,1} < \infty, \qquad (\lambda < \beta),$$

then for $s = \text{const}$, by (A.169) we have

$$\|Q_{\lambda,s} Q_{\lambda,n}\| = \|Q_{\lambda,n} Q_{\lambda,s}\| \to 0, \qquad \text{as } n \to \infty. \tag{A.171}$$

To get a contradiction for $m_\lambda > 0$, we will find n and s, $n > s > N$, such that

$$\forall x_n \in H_{\lambda,n} \subset \mathrm{D}(a_n) \subseteq \mathrm{D}(a_s), \qquad x_n \neq 0,$$

and the following inequality holds:

$$a_s[x_n] > a_n[x_n], \tag{A.172}$$

contrary to (A.160).

By (A.164) we have

$$a_s[x_n] \geq (\lambda + \delta) \cdot \|(I - E_{\lambda+\delta,s})x_n\|^2 = (\lambda + \delta)[\|x_n\|^2 - \|E_{\lambda,s}x_n\|^2],$$

and this, together with (A.166), gives

$$a_s[x_n] - a_n[x_n] \geq 2\delta\|x_n\|^2 - (\lambda + \delta)\|E_{\lambda,s}x_n\|^2. \tag{A.173}$$

Let $\epsilon > 0$. By (A.170) there exists N_ϵ such that for all $s > N_\epsilon$, the following inequality holds:

$$\|F_{\lambda,s} - E_{\lambda,\omega}\| < \epsilon.$$

Since, by (A.165), $E_{\lambda,\omega}x_n = 0$, then

$$\|F_{\lambda,s}x_n\| < \epsilon\|x_n\|, \qquad (s > N_\epsilon). \tag{A.174}$$

By (A.171) there exists $n = n(\epsilon, s) > s$ such that

$$\|Q_{\lambda,s}Q_{\lambda,n}\| < \epsilon, \qquad (n > n(\epsilon, s) > s > N_\epsilon),$$

and, hence, $\|Q_{\lambda,s}x_n\| < \epsilon\|x_n\|$ because $Q_{\lambda,n}x_n = x_n$.

This, together with (A.174) and (A.167) gives

$$\|E_{\lambda,s}x_n\|^2 = \|F_{\lambda,s}x_n\|^2 + \|Q_{\lambda,s}x_n\|^2 \leq 2\epsilon^2\|x_n\|^2.$$

Therefore, for sufficiently small ϵ and $x_n \neq 0$, the inequality (A.172) follows from (A.173), and the Theorem is proven. □

Using a result of B. Simon [860] (see Theorem A.17 below), Theorem A.16, whose proof we have just given, together with Proposition A.5, enables us to establish the continuity and stability with respect to ξ of eigenvalues $\lambda_j(A_\xi)$ for a family of operators $\{A_\xi\}$ (forms, relations) monotonely depending on a continuous parameter ξ or for a family of operators (forms, relations) which is, in a certain sense, close to a monotone one.

Theorem A.17 **(B. Simon [860])** *Assume that $\{a_\xi\}_{\xi=0}^{\infty}$, $a \geq \gamma$, $a_\xi \subset a$ are closed symmetric forms, not necessarily densely defined. Let $A_\xi \sim \theta_\xi \sim a_\xi$ be the operators and Hermitian relations associated to a_ξ, let $D(a_\xi) \subset D(a_\eta)$ for $\xi < \eta$, and let*

$$\bigcap_{\xi>\xi_0} D(a_\xi) = \left[\bigcup_{\xi<\xi_0} D(a_\xi)\right]_a^{\sim}, \tag{A.175}$$

where $[\cdot]_a^{\sim}$ is the closure in the a-metric. Then ξ_0 is a point of $s-R$-continuity of A_ξ and θ_ξ.

Theorem A.18 **([782])** *Under the hypotheses of Theorem A.17, if the spectrum of the form a is discrete for $\lambda < \beta$, so are the spectra $\sigma(A_\xi)$. Moreover, $\lambda_j(A_\xi)(< \beta)$ are continuous and stable for $\xi = \xi_0$, $(j = 1, 2, \ldots)$.*

Proof. Let

$$a_{\xi_0+} := a|_{\bigcap_{\xi>\xi_0} \mathrm{D}(a_\xi)}, \qquad a_{\xi_0-} := a|_{\bigcup_{\xi<\xi_0} \mathrm{D}(a_\xi)}.$$

Then, by (A.175) and since a_{ξ_0} is closed, we have $a_{\xi_0} = a_{\xi_0+} = a^{\sim}_{\xi_0-}$, where $a^{\sim}$ is the closure of a.

The remaining assertions of the Theorem follow from Theorems A.16, A.17 and Proposition A.5 after the corresponding change of indices (instead of $n \to \infty$, take $\xi \nearrow \xi_0$ or $\xi \searrow \xi_0$). □

Corollary A.7 **([782])** *Let $A \geq \gamma$ be a symmetric operator, and let $D_\xi \subseteq D_\eta \subseteq D(A)$ for all $\xi < \eta$. If $\left[\bigcup_{\xi<\xi_0} D_\xi\right]_A^{\sim} = \bigcap_{\xi>\xi_0} [D_\xi]_A^{\sim}$, then the family $A_\xi^F := \left(A|_{D_\xi}\right)^F$ is $s-R$-continuous at ξ_0.*

If $\dim E_\beta(A^F) < \infty$, then $\lambda_j(A_\xi^F)\,(< \beta)$ are continuous at $\xi = \xi_0$ and stable.

The proof is carried out by taking $a \sim A^F$ and $a_\xi \sim A_\xi^F$ as the forms a and a_ξ in Theorems A.17 and A.18.

The following Theorem allows us to weaken the monotonicity requirement in Theorem A.17.

Theorem A.19 **([782])** *Assume that a family of Hermitian relations $\{\theta_\xi\}$ is nested between two monotone in ξ families $\{\theta_\xi'\}$ and $\{\theta_\xi''\}$:*

$$\theta_\xi'' \geq \theta_\xi \geq \theta_\xi' \quad (\geq \gamma).$$

Assume also that

$$s - R - \lim \theta'_\xi = s - R - \lim \theta''_\xi (=: \theta), \qquad as\ \xi \to \infty. \tag{A.176}$$

Then the limit $s - R - \lim \theta_\xi$, *as* $\xi \to \infty$, *exists and is equal to* θ.

If, in addition,

$$\dim E_\beta(\theta) < \infty \quad and \quad \dim E_\beta(\theta'_\xi) < \infty,$$

then $\dim E_\beta(\theta_\xi) < \infty$, *and, as* $\xi \to \infty$,

$$\lambda_j(\theta_\xi) \to \lambda_j(\theta)\ (< \beta), \qquad j = 1, 2, \ldots, \tag{A.177}$$

and $\lambda_j(\theta)$ *are stable for* $\lambda_j < \beta$ *in the sense of Definition A.6.*

The proof of Theorem A.19 easily follows from previous results.

Example A.18 Dirichlet problems for the equation

$$Lu := -\Delta u + (x^2 + y^2)u = \lambda u \tag{A.178}$$

have discrete spectra in each of the domains

$$\Omega_\xi := (-\infty, \xi) \times \mathbb{R} \quad \text{if } \xi \leq 0 \tag{A.179}$$

$$\Omega_\xi := \mathbb{R}^2 \backslash \left([0, \infty) \times \bigcup_{n=-\infty}^{\infty} n2^{-k} \right), \quad \text{if } \xi > 0 \tag{A.180}$$

(where $k = [1/\xi]$, i.e. k is the greatest integer $\leq 1/\xi$), and every eigenvalue $\lambda_j(\xi)$ of those problems is a continuous function of ξ at $\xi = 0$ (and, evidently, at $\xi < 0$ as well).

Let us prove the continuity of eigenvalues at $\xi = 0$. Let f be an arbitrary function belonging to $[C_0^\infty(\Omega_\xi)]_{\tilde{L}}$. Then f also belongs to $[C_0^\infty(\mathbb{R}^2)]_{\tilde{L}}$, and, *a fortiori*, f belongs to Sobolev space $W_2^1(\mathbb{R}^2)$ of square integrable functions with square integrable first derivatives; see [118], [134], [148] or [184]. Therefore, by Sobolev's theorem,

$$F(y) := \int_0^\infty |f(x, y)|^2 \, dx$$

is a continuous function of y. Using boundary conditions we also have $F(n2^{-k}) = 0$ for all integers n and k such that $0 < \xi \leq k^{-1}$.

Therefore, if $f \in [C_0^\infty(\Omega_\xi)]_{\tilde{L}}$ for $\xi = 0$ (hence, for all $\xi > 0$), then $F(y) \equiv 0$, and we have $f(x, y) = 0$ at $x \geq 0$. Therefore, the condition (A.175) of Theorems A.17 and A.18 and conditions of Corollary A.7 are satisfied because both parts of (A.175) at $\xi_0 = 0$ are equal to the same set of functions—the set coinciding at $x \leq 0$ with the set $[C_0^\infty]_{\tilde{L}}$ which is contained in $\dot{W}_2^1(\mathbb{R}_- \times \mathbb{R})$. (This is the Sobolev space of functions in W_2^1 that are equal to 0 at $x = 0$.) At $x > 0$ both parts of (A.175) are functions identically equal to zero. By Theorem A.18 and Corollary A.7 the eigenvalues $\lambda_j(\xi)$ are continuous with respect to ξ at $\xi = 0$.

Thus, Example A.18 shows that using our theorems, it is possible, in some cases, to establish continuous dependence of eigenvalues of Dirichlet problems on perturbation of the domain without Courant's condition [48] of closeness of the boundaries of perturbed and unperturbed domains and without assuming boundedness of the considered domains. Our theorems can be applied to the case of strongly elliptic systems. (Note that the monograph [84] is devoted to boundary problems in bounded domains with smooth boundaries and boundaries smoothly depending on parameter. Refined classical results are contained in [88].)

Open Question A.1 What conditions guarantee continuous dependence of eigenvalues of the Dirichlet problem on perturbation of the domain (or on the parameter that determines the corresponding perturbation, as in Example A.18) if we do not require (unlike theorems from [48]) that to small perturbations corresponds closeness of boundaries of perturbed and unperturbed domains?

A.4.3 *Characteristic Operator and Boundary-Value Problems with Separated Boundary Conditions*

In this Section we mention the results of V. I. Khrabustovkiy [552] concerning the properties of characteristic operators for the boundary value problems with self-adjoint or dissipative boundary conditions.

Many dissipative differential and difference equations can be reduced to a dissipative Hamiltonian equation in a separable Hilbert space H (see [13], [73], [74], [496], [779] and [809]), with a spectral parameter λ entering the equation linearly or non-linearly (for example, through Nevanlinna type). For a fixed non-real parameter λ (for definiteness, $\lambda = i$), the corresponding Hamiltonian equation has the following form:

$$iGx'(t) - H(t)x(t) = [\operatorname{Im} H(t)]f(t), \qquad t \in (a,b) \subseteq \mathbb{R}^1, \tag{A.181}$$

where $G = G^*, G^{-1}, H(t) \in B(H)$; weight $\operatorname{Im} H(t) \geq 0$; the operator–function $H(t)$ is locally Bochner integrable.

By $X(t)$ we denote the operator solution of the homogeneous equation corresponding to (A.181) (i.e. the equation with $f(t) = 0$), normalized by the condition $X(c) = I$, where $c \in \overline{(a,b)}$, and by $\widetilde{X}(t) = GX^{-1}(t)G^{-1}$ we denote the solution of the equation

$$i\widetilde{X}'(t)G + \widetilde{X}(t)H(t) = 0.$$

Definition A.7 An operator $M \in B(H)$ is said to be the characteristic operator of the equation (A.181) if for all piecewise continuous vector-functions $f(t) \in H$ with compact support, the solution $x(t)$ of (A.181) of the form

$$x(t) = \int_a^b X(t)\left\{M - \frac{1}{2}\text{sgn}\,(s-t)(iG)^{-1}\right\}\widetilde{X}(s)[\text{Im}\ H(s)]f(s)\,ds \tag{A.182}$$

satisfies the inequality

$$\lim_{\alpha\downarrow a,\beta\uparrow b}[(Gx(\beta), x(\beta)) - (Gx(\alpha), x(\alpha))] \leq 0. \tag{A.183}$$

Note that the limit in (A.183) exists if the expression in brackets in (A.183) is bounded above uniformly in $\alpha, \beta \in (a,b)$ because, the function $X^*(t)GX(t)$ is nondecreasing. This can be seen as follows: for any finite subinterval $(\alpha,\beta) \subseteq (a,b)$ we have

$$X^*(\beta)GX(\beta) - X^*(\alpha)GX(\alpha) = 2\Delta(\alpha,\beta),$$

where

$$\Delta(\alpha,\beta) \equiv \int_\alpha^\beta X^*(t)[\text{Im}\ H(t)]X(t)\,dt \geq 0. \tag{A.184}$$

In what follows, we assume that the subspace of vectors belonging to the kernel of $\Delta(\alpha,\beta)$ for all finite $(\alpha,\beta) \subseteq (a,b)$ is trivial, i.e. equals $\{0\}$.

Definition A.8 The inequality (A.183) is said to be separated if, under the considered conditions, it is equivalent to the following system of inequalities:

$$\lim_{\alpha\downarrow a}(Gx(\alpha), x(\alpha)) \geq 0, \qquad \lim_{\beta\uparrow b}(Gx(\beta), x(\beta)) \leq 0. \tag{A.185}$$

Theorem A.20 *([552]) The inequality (A.183) is separated with respect to the solutions (A.182) of the equation (A.181) if and only if the charac-*

teristic operator M admits the following representation:

$$M = \left(P - \frac{1}{2}I\right)(iG)^{-1}, \tag{A.186}$$

where P is a projection, i.e.

$$P^2 = P \in B(H). \tag{A.187}$$

It is well known that, in general, to every self-adjoint, dissipative or accumulative regular boundary-value problem for the equation of type (A.181) corresponds some characteristic operator, and, conversely, to the characteristic operator of the equation (A.181) corresponds some boundary-value problem for this equation; see, for example, [13], [46], [73], [74], [113], [145], [197], [316], [317], [320], [321], [322], [360], [496], [552], [719], [720] [828], [850], and [851] (see also Chapter 5 and the Appendix of the present book). When a finite interval is extended to a half-axis or the whole axis, to a sequence of boundary-value problems on the corresponding intervals corresponds a convergent (sub)sequence of the corresponding characteristic operators (see the works cited above). Defining a characteristic operator for (A.181) on an infinite interval may be considered as a way of specifying the boundary condition at infinity. (Of course, in the Weyl limit point case, the boundary condition at the corresponding endpoint of (a, b) is reduced to requiring the square integrability of solutions in a neighborhood of the corresponding point with respect to the considered weight.)

There is a clear connection between characteristic operators and regular boundary-value problems on a finite interval (a, b) obtained by imposing boundary conditions on the equation (A.181). We will consider such conditions in the following form (see [13]):

$$\exists v = v(f) \in H\colon\ x(a) = Av, \qquad x(b) = Bv, \tag{A.188}$$

where the operators A, $B \in B(H)$ satisfy the condition $\|Ah\| + \|Bh\| \neq 0$ for all $h \neq 0$, and the lineal $\{Ah \oplus Bh\colon h \in H\} \subset H^2$ is maximal $\begin{pmatrix} G & 0 \\ 0 & -G \end{pmatrix}$-nonnegative subspace (see [14]), and, hence,

$$B^*GB - A^*GA \leq 0.$$

Proposition A.7 *([552]) If there exists an interval $(\alpha, \beta) \subseteq (a, b)$ such that $\Delta(\alpha, \beta) >> 0$, then the boundary-value problem (A.181), (A.188) has a unique solution $x(t)$ of the form (A.182), where M is the characteristic* *operator. Besides, if the coefficients A, B of the boundary condition (A.188) satisfy, respectively, the inequalities*

$$A^*GA \geq 0, \qquad B^*GB \leq 0, \tag{A.189}$$

then the inequality (A.183) is separated with respect to the solution $x(t)$ of the form (A.182), and, then, the characteristic operator M has representation (A.186), where P is a projection as in (A.187).

Conversely, if M is the characteristic operator of (A.181), then the vector function $x(t)$ in (A.182) is a solution of some boundary-value problem of the form (A.181), (A.188). If, in addition, M has the representation as in (A.186), (A.187), then the boundary condition (A.188) for this equation satisfies the property (A.189).

Remark A.8 *([552])* In Proposition A.7 the characteristic operator M and the projection P are respectively equal to:

$$M = -\frac{1}{2}\left(X^{-1}(a)A + X^{-1}(b)B\right)\left(X^{-1}(a)A - X^{-1}(b)B\right)^{-1}(iG)^{-1}, \tag{A.190}$$

$$P = -X^{-1}(b)B\left(X^{-1}(a)A - X^{-1}(b)B\right)^{-1},$$

where

$$\left(X^{-1}(a)A - X^{-1}(b)B\right)^{-1} \in B(H).$$

For $\dim H < \infty$, the problem (A.181), (A.188) was studied in [13], [320], [720] and [851], and, in particular, the construction of its solution (A.182), (A.190) was given there. However, the fact that P is a projection in the case of separated boundary conditions was not established in those references.

For a broad class of eigenvalue problems in the equation (A.181), the Hamiltonian $H(t) = H_\lambda(t)$ exhibits Nevanlinna-type dependence on the non-real spectral parameter λ: $\text{Im}\ H_\lambda(t)/\text{Im}\ \lambda \geq 0$, $\text{Im}\ \lambda \neq 0$. In this case, the equation (A.181) has the form:

$$iGx'(t) - H_\lambda(t)x(t) = w_\lambda(t)f(t), \tag{A.191}$$

where the weight is $w_\lambda(t) = \text{Im}\ H_\lambda(t)/\text{Im}\ \lambda$ (instead of $\text{Im}\ H_\lambda(t)$), and one considers characteristic operators $M = M(\lambda)$ depending analytically on non-real λ. The conditions for the existence of such operators are given in [552]. For $\dim H < \infty$, the resolvents of type (A.182), (A.183) of the equation (A.191) with analytic characteristic operator $M = M(\lambda)$ are constructed in [719], [720], and [779] when one of the endpoints of (a, b) is regular.

Bibliography

Monographs

[1] Agranovich, Z. S. and Marchenko, V. A. (1960). The inverse problem of scattering theory, Khar'kov State Univ., 268pp. (in Russian). (English Translation: (1963). Gordon and Breach Science Publishers, New York-London, xiii+291 pp.)

[2] Akhiezer, N. I. (1962). The classical moment problem and some related questions in analysis, Gosudarstv. Izdat. Fiz.-Mat. Lit., Moscow, 312pp. (in Russian). (English Translation: (1965). Translated by N. Kemmer, Hafner Publishing Co., New York, x+253 pp.)

[3] Akhiezer, N. I. (1970). Elements of the theory of elliptic functions, Second edition, Nauka, Moscow, 304 pp. (in Russian). (English Translation: (1990). Translated from the second Russian edition by H. H. McFaden, *Translations of Mathematical Monographs*, Vol. 79, American Mathematical Society, Providence, RI, 1990. viii+237 pp.)

[4] Akhiezer, N. I. (1978). Some inverse problems of spectral analysis related to hyperelliptic integrals (appendix to [6]), 3rd ed., vol. 2, Vyshcha Shkola, Khar'kov, pp. 242–283. (in Russian and in German [6]).

[5] Akhiezer, N. I. (2001). Selected works on the theory of functions and mathematical physics, vol. 1, vol. 2, Akta, Khar'kov. (in Russian).

[6] Akhiezer, N. I. and Glazman, I. M. (1966). Theory of linear operators in Hilbert spaces, Nauka, Moscow, 544pp. (in Russian). (1977). 3rd ed., vol.1, Vyshcha Shkola, Khar'kov. (in Russian). (1978). 3rd ed., vol.2, Vyshcha Shkola, Khar'kov. (in Russian). (English Translation: (1981). Translated from the third Russian edition by E. R. Dawson, Translation edited by W. N. Everitt, *Monographs and Studies in Mathematics* **9**, Pitman (Advanced Publishing Program), Boston, Mass.-London.) (German Translation: N. I. Achieser and I. M. Glassmann. (1981). Theorie der Linearen Operatoren in Hilbertraum, Academie Verlag, Berlin.) (Japanese Translation: (1972). Shuppan Kyoritsu, Tokyo.)

[7] Albeverio, S., Gesztesy F., Hoegh-Kron, R. and Holden, H. (1988). Solvable

models in quantum mechanics, Springer, New York–Berlin, xiv+452 pp. (Russian Translation: (1991). Mir, Moscow, 556 pp.)

[8] Amrein, W. O., Hinz, A. M. and Pearson, D. B. (Editors) (2005). Sturm–Liouville theory: past and present (Invited papers by participants of Sturm-200 Colloquium in Geneva, 2003), Birkhäuser, Basel.

[9] Anosov, D. V. and Bolibrukh, A. A. (1994). The Riemann-Hilbert problem, *Aspects of Mathematics*, E22, Friedr. Vieweg and Sohn, Braunschweig, x+190 pp.

[10] Arnold, V. I. (1974). Mathematical methods of classical mechanics, Nauka, Moscow, 432pp. (in Russian). (English Translation: (1989). Translated from the 1974 Russian original by K. Vogtmann and A. Weinstein, *Graduate Texts in Mathematics*, No. 60, Springer-Verlag, New York, xvi+516 pp.) (German Translation: (1988). Mathematische Methoden der klassischen Mechanik, Translated from the second Russian edition by Peter Möbius, VEB Deutscher Verlag der Wissenschaften, Berlin, 550 pp.)

[11] Arnold, V. I. (1984). Ordinary differential equations, Nauka, Moscow, (in Russian). (English Translation: (1978). Translated from the Russian and edited by Richard A. Silverman, MIT Press, Cambridge, Mass.-London, ix+280 pp.) (German Translation: (1991). Gewöhnliche Differentialgleichungen, translated from the third Russian edition by B. Mai and W. Plischke, Second edition, *Hochschulbücher für Mathematik [University Books for Mathematics]*, Bd. 83, Deutscher Verlag der Wissenschaften, Berlin, 340 pp.)

[12] Aslanyan, A. G. and Lidskiy, V. B. (1974). The distribution of characteristic frequneces of thin elastic shells, Nauka, Moscow, 156 pp. (in Russian).

[13] Atkinson, F. (1964). Discrete and continuous boundary problems, Acad. Press, New York–London. (Russian Translation: (1968). With supplements to the Russian edition by I. S. Kats and M. G. Krein: I. R-finctions—analytic functions mapping the upper half plane into itself; (English Translation of the supplement I: (1974). *Amer. Math. Soc. Transl. Ser. II* **103**, pp. 1–18.; II. On spectral functions of a string, Mir, Moscow, 750 pp.)

[14] Azizov, T. Ya. and Iokhvidov, I. S. (1986). Linear operators in spaces with an indefinite metric, Nauka, Moscow, 352pp. (in Russian). (English Translation: (1989). *Pure and Applied Mathematics*, Chichester, John Wiley and Sons Ltd., xii+304 pp.)

[15] Babenko, V. F., Korneychuk, N. P., Kofanov, V. A. and Pichugov, S. A. (2003). Inequalities for derivatives and their applications, Naukova Dumka, Kiev, 591 pp. (in Russian).

[16] Babikov, V. V. (1968). The method of phase functions in quantum mechanics, 1st ed., Nauka, Moscow, 224 pp. (in Russian). ((1976). 2nd ed., Nauka, Moscow, 288 pp. (in Russian); (1988). 3rd ed., Nauka, Moscow, 256 pp. (in Russian)).

[17] Babskiy, V. G., Kopachevskiy, N. D., Myshkis A. D., Slobozhanin, L. A. and Tyuptsov A. D. (Ed. Myshkis A. D.) (1976). Hydromechanics of ungravity, Nauka, Moscow, 504 pp. (in Russian). (For an augmented English version see [144].)

[18] Bakhvalov, N. S. and Panasenko, G. P. (1989). Homogenisation: averaging processes in periodic media, Mathematical problems in the mechanics of composite materials, Translated from the Russian by D. Leites, *Mathematics and its Applications (Soviet Series)* **36**, Kluwer Academic Publishers Group, Dordrecht, xxxvi+366 pp.

[19] Bar'yakhtar, V. G., Zakharov, V. E. and Chernousenko, V. M. (Editors). (1990). Integrability and kinetic equations for solitons, Collections of scientific works, Nukova Dumka, Kiev, 472 pp. (in Russian).

[20] Beckenbach, E. and Bellman, R. (1961). Inequalities, *Ergebnisse der Mathematik und ihrer Grenzgebiete. N. F.*, Bd. 30, Springer-Verlag, Berlin-Göttingen-Heidelberg, xii+198 pp. (Russian Translation: (1965). Mir, Moscow, 276 pp.)

[21] Bellman, R. (1953). Stability theory of differential equations, McGraw-Hill Book Company, Inc., New York-Toronto-London, xiii+166 pp. (Russian Translation: (1954). Foreign Literature Press, Moscow, 216 pp.)

[22] Belokolos, E. D., Bobenko, A. I., Enol'skiy, V. Z., Its, A. R. and Matveev, V. B. (1994). Algebro-geometric approach to non-linear integrable equations, Springer-Verlag, Berlin.

[23] Bensoussan, A. Lions, J.-L. and Papanicolaou, G. (1978). Asymptotic analysis for periodic structures, *Studies in Mathematics and its Applications* **5**, North-Holland Publishing Co., Amsterdam-New York, xxiv+700 pp.

[24] Berezanskiy, Yu. M. (1965). Expansions in eigenfunctions of self-adjoint operators, Naukova Dumka, Kiev, 800 pp. (in Russian). (English Translation: (1968). Translated from the Russian by R. Bolstein, J. M. Danskin, J. Rovnyak and L. Shulman, *Translations of Mathematical Monographs*, Vol. 17, American Mathematical Society, Providence, R.I., ix+809 pp.)

[25] Berezanskiy, Yu. M. (1978). Selfadjoint operators in spaces of functions of an infinite number of variables, Naukova Dumka, Kiev, 360 pp. (in Russian). (English Translation: (1986). Translated from the Russian by H. H. McFaden, Translation edited by Ben Silver, *Translations of Mathematical Monographs*, Vol. 63, American Mathematical Society, Providence, RI, xiv+383 pp.)

[26] Berezanskiy, Yu. M., Sheftel, Z. G. and Us, G. F. (1990). Functional analysis, Vyshcha Shkola, Kiev, 600 pp. (in Russian). (English Translation: (1996). Translated from the 1990 Russian original by Peter V. Malyshev, Vol. I, *Oper. Theory Adv. Appl.* **85**, Birkhäuser Verlag, Basel, xx+423 pp; Vol. II, *Oper. Theory Adv. Appl.* **86**, Birkhäuser Verlag, Basel, xvi+293 pp.)

[27] Berezin, F. A. and Shubin, M. A. (1983). Schrödinger equation, Izdat. Moskov. Gos. Univ., 393 pp. (in Russian). (English Translation: (1991). Translated from the 1983 Russian edition by Yu. Rajabov, D. A. Leites and N. A. Sakharova and revised by M. A. Shubin. With contributions by G. L. Litvinov and Leites, *Mathematics and its Applications (Soviet Series)* **66**, Kluwer Academic Publishers Group, Dordrecht, 1991. xviii+555 pp.)

[28] Bobik, O. I., Bodnarchuk, P. I., Ptashnik, B. Y. and Skorobogat'ko, V. Ya. (1972). Elements of the qualitative theory of partial differential equations (Ed. Skorobogat'ko V. Ya.), Naukova Dumka, Kyiv, 176 pp. (in Ukrainian).

[29] Bôcher, M. (1917). Leçons sur les méthodes de *Sturm* dans la théorie des équations différentielles linéaires et leurs développements modernes, Gauthier-Villars et Cie., Paris. (in French).

[30] Bogolyubov, N. N. (Jr.), Mitropol'skiy, Yu. A., Prikarpatskiy, A. K. and Samoilenko, V . G. (1987). Integrable dynamical systems: spectral and differential-geometric aspects, Naukova Dumka, Kiev, 292 pp. (in Russian).

[31] Bohr, H. (1932). Fastperiodische Funktionen, Springer, Berlin, v+95 pp. (in German). (English Translation: (1947). Almost Periodic Functions, Chelsea Publishing Company, New York, ii+114 pp.) (Russian Translation: (1934). GTTI, Moscow and Leningrad, 128 pp.)

[32] Bolibrukh, A. A. (1994). Hilbert's twenty-first problem for linear Fuchsian systems, *Trudy Mat. Inst. Steklov.* **206**, 160 pp. (in Russian). (English Translation: (1995). *Proc. Steklov Inst. Math.*, no. 5 (206), viii + 145pp.)

[33] Burskiy, V. P. (2002). Methods of investigation of boundary-value problems for general differential equations, Naukova Dumka, Kiev. (in Russian).

[34] Calogero, F. (1967). Variable phase approach to potential sattering, Academic Press, New York-London, xii+244 pp. (Russian Translation: (1972). Mir, Moscow, 296 pp.)

[35] Calogero, F. and Degasperis, A. (1982). Spectral transform and solitons, Vol. I. Tools to solve and investigate nonlinear evolution equations, *Studies in Mathematics and its Applications* **13**, *Lecture Notes in Computer Science* **144**, North-Holland Publishing Co., Amsterdam-New York, xv+516 pp. (Russian Translation: (1985). Mir, Moscow, 471 pp.)

[36] Carleman, T. (1960). Édition complète des articles de Torsten Carleman, Malmö: Institut Mathématique Mittag-Leffler, Djursholm, Suède, 552 pp. (in French).

[37] Carmona, R. and Lacroix, J. (1990). Spectral theory of random Schrödinger operators, Birkhäuser, Boston, MA, xxvi+587 pp.

[38] Cercignani, C., Gerasimenko, V. I. and Petrina, D. Ya. (1997). Many-particle dynamics and kinetic equations, Translated from the Russian manuscript by K. Petrina and V. Gredzhuk, *Mathematics and its Applications* **420**, Kluwer Academic Publishers Group, Dordrecht, 1997. viii+244 pp.

[39] Cesari, L. (1959). Asymptotic behavior and stability problems in ordinary differential equations, *Ergebnisse der Mathematik und ihrer Grenzgebiete. N.F.*, Heft 16, Springer-Velag, Berlin-Göttingen-Heidelberg, vii+271 pp. (Russian Translation: (1964). Mir, Moscow, 478 pp.)

[40] Chadan, K. and Sabatier, P. (1977). Inverse problems in quantum scattering theory, With a foreword by R. G. Newton, *Texts and Monographs in Physics*, Springer-Verlag, New York-Berlin, xxii+344 pp. (Russian Translation: (1980). Translated from the English by Ju. Muzyčka, B. V. Rudjak and A. A. Suz'ko, With a foreword by B. N. Zahar' ev and A. A. Suz'ko, Mir, Moscow, 408 pp.)

[41] Chaplygin, S. A. (1954). Selected works in mechanics and mathematics, Gosudarstv. Izdat. Tehn.-Teor. Lit., Moscow, 568 pp. (in Russian).

[42] Chaplygin, S. A. (1976). New method for approximate integration of differential equations, Selected works, Nauka, Moscow, pp. 307–362. (in Russian).

[43] Chueshov, I. (2002). Monotone random systems theory and applications, *Lecture Notes in Math.* **1779**, Springer-Verlag, Berlin, viii+234 pp.

[44] Cioranescu, D. and Saint Jean Paulin, J. (1999). Homogenization of reticulated structures, *Applied Mathematical Sciences* **136**, Springer-Verlag, New York, xx+346 pp.

[45] Coddington, E. A. (1973). Extension theory of formally normal and symmetric subspaces, *Memoirs of the American Mathematical Society*, No. 134, American Mathematical Society, Providence, R.I., iv+80 pp.

[46] Coddington, E. A. and Levinson, N. (1955). Theory of ordinary differential equations, McGraw-Hill Book Company, Inc., New York-Toronto-London, xii+429 pp. (Russian Translation: (1958). Foreign Literature Press, Moscow.)

[47] Collatz, L. (1963). Eigenwertaufgaben mit technischen Anwendungen. Zweite, durchgesehene Auflage Akademische Verlagsgesellschaft Geest Portig K.-G., Leipzig, xiv+500 pp. (in German). (Russian Translation: (1968). Eigenvalue problems with applications to engineering, Translated from the second German edition under the general editorship of V. V. Nikol'skiy, Nauka, Moscow, 503 pp.)

[48] Courant, R. and Hilbert, D. (1943). Methoden der Mathematischen Physik, Vols. I, II, Interscience Publishers, Inc., N.Y., xiv+469 pp., xiv+549 pp. (in German). (English Translation: (1953). Methods of mathematical physics. Vol. I, Interscience Publishers, Inc., New York, N.Y., xv+561 pp; (1962). Vol. II: Partial differential equations, Interscience Publishers, New York-London, xxii+830 pp.)

[49] Cross, R. (1998). Multivalued linear operators, *Monographs and Textbooks in Pure and Applied Mathematics* **213**, Marcel Dekker, Inc., New York, x+335 pp.

[50] Cycon, H., Froese, R., Kirch, W. and Simon, B. (1987). Schrödinger operators with application to quantum mechanics and global geometry, *Texts and Monographs in Physics*, Springer Study Edition, Springer-Verlag, Berlin, x+319 pp. (Russian Translation: (1990). Translated from the English by A. V. Sobolev, Translation edited and with a preface by D. R. Yafaev, Mir, Moscow, 408 pp.)

[51] Daletskiy, Yu. L. and Krein, M. G. (1970). Stability of solutions of differential equations in Banach space, Nauka, Moscow, 534pp. (in Russian). (English Translation: (1974). Translated from the Russian by S. Smith, *Translations of Mathematical Monographs*, Vol. 43, Providence, R.I., American Mathematical Society, vi+386 pp.)

[52] Dekkers, W. (1976). Moduli spaces for pairs of 2×2-matrices and for certain connections on $P^1(\mathrm{C})$, *Dissertation, Catholic University of Nijmegen, Nijmegen, 1976*, With a Dutch summary, Krips Repro B.V., Meppel, 146 pp.

[53] Deligne, P. (1970). Équations différentielles à points singuliers réguliers, *Lecture Notes in Math.* **163**, Springer-Verlag, Berlin-New York, iii+133 pp. (in French). (See also N. M. Katz in the book [80], pp. 537–559.)

[54] Dezin, A. A. (1980). General questions of the theory of boundary value prob-

lems, Nauka, Moscow, 208 pp. (in Russian).

[55] Dubovoy, V. K., Fritzsche, B. and Kirstein, B. (1992). Matricial version of the classical Schur problem. With German, French and Russian summaries, *Teubner-Texte zur Mathematik [Teubner Texts in Mathematics]*, No. 129, B. G. Teubner Verlagsgesellschaft mbH, Stuttgart, 355 pp.

[56] Dubrovin, B. A. (1986). Riemann surfaces and nonlinear equations, Part 1, Moscow University, Moscow, 92 pp. (in Russian).

[57] Dubrovin, B. A., Novikov S. P. and Fomenko, A. T. (1986). Modern Geometry–methods and applications, 2nd edition, Nauka, Moscow, 760pp. (in Russian). (English Translation: Translated from the Russian by Robert G. Burns, (1992). Part I, The geometry of surfaces, transformation groups, and fields, Second edition, *Graduate Texts in Mathematics*, No. 93, Springer-Verlag, New York, xvi+468 pp; (1985). Part II, The geometry and topology of manifolds, *Graduate Texts in Mathematics*, No. 104, Springer-Verlag, New York, xv+430 pp; (1990). Part III, Introduction to homology theory, *Graduate Texts in Mathematics*, No. 124, Springer-Verlag, New York, 1990, x+416 pp.)

[58] Dunford, N. and Schwartz, J. T. (1963). Linear operators. Part II: Spectral theory. Self adjoint operators in Hilbert space, With the assistance of William G. Bade and Robert G. Bartle, Interscience Publishers [John Wiley and Sons, Inc.], New York-London, ix+pp. 859–1923+7 pp. (Russian Translation: (1966). Mir, Moscow, 1064 pp.)

[59] Dunford, N. and Schwartz, J. T. (1971). Linear operators. Part III: Spectral operators, With the assistance of William G. Bade and Robert G. Bartle, *Pure and Applied Mathematics*, Vol. VII, Interscience Publishers [John Wiley and Sons, Inc.], New York-London-Sydney, i–xx and 1925–2592 pp. (Russian Translation: (1974). Mir, Moscow, 663 pp.)

[60] Eastham, M. S. P. (1973). The spectral theory of periodic differential equations, Scottish Acad. Press, Edinburgh-London, 130 pp.

[61] Eastham, M. S. P. (1989). The asymptotic solution of linear differential systems. Applications of the Levinson theorem, *London Mathematical Society Monographs. New Series* **4**, Oxford Science Publications, The Clarendon Press, Oxford University Press, New York, x+241 pp.

[62] Eastham, M. S. P. and Kalf, H. (1982). Schrödinger-type operators with continuous spectra, *Research Notes in Mathematics*, No. 65, Pitman (Advanced Publishing Program), Boston, Mass.-London, vii+281 pp.

[63] Erdélyi, A., Magnus, W., Oberhettinger, F. and Tricomi, F. G. (1955). Higher transcendental functions. Vol. III, Based, in part, on notes left by Harry Bateman, McGraw-Hill Book Company, Inc., New York-Toronto-London, xvii+292 pp. (Russian Translation: (1967). Nauka, Moscow, 300 pp.)

[64] Erugin, N. P. (1982). The Riemann problem, *Lecture Notes in Mathematics* **936**, Nauka i Tekhnika, Minsk, 336 pp. (in Russian).

[65] Favini, A. and Yagi, A. (1999). Degenerate differential equations in Banach spaces, Monographs and *Textbooks in Pure and Applied Mathematics* **215**, Marcel Dekker, Inc., New York, xii+313 pp.

[66] Fisher, R. A. (1930). The genetical theory of natural selection, Oxford Uni-

versity Press, Oxford.

[67] Gabov, S. A. and Sveshnikov, A. G. (1986). Problems of the dynamics of stratified fluids, Nauka, Moscow, 288 pp. (in Russian).

[68] Gabov, S. A. and Sveshnikov, A. G. (1990). Linear problems in the theory of unsteady internal waves, Nauka, Moscow, 344 pp. (in Russian).

[69] Gantmakher, F. R. and Krein, M. G. (1950). Oscillatory matrices and kernels and small vibrations of mechanical systems, Gosudarstv. Izdat. Tehn.-Teor. Lit., Moscow-Leningrad, 360 pp. (in Russian). (English Translation: (2002). Oscillation matrices and kernels and small vibrations of mechanical systems, Revised edition, Translation based on the 1941 Russian original, Edited and with a preface by Alex Eremenko, AMS Chelsea Publishing, Providence, RI, viii+310 pp.)

[70] Glazman, I. M. (1963). Direct methods of qulitative spectral analysis of singular differential operators, Fizmatgiz, Moscow, 340 pp. (in Russian). (English Translation: (1965). Translated from the Russian by the IPST staff Israel Program for Scientific Translations, Jerusalem; (1966). Daniel Davey and Co., Inc., New York 1966 ix+234 pp.)

[71] Gol'dberg, A. A. and Ostrovskiy, I. V. (1970). The distribution of values of meromorphic functions, Nauka, Moscow, 592 pp. (in Russian).

[72] Gol'denveizer, A. L., Lidskiy, V. B. and Tovstik, P. E. (1979). Free oscillations of thin elastic shells, Nauka, Moscow, 384 pp. (in Russian).

[73] Gorbachuk, V. I. and Gorbachuk, M. L. (1984). Boundary value problems for operator-differential equations, Naukova Dumka, Kiev, 283 pp. (in Russian).

[74] Gorbachuk, V. I. and Gorbachuk, M. L. (1991). Boundary value problems for operator differential equations, Translated and revised from the 1984 Russian original, *Mathematics and its Applications (Soviet Series)* **48**, Kluwer Academic Publishers Group, Dordrecht, xii+347 pp.

[75] Griffiths, Ph. A. (1989). Introduction to algebraic curves, Translated from the Chinese by Kuniko Weltin, *Translations of Mathematical Monographs*, Vol. 76, American Mathematical Society, Providence, RI, 1989. x+221 pp.

[76] Gromoll, D., Klingenberg, W. and Meyer, W. (1968). Riemannsche Geometrie im Grossen, *Lecture Notes in Math.* **55**, Springer-Verlag, Berlin, vi+287 pp. (in German). (Russian Translation: (1971). Mir, Moscow, 343 pp.)

[77] Hardy, G. H., Littlewood, J. E. and Pólya, G. (1934, 1952). Inequalities, Cambridge University Press. (Russian Translation: (1948). Foreign Literature Press.)

[78] Hartman, P. (2002). Ordinary differential equations, Corrected reprint of the second (1982) edition [Birkhäuser, Boston, MA], With a foreword by Peter Bates, *Classics in Applied Mathematics*, No. 38, Society for Industrial and Applied Mathematics (SIAM), Philadelphia, PA, 2002, xx+612 pp. (Russian Translation: (1970). Translated from the English by I. H. Sabitov and Ju. V. Egorov, Edited by V. M. Alekseev, Mir, Moscow, 720 pp.)

[79] (1969). Hilbert's problems, (Collection edited by P. S. Alexandrov), Nauka, Moscow, 239 pp.

[80] (1976). Mathematical developments arising from Hilbert problems. Proceed-

ings of the Symposium in Pure Mathematics of the American Mathematical Society held at Northern Illinois University, De Kalb, Ill., May, 1974, Edited by Felix E. Browder, *Proceedings of Symposia in Pure Mathematics*, Vol. XXVIII, American Mathematical Society, Providence, R. I., xii+628 pp.

[81] Hille, E. and Phillips, R. S. (1957). Functional analysis and semi-groups, *American Mathematical Society Colloquium Publications*, Vol. 31, American Mathematical Society, Providence, R. I., xii+808 pp. (Russian Translation: (1962). Translated from the English by D. A. Vasil'kov, Edited by V. M. Alekseev and S. V. Fomin, Izdat. Inostran. Lit., Moscow 829 pp.)

[82] Hurwitz, A. and Courant, R. (1964). Vorlesungen über allgemeine Funktionentheorie und elliptische Funktionen, *Die Grundlehren der Mathematischen Wissenschaften*, Band 3, Springer-Verlag, Berlin-New York, xiii+706 pp. (in German). (Russian Translation: (1968). Translated and adapted from the fourth German edition by M. A. Evgrafov, Nauka, Moscow, 648 pp.)

[83] Iohvidov, I. S., Krein, M. G. and Langer, H. (1982). Introduction to the spectral theory of operators in spaces with an indefinite metric, *Mathematical Research*, No. 9, Akademie-Verlag, Berlin, 120 pp.

[84] Ivanov, L., Kotko, L. and Krein, S. (1977). Boundary value problems in variable domains, *Differentsial'nye Uravneniya i Primenen.—Trudy Sem. Processy Optimal. Upravleniya*, Issue 19, pp. 7–160. (in Russian).

[85] Jörgens, K. and Weidmann, J. (1973). Spectral properties of Hamiltonian operators, *Lecture Notes in Math.* **313**, Springer-Verlag, Berlin-New York, iii+140 pp. (Russian Translation: (1976). With a supplement "The discrete spectrum of n-particle Hamiltonians" by M. A. Antonets, G. M. Zhislin and I. A. Shereshevskiy, Translated from the English by G. M. Zhislin, Mir, Moscow, 152 pp.)

[86] Kamke, E. (1944). Differentialgleichungen. Lösungsmethoden und Lösungen. Band I. Gewöhnliche Differentialgleichungen, 3d ed, *Mathematik und ihre Anwendungen in Physik und Technik*, Band 18_1, Akademische Verlagsgesellschaft, Leipzig, xxvi+666 pp. (in German). (Russian Tranlslation: (1976). Manual of ordinary differential equations, Supplemented by a translation by N. H. Rozov of excerpts of articles by D. S. Mitrinović and J. Zbornik, Fifth unrevised edition, Translated by S. V. Fomin from the sixth German edition and revised in accordance with the eighth German edition, Nauka, Moscow, 576 pp.)

[87] Kato, T. (1995). Perturbation theory for linear operators, Reprint of the 1980 edition, *Classics in Mathematics*, Springer-Verlag, Berlin, xxii+619 pp. (Russian Translation: (1972). Mir, Moscow, 740 pp.)

[88] Keldysh, M. V. (1985). Selected works. Mathematics, Edited and with a preface by K. I. Babenko, N. N. Bogolyubov, and N. N. Chentsov, Nauka, Moscow, 448 pp. (in Russian).

[89] Kiguradze, I. T. and Chanturiya, T. A. (1990). Asymptotic properties of solutions of non-autonomous ordinary differential equations, Nauka, Moscow, 432 pp. (in Russian). (English Translation: (1993). *Mathematics and its*

Applications (Soviet Series) **89**, Kluwer Academic Publishers Group, Dordrecht, xiv+331 pp.)

[90] Kolmogorov, A. N. (1985). Selected works of A. N. Kolmogorov. Vol. I. Mathematics and mechanics, With commentaries by V. I. Arnold, V. A. Skvortsov, P. L. Ul'yanov, et al., Nauka, Moscow, 470 pp. (in Russian). (English Translation: (1991). Translated from the Russian original by V. M. Volosov, Edited and with a preface, foreword and brief biography by V. M. Tikhomirov, *Mathematics and its Applications (Soviet Series)* **25**, Kluwer Academic Publishers Group, Dordrecht, xx+551 pp.)

[91] Kopachevskiy, N. D. and Krein, S. G. (2001). Operator approach to linear problems of hydrodynamics. Vol. 1. Self-adjoint problems for an ideal fluid, *Oper. Theory Adv. Appl.* **128**, Birkhäuser Verlag, Basel, xxiv+384 pp.

[92] Kopachevskiy, N. D. and Krein, S. G. (2003). Operator approach to linear problems of hydrodynamics. Vol. 2. Nonself-adjoint problems for viscous fluids, *Oper. Theory Adv. Appl.* **146**, Birkhäuser Verlag, Basel, xxiv+444 pp.

[93] Kopachevskiy, N. D., Krein, S. G. and Ngo Zuy Kan (1989). Operator methods in linear hydrodynamics, Nauka, Moscow, 416 pp. (in Russian).

[94] Kostyuchenko, A. G. and Sargasyan, I. S. (1979). The distribution of eigenvalues, Nauka, Moscow, 400 pp. (in Russian).

[95] Kozlov, V. and Maz'ya, V. (1999). Differential equations with operator coefficients with applications to boundary value problems for partial differential equations, *Springer Monographs in Mathematics*, Springer-Verlag, Berlin, xx+442 pp.

[96] Krein, M. G. and Nudel'man, A. A. (1973). The Markov moment problems and extremal problems, Nauka, Moscow, 552 pp. (in Russian). (English Translation: (1977). Ideas and problems of P. L. Čebyšev and A. A. Markov and their further development, Translated from the Russian by D. Louvish, *Translations of Mathematical Monographs*, Vol. 50, American Mathematical Society, Providence, R.I., v+417 pp.)

[97] Krein, S. G. (1967). Linear differential equations in Banach space, Nauka, Moscow, 464 pp. (in Russian). (English Translation: (1971). Translated from the Russian by J. M. Danskin, *Translations of Mathematical Monographs*, Vol. 29, American Mathematical Society, Providence, R.I., v+390 pp.)

[98] Kreith, K. (1974). Oscillation Theory, *Lecture Notes in Math.* **324**, Springer, Berlin.

[99] Krylov, A. N. (1932). On some differential equations of mathematical physics with applications to technical questions, Edition of Akad. Sci. USSR. (in Russian).

[100] Kuchment, P. A. (1993). Floquet theory for partial differential equations, *Oper. Theory Adv. Appl.* **60**, Birkhäuser Verlag, Basel, xiv+350 pp.

[101] Kurpel', N. S. and Shuvar, B. A. (1980). Two-sided operator inequalities and their applications, Naukova Dumka, Kiev, 268 pp. (in Russian).

[102] Kuzhel', A. V. (1989). Extensions of Hermitian operators, Vyshcha Shkola, Kiev, 56 pp. (in Russian).

[103] Kuzhel', A. V. and Kuzhel', S. A. (1998). Regular extensions of Hermitian operators, Translated from the Russian by P. Malyshev and D. Malyshev, VSP, Utrecht, xii+273 pp.

[104] Ladyzhenskaya, O. A. (1970). Mathematical questions of the dynamics of a viscous incompressible fluid, Nauka, Moscow, 288 pp. (in Russian). (English Translation: (1969). 2nd English ed., rev. and enl., Translated from the Russian by Richard A. Silverman and John Chu, Gordon and Breach, New York, xviii+224 pp.).

[105] Ladyzhenskaya, O. A. (1991). Attractors for semigroups and evolution equations, *Lezioni Lincee [Lincei Lectures]*, Cambridge University Press, Cambridge, xii+73 pp.

[106] Lang, S. (1973). Elliptic functions, With an appendix by J. Tate, Addison-Wesley Publishing Co., Inc., Reading, Mass.-London-Amsterdam, xii+326 pp. (Russian Translation: (1984). Translated from the English and with a preface by S. A. Stepanov, Nauka, Moscow, 312 pp.)

[107] Lappo-Danilevskiy, I. A. (1957). Application of matrix functions to the theory of linear systems of ordinary differential equations, Edited by V. I. Smirnov with an introductory essay, Gosudarstv. Izdat. Tehn.-Teor. Lit., Moscow, 456 pp. (in Russian).

[108] Lax, P. D. and Phillips, R. S. (1967). Scattering theory, *Pure and Applied Mathematics*, Vol. 26, Academic Press, New York-London, xii+276 pp. (Russian Translation: (1971). Translated from the English by N. K. Nikol'skiy and B. S. Pavlov, Edited by M. Sh. Birman, Mir, Moscow, 312 pp.)

[109] Lebowitz, A. and Rauch, H. E. (1973). Elliptic functions, theta functions and Riemann surfaces, The Williams and Wilkins Co., Baltimore, Maryland, xii+292 pp.

[110] Levin, B. Ya. (1956). Distribution of zeros of entire functions, Gosudarstv. Izdat. Tehn.-Teor. Lit., Moscow, 632 pp. (in Russian). (English Translation: (1964). *Translations of Mathematical Monographs*, Vol. 5, American Mathematical Society, Providence, R.I., xii+523 pp.)

[111] Levitan, B. M. (1953). Almost periodic functions, Gosudarstv. Izdat. Tehn.-Teor. Lit., Moscow, 396pp. (in Russian).

[112] Levitan, B. M. (1984). Inverse Sturm-Liouville problems, Nauka, Moscow, 240 pp. (in Russian). (English Translation: (1987). Translated from the Russian by O. Efimov, VSP, Zeist, 1987. x+240 pp.)

[113] Levitan, B. M. and Sargasyan, I. S. (1970). Introduction to spectral theory, Nauka, Moscow, 672 pp. (in Russian). (English Translation: (1975). Translated from the Russian by Amiel Feinstein, *Translations of Mathematical Monographs*, Vol. 39, American Mathematical Society, Providence, R.I., xi+525 pp.)

[114] Levitan, B. M. and Sargsyan, I. S. (1988). Sturm-Liouville and Dirac operators, Nauka, Moscow, 432 pp. (in Russian). (English Translation: *Mathematics and its Applications (Soviet Series)* **59**, Kluwer Academic Publishers Group, Dordrecht, 1991. xii+350 pp.)

[115] Levitan, B. M. and Zhikov, V. V. (1978). Almost periodic functions and

differential equations, Moscow State Univ., 205 pp. (in Russian). (English Translation: (1982). Translated from the Russian by L. W. Longdon, Cambridge University Press, Cambridge-New York, xi+211 pp.)

[116] Lifschitz, A. E. (1989). Magnetohydrodynamics and spectral theory, *Developments in Electromagnetic Theory and Applications*, No. 4, Kluwer Academic Publishers Group, Dordrecht, xii+446 pp.

[117] Lifschitz, I. M., Gredeskul, S. A. and Pastur, L. A. (1982). Introduction to the theory of disordered systems, Nauka, Moscow, 360 pp. (English Translation: (1988). Translated from the Russian by Eugene Yankovsky, A Wiley-Interscience Publication, John Wiley and Sons, Inc., New York, xviii+462 pp.)

[118] Lions, J.-L. and Magenes, E. (1968). Problèmes aux limites non homogènes et applications, Vol. 1, *Travaux et Recherches Mathématiques*, No. 17, Dunod, Paris, xx+372 pp. (in French). (English Translation: (1972). Translated from the French by P. Kenneth, *Die Grundlehren der mathematischen Wissenschaften*, Band 181, Springer-Verlag, New York-Heidelberg, xvi+357 pp.) (Russian Translation: (1971). Translated from the French by L. S. Frank, Mir, Moscow, 371 pp.)

[119] Lyantse, V. E. and Storozh, O. G. (1983). Methods of the theory of unbounded operators, Naukova Dumka, Kiev, 210 pp. (in Russian).

[120] Lyapunov, A. M. (1956). Collected works, Vol. II, Izdat. Akad. Nauk SSSR, Moscow, pp. 401, 404–406, 409. (in Russian).

[121] Marchenko, V. A. (1977). Sturm-Liouville operators and their applications, Naukova Dumka, Kiev, 331 pp. (in Russian). (English Translation: (1986). Translated from the Russian by A. Iacob, *Oper. Theory Adv. Appl.* **22**, Birkhäuser Verlag, Basel, xii+367 pp.)

[122] Marchenko, V. A. (1986). Non-linear equations and operator algebras, Naukova Dumka, Kiev, 155 pp. (in Russian). (English Translation: (1988). Nonlinear equations and operator algebras, Translated from the Russian by V. I. Rublinetskiy, *Mathematics and its Applications (Soviet Series)* **17**, D. Reidel Publishing Co., Dordrecht, xvi+157 pp.)

[123] Marchenko, V. A. (2005). Introduction to the theory of inverse problems of spectral analysis, Acta, Kharkov. (in Russian), to appear.

[124] Marchenko, V. A. and Khruslov, E. Ya. (1974). Boundary value problems in domains with a fine-grained boundary, Naukova Dumka, Kiev, 280 pp. (in Russian).

[125] Marchenko, V. A. and Khruslov, E. Ya. (2005). Homogenization of partial differential equations, Birkhäuser Verlag, to appear.

[126] Marchenko, V. A. and Khruslov, E. Ya. (2005). Homogenized models of microinhomogeneous media, Naukova Dumka, Kiev, (in Russian), to appear.

[127] Martynyuk, A. A. and Gutovski, R. (1977). Integral inequalities and stability of motion, Naukova Dumka, Kiev, 272 pp. (in Russian).

[128] Maslov, V. P. (1965). WKB method in many-dimensional case. Appendix to the book by J. Heading "Introduction to pahse-integral methods," Methuen and Co., Ltd., London, John Wiley and Sons, Inc., New York (1962); the Appendix published by Mir, Moscow, pp. 177–237. (in Russian).

[129] Maslov, V. P. (1973). Operator methods, Nauka, Moscow, 544 pp. (in Russian). (French Translation: (1987). Translated from the Russian by Djilali Embarek, Editions Mir, Moscow, 708 pp.) (Spanish Translation: (1982). Translated from the Russian by Andrés Fraguela Collar, Mir, Moscow, 621 pp.)

[130] Maslov, V. P. (1987). Asymptotic methods for solving pseudodifferential equations, Nauka, Moscow, 408 pp. (in Russian).

[131] Maslov, V. P. (1988). Asymptotic methods and perturbation theory, Nauka, Moscow, 311 pp. (in Russian).

[132] Massera, J. and Schäffer, J. (1966). Linear differential equations and function spaces, *Pure and Applied Mathematics*, Vol. 21, Academic Press, New York-London, xx+404 pp. (Russian Translation: (1970). Translated from the English by A. M. Zverkin and G. A. Kamenskiy, Mir, 456 pp.)

[133] Matveev, V. B. and Salle, M. A. (1991). Darboux transformations and solitons, *Springer Series in Nonlinear Dynamics*, Springer-Verlag, Berlin, x+120 pp.

[134] Maz'ya (=Maz'ja), V. G. (1985). Sobolev spaces, Translated from the Russian by T. O. Shaposhnikova, *Springer Series in Soviet Mathematics*, Springer-Verlag, Berlin, xix+486 pp.

[135] Merkur'ev, S. P. and Faddeev, L. D. (1985). Quantum scattering theory for several particle systems, Nauka, Moscow, 400 pp. (in Russian). (English Translation: (1993). *Mathematical Physics and Applied Mathematics* **11**, Kluwer Academic Publishers Group, Dordrecht, xiv+404 pp.)

[136] Milnor, J. (1963). Morse theory, Based on lecture notes by M. Spivak and R. Wells, *Annals of Mathematics Studies*, No. 51, Princeton University Press, Princeton, N.J., vi+153 pp. (Russian Translation: (1965). Translated from the English by V. I. Arnol'd, Mir, Moscow, 184 pp.)

[137] Mishchenko, A. S., Sternin, B. Yu. and Shatalov, V. E. (1978). Lagrangian manifolds and the method of the canonical operator, Nauka, Moscow, 352 pp. (in Russian). (English Translation: (1990). With an appendix by V. E. Nazaikinskiy, V. G. Oshchmian, Sternin and Shatalov, Translated from the Russian by Dana Mackenzie, *Springer Series in Soviet Mathematics*, Springer-Verlag, Berlin, x+395 pp.)

[138] Mitropol'skiy, Yu. A., Samoilenko, A. M. and Martynyuk, D. I. (1984). Systems of evolution equations with periodic and quasiperiodic coefficients, Naukova Dumka, Kiev, 214 pp. (in Russian). (English Translation: (1993). Translated from the 1984 Russian original, *Mathematics and its Applications (Soviet Series)* **87**, Kluwer Academic Publishers Group, Dordrecht, xiv+280 pp.)

[139] Mizohata, S. (1965). The theory of partial differential equations, *Contemporary Mathematics*, No. 9, Iwanami Shoten, viii+462 pp. (in Japanese). (English Translation: (1973). Translated from the Japanese by Katsumi Miyahara, Cambridge University Press, New York, xii+490 pp.)

[140] Modevich, V. S., Nikulin, V. I. and Stetsenko, A. G. (1988). Dynamics of the interior intermixing in the stratified medium, Hydromechanics Inst. Acad. Sci. Ukrain. SSR, Naukova Dumka, Kiev. (in Russian).

[141] Morse, M. (1996). The calculus of variations in the large, Reprint of the 1932 original, *American Mathematical Society Colloquium Publications*, Vol. 18, American Mathematical Society, Providence, R.I., xii+368 pp.

[142] Murtazin, Kh. Kh. and Sadovnichiy, V. A. (1988). Spectral analysis of the multiparticle Schrödinger operator, Moscow University, Moscow. (in Russian).

[143] Muskhelishvili, N. I. (1962, 1968). Singular integral equations. Boundary problems of functions theory and their applications to mathematical physics; Second Edition, 1962; Third, corrected and augmented edition, With an appendix by B. Bojarski, Nauka, Moscow, 1968, 511 pp. (in Russian). (English Translation: (1972). Revised translation from the Russian, edited by J. R. M. Radok, Reprinted, Wolters-Noordhoff Publishing, Groningen, xii+7–447 pp.)

[144] Myshkis, A. D., Babskiy, V. G., Kopachevskiy, N. D., Slobozhanin, L. A. and Tyuptsov, A. D. (1987). Low-gravity fluid mechanics. Mathematical theory of capillary phenomena, Translated from the Russian by R. S. Wadhwa, Springer-Verlag, Berlin, xx+583 pp.

[145] Naimark, M. A. (1967). Linear differential operators, Part I: Elementary theory of linear differential operators, Frederick Ungar Publishing Co., New York, xiii+144 pp; (1968). Part II: Linear differential operators in Hilbert space, With additional material by the author, and a supplement by V. E. Ljance, Translated from the Russian by E. R. Dawson, English translation edited by W. N. Everitt Frederick Ungar Publishing Co., New York, xv+352 pp.

[146] Newton, R. G. (1982). Scattering theory of waves and particles, Second edition, *Texts and Monographs in Physics*, Springer-Verlag, New York-Berlin, xx+743 pp.

[147] Newton, R. G. (1989). Inverse Schrödinger scattering in three dimensions, *Texts and Monographs in Physics*, Springer-Verlag, Berlin, x+170 pp.

[148] Nikol'skiy, S. M. (1977). Approximation of functions of several variables and imbedding theorems, Second edition, revised and supplemented, Nauka, Moscow, 455 pp. (in Russian). (English Translation: (1975). Approximation of functions of several variables and imbedding theorems, Translated from the Russian by John M. Danskin, Jr, *Die Grundlehren der Mathematischen Wissenschaften*, Band 205, Springer-Verlag, New York-Heidelberg, viii+418 pp.)

[149] Novikov, S. P. and Fomenko, A. T. (1987). Elements of differential geometry and topology, Nauka, Moscow, 432 pp. (English Translation: (1990). Basic elements of differential geometry and topology, Translated from the Russian by M. V. Tsaplina, *Mathematics and its Applications (Soviet Series)* **60**, Kluwer Academic Publishers Group, Dordrecht, 1990. x+490 pp.)

[150] Oleinik, O. A., Iosifyan, G. A. and Shamaev, A. S. (1990). Mathematical problems in the theory of strongly non-homogeneous elastic media, Moscow University, Moscow, 312 pp. (in Russian).

[151] Otelbaev, M. O. (1990). Estimates of spectrum of the Sturm-Liouville operator (with appendices by Minkin, A. M., Shuster, L. A., Turmukhambetov,

D. N., Rofe-Beketov, F. S. and Grinshpun, E. Z.), Gylym, Alma-Ata, 192 pp. (in Russian).

[152] Pankov, A. A. (1985). Bounded and almost periodic solutions of nonlinear operator-differential equations, Naukova Dumka, Kiev, 182 pp. (in Russian). (English Translation: (1990). Translated from the Russian by V. S. Zajačkovskiy and the author, *Mathematics and its Applications (Soviet Series)* **55**, Kluwer Academic Publishers Group, Dordrecht, x+221 pp.)

[153] Pastur, L. A. (1987). Spectral properties of random and almost-periodic operators, *Mathematical physics reviews*, Vol. 6, *Soviet Sci. Rev. Sect. C Math. Phys. Rev.* **6**, Harwood Academic Publ., Chur, pp. 1–112.

[154] Pastur, L. A. and Figotin, A. L. (1991). Random and almost periodic self-adjoint operators. General spectral properties and distribution of eigenvalues, Nauka, Moscow, 336 pp. (in Russian).

[155] Pastur, L. A. and Figotin, A. L. (1992). Spectra of random and almost-periodic operators, *Grundlehren der Mathematischen Wissenschaften*, Vol. 297, Springer-Verlag, Berlin, viii+587 pp.

[156] Petrovskiy, I. G. (1961). Lectures on partial differential equations, Gosudarstv. Izdat. Fiz.-Mat. Lit., Moscow, Augmented third edition, 400 pp. (in Russian). (English Translation: (1967). Translated from the Russian by Scripta Technica, Ltd. Iliffe Books Ltd., London (distributed by W. B. Saunders Co., Philadelphia, Pa.).)

[157] Petrovskiy, I. G. (1984). Lectures on the theory of ordinary differential equations, Seventh edition, Edited and with a preface by A. D. Myshkis and O. A. Oleinik, Moskov. Gos. Univ., Moscow, 296 pp. (in Russian). (English Translation: (1966). Ordinary differential equations, Revised English edition, Translated from the Russian and edited by Richard A. Silverman, Prentice-Hall, Inc., Englewood Cliffs, N.J. x+232 pp.) (French Translation: (1988). Translated from the Russian by Irina Petrova, Mir, Moscow, 359 pp.)

[158] Plemelj, J. (1964). Problems in the sense of Riemann and Klein, Edited and translated by J. R. M. Radok, *Interscience Tracts in Pure and Applied Mathematics*, No. 16, Interscience Publishers John Wiley and Sons Inc., New York-London-Sydney, vii+175 pp.

[159] Postnikov, M. M. (1965). Variational theory of geodesics, Nauka, Moscow, 248 pp. (in Russian). (English Translation: (1967). Translated from the Russian by Scripta Technica, Inc. Translation edited by Bernard R. Gelbaum W. B. Saunders Co., Philadelphia, Pa.-London, viii+200 pp.)

[160] Postnikov, M. M. (1971). Introduction to Morse theory, Nauka, Moscow, 568 pp. (in Russian).

[161] Prilepko, A. I., Orlovskiy, D. G. and Vasin, I. A. (2000). Methods for solving inverse problems in mathematical physics, *Monographs and Textbooks in Pure and Applied Mathematics* **231**, Marcel Dekker, Inc., New York, xiv+709 pp.

[162] Ptashnik, B. I. (1984). Ill-posed boundary value problems for partial differential equations, Naukova Dumka, Kiev, 264 pp. (in Russian).

[163] Ramm, A. G. (1992). Multidimensional inverse scattering problems, *Pitman*

Monographs and Surveys in Pure and Applied Mathematics **51**, Longman Scientific and Technical, Harlow; copublished in the United States with John Wiley and Sons, Inc., New York, viii+379 pp. (Expanded Russian Edition: (1994). Mir, Moscow.)

[164] Rapoport, I. M. (1954). On some asymptotic methods in the theory of differential equations, Izdat. Akad. Nauk Ukrain. SSR, Kiev, 292 pp. (in Russian).

[165] Reed, M. and Simon, B. (1972). Methods of modern mathematical physics I. Functional analysis, Academic Press, Inc. [Harcourt Brace Jovanovich, Publishers], New York, xv+400 pp. (Russian Translation: (1977). Translated from the English by A. K. Pogrebkov and V. N. Sushko, With a preface by N. N. Bogolyubov, Edited by M. K. Polivanov, Mir, Moscow, 357 pp.)

[166] Reed, M. and Simon, B. (1975). Methods of modern mathematical physics II. Fourier analysis, self-adjointness, Academic Press [Harcourt Brace Jovanovich, Publishers], New York-London, xv+361 pp. (Russian Translation: (1978). Translated from the English by A. K. Pogrebkov and V. N. Sushko, Edited by M. K. Polivanov, Mir, Moscow, 395 pp.)

[167] Reed, M. and Simon, B. (1978). Methods of modern mathematical physics IV. Analysis of operators, Academic Press [Harcourt Brace Jovanovich, Publishers], New York-London, xv+396pp. (Russian Translation: (1982). Translated from the English by A. K. Pogrebkov and V. N. Sushko, Mir, Moscow, 430 pp.)

[168] Reed, M. and Simon, B. (1979). Methods of modern mathematical physics III. Scattering theory, Academic Press [Harcourt Brace Jovanovich, Publishers], New York-London, xv+463 pp. (Russian Translation: (1982). Translated from the English by A. K. Pogrebkov and V. N. Sushko, Mir, Moscow, 445 pp.)

[169] Riesz, F. and Sz.-Nagy, B. (1965). Leçons d'analyse fonctionnelle, Quatrième édition, Académie des Sciences de Hongrie Gauthier-Villars, Editeur-Imprimeur-Libraire, Paris; Akadémiai Kiadó, Budapest, viii+490 pp. (in French). (English Translation: (1990). Functional Analysis, Translated from the second French edition by Leo F. Boron, Reprint of the 1955 original, Dover Books on Advanced Mathematics, Dover Publications, Inc., New York, xii+504 pp.) (German Translation: (1956). Translated by Siegfried Brehmer and Brigitte Mai, *Hochschulbücher für Mathematik*, Bd. 27, VEB Deutscher Verlag der Wissenschaften, Berlin, xi+482 pp.) (Russian Translation: (1979). Translated from the French by D. A. Vasil'kov, Second edition, revised and supplemented by S. A. Teljakovskiy, Mir, Moscow, 590 pp.)

[170] Roseau, F. (1984). Vibrations des systèmes mécaniques. Méthodes analytiques et applications, Masson, Paris, 489 pp. (in French). (English Translation: (1987). Vibrations in mechanical systems. Analytical methods and applications, Translated from the French by H. L. S. Orde, Springer-Verlag, Berlin, xiv+515 pp.)

[171] Sakhnovich, L. A. (1997). Interpolation theory and its applications, *Math-*

ematics and its Applications, Vol. 428, Kluwer Academic Publishers, Dordrecht, xviii+197 pp.

[172] Sakhnovich, L. A. (1999). Spectral theory of canonical differential systems. Method of operator identities, Translated from the Russian manuscript by E. Melnichenko, *Oper. Theory Adv. App.* **107**, Birkhäuser Verlag, Basel, vi+202 pp.

[173] Samarskiy, A. A., Galaktionov, V. A., Kurdyumov, S. P. and Mikhailov, A. P. (1995). Blow-up in quasilinear parabolic equations, Translated from the 1987 Russian original by Michael Grinfeld and revised by the authors, *de Gruyter Expositions in Mathematics* **19**, Walter de Gruyter and Co., Berlin, xxii+535 pp.

[174] Samoilenko, A. M. , Shkil', M. I. and Yakovets, V. P. (2000). Linear systems of differential equations with degenerations, Vyshcha Shkola, Kyiv, 294 pp. (in Ukrainian).

[175] Sánchez-Palencia, E. (1980). Nonhomogeneous media and vibration theory, *Lecture Notes in Physics* **127**, Springer-Verlag, Berlin-New York, ix+398 pp.

[176] Sansone, G. (1948). Equazioni Differenziali nel Campo Reale, Vol. 1, 2d ed, Nicola Zanichelli, Bologna, xvii+400 pp; (1949). Vol. 2, 2d ed, Nicola Zanichelli, Bologna, xvi+475 pp. (in Italian). (Russian Translation: (1953). vol 1, Foreign Literature Press, Moscow, 347 pp; (1954). vol 2, Foreign Literature Press, Moscow, 416pp.)

[177] Schmid, E. W. and Ziegelmann, H. (1974). The quantum mechanical three-body problem, *Vieweg Tracts in Pure and Applied Physics* **2**, Pergamon Press, New York, 217 pp. (Russian Translation: (1979). Nauka, Moscow.)

[178] Shevelo, V. N. (1978). Oscillation of solutions of differential equations with deviating argument, Naukova Dumka, Kiev, 155 pp. (in Russian).

[179] Shkil', N. I., Starun, I. I. and Yakovets, V. P. (1989). Asymptotic integration of linear systems of ordinary differential equations, Vyshcha Shkola, Kiev, 287 pp. (in Russian).

[180] Showalter, R. E. (1997). Monotone operators in Banach space and nonlinear partial differential equations, *Mathematical Surveys and Monographs* **49**, American Mathematical Society, Providence, RI, xiv+278 pp.

[181] Simon, B. (2004). Orthogonal polynomials on the unit circle, Part 1: Classical theory, Part 2: Spectral theory, *American Mathematical Society Colloquium Publications*, Vol. 54, American Mathematical Society, Providence, RI, 1044 pp.

[182] Skorobogat'ko, V. Ya. (1980). Investigation in the qualitative theory of partial differential equations, Naukova Dumka, Kiev, 244 pp. (in Russian).

[183] Skrypnik, I. V. (1994). Methods for analysis of nonlinear elliptic boundary value problems, Translated from the 1990 Russian original by Dan D. Pascali, *Translations of Mathematical Monographs*, Vol. 139, American Mathematical Society, Providence, RI, xii+348 pp.

[184] Sobolev, S. L. (1950). Some applications of functional analysis in mathematical physics, Izdat. Leningrad. Gos. Univ., Leningrad, 255 pp. (in Russian). (English Translation: (1963). Applications of functional analysis in math-

ematical physics, Translated from the Russian by F. E. Browder, *Translations of Mathematical Monographs*, Vol. 7, American Mathematical Society, Providence, R.I. vii+239 pp.)

[185] Stepanov, V. V. (1950). A course in differential equations, Gosudarstv. Izdat. Tehn.-Teor. Lit., Moscow-Leningrad, 468 pp. (in Russian).

[186] Strizhak, T. G. (1984). An asymptotic normalization method, Vyshcha Shkola, Kiev, 280 pp. (in Russian).

[187] Swanson, C. A. (1968). Comparison and oscillation theory of linear differential equations, *Mathematics in Science and Engineering*, Vol. 48, Academic Press, New York-London, viii+227 pp.

[188] Sz.-Nagy, B. and Foiaş, C. (1967). Analyse harmonique des opérateurs de l'espace de Hilbert, Masson et Cie, Paris; Akadémiai Kiadó, Budapest, xi+373 pp. (in French). (English Translation: (1970). Harmonic analysis of operators on Hilbert space, Translated from the French and revised North-Holland Publishing Co., Amsterdam-London; American Elsevier Publishing Co., Inc., New York; Akadémiai Kiadó, Budapest, xiii+389 pp.) (Russian Translation: (1970). Translated from the French by Jr. L. Šmul'jan, Edited by Ju. P. Ginzburg, With a foreword by M. G. Krein, Mir, Moscow 431 pp.)

[189] Takhtajan, L. A. and Faddeev, L. D. (1986). Hamiltonian methods in the theory of solitons, Nauka, Moscow, 528pp. (in Russian). (English Translation: (1987). Translated from the Russian by A. G. Reyman, *Springer Series in Soviet Mathematics*, Springer-Verlag, Berlin, x+592 pp.)

[190] Timoshenko, S. P. (1975). Strength and vibrations of elementary constructions, Nauka, Moscow. (in Russian).

[191] Titchmarsh, E. Ch. (1962). Eigenfunction expansions associated with second-order differential equations, Vol. 1, Second Edition, Clarendon Press, Oxford, vi+203 pp; (1958). Vol. 2, Clarendon Press, Oxford, xi+404 pp. (Russian Translation: Translated from the English by V. B. Lidskiy, edited by B. M. Levitan, with appendices by B. M. Levitan, V. B. Lidskiy, and A. I. Baz', Izdat. Inostran. Lit., Moscow, (1960). Vol. 1, 278 pp; (1961). Vol. 2, 556 pp.)

[192] Tricomi, F. (1961). Equazioni differenziali, 3a ed. riveduta e ampliata, Manuali: Serie di matematica Paolo Boringhieri, Torino, 351 pp. (in Italian). (English Translation: (1961). Differential equations, Translated by Elizabeth A. McHarg, Hafner Publishing Co., New York, x+273 pp.) (Russian Translation: (1962). Foreign Literarure Press, Moscow.)

[193] Vekua, N. P. (1970). Systems of singular integral equations and certain boundary value problems, Second revised and supplemented edition, Nauka, Moscow, 379 pp. (in Russian). (English Translation: (1967). Translated from an earlier Russian edition by A. G. Gibbs and G. M. Simmons, Edited by J. H. Ferziger P. Noordhoff, Ltd., Groningen, 216 pp.)

[194] Vishik, M. I. and Fursikov, A. V. (1980). Mathematical problems of statistical hydromechanics, Nauka, Moscow, 440 pp. (in Russian).

[195] Wasow, W. (1987). Asymptotic expansions for ordinary differential equations, Reprint of the 1976 edition, Dover Publications, Inc., New York,

x+374 pp. (Russian Translation: (1968). Mir, Moscow.)

[196] Weidmann, J. (1987). Spectral theory of ordinary differential operators, *Lecture Notes in Math.* **1258**, Springer Verlag, Berlin.

[197] Weidmann, J. (2003). Lineare Operatoren in Hilberträumen, Teil II: Anwendungen, B. G. Teubner, Stuttgart, 475 pp. (in German).

[198] Whittaker, E. T. and Watson, G. N. (1996). A course of modern analysis, An introduction to the general theory of infinite processes and of analytic functions, with an account of the principal transcendental functions, Reprint of the fourth (1927) edition, *Cambridge Mathematical Library*, Cambridge University Press, Cambridge, vi+608 pp. (Russian Translation: (1963). Part II, Gosudarstv. Izdat. Fiz.-Mat. Lit., Moscow, 515 pp.)

[199] Yakubovich, V. A. and Starzhinskiy, V. M. (1972). Linear differential equations with periodic conditions and their applications, Nauka, Moscow, 720 pp. (in Russian). (English Translation: (1975). Linear differential equations with periodic coefficients. 1, 2. Translated from Russian by D. Louvish, Halsted Press [John Wiley and Sons] New York-Toronto, Ont.; Israel Program for Scientific Translations, Jerusalem-London, Vol. 1: xii+386 pp.; Vol. 2: xii+pp. 387–839.)

[200] Yakubovich, V. A. and Starzhinskiy, V. M. (1987). Parametric resonance in linear systems, Nauka, Moscow, 328 pp. (in Russian).

[201] Yosida, K. (1995). Functional analysis, Reprint of the sixth (1980) edition, *Classics in Mathematics*, Springer-Verlag, Berlin, xii+501 pp. (Russian Translation: (1967). Mir, Moscow, 624 pp.)

[202] Yurko, V. A. (1989). The inverse problem for differential operators, Saratov University Press, 175 pp. (in Russian).

[203] Yurko, V. A. (2000). Inverse spectral problems for differential operators and their applications, Analytical Methods and Special Functions, 2, Gordon and Breach Science Publishers, Amsterdam, 2000, xiv+253 pp.

[204] Yurko, V. A. (2001). The inverse spectral problems and their applications, Saratov Pedagogical Institute Press, 499 pp. (in Russian).

[205] Yurko, V. A. (2002). Method of spectral mappings in the inverse problems theory, *Inverse and Ill-posed Problems Series*, VSP, Utrecht.

[206] Zakharov, V. E., Manakov, S. V., Novikov, S. P. and Pitaevskiy, L. P. (ed. Novikov, S. P.) (1980). Theory of solitons. The inverse scattering method, Nauka, Moscow, 320 pp. (in Russian). (English Translation: (1984). *Contemporary Soviet Mathematics*, Consultants Bureau [Plenum], New York, xi+276 pp.)

[207] Zhikov, V. V., Kozlov, S. M. and Oleinik, O. A. (1994). Homogenization of differential operators and integral functionals, Translated from the Russian by G. A. Yosifian, Springer-Verlag, Berlin, xii+570 pp.

[208] Ziman, J. M. (1960). Electrons and phonons, *The International Series of Monographs on Physics*, Oxford University Press, London, xiv+554 pp. (Russian Translation: (1962). Foreign Literature Press, Moscow).

[209] Zolotarev, V. A. (2003). Analytic methods of spectral representations of non-self-adjoint and non-unitary operators, Kharkov National University, 342 pp. (in Russian).

Articles

[210] Abramov, A. A., Balla, K., Ul'yanova, V. I. and Yukhno, L. F.(2003). On a nonlinear selfadjoint spectral problem for a class of differential algebraic equations, *Zh. Vychisl. Mat. Mat. Fiz.* **43**, no. 3, pp. 410–421. (in Russian). (English Translation: (2003). *Comput. Math. Math. Phys.* **43**, no. 3, pp. 391–402.)

[211] Akhiezer, A. I., Belozorov, D. P., Rofe-Beletov, F. S., Davydov, L. N. and Spolnik, Z. A. (1999). On the theory of propagation of chain nuclear reaction, *Physica A* **273**, pp. 272–285.

[212] Akhiezer, A. I., Belozorov, D. P., Rofe-Beketov, F. S., Davydov, L. N. and Spolnik, Z. A. (1999). On the theory of propagation of chain nuclear reaction in diffusion approximation, *Yadernaya Fizika* **62**, No. 9, pp. 1567–1575. (in Russian). (English Translation: (1999). *Physics of Atomic Nuclei* **62**, No. 9, pp. 1474–1481.)

[213] Akhiezer, A. I., Belozorov, D. P., Rofe-Beketov, F. S., Davydov, L. N. and Spolnik, Z. A. (2001). The velocity of slow nuclear burning in the two-group approximation, *Problems of Atomic Science and Technology (Special issue dedicated to the 90-th birthday anniversary of A. I. Akhiezer)*, No. 6(2), pp. 276–278.

[214] Akhiezer, A. I., Davydov, L. N. and Peletminskiy, S. V. (2000). On the theory of magnetohydrodynamic waves in nuclear matter, *Dopov. Nats. Akad. Nauk Ukr. (Mat. Prirodozn. Tekh. Nauki)*, No. 7, pp. 88–92.

[215] Akhiezer, A. I., Davydov, L. N. and Peletminskiy, S. V. (2000). On the kinetic theory of the magnetohydrodynamics of neutron matter, *Teoret. Mat. Fiz.* **125**, no. 3, pp. 471–490. (in Russian). (English Translation: (2000). *Theoret. and Math. Phys.* **125**, no. 3, pp. 1702–1720.)

[216] Akhiezer, N. I. (1961). Continuous analogue of orthogonal polynomials on a system of intervals, *Dokl. Akad. Nauk SSSR* **141**, no. 2, pp. 262–266. (in Russian). (English Translation: (1961). *Soviet Math. Dokl.* **2**, pp. 1409–1412.)

[217] Akhtyamov, A. M. (2004). On the coefficients of eigenfunction expansions of boundary value problems with a parameter in the boundary conditions, *Mat. Zametki* **75**, no. 4, pp. 493–506. (in Russian).

[218] Akkermans, E., Comtet, A., Desbois, J., Montambaux, G. and Texier, C. (2000). Spectral determinant on quantum graphs, *Ann. Physics* **284**, no. 1, pp. 10–51.

[219] Aleksandryan, R. A. (1960). Spectral properties of operators generated by a system of S. L. Sobolev-type differential equations, *Tr. Mosk. Mat. Obs.* **9**, pp. 455–505. (in Russian).

[220] Aleksandryan, R. A., Berezanskiy, Yu. M., Il'in, V. A. and Kostyuchenko, A. G. (1970). Some questions in spectral theory of partial differential operators, *Partial Differential Equations, Proceedings of the Conference on the Occasion of the 60-th Birthday of Academician S. L. Sobolev*, Nauka, Moscow, pp. 3–32. (in Russian).

[221] Allakhverdiev, B. P. (1995). On extensions of symmetric Schrödinger operators with matrix-valued potentials, *Izv. Ross. Akad. Nauk Ser. Mat.* **59**, no. 1, pp. 49–64. (in Russian). (English Translation: (1995). *Izv. Math.* **59**, no. 1, pp. 45–62.)

[222] Alonso, A. and Simon, B. (1980). The Birman-Krein-Vishik theory of selfadjoint extensions of semibounded operators, *J. Operator Theory* **4**, no. 2, pp. 251–270.

[223] Anders, I. A., Khruslov, E. Ya. and Kotlyarov, V. P. (1997). Soliton asymptotics of non-localized solutions of non-linear evolutionary equations, *Proceedings of the Second World Congress of Nonlinear Analysts, Part 7 (Athens, 1996)*, *Nonlinear Anal.* **30**, no. 7, pp. 3951–3961.

[224] Anders, I. A. and Kotlyarov, V. P. (1991). Characterization of scattering data for Schrödinger and Dirac operators, *Teoret. Mat. Fiz.* **88**, no. 1, pp. 72–84 (in Russian). (English Translation: (1992). *Theoret. and Math. Phys.* **88**, no. 1, pp. 725–734.)

[225] Ando, T. and Nishio, K. (1970). Positive selfadjoint extensions of positive symmetric operators, *Tohoku Math. J.* **22**, no. 1, pp. 65–75.

[226] Androshchuk, A. A. (1971). Transformation operators and a uniqueness theorem in the inverse problem for a second order differential equation with operator-valued coefficients, *Dokl. Akad. Nauk SSSR* **198**, no. 1, pp. 9–12. (in Russian). (English Translation: (1971). *Soviet Math. Dokl.* **12**, pp. 687–691.)

[227] Anikeeva, L. I. (1977). The deficiency index of a certain higher order differential operator, *Uspekhi Mat. Nauk* **32**, no. 1 (193), pp. 179–180. (in Russian).

[228] Ao, P. (1990). Abscence of localization in energy space of a Bloch electron driven by a constant electric force, *Phys. Rev. B.* **41**, pp. 2998–4001.

[229] Arens, R. (1961). Operational calculus of linear relations, *Pacific J. Math.* **11**, no. 1, pp. 9–23.

[230] Arkeryd, L. and Nouri, A. (1999). On the stationary Povzner equation in R^n, *J. Math. Kyoto Univ.* **39**, no. 1, pp. 115–153.

[231] Arlinskiy, Yu. M. (1988). Positive spaces of boundary values and sectorial extensions of a nonnegative symmetric operator, *Ukraïn. Mat. Zh.* **40**, no. 1, pp. 8–14. (in Russian). (English Translation (1988). *Ukrainian Math. J.* **40**, no. 1, pp. 5–10.)

[232] Arlinskiy, Yu. M. (1998). On m-accretive extensions and restrictions, *Methods Funct. Anal. and Topology* **4**, no. 3, pp. 1–26.

[233] Arlinskiy, Yu. M. (1999). On functions connected with sectorial operators and their extensions, *Integral Equations Operator Theory* **33**, pp. 125–152.

[234] Arlinskiy, Yu. M. (2000). Abstract boundary conditions for maximal sectorial extensions of sectorial operators, *Math. Nachr.* **209**, pp. 5–36.

[235] Arlinskiy, Yu. M. (2000). M-accretive extensions of sectorial operators and Krein spaces, *Oper. Theory Adv. Appl.* **118**, pp. 67–82.

[236] Arlinskiy, Yu. M., Hassi, S., Sebestyén, Z. and de Snoo, H. S. V. (2001). On the class of extremal extensions of a nonnegative operator, *Recent advances in operator theory and related topics (Szeged, 1999)*, *Oper. Theory Adv.*

Appl. **127**, Birkhäuser, Basel, pp. 41–81.

[237] Arnold, V. I. (1967). On a characteristic class entering into conditions of quantization, *Funktsional Anal. i Prilozhen.* **1**, no. 1, pp. 1–14. (in Russian). (English Translation: (1967). *Funct. Anal. Appl.* **1**, pp. 1–13.)

[238] Arnold, V. I. (1985). Sturm theorems and symplectic geometry, *Funktsional Anal. i Prilozhen.* **19**, no. 4, pp. 1–10. (in Russian). (English Translation: (1985). *Funct. Anal. Appl.* **19**, pp. 251–259.)

[239] Arnold, V. I. and Ilyashenko, Yu. S. (1985). Ordinary differential equations, *Current problems in mathematics. Fundamental directions*, Vol. 1, *Itogi Nauki i Tekhniki*, Akad. Nauk SSSR, Vsesoyuz. Inst. Nauchn. i Tekhn. Inform., Moscow, pp. 7–149. (in Russian). (English Translation: (1988). Translated from the Russian by E. R. Dawson and D. O'Shea, *Encyclopaedia Math. Sci., 1, Dynamical systems*, I, pp. 1–148, Springer, Berlin.)

[240] Arnold, V., Kalf, H. and Schneider, A. (1997). Separated Dirac operators and asymptotically constant linear systems, *Math. Proc. Cambridge Philos. Soc.* **121**, no. 1, pp. 141–146.

[241] Arov, D. Z. (1979). Passive linear steady-state dynamical systems, *Sibirsk. Mat. Zh.* **20**, no. 2, pp. 211–228. (in Russian). (English Translation: (1979). *Siberian Math. J.* **20**, pp. 149–162.)

[242] Atkinson, F. V., Everitt, W. N. and Ong, K. S. (1974). On the m-coefficient of Weyl for a differential equation with an indefinite weight function, *Proc. London Math. Soc. (3)* **29**, pp. 368–384.

[243] Atkinson, F. V., Langer, G., Mennicken, R. and Shkalikov, A. A. (1994). The essential spectrum of some matrix operators, *Math. Nachr.* **167**, pp. 5–20.

[244] Avron, J., Gunter, L. and Zak, J. (1975). Energy uncertainty in "Stark ladder," *Solid State Commun.* **16**, no. 2, pp. 189–191.

[245] Avron, J. and Simon, B. (1981/82). Almost periodic Schrödinger operators. I. Limit periodic potentials, *Comm. Math. Phys.* **82**, no. 1, pp. 101–120.

[246] Bailey P. B., Everitt, W. N. and Zettl, A. (2001). The SLEIGN2 Sturm–Liouville code, *ACM Trans. Math. Software* **27**, pp. 143–192.

[247] Bakh, S. N. and Pankov, A. A. (2004). On periodic oscillations of infinite chain of linear connected non-linear oscillators, *Dopov. Nats. Akad. Nauk Ukr. (Mat. Prirodozn. Tekh. Nauki)*, No. 9, pp. 13–16. (in Russian).

[248] Baskakov, A. G. and Chernyshov, K. I. (2002). Linear relations, differential inclusions, and degenerate semigroups, *Funktsional. Anal. i Prilozhen.* **36**, no. 4, pp. 65–70. (in Russian). (English Translation: (2002). *Funct. Anal. Appl.* **36**, no. 4, pp. 306–310.)

[249] Batchenko, V. and Gesztesy, F. (2003). The spectrum of Schrödinger operators with quasi-periodic algebro-geometric KdV potentials, *Mat. Fiz. Anal. Geom.* **10**, no. 4, pp. 447–468.

[250] Batchenko, V. and Gesztesy, F. (2003). The spectrum of Schrödinger operators with quasi-periodic algebro-geometric KdV potentials, Preprint.

[251] Bayasgalan, Ts. (2000). The numerical range of linear relations in spaces with an indefinite metric, *Mat. Fiz. Anal. Geom.* **7**, no. 1, pp. 115–118.

[252] Bayramogly, M. (1976). Selfadjoint extensions of an operator equation with a singularity, *Izv. Akad. Nauk Azerbaidzhan. SSR Ser. Fiz.-Tehn. Mat.*

Nauk **1976**, no. 2, pp. 140–143. (in Russian).

[253] Belokolos, E. D. (1976). A quantum particle in a one-dimensional deformed lattice. Dependence of the energy on the quasimomentum, *Teoret. Mat. Fiz.* **26**, no. 1, pp. 35–41. (in Russian). (English Translation: (1976). *Theoret. and Math. Phys.* **26**, no. 1, pp. 21–25.)

[254] Belokolos, E. D., Bobenko, A. I., Matveev, V. B. and Enol'skiy, V. Z. (1986). Algebro-geometric superposition principles of finite-zone solutions of integrable non-linear equations, *Uspekhi Mat. Nauk* **41**, no. 2, pp. 3–42. (in Russian). (English Translation: (1986). *Russian Math. Surveys* **41**, no. 2, pp. 1–49.)

[255] Belokolos E. D., Gesztesy F., Makarov K. A. and Sakhnovich L. A. (2002). Matrix-valued generalizations of the theorems of Borg and Hochstadt, Preprint, arXiv:math.SP/0201019 v1 3 Jan 2002, 28 pp.

[256] Ben Amara, Zh. and Shkalikov, A. A. (1999). The Sturm-Liouville problem with physical and spectral parameters in the boundary condition, *Mat. Zametki* **66**, no. 2, pp. 163–172. (in Russian). (English Translation: (1999). *Math. Notes* **66**, no. 1-2, pp. 127–134.)

[257] Ben Amara, Zh. and Vladimirov, A. A. (2004). On a fourth-order problem with spectral and physical parameters in the boundary condition, *Izv. Ross. Akad. Nauk Ser. Mat.* **68**, no. 4, pp. 3–18. (in Russian).

[258] Benderskiy, M. M. and Pastur, L. A. (1970). The spectrum of the one-dimensional Schrödinger operator with random potential, *Mat. Sb.* **82**, no. 2, pp. 273–284. (in Russian). (English Translation: (1970). *Sb. Math.* **11**, pp. 245–256.)

[259] Bennewitz, Ch. (1972). Symmetric relations on a Hilbert space, *Conference on the Theory of Ordinary and Partial Differential Equations (Univ. Dundee, Dundee, 1972), Lecture Notes in Math.*, Vol. 280, Springer, Berlin, pp. 212–218.

[260] Bentosela, F., Carmona, R., Duclos, P., Simon, B., Souillard, B. and Weder, R. (1983). Schrödinger operators with an electric field and random or deterministic potentials, *Comm. Math. Phys.* **88**, no. 3, pp. 387–397.

[261] Bentosela, F., Briet, Ph. and Pastur, L. (2004). Spectral analysis of the generalized surface Maryland model, *Algebra i Analiz* **16**, no. 6, pp. 1–27.

[262] Berezanskiy, Yu. M. (1956). Expansion in eigenfunctions of second order partial difference equations, *Tr. Mosk. Mat. Obs.* **5**, pp. 203–268. (in Russian).

[263] Berezanskiy, Yu. M. (1956). On expansion according to eigenfunctions of general self-adjoint differential operators, *Dokl. Akad. Nauk SSSR (N.S.)* **108**, pp. 379–382. (in Russian).

[264] Berezanskiy, Yu. M. (1958). The uniqueness theorem in the inverse problem of spectral analysis for the Schrödinger equation, *Tr. Mosk. Mat. Obs.* **7**, pp. 1–62. (in Russian). (English Translation: (1964). *Amer. Math. Soc. Transl. Ser. II* **35**, pp. 167–235.)

[265] Berezanskiy, Yu. M. (1985). Integration of nonlinear difference equations by the method of the inverse spectral problem, *Dokl. Akad. Nauk SSSR* **281**, no. 1, pp. 16–19. (in Russian). (English Translation: (1985). *Soviet Math.*

Dokl. **31**, no. 2, pp. 264–267.)

[266] Berezanskiy, Yu. M. (1986). The integration of semi-infinite Toda chain by means of inverse spectral problem, *Rep. Math. Phys.* **24**, no. 1, pp. 21–47.

[267] Berezanskiy, Yu. and Brasche, J. (2002). Generalized selfadjoint operators and their singular perturbations, *Methods Funct. Anal. Topology* **8**, no. 4, pp. 1–14.

[268] Berezanskiy, Yu. M. and Samoilenko, V. G. (1981). Self-adjointness of differential operators with a finite and infinite number of variables and evolution equations, *Uspekhi Mat. Nauk* **36**, no. 5, pp. 3–56. (in Russian). (English Translation: (1981). *Russian Math. Surveys* **36**, no. 5, pp. 1–62.)

[269] Berezanskiy, Yu. M. and Zhernakov, N. V. (1987). Integration of the Toda lattice in the class of Hilbert-Schmidt operators by means of inverse spectral problem, *Uspekhi Mat. Nauk* **42**, no. 4, pp. 171–172. (in Russian).

[270] Berezin, F. A. and Faddeev, L. D. (1961). Remark on the Schrödinger equation with singular potential, *Dokl. Akad. Nauk SSSR* **137**, no. 5, pp. 1011–1014. (in Russian). (English Translation: (1961). *Soviet Math. Dokl.* **2**, pp. 372–375.)

[271] Binding, P. A., Brown, P. J. and Seddighi, K. (1994). Sturm-Liouville problems with eigenparameter dependent boundary conditions, *Proc. Edinburgh Math. Soc. (2)* **37**, no. 1, pp. 57–72.

[272] Birger, E. S. and Kalyabin, G. A. (1976). Theory of Weyl circles in the case of a nonselfadjoint system of second order differential equations, *Differ. Uravn.* **12**, no. 9, pp. 1531–1540. (in Russian). (English Translation: (1976). *Differ. Equ.* **12**, no. 9, pp. 1077–1084.)

[273] Birkhoff, G. D. (1913). The generalized Riemann problem for linear differential equations and the allied problems for linear difference and q-difference equations, *Proc. Natl. Acad. Sci. USA* **49**, pp. 521–568; *Bull. Amer. Math. Soc. (2)* **19**, pp. 508–509.

[274] Birkhoff, G. D. and Hestens, M. R. (1935). Natural isoperimetric conditions in the calculus of variations, *Duke Math. J.* **1**, pp. 198–286.

[275] Birman, M. Sh. (1956). On the theory of self-adjoint extensions of positive definite operators, *Mat. Sb.* **38**, no. 4, pp. 431–450. (in Russian).

[276] Birman, M. Sh. (1961). On the spectrum of singular boundary problems, *Mat. Sb.* **55**, no. 2, pp. 125–174. (in Russian). (English Translation: (1966). *Amer. Math. Soc. Transl. Ser. II* **53**, pp. 23–80.)

[277] Birman, M. Sh. (1991). On the discrete spectrum in spectral gaps of a perturbed periodic second order operator, *Funktsional Anal. i Prilozhen.* **25**, no. 2, pp. 89–92. (in Russian). (English Translation: (1991). *Funct. Anal. Appl.* **25**, no. 2, pp. 158–161.)

[278] Birman, M. Sh. (1996). Discrete spectrum of the Schrödinger operator perturbed by a decreasing potential, *Algebra i Analiz* **8**, no. 1, pp. 3–20. (in Russian). (English Translation: (1996). *St. Petersburg Math. J.* **8**, no. 1, pp. 1–14.)

[279] Birman, M. Sh. and Glazman, I. M. (1964). Spectra of singular differential operators, *Trudy IV vsesoyuznogo sjezda*, No. 2, pp. 253–261. (in Russian).

[280] Birman, M. Sh., Laptev, A. and Suslina, T. A. (2000). Discrete spectrum of

a two-dimensional elliptic second-order operator perturbed by a decreasing potential. I. Semi-infinite spectral gap, *Algebra i Analiz* **12**, no. 4, pp. 36–78. (in Russian). (English Translation: (2000). *St. Petersburg Math. J.* **12**, no. 4, pp. 535–567.)

[281] Birman, M. Sh. and Suslina, T. A. (2003). Second order periodic differential operators. Threshold properties and homogenization, *Algebra i Analiz* **15**, no. 5, pp. 1–108. (in Russian). (English Translation: (2004). *St. Petersburg Math. J.* **15**, no. 5, pp. 639–714.)

[282] Birman, M. Sh., Suslina, T. A. and Shternberg, R. G. (2000). Absolute continuity of the two-dimensional Schrödinger operator with delta-potential concentrated on a periodic system of curves, *Algebra i Analiz* **12**, no. 6, pp. 140–177. (in Russian). (English Translation: (2000). *St. Petersburg Math. J.* **12**, no. 6, pp. 983–1012.)

[283] Birnir, B. (1986). Complex Hill's equation and the complex periodic Korteweg-de Vries equations, *Comm. Pure Appl. Math.* **39**, no. 1, pp. 1–49.

[284] Birnir, B. (1986). Singularities of the complex Korteweg-de Vries flows, *Comm. Pure Appl. Math.* **39**, no. 3, pp. 283–305.

[285] Bliss, G. A. and Schoenberg, I. J. (1931). On separation, comparison and oscillation theorems for self-adjoint systems of linear second-order differential equations, *Amer. J. Math.* **53**, pp. 781–800.

[286] Blokh, A. Sh. (1953). On the determination of the differential equation by its spectral matrix-function, *Dokl. Akad. Nauk SSSR* **92**, no. 2, pp. 209–212. (in Russian).

[287] Bobylev, A. V. (1984). Exact solutions of the nonlinear Boltzmann equation and the theory of relaxation of a Maxwell gas, *Teoret. Mat. Fiz.* **60**, no. 2, pp. 280–310. (in Russian). (English Translation: (1984). *Theoret. and Math. Phys.* **60**, no. 2, pp. 820–841.)

[288] Bobylev, A. V. and Cercignani, C. (2002). Exact eternal solutions of the Boltzmann equation, *J. Statist. Phys.* **106**, no. 5-6, pp. 1019–1038.

[289] Bobylev, A. V. and Cercignani, C. (2002). Self-similar solutions of the Boltzmann equation and their applications, *J. Statist. Phys.* **106**, no. 5-6, pp. 1039–1071.

[290] Bolibrukh, A. A. (1989). On the Riemann-Hilbert problem for the system of three equations, *Modern mathematics problems in physics and mechanics problems*, Moskov. Fiz.-Tekhn. Inst., Moscow, pp. 10–24. (in Russian).

[291] Bolibrukh, A. A. (1989). The Riemann-Hilbert problem on the complex projective line, *Mat. Zametki* **46**, no. 3, pp. 118–120. (in Russian).

[292] Bolibrukh, A. A. (1990). The Riemann-Hilbert problem, *Uspekhi Mat. Nauk* **45**, no. 2(272), pp. 3–47. (in Russian). (English Translation: (1990). *Russian Math. Surveys* **45**, no. 2, pp. 1–58.)

[293] Bolibrukh, A. A. (1992). Sufficient conditions for the positive solvability of the Riemann-Hilbert problem, *Mat. Zametki* **51**, no. 2, pp. 9–19. (in Russian). (English Translation: (1992). *Math. Notes* **51**, no. 1-2, pp. 110–117.)

[294] Bolibrukh, A. A. (2003). The monodromy inverse problems of the analytical theory of differential equations, *Mathematical events of the XX century*,

Fazis, Moscow, pp. 53–59. (in Russian).

[295] Bolibrukh, A. A., Its, A. R. and Kapaev, A. A. (2004). On the Riemann-Hilbert-Birkhoff inverse monodromy problem and the Painlevé equations, *Algebra i Analiz* **16**, no. 1, pp. 121–162.

[296] Borg, G. (1946). Eine Umkehrung der Sturm-Liouvilleschen Eigenwertaufgabe. Bestimmung der Differentialgleichung durch die Eigenwerte, *Acta Math.* **78**, pp. 1–96. (in German).

[297] Bott, R. (1956). On the iterations of closed geodesics and the Sturm intersection theory, *Comm. Pure Appl. Math.* **9**, no. 2, pp. 171–206.

[298] Boutet de Monvel, A. and Egorova, I. E. (1997). On solutions of nonlinear Schrödinger equations with Cantor-type spectrum, *J. Anal. Math.* **72**, pp. 1–20.

[299] Boutet de Monvel, A., Khruslov, E. Ya. and Kotlyarov, V. P. (1998). The Cauchy problem for the sinh-Gordon equation and regular solitons, *Inverse Problems* **14**, no. 6, pp. 1403–1427.

[300] Boutet de Monvel, A., Khruslov, E. Ya. and Kotlyarov, V. P. (2002). Soliton asymptotics of rear part of non-localized solutions of the Kadomtsev-Petviashvili equation, *J. Nonlinear Math. Phys.* **9**, no. 1, pp. 58–76.

[301] Boutet de Monvel, A. and Kotlyarov, V. P. (2000). Scattering problem for the Zakharov-Shabat equations on the semi-axis, *Inverse Problems* **16**, no. 6, pp. 1813–1837.

[302] Boutet de Monvel, A. and Kotlyarov, V. P. (2003). Generation of asymptotic solitons of the nonlinear Schrödinger equation by boundary data, Integrability, topological solitons and beyond, *J. Math. Phys.* **44**, no. 8, pp. 3185–3215.

[303] Boutet de Monvel, A. and Kotlyarov, V. P. (2004). Characteristic properties of the scattering data for modified Koretweg-de Vries equation on the half-line, submitted.

[304] Boutet de Monvel, A. and Marchenko, V. (1994). Generalization of the Darboux transform, *Mat. Fiz. Anal. Geom.* **1**, no. 3-4, pp. 479–504.

[305] Brander, O. (1980). High energy behavior of phase shifts for scattering from singular potentials, *Preprint LPTHE 80/3*, Paris, Jan., pp. 1–28; (1981). *J. Math. Phys.* **22**, no. 6, pp. 1229–1235.

[306] Brasche, J. (1985). Perturbation of Schrödinger Hamiltonians by measures—selfadjointness and lower semiboundedness, *J. Math. Phys.* **26**, no. 4, pp. 621–626.

[307] Brasche, J. F., Koshmanenko, V. and Neidhardt, H. (1994). New aspects of Krein's extension theory, *Ukraïn. Mat. Zh.* **46**, no. 1-2, pp. 37–54. (in Russian). (English Translation: (1994). *Ukrainian Math. J.* **46**, no. 1-2, pp. 34–53.)

[308] Brasche, J. F., Malamud, M. and Neidhardt, H. (2002). Weyl function and spectral properties of self-adjoint extensions, *Integral Equations Operator Theory* **43**, no. 3, pp. 264–289.

[309] Braverman, M., Milatovic, O. and Shubin, M. (2002). Essential selfadjointness of Schrödinger-type operators on manifolds, *Uspekhi Mat. Nauk* **57**, no. 4(346), pp. 3–58. (in Russian). (English Translation: (2002). *Russian*

Math. Surveys **57**, no. 4, pp. 641–692.)

[310] Breimesser, S. V. and Pearson, D. B. (2003). Geometrical aspects of spectral theory and value distribution for Herglotz functions, *Math. Phys. Anal. Geom.* **6**, 1, pp. 29–57.

[311] Brézis, H. (1988). Points critiques dans les problèmes variationnels sans compacité, *Séminaire Bourbaki*, Vol. 1987/88, *Astérisque*, No. 161-162, Exp. No. 698, 5, pp. 239–256. (in French). (Russian Translation: (1990). *Proc. Sem. N. Bourbaki (1988)*, Collection of papers, Mir, Moscow, pp. 252–269.)

[312] Brown, B. M., Eastham, M. S. P., Hinz, A. M., Kreicherbauer, T., McCormack, D. K. R. and Schmidt, K. M. (1991). Welsh eigenvalues of radially periodic Schrödinger operators, *J. Math. Anal. Appl.* **225**, pp. 347–357.

[313] Brown, R. C. (1979). Notes on generalized boundary value problems in Banach spaces. I. Adjoint and extension theory, *Pacific J. Math.* **85**, no. 2, pp. 295–322.

[314] Bruk, V. M. (1974). Dissipative extensions of a differential operator of elliptic type, *Functional Analysis*, No. 3, Ulyanovsk, pp. 35–43. (in Russian).

[315] Bruk, V. M. (1974). Generalized resolvents and spectral functions of odd order differential operators in a space of vector-valued functions, *Functional Analysis*, No. 3, Ul'yanovsk, pp. 44-50. (in Russian).

[316] Bruk, V. M. (1974). The generalized resolvents and spectral functions of even order differential operators in a space of vector-valued functions, *Mat. Zametki* **15**, no. 6, pp. 945–954. (in Russian). (English Translation: (1974). *Math. Notes* **15**, pp. 563–568.)

[317] Bruk, V. M. (1975). On the number of linearly independent square integrable solutions of systems of differential equations, *Functional Analysis*, No. 5, Ul'yanovsk, pp. 25–33. (in Russian).

[318] Bruk, V. M. (1976). A certain class of boundary value problems with a spectral parameter in the boundary condition, *Mat. Sb.* **100**, no. 2, pp. 210–216. (in Russian). (English Translation: (1976). *Sb. Math.* **29**, no. 2, pp. 186–192.)

[319] Bruk, V. M. (1977). Extensions of symmetric relations, *Mat. Zametki* **22**, no. 6, 825–834. (in Russian). (English Translation: (1977). *Math. Notes* **22**, no. 5–6, pp. 953–958.)

[320] Bruk, V. M. (1978). Linear relations in a space of vector functions, *Mat. Zametki* **24**, no. 6, pp. 499–511. (in Russian). (English Translation: (1979). *Math. Notes* **24**, pp. 767–773).

[321] Bruk, V. M. (1980). On symmetric relations generated by a differential expression and a nonnegative operator function, *Functional Analysis*, No. 15, Ul'yanovsk, pp. 33–44. (in Russian).

[322] Bruk, V. M. (1986). Generalized resolvents of symmetric relations in a space of vector functions, *Functional Analysis*, No. 26, Ul'yanovsk, pp. 52–61. (in Russian).

[323] Bruk, V. M. and Krys'ko, V. A. (2004). On the spectrum of operators associated with uniformly well-posed problems, *Differ. Uravn.* **40**, no. 10, pp. 1417–1418. (in Russian). (English Translation: (2004). *Differ. Equ.* **21**, no. 10, pp. 1496–1497.)

[324] Brusentsev, A. G. (1985). Selfadjointness in essentially semibounded higher-order elliptic operators, *Differ. Uravn.* **21**, no. 4, pp. 668–677. (in Russian). (English Translation: (1985). *Differ. Equ.* **21**, pp. 450–458.)

[325] Brusentsev, A. G. (2003). The multidimensional theorem of H. Weyl and covering families. *Mat. Zametki* **73**, no. 1, pp. 38–48. (in Russian). (English Translation: (2003). *Math. Notes* **73**, no. 1, pp. 36–45.)

[326] Brusentsev, A. G. (2004). Self-adjointness of elliptic differential operators in $L_2(G)$ and the correcting potentials, *Tr. Mosk. Mat. Obs.* **65**, pp. 35–68. (in Russian).

[327] Brusentsev, A. G. and Rofe-Beketov, F. S. (1974). Self-adjointness conditions for elliptic systems of arbitrary order, *Mat. Sb.* **95**, no. 1, pp. 108–129. (in Russian). (English Translation: (1974). *Sb. Math.* **24**, no. 1, pp. 103–126.)

[328] Burnat, M. (1961). Stability of eigenfunctions and the spectrum for Hill's equation, *Bull. Acad. Polon. Sci. Sér. Sci. Math. Astronom. Phys.* **9**, No. 11, pp. 795–798.

[329] Burnat, M. (1964). Die Spektraldarstellung einiger Differentialoperatoren mit periodischen Koeffiezienten im Raume der fastperiodischen Funktionen, *Studia Math.* **25**, pp. 33–65. (in German).

[330] Buslaev, V. S. (1987). Quasi-classical approximation for equations with periodic coefficients, *Uspekhi Mat. Nauk* **42**, no. 6, pp. 77–98. (in Russian). (English Translation: (1987). *Russian Math. Surveys* **42**, no. 6, pp. 97–125.)

[331] Buslaev, V. S. (1999). Kronig-Penney electron in a homogeneous electric field, *Amer. Math. Soc. Transl. Ser. II* **189**, Amer. Math. Soc., Providence, RI, pp. 45–57.

[332] Buslaev, V. S. and Dmitrieva, L. A. (1989). A Bloch electron in an external field, *Algebra i Analiz* **1**, no. 2, pp. 1–29. (in Russian). (English Translation: (1990). *Leningrad Math. J.* **1**, no. 2, pp. 287–320.)

[333] Buslaev, V. S. and Faddeev, L. D. (1960). Formulas for traces for a singular Sturm-Liouville differential operator, *Dokl. Akad. Nauk SSSR* **132**, pp. 13–16. (in Russian). (English Translation: (1960). *Soviet Math. Dokl.* **1**, pp. 451–454.)

[334] Calkin, J. W. (1939). Abstract symmetric boundary conditions, *Trans. Amer. Math. Soc.* **45**, no. 3, pp. 369–442.

[335] Cantero, M. J., Moral, L. and Velázquez, L. (2003). Five-diagonal matrices and zeros of orthogonal polynomials on the unit circle, *Linear Algebra Appl.* **362**, pp. 29–56.

[336] Carleman, T. (1933). On the theory of the integro-differential Boltzmann equation, *Acta Math.* **60**, pp. 91–146.

[337] Carleman, T. (1939). Sur un probléme d'unicité pour les systèmes d'équations aux dérivées partielles à deux variables indépendantes, *Ark. Mat. Astron. Fysik*, Bd. 26 B, No. 17, pp. 1–9. (in French). (See also the book [36], pp. 497–505.)

[338] Chaudhuri, J. and Everitt, W. N. (1969). On the square of a formally self-adjoint differential expression, *J. London Math. Soc. (2)* **1**, pp. 661–673.

[339] Chien, M.-T., Tso, S.-H. and Wu, P. Y. (2003). Higher-dimensional numerical ranges of quadratic operators, *J. Operator Theory* **49**, no. 1, pp. 153–171.
[340] Chulaevskiy, V. A. (1981). On perturbations of a Schrödinger operator with periodic potential, *Uspekhi Mat. Nauk* **36**, no. 35, pp. 203–204. (in Russian). (English Translation: (1981). *Russian Math.Surveys* **36**, no. 5, pp. 143–144.)
[341] Clark, S. and Gesztesy F. (2001). Weyl-Titchmarsh M-function asymptotics for matrix-valued Schrödinger operators, *Proc. London. Math. Soc.* **82**, pp. 701–724.
[342] Clark, S. and Gesztesy F. (2003). On Povzner-Wienholtz-type self-adjointness results for matrix-valued Sturm-Liouville operators, *Proc. Roy. Soc. Edinburgh Sect. A* **133**, no. 4, pp. 747–758.
[343] Clark, S., Gesztesy, F., Holden, H. and Levitan, B. M. (2000). Borg-type theorems for matrix-valued Schrödinger operators, *J. Differential Equations* **167**, no. 1, pp. 181–210.
[344] Clark, S., Gesztesy, F. and Renger, W. (2004). Trace formulas and Borg-type theorems for matrix-valued Jacobi and Dirac finite difference operators, Preprint, arXiv: math.SP/0408074, 4 Aug. 2004, 28 pp.
[345] Coddington, E. A. and de Snoo, H. S. V. (1978). Positive self-adjoint extensions of positive symmetric subspaces, *Math. Z.* **159**, 3, pp. 203–214.
[346] Crooks, E. C. M. and Toland, J. F. (1998). Travelling waves for reaction-diffusion-convection systems, *Topol. Methods Nonlinear Anal.* **11**, no. 1, pp. 19–43.
[347] Danielyan, A. A. (1987). On the width of the strip of regularity of finite-zone potentials of the Dirac operator, *Vestnik Moskov. Univ. Ser. I Mat. Mekh.* **1987**, no. 3, pp. 81–82. (in Russian). (English Translation: (1987). *Moscow Univ. Math. Bull.* **42**, no. 3, pp. 77–79.)
[348] Danilov, L. I. (2003). On the spectrum of the two-dimensional periodic Schrödinger operator, *Teoret. Mat. Fiz.* **134**, no. 3, pp. 447–459. (in Russian). (English Translation: (2003). *Theoret. and Math. Phys.* **134**, no. 3, pp. 392–403.)
[349] Darboux, G. (1882). Sur une proposition relative a une equation linéaire, *C. R. Acad. Sci. Paris*, v. 54. (in French).
[350] Davydov, R. N. (1978). On the spectral function of a class of non-self-adjoint Sturm-Liouville operators, *Differential Equations and Methods of Functional Analysis*, Naukova Dumka, Kiev, pp. 77–88. (in Russian).
[351] Deift, P. and Killip, R. (1999). On the absolutely continuous spectrum of one-dimensional Schrödinger operators with square summable potentials, *Comm. Math. Phys.* **203**, no. 2, pp. 341–347.
[352] Dekkers, W. (1979). The matrix of a connection having regular singularities on a vector bundle of rank 2 on $P^1(C)$, *Équations différentielles et systèmes de Pfaff dans le champ complexe (Sem., Inst. Rech. Math. Avance, Strasbourg, 1975)*, *Lecture Notes in Math.* **712**, Springer, Berlin, pp. 33–43.
[353] Delyon, F., Simon, B. and Souillard, B. (1985). From power pure point to continuous spectrum in disordered systems, *Ann. Inst. H. Poincaré Phys.*

Théor. **42**, no. 3, pp. 283–309.

[354] Derkach, V. (1995). On Weyl function and generalized resolvents of a Hermitian operator in a Krein space, *Integral Equations Operator Theory* **23**, no. 4, pp. 387–415.

[355] Derkach, V. A. (1999). On generalized resolvents of Hermitian relations in Krein spaces. Functional analysis, *J. Math. Sci. (New York)* **97**, no. 5, pp. 4420–4460.

[356] Derkach, V. A. (2001). On characteristic functions of linear relations and unitary colligations, *Dopov. Nats. Akad. Nauk Ukr. (Mat. Prirodozn. Tekh. Nauki)*, no. 11, pp. 28–33.

[357] Derkach, V. A., Hassi, S., Malamud, M. M. and de Snoo, H. S. V. (2000). Generalized resolvents of symmetric operators and admissibility, *Methods Funct. Anal. Topology* **6**, no. 3, pp. 24–55.

[358] Derkach, V. A., Hassi, S., Malamud, M. M. and de Snoo, H. S. V. (2001). Weyl functions and intermediate extensions of symmetric operators, *Dopov. Nats. Akad. Nauk Ukr. (Mat. Prirodozn. Tekh. Nauki)*, no. 10, pp. 33–39.

[359] Derkach, V. A. and Malamud, M. M. (1987). On the Weyl function and Hermite operators with lacunae, *Dokl. Akad. Nauk SSSR* **293**, no. 5, pp. 1041–1046. (in Russian). (English Translation: (1987). *Soviet Math. Dokl.* **35**, pp.. 393–398.)

[360] Derkach, V. A. and Malamud, M. M. (1991). Generalized resolvents and the boundary value problems for Hermitian operators with gaps, *J. Funct. Anal.* **95**, no. 1, pp. 1–95.

[361] Derkach, V. A. and Malamud, M. M. (1995). The extension theory of Hermitian operators and the moment problem. Analysis. 3, *J. Math. Sci.* **73**, no. 2, pp. 141–242.

[362] Derkach V. A., Malamud, M. M., de Snoo, H. S. V. and Hassi, S. (2004). Boundary relations and Weyl families, *Dokl. Akad. Nauk* **399**, no. 2, pp. 151–156. (in Russian).

[363] Derkach, V. A., Malamud, M. M. and Tsekanovskiy, E. R. (1988). Sectorial extensions of a positive operator, and the characteristic function, *Dokl. Akad. Nauk SSSR* **298**, no. 3, 537–541. (in Russian). (English Translation: (1988). *Soviet Math. Dokl.* **37**, no. 1, pp. 106–110.)

[364] Dijksma, A. and de Snoo, H. S. V. (1974). Self-adjoint extensions of symmetric subspaces, *Pacific J. Math.* **54**, pp. 71–100.

[365] Dikhamindzhiya, G. V. (1976). On the two-sided asymptotics of the number of eigenvalues of differential operators, *Soobshch. Akad. Nauk Gruzin. SSR* **81**, no. 1, pp. 25–28. (in Russian).

[366] Dinaburg, E. I. (1997). Some questions of the spectral theory of discrete operators with quasi-periodic coefficients, *Uspekhi Mat. Nauk* **52**, no. 3, pp. 3–52. (in Russian). (English Translation: (1997). *Russian Math. Surveys* **52**, no. 3, pp. 451–499.)

[367] Dinaburg, E. I. and Sinai, Ya. G. (1975). The one-dimensional Schrödinger equation with quasiperiodic potential, *Funktsional Anal. i Prilozhen.* **9**, no. 4, pp. 8–21. (in Russian). (English Translation: (1975). *Funct. Anal. Appl.* **9**, pp. 279–289.)

[368] Di Perna, R. J. and Lions, P. L. (1991). Global solutions of Boltzmann's equation and the entropy inequality, *Arch. Rational Mech. Anal.* **114**, no. 1, pp. 47–55.

[369] Djakov, P. and Mityagin, B. (2002). Smoothness of Schrödinger operator potential in the case of Gevrey type asymptotics of the gaps, *J. Funct. Anal.* **195**, no. 1, pp. 89–128.

[370] Djakov, P. and Mityagin, B. (2002). Smoothness of solutions of nonlinear ODE, *Math. Ann.* **324**, no. 2, pp. 225–254.

[371] Djakov, P. and Mityagin, B. (2002). Smoothness of solutions of a nonlinear ODE, *Integral Equations Operator Theory* **44**, no. 2, pp. 149–171.

[372] Djakov, P. and Mityagin, B. (2003). Spectra of 1-D periodic Dirac operators and smoothness of potentials, *C. R. Math. Acad. Sci. Soc. R. Can.* **25**, no. 4, pp. 121–125.

[373] Djakov, P. and Mityagin, B. (2003). Spectral triangles of Schrödinger operators with complex potentials, *Selecta Math. (N.S.)* **9**, no. 4, pp. 495–528.

[374] Djakov, P. and Mityagin, B. (2003). Spectral gaps of the periodic Schrödinger operator when its potential is an entire function, *Adv. in Appl. Math.* **31**, no. 3, pp. 562–596.

[375] Djakov, P. and Mityagin, B. (2003). The asymptotics of spectral gaps of a 1-D Dirac operator with cosine potential, *Lett. Math. Phys.* **65**, no. 2, pp. 95–108.

[376] Djakov, P. and Mityagin, B. (2004). Multiplicities of the eigenvalues of periodic Dirac operators, *The Ohio State Univ. Math. Res. Inst. Preprints*, No. 04-1, January 2004, pp. 1–32. (Online: http://www.math.ohio-state.edu/mri/preprints). (to appear in *J. Differential Equations*).

[377] Djakov, P. and Mityagin, B. (2004). Simple and double eigenvalues of the Hill operator with a two term potential, *The Ohio State Univ. Math. Res. Inst. Preprints*, No. 04-9, July 2004, pp. 1–40. (Online: http://www.math.ohio-state.edu/mri/preprints).

[378] Djakov, P. and Mityagin, B. (2004). Asymptotics of instability zones of the Hill operator with a two term potential, *The Ohio State Univ. Math. Res. Inst. Preprints*, No. 04-10, July 2004, pp. 1–39. (Online: http://www.math.ohio-state.edu/mri/preprints).

[379] Djakov, P. and Mityagin, B. (2004). Asymptotics of instability zones of Hill operators with a two term potential, *C. R. Acad. Sci. Paris Sér. I Math.* **339**, pp. 351–354.

[380] Djakov, P. and Mityagin, B. (2004). Instability zones of a periodic 1D Dirac operator and smoothness of its potential, *The Ohio State Univ. Math. Res. Inst. Preprints*, No. 04-12, August 2004, pp. 1–44. (Online: http://www.math.ohio-state.edu/mri/preprints).

[381] Dobrokhotov, S. Yu. and Zhevandrov, P. N. (1985). Nonstandard characteristics and the Maslov operator method in linear problems on time-dependent waves in water, *Funktsional. Anal. i Prilozhen.* **19**, no. 4, pp. 43–54. (in Russian). (English Translation: (1985). *Funct. Anal. Appl.* **19**, pp. 285–295.)

[382] Domshlak, Yu. I. (1970). On the oscillation of solutions to vector differential

equations, *Dokl. Akad. Nauk SSSR* **193**, no. 1, pp. 21–23. (in Russian). (English Translation: (1970). *Soviet Math. Dokl.* **11**, pp. 839–841.)

[383] Domshlak, Yu. I. (1971). On the oscillation and non-oscillation of the solutions of vector differential equations, *Differ. Uravn.* **7**, no. 6, pp. 961–969. (in Russian).

[384] Domshlak, Yu. I. (1979). A Sturm-type comparison theorem for a first-order equation in Hilbert space with a cone and its applications to systems of ordinary differential equations, *Akad. Nauk Azerbaidzhan. SSR Dokl.* **35**, no. 5, pp. 7–12. (in Russian).

[385] Dubovoy, V. K. (1983). Multiplicities of one-sided shifts contained in a contraction operator, *Funktsional. Anal. i Prilozhen.* **17**, no. 1, pp. 70. (English Translation: (1983). *Funct. Anal. Appl.* **15**, pp. 55–56.)

[386] Duistermaat, J. J. (1976). On the Morse index in variational calculus, *Advances in Math.* **21**, no. 2, pp. 173–195.

[387] Dymarskiy, Ya. M. (2000). The periodic Choquard equation, *Differential operators and related topics, Vol. I (Odessa, 1997), Oper. Theory Adv. Appl.* **117**, Birkhäuser, Basel, pp. 87–99.

[388] Dymarskiy, Ya. M. (2002). Manifolds generated by a family of periodic boundary value problems, *Dopov. Nats. Akad. Nauk Ukr. (Mat. Prirodozn. Tekh. Nauki)*, no. 2, pp. 16–19. (in Russian).

[389] Dymarskiy, Ya. M. (2002). Manifolds of eigenfunctions and potentials for a family of periodic Sturm-Liouville problems, *Ukraïn. Mat. Zh.* **54**, no. 8, pp. 1042–1052. (in Russian). (English Translation: (2002). *Ukrainian Math. J.* **54**, no. 8, pp. 1251–1263.)

[390] Dyukarev, Yu. M. (2004). Indeterminacy criteria for the Stieltjes matrix moment problem, *Mat. Zametki* **75**, no. 1, pp. 71–88. (in Russian). (English Translation: (2004). *Math. Notes* **75**, no. 1-2, pp. 66–82.)

[391] Dyukarev, Yu. M. (2004). Nevanlinna–Pick problem for the Stiltjes matrix-functions, *Ukraïn. Mat. Zh.* **56**, no. 3, pp. 366–380. (in Russian).

[392] Edwards, H. M. (1964). A generalized Sturm theorem, *Ann. of Math. (2)* **80**, pp. 22–57.

[393] Efimov, A. V. and Potapov, V. P. (1973). J-expanding matrix-valued functions, and their role in the analytic theory of electrical circuits, *Uspekhi Mat. Nauk* **28**, no. 1(169), pp. 65–130. (English Translation: (1973). *Russian Math. Surveys* **28**, no. 1, pp. 69–140.)

[394] Egorova, I. E. (1986). Spectral properties of the Dirac operator with limit-periodic coefficients, *Dokl. Akad. Nauk Ukrain. SSR Ser. A*, no. 5, pp. 10–13. (in Russian).

[395] Egorova, I. E. (1987). Spectral analysis of Jacobi limit-periodic matrices, *Dokl. Akad. Nauk Ukrain. SSR Ser. A*, no. 3, pp. 7–9. (in Russian).

[396] Egorova, I. E. (1987). The complete set of spectral parameters of the Dirac operator with limit-periodic coefficients, *Teor. Funktsii Funktsional. Anal. i Prilozhen.*, No. 47, pp. 25–35. (in Russian). (English Translation: (1990). *J. Soviet Math.* **48**, no. 6, pp. 636–643.)

[397] Egorova, I. E. (1992). On a class of almost periodic solutions of the KdV equation with a nowhere dense spectrum, *Dokl. Akad. Nauk* **323**, no. 2,

pp. 219–222. (in Russian). (English Translation: (1992). *Russian Acad. Sci. Dokl. Math.* **45**, no. 2, pp. 290–293.)

[398] Egorova, I. E. (1994). The Cauchy problem for the KdV equation with almost periodic initial data whose spectrum is nowhere dense, *Spectral operator theory and related topics, Adv. Soviet Math.* **19**, Amer. Math. Soc., Providence, RI, pp. 181–208.

[399] Egorova, I. E. and Surkova A. B. (1995). On the almost periodicity of "non-reflective" Dirac operators with the Cantor spectrum, *Dopov. Nats. Akad. Nauk Ukraïni*, no. 12, pp. 13–15. (in Russian).

[400] Eliasson, L. H. (1992). Floquet solution for one-dimensional quasi-periodic Schrödinger equation, *Comm. Math. Phys.* **146**, no. 3, pp. 447–482.

[401] Er'omenko, V. O. (1998). Periodic solutions of systems of two first-order linear ordinary differential equations with a degenerate asymmetric matrix coefficient of the derivatives., *Ukraïn. Mat. Zh.* **50**, no. 3, pp. 350–356. (in Ukrainian). (English Translation: (1998). *Ukrainian Math. J.* **50**, no. 3, pp. 400–407.)

[402] Etgen, G. J. and Lewis, R. T. (1980). A Hille-Wintner comparison theorem for second order differential systems, *Czechoslovak Math. J.* **30**, no. 1, pp. 98–107.

[403] Etgen, G. J. and Pawlowski, J. F. (1977). A comparison theorem for second order differential systems, *Pacific J. Math.* **72**, no. 1, pp. 59–69.

[404] Etkin, A. (2000). On an abstract boundary value problem with the eigenvalue parameter in the boundary condition, *Operator theory and its applications (Winnipeg, MB, 1998), Fields Inst. Commun.* **25**, Amer. Math. Soc., Providence, RI, 2000, pp. 257–266.

[405] Evans, W. D. and Zettl, A. (1977). Interval limit-point criteria for differential expressions and their powers, *J. London Math. Soc. (2)* **15**, no. 1, pp. 119–133.

[406] Everitt, W. N. (1959). Integrable-square solutions of ordinary differential equations, *Quart. J. Math. Oxford Ser. (2)* **10**, pp. 145–155.

[407] Everitt, W. N. (1962). Two theorems in matrix theory, *Amer. Math. Monthly* **69**, pp. 856–859.

[408] Everitt, W. N. (1964). Singular differential equations. I. The even order case, *Math. Ann.* **156**, pp. 9–24.

[409] Everitt, W. N. (1977). On the deficiency index problem for ordinary differential operators 1910–1976, *Differential equations (Proc. Internat. Conf., Uppsala, 1977), Sympos. Univ. Upsaliensis Ann. Quingentesimum Celebrantis*, No. 7, Almqvist and Wiksell, Stockholm, pp. 62–81.

[410] Everitt, W. N. (2003). A Catalogue of Sturm-Liouville differential equations, Preprint 2003/33, The School of Mathematics and Statistics, University of Birmingham, pp. 1–61.

[411] Everitt, W. N. (2004/2005). Charles Sturm and the development of Sturm–Liouville theory in the years 1900–1950, (A Lecture given at the *Sturm Bicentennial Meeting* held at the University of Geneva, Switzerland, in September 2003), pp. 1–30, in print.

[412] Everitt, W. N. and Giertz, M. (1975). On the deficiency indices of powers of

formally symmetric differential expressions, With addenda by T. T. Reed, R. M. Kaufman and Anton Zettl, *Spectral theory and differential equations (Proc. Sympos., Dundee, 1974; dedicated to Konrad Jörgens)*, *Lecture Notes in Math.*, Vol. 448, Springer, Berlin, pp. 167–181.

[413] Everitt, W. N. and Markus, L. (1999). Complex symplectic geometry with applications to ordinary differential operators, *Trans. Amer. Math. Soc.* **351**, no. 12, pp. 4905–4945.

[414] Everitt, W. N. and Markus, L. (2003). Elliptic partial differential operators and symplectic algebra, *Mem. Amer. Math. Soc.* **162**, no. 770, x+111 pp.

[415] Everitt, W. N. and Zettl, A. (1977). The number of integrable-square solutions of products of differential expressions, *Proc. Roy. Soc. Edinburgh Sect. A* **76**, pp. 215–226.

[416] Exner, P. (1995). The absence of the absolutely continuous spectrum for δ' Wannier-Stark ladders, *J. Math. Phys.* **36**, no. 9, pp. 4561–4570.

[417] Faddeev, L. D. (1964). Properties of the S-matrix of the one-dimensional Schrödinger equation, *Trudy Mat. Inst. Steklov.* **73**, pp. 314–336. (in Russian).

[418] Faddeev, L. D. (1974). The inverse problem in the quantum theory of scattering. II, *Current problems in mathematics*, Vol. 3, Akad. Nauk SSSR Vsesoyuz. Inst. Nauchn. i Tekhn. Informatsii, Moscow, pp. 93–180. (in Russian). (English Translation: (1976). *J Soviet Math.* **5**, pp. 334–396.)

[419] Favini, A. and Rutkas, A. (1999). Existence and uniqueness of solutions of some abstract degenerate nonlinear equations, *Differential Integral Equations* **12**, no. 3, pp. 373–394.

[420] Favini, A. and Vlasenko, L. (2003). Degenerate non-stationary differential equations with delay in Banach spaces, *J. Differential Equations* **192**, no. 1, pp. 93–110.

[421] Fedoryuk, M. V. (1966). Asymptotic methods in the theory of one-dimensional singular differential operators *Tr. Mosk. Mat. Obs.* **15**, pp. 296–345. (in Russian). (English Translation: (1966). *Trans. Moscow Math. Soc.* **1966**, pp. 333–386.)

[422] Fedoryuk M. V. (1969). Asymptotic methods in the theory of ordinary differential equations *Mat. Sb.* **79**, pp. 477–516. (in Russian). (English Translation: (1969). *Sb. Math.* **8**, no. 4, pp. 451–491.)

[423] Feoktistov, L. P. (1989). Neutron-fissioning wave, *Dokl. Akad. Nauk SSSR* **309**, pp. 864–867. (in Russian).

[424] Firsov, O. B. (1975). On calculation of the effective mass of an electron in a periodic field near the zonal boundary, *Zh. Eksper. i Teor. Fiz.* **68**, no. 2, pp. 568–575. (in Russian). (English Translation: (1975). *JETP* **41**, no. 2, pp. 280–289.)

[425] Firsova, N. E. (1974). A trace formula for a perturbed one-dimensional Schrödinger operator with a periodic potential I., *Problems in mathematical physics*, No. 7, Izdat. Leningrad. Univ., Leningrad, pp. 162–177. (in Russian).

[426] Firsova, N. E. (1975). The Riemann surface of a quasimomentum, and scattering theory for a perturbed Hill operator, Mathematical questions in

the theory of wave propagation, *Zap. Nauchn. Sem. Len. Otdel. Mat. Inst. Steklov. (LOMI)* **51**, no. 7, pp. 183–196. (in Russian). (English Translation: (1979). *J. Soviet Math.* **11**, no. 3, pp. 487–510.)

[427] Firsova, N. E. (1976). A trace formula for a perturbed one-dimensional Schrödinger operator with a periodic potential II., *Problems in mathematical physics*, No. 8, Izdat. Leningrad. Univ., Leningrad, pp. 158–171. (in Russian).

[428] Firsova, N. E. (1979). Effective mass of the one-dimensional Hill operator, *Vestnik Leningrad. Univ. Fiz. Khim.*, no. 2, pp. 13–18. (in Russian).

[429] Firsova, N. E. (1985). The Levinson formula for a perturbed Hill operator, *Teoret. Mat. Fiz.* **62**, no. 2, pp. 196–209. (in Russian). (English Translation: (1985). *Theoret. and Math. Phys.* **62**, pp. 130–140.)

[430] Firsova, N. E. (1986). A direct and inverse scattering problem for a one-dimensional perturbed Hill operator, *Mat. Sb.* **130**, no. 3, pp. 349–385. (in Russian). (English Translation: (1986). *Sb. Math.* **58**, no. 2, pp. 351–388.)

[431] Fisher, R. A. (1937). The wave of advance of advantageous genes, *Ann. Eugenics* **7**, pp. 355–369.

[432] Foliadova, E. V. (1984). Oscillatory theorem for Dirac's system with a parameter in the boundary condition, *Functional Analysis. Spectral Theory.*, No. 22, Ul'yanovsk, pp. 111–122. (in Russian).

[433] Foliadova, E. V. (1984). Oscillation of eigenfunctions of the Sturm-Liouville problem with weight and with a spectral parameter in the boundary condition, *Functional Analysis. Spectral Theory.*, No. 23, Ul'yanovsk, pp. 125–135. (in Russian).

[434] Foliadova, E. V. (1985). Oscillation of solutions of the Sturm-Liouville problem with operator potential and parameter in the boundary condition, *Functional Analysis. Linear Spaces.*, Ul'yanovsk, No. 25, pp. 129–141. (in Russian).

[435] Foliadova, E. V. (1985). Oscillatory properties of solutions of certain multiparameter boundary value problems, *Functional Analysis. Operator Theory.*, Ul'yanovsk, No. 24, pp. 115–122. (in Russian).

[436] Foliadova, E. V. (1994). Perturbations of the spectrum of Sturm-Liouville boundary value problems of boundary conditions containing a spectral parameter, *Functional Analysis*, Ul'yanovsk, No. 35, pp. 103–112. (in Russian).

[437] Frantzeskakis, D. J., Stratis, I. S. and Yannacopoulos, A. N. (2004). On equilibria of the two-fluid model in magnetohydrodynamics, *Math. Phys. Anal. Geom.* **7**, no. 2, pp. 97–117.

[438] Fulton, C. T. (1977). Parametrizations of Titchmarsh $m(\lambda)$-functions in the limit circle case, *Trans. Amer. Math. Soc.* **229**, pp. 51–63.

[439] Gardner, G. S. (1971). Korteweg-de Vries equation and generalization. IV. The Korteweg-de Vries equation as a Hamiltonian system, *J. Math. Phys.* **12**, pp. 1548–1551.

[440] Gasymov, M. G. (1968). Expansion in terms of the solutions of a scattering theory problem for the nonselfadjoint Schrödinger equation, *Dokl. Akad. Nauk SSSR* **179**, no. 3, pp. 518–521. (in Russian). (English Translation:

(1968). *Soviet Math. Dokl.* **9**, pp. 390–393.)

[441] Gasymov, M. G. (1980). Spectral analysis of a class of second order non-selfadjoint differential operators, *Funktsional. Anal. i Prilozhen.* **14**, no. 1, pp. 14–19. (in Russian). (English Translation: (1980). *Funct. Anal. Appl.* **14**, pp. 11-15.)

[442] Gasymov, M. G. and Guseinov, G. Sh. (1989). Some uniqueness theorems in inverse problems of spectral analysis for Sturm-Liouville operators in the case of the Weyl limit circle, *Differ. Uravn.* **25**, no. 4, pp. 588–599. (in Russian). (English Translation: (1989). *Differ. Equ.* **25**, no. 4, pp. 394–402.)

[443] Gasymov M. G., Orudgev, E. G. and Gasymova, D. G. (1998). Spectral analysis of a class of non-selfadjoint differential operators with matrix periodic coefficients, *Trans. Acad. Sci. Azerb. Ser. Phys.-Tech. Math. Sci.* **18**, no. 3-4, pp. 23–26.

[444] Gasymov, M. G., Zhikov, V. V. and Levitan, B. M. (1967). Conditions for the discreteness and finiteness of the negative spectrum of the second order operator Schrödinger equation, *Mat. Zametki* **2**, no. 5, pp. 531–538. (in Russian). (English Translation: (1967). *Math. Notes* **2**, pp. 813-817.)

[445] Gefter, S. L. and Mokrenyuk, V. N. (2004/2005). Power series $\sum_{n=0}^{\infty} n!z^n$ and holomorphic solutions of some differential equations in Banach space, submitted to *Mat. Fiz. Anal. Geom.* (in Russian).

[446] Gel'fand, I. M. (1950). Expansion in eigenfunctions of equations with periodic coefficients, *Dokl. Akad. Nauk SSSR* **73**, 6, pp. 1117–1120. (in Russian). (English Translation: (1987). *I. M. Gel'fand. Collected Papers*, Vol. 1, Part III, No. 1, Springer-Verlag, New York, pp. 401–404.)

[447] Gel'fand, I. M. and Dikiy, L. A. (1975). Asymptotic properties of the resolvent of Sturm-Liouville equations, and the algebra of Korteweg-de Vries equations, *Uspekhi Mat. Nauk* **30**, no. 5(185), pp. 67–100. (in Russian). (English Translation: *Russian Math. Surveys* **30**, no. 5, pp. 77–113.)

[448] Gel'fand, I. M. and Kostyuchenko, A. G. (1955). Expansion in eigenfunctions of differential and other operators, *Dokl. Akad. Nauk SSSR* **103**, no. 3, pp. 349–352. (in Russian). (English Translation: (1987). *I. M. Gel'fand. Collected Papers*, Vol. 1, Part III, No. 6, Springer-Verlag, New York, pp. 505–509.)

[449] Gel'fand, I. M. and Levitan, B. M. (1951). On the determination of the differential equation by its spectral function, *Izv. Akad. Nauk SSSR Ser. Mat.* **15**, pp. 309–360. (in Russian).

[450] Gel'fand, I. M. and Lidskiy, V. B. (1955). On the structure of the regions of stability of linear canonical systems of differential equations with periodic coefficients, *Uspekhi Mat. Nauk* **10**, no. 1(63), pp. 3–40. (in Russian). (English Translation: (1987). *I. M. Gel'fand. Collected Papers*, Vol. 1, Part III, No. 5, Springer-Verlag, New York, pp. 466–504.)

[451] Gel'fand, I. M. and Ponomarev, V. A. (1968). Indecomposable representations of the Lorenz group, *Uspekhi Mat. Nauk* **23**, no. 2, pp. 3–60. (in Rus-

sian). (English Translation: (1968). *Russian Math. Surveys* **23**, 2, pp. 1–58.)

[452] Gérard, P. (1988) Solutions globales du problème de Cauchy pour l'équation de Boltzmann (d'après R. J. DiPerna et P.-L. Lions), *Séminaire Bourbaki, Vol. 1987/88, Astérisque*, No. 161-162, Exp. No. 699, 5, pp. 257–281. (in French). (Russian Translation: (1990). *Proc. Sem. N. Bourbaki (1988)*, Collection of papers, Mir, Moscow, pp. 252–269.)

[453] Gerasimenko, V. I. and Petrina, D. Ya. (1997). Generalized kinetic equation generated by BBGKY hierarchy, *Ukrainian J. of Physics* **43**, no. 6-7, pp. 697–702.

[454] Geronimo, J. S., Gesztesy, F. and Holden, H. (2004). Algebro-geometric solutions of a discrete system related to the trigonometric moment problem, Preprint, arXiv: math.SP/0408073, 4 Aug. 2004, 22 pp.

[455] Gestrin, G. N. (1979). On a uniqueness theorem in a multidimensional inverse problem of spectral analysis of the Schrödinger operator in a class of analytic potentials, *Teor. Funktsii Funktsional. Anal. i Prilozhen.*, No. 31, pp. 28–32. (in Russian).

[456] Gesztesy, F. (2001). Integrable systems in the infinite genus limit, *Differential Integral Equations* **14**, pp. 671–700.

[457] Gesztesy, F., Kiselev, A. and Makarov, K. A. (2002). Uniqueness results for matrix-valued Schrödinger, Jacobi and Dirac type operators, *Math. Nachr.* **239/240**, pp. 103–145.

[458] Gesztesy, F. and Sakhnovich, L. A. (2001). A class of the matrix-valued Schrödinger operators with prescribed finite-band spectra, Preprint, University of Missouri, Columbia, USA.

[459] Gesztesy, F. and Sakhnovich, L. A. (2003). Schrödinger operators with prescribed finite-band spectra, In: Reproducing kernel spaces and applications, *Oper. Theory Adv. Appl.* **143**, pp. 213–253.

[460] Gesztesy, F. and Simon, B. (1993). A short proof of Zheludev's theorem, *Trans. Amer. Math. Soc.* **335**, no. 1, pp. 329–340.

[461] Gesztesy, F. and Simon, B. (2000). A new approach to inverse spectral theory. II. General real potentials and the connection to the spectral measure, *Ann. of Math. (2)* **152**, no. 2, pp. 593–643.

[462] Gesztesy, F. and Simon, B. (2003). Connectedness of the isospectral manifold for one-dimensional half-line Schrödinger operators, Preprint, arXiv: math.SP/0307007, v 1, pp. 1–5.

[463] Gesztesy, F., Simon, B. and Teschl, G. (1996). Zeros of the Wronskian and renormalized oscillation theory, *Amer. J. Math.* **118**, pp. 571–594.

[464] Gesztesy, F. and Ünal, M. (1998). Perturbative oscillation criteria and Hardy-type inequalities, *Math. Nachr.* **189**, pp. 121–144.

[465] Gesztesy, F. and Weikard, R. (1996). Picard potentials and Hill's equation on a torus, *Acta Math.* **176**, no. 1, pp. 73–107.

[466] Gesztesy, F. and Zinchenko, M. (2005). A Borg-type theorem associated with orthogonal polynomials on the unit circle, Preprint arXiv:math.SP/0501212, v 1, 14 Jan 2005, pp. 1–29.

[467] Gesztesy, F. and Zinchenko, M. (2005). Weyl–Titchmarsh theory for CMV

operators associated with orthogonal polynomials on the unit circle, Preprint, arXiv: math.SP/0501210, v 1, 14 Jan 2005, pp. 1–39.

[468] Geyler, V. A. and Senatorov, M. M. (1997). The structure of the spectrum of the Schrödinger operator with a magnetic field in a strip, and finite-gap potentials, *Mat. Sb.* **188**, no. 5, pp. 21–32. (in Russian). (English Translation: (1997). *Sb. Math.* **188**, no. 5, pp. 657–669.)

[469] Geyler, V. A. and Senatorov, M. M. (1997). Periodic potentials for which all the gaps are nonzero, *Funktsional. Anal. i Prilozhen.* **31**, no. 1, pp. 67–70. (in Russian). (English Translation: (1997). *Funct. Anal. Appl.* **31**, no. 1, pp. 52–54.)

[470] Gilbert, D. J. and Pearson, D. B. (1987). On subordinacy and analysis of the spectrum of one-dimensional Schrödinger operators, *J. Math. Anal. Appl.* **128**, no. 1, pp. 30–56.

[471] Gilbert, R. C. (1977). A class of formally selfadjoint ordinary differential operators whose deficiency numbers differ by an arbitrary pre-assigned positive integer, *The Uppsala International Conference on Differential Equations (April 18–22, 1977) Abstracts*, pp. 36–37.

[472] Gilbert, R. C. (1977). The deficiency index of a symmetric ordinary differential operator with complex coefficients, *J. Differential Equations* **25**, 3, pp. 425–459.

[473] Gilbert, R. C. (1977). The deficiency index of a third order operator, *Pacific J. Math.* **68**, no. 2, pp. 369–392.

[474] Gilbert, R. C. (1978). Asymptotic formulas for solutions of a singular linear ordinary differential equation, *Proc. Roy. Soc. Edinburgh Sect. A* **81**, no. 1-2, pp. 57–70.

[475] Gilbert, R. C. (1978/79). A class of symmetric ordinary differential operators whose deficiency numbers differ by an integer, *Proc. Roy. Soc. Edinburgh Sect. A* **82**, no. 1-2, pp. 117–134.

[476] Gilbert, R. C. (1981). Integrable-square solutions of a singular ordinary differential equation, *Proc. Roy. Soc. Edinburgh Sect. A* **89**, no. 3-4, pp. 267–279.

[477] Gimadislamov, M. G. (2005). On the spectrum of a higher order differential operator, *Mat. Zametki* **77**, no. 2, pp. 188–193. (in Russian).

[478] Glazman, I. M. (1949). On the deficiency index of differential operators, *Dokl. Akad. Nauk SSSR* **64**, pp. 151–154. (in Russian).

[479] Glazman, I. M. (1950). On the theory of singular differential operators, *Uspekhi Mat. Nauk* **5**, no. 6(40), pp. 102–135. (in Russian). (English Translation: (1953). *Amer. Math. Soc. Transl. Ser. I*, No. 96, pp. 1–43.)

[480] Gohberg, I., Kaashoek, M. A. and Sakhnovich, A. L. (1998). Sturm-Liouville systems with rational Weyl functions: explicit formulas and applications. Dedicated to the memory of Mark Grigorievich Krein (1907–1989), *Integral Equations Operator Theory* **30**, no. 3, pp. 338–377.

[481] Gohberg, I., Kaashoek, M. A. and Sakhnovich, A. L. (2001). Bound states of a canonical system with a pseudo-exponential potential, *Integral Equations Operator Theory* **40**, 3, pp. 268–277.

[482] Gohberg, I., Kaashoek, M. A. and Sakhnovich, A. L. (2002). Scattering prob-

lems for a canonical system with a pseudo-exponential potential, *Asymptot. Anal.* **29**, no. 1, pp. 1–38.

[483] Gol'din, V. Ya. and Anistratov, V. Yu. (1995). Fast neutron reactor in the self-adjusting neutron-nucleade regime, *Mat. Model.* **7**, no. 10, pp. 12–32. (in Russian).

[484] Gol'din, V. Ya., Sosnin, N. V. and Troshchiev, Yu. V. (1998). Fast neutron reactor in the self-adjusting regime of the second kind, *Dokl. Akad. Nauk* **358**, no. 6, pp. 747–748. (in Russian). (English Translation: (1998). *Russian Acad. Sci. Dokl. Math.* **57**, no. 1, pp. 147–148.)

[485] Gomilko, A. M. (2003). On the factorization of operator functions in a Hilbert space, *Funktsional. Anal. i Prilozhen.* **37**, no. 1, pp. 19–24. (in Russian). (English Translation: (2003). *Funct. Anal. Appl.* **37**, no. 1, pp. 16–20.)

[486] Gorbachuk, M. L. (1966). On spectral functions of a second order differential operator with operator coefficients, *Ukraïn. Mat. Zh.* **18**, no. 2, pp. 3–21. (in Russian). (English Translation: (1968). *Amer. Math. Soc. Transl. Ser. II* **72**, pp. 177–202.)

[487] Gorbachuk, M. L. (1971). Selfadjoint boundary value problems for a second order differential equation with an unbounded operator coefficient, *Funktsional. Anal. i Prilozhen.* **5**, no. 1, pp. 10–21. (in Russian). (English Translation: (1971). *Funct. Anal. Appl.*, 5, pp. 9–18.)

[488] Gorbachuk, M. L. (2000). On the well-posedness of the Cauchy problem for operator-differential equations in classes of analytic vector functions, *Dokl. Akad. Nauk* **374**, no. 1, pp. 7–9. (in Russian). (English Translation: (2000). *Russian Acad. Sci. Dokl. Math.* **62**, no. 2, pp. 149–151.)

[489] Gorbachuk, M. L., Kochubei, A. N. and Rybak, M. A. (1972). Dissipative extensions of differential operators in a space of vector-valued functions, *Dokl. Akad. Nauk SSSR* **205**, no. 5, pp. 1029–1032. (in Russian). (English Translation: (1972). *Soviet Math. Dokl.* **13**, pp. 1063–1067.)

[490] Gorbachuk, M. L., Kochubei, A. N. and Rybak, M. A. (1973). On some classes of extensions of differential operators in a space of vector-functions, in Collection: *Primen. funkts. analiza k zadacham matem. fiziki*, Kiev, pp. 56–82. (in Russian).

[491] Gorbachuk, M. L. and Mikhailets, V. A. (1976). Semi-bounded self-adjoint extensions of symmeteric operators, *Dokl. Akad. Nauk SSSR* **226**, no. 4, pp. 765–767. (in Russian). (English Translation: (1976). *Soviet Math. Dokl.* **17**, pp. 185–187.)

[492] Gorbachuk, V. I. and Gorbachuk, M. L. (1969). Expansion in the eigenfunctions of second order differential equation with operator coefficients, *Dokl. Akad. Nauk SSSR* **184**, no. 4, pp. 774–777. (in Russian). (English Translation: (1969). *Soviet Math. Dokl.* **10**, pp. 158–162.)

[493] Gorbachuk, V. I. and Gorbachuk, M. L. (1972). Certain classes of boundary value problems for the Sturm–Liouville equation with an operator-valued potential, *Ukraïn. Mat. Zh.* **24**, no. 3, pp. 291–304. (in Russian).

[494] Gorbachuk, V. I. and Gorbachuk, M. L. (1972). On the spectrum of self-adjoint extensions of the minimal operator generated by Sturm–Liouville

expression with operator-potential, *Ukraïn. Mat. Zh.* **24**, no. 6, pp. 726–734. (in Russian). (English Translation: (1972). *Ukrainian Math. J.* **24**, no. 6, pp. 582–588.)

[495] Gorbachuk, V. I. and Gorbachuk, M. L. (1973). Some questions of the spectral theory of first order differential equations with operator coefficients, *Application of functional analysis to problems of mathematical physics*, Institute for Mathematics of the Academy of Sciences of Ukrainian SSR, pp. 83–124. (in Russian).

[496] Gorbachuk, V. I., Gorbachuk, M. L. and Kochubei, A. N. (1989). The theory of extensions of symmetric operators, and boundary value problems for differential equations, *Ukraïn. Mat. Zh.* **41**, no. 10, pp. 1299–1313, 1436. (in Russian). (English Translation: (1990). *Ukrainian Math. J.* **41**, no. 10, pp. 1117–1129.)

[497] Gordevskiy, V. D. (2000). Approximate biflow solutions of the kinetic Bryan-Pidduck equation, *Math. Methods Appl. Sci.* **23**, no. 13, pp. 1121–1137.

[498] Gordevskiy, V. D. (2003). Bimodal distributions with the spiral-type Maxwellians for the Bryan-Pidduck model, *J. Math. Anal. Appl.* **283**, no. 1, pp. 192–201.

[499] Gordevskiy, V. D. (2003). Transitional regime between vortical states of a gas, *Nonlinear Anal.* **53**, no. 3-4, pp. 481–494.

[500] Gordevskiy, V. D. (2004). On the non-stationary Maxwellians, *Math. Methods Appl. Sci.* **27**, no. 2, pp. 231–247.

[501] Graff, A. A. (1946–1947). On the theory of differential systems in one dimensional domain I, II, *Mat. Sb.* **18**, no. 2, pp. 305–328; **21**, no. 1, pp. 143–159. (in Russian).

[502] Guseinov, I. M. (1982). Reconstruction of a second-order difference equation from a spectral matrix, *Mat. Zametki* **32**, no. 5, pp. 737–746. (in Russian). (English Translation: (1983). *Math. Notes* **32**, no. 5–6, pp. 853–858.)

[503] Guseinov, I. M. and Pashaev, R. T. (1983). Description of selfadjoint extensions of a class of differential operators of order $2n$ with defect indices $(n+k, n+k), 0 < k < n, 0 < k < n$, *Izv. Akad. Nauk Azerb. Ser. Fiz.-Tekh. Mat. Nauk*, No. 2, pp. 15–19. (in Russian).

[504] Hartman, P. (1979). Oscillation criteria for selfadjoint second-order differential systems and "principal sectional curvatures," *J. Differential Equations* **34**, no. 2, pp. 326–338.

[505] Hassi, S., Malamud, M. and de Snoo, H. (2002). On extension theory of nonnegative operators, *Preprint*, University of Vaasa, Finland, 32 pp.

[506] Heinz, E. (1958). Halbbeschränktheit gewöhnlicher Differentialoperatoren höherer Ordnung, *Math. Ann.* **135**, no. 1, pp. 1–49 (Auf Graund des Nachlasses von F. Rellich). (in German).

[507] Hempel, R., Herbst, I., Hinz, A. M. and Kalf, H. (1991). Intervals of dense point-spectrum for spherically symmetric Schrödinger operators of the type $-\Delta + \cos|x|$, *J. London Math. Soc. (2)* **43**, no. 2, pp. 295–304.

[508] Hempel, R., Hinz, A. M. and Kalf, H. (1987). On the essential spectrum of Schrödinger operators with spherically symmetric potentials. With a comment by J. Weidmann, *Math. Ann.* **277**, no. 2, pp. 197–211.

[509] Henniart, G. (1985). Les inégalités de Morse (d'après E. Witten). *Séminaire Bourbaki*, Vol. 1983/84, *Astérisque*, No. 121-122, pp. 43–61. (in French). (Russian Translation: (1990). *Mathematical Analysis and Geometry, Selected Works of the Seminar N. Bourbaki*, Collection of papers, Mir, Moscow, pp. 80–101.)

[510] Hilbert, D. (1900). Mathematische Probleme. Vortrag, gehalten auf dem internationalen Mathematiker-Congress zu Paris, *Gött. Nachr.* **1900**, pp. 253–297. (in German). (English Translation: see [511].) (Russian Translation: see [79], pp. 11–64.)

[511] Hilbert, D. (1902). Mathematical problems. Lecture delivered before the international congress of mathematicians at Paris in 1900, Translated by Mary Winston Newson, *Bull. Amer. Math. Soc. (2)* **8**, pp. 437–479.

[512] Hilbert, D. (1905). Über eine Anwendung der Integralgleichungen auf ein Problem der Funktionentheorie, *Verh. d. 3. intern. Math. Kongr. Heidelb.*, pp. 233–240. (in German).

[513] Hilbert, D. (1905). Grundzüge einer allgemeinen Theorie der linearen Integralgleichungen, (Dritte Mitteilung), *Gött. Nachr.*, pp. 307–338. (in German).

[514] Hinton, D. (1970). Continuous spectra of second-order differential operators. *Pacific J. Math.* **33**, pp. 641–643.

[515] Hinton, D. B. (1979). An expansion theorem for an eigenvalue problem with eigenvalue parameter in the boundary condition, *Quart. J. Math. Oxford Ser. (2)* **30**, no. 117, pp. 33–42.

[516] Hryniv, R. and Mykytyuk, Ya. (2001). 1-D Schrödinger operators with periodic singular potentials, *Methods Funct. Anal. Topology* **7**, no. 4, pp. 31–42.

[517] Hryniv, R. O. and Mykytyuk, Ya. V. (2004). Transformation operators for Sturm-Liouville operators with singular potentials, *Math. Phys. Anal. Geom.* **7**, no. 2, pp. 119–149.

[518] Ismagilov, R. S. (1971). The asymptotic behavior of the spectrum of a differential operator in a space of vector valued fnctions, *Mat. Zametki* **9**, pp. 667–676. (in Russian). (English Translation: (1971). *Math. Notes* **9**, pp. 387–392.)

[519] Ismagilov, R. S. (1985). The spectrum of the Sturm-Liouville equation with oscillating potential, *Mat. Zametki* **37**, no. 6, pp. 869–879. (in Russian). (English Translation: (1985). *Math. Notes* **37**, no. 5-6, pp. 476–482.)

[520] Its, A. R. and Matveev, V. B. (1975). Hill operators with a finite number of lacunae, *Funkts. Anal. i Prilozhen.* **9**, no. 1, pp. 69–70. (in Russian). (English Translation: (1975). *Funct. Anal. Appl.* **9**, pp. 65–66.)

[521] Its, A. R. and Matveev, V. B. (1976). A class of solutions of Korteweg-de Vries equation, *Problems of mathematical physics*, No. 8, Izdat. Leningrad. Univ., Leningrad, pp. 70–92. (in Russian).

[522] Its, A. R. and Novokshenov, V. Yu. (1988). Effective sufficient conditions for the solvability of an inverse problem of monodromy theory for systems of linear ordinary differential equations, *Funktsional. Anal. i Prilozhen.* **22**, no. 3, pp. 25–36. (in Russian). (English Translation: (1988). *Funct. Anal. Appl.* **22**, no. 3, pp. 190–200.)

[523] Jodeit, M. Jr. and Levitan B. M. (1998). Isospectral vector-valued Sturm-Liouville problems, *Lett. Math. Phys.* **43**, no. 2, pp. 117–122.

[524] Johnson, R. and Moser, J. (1982). The rotation number for almost periodic potentials, *Comm. Math. Phys.* **84**, no. 3, pp. 403–438.

[525] Kabalyants, P. S. and Rutkas, A. G. (1999). On an quasilinear differential equation with the singulyarity of the Kronecker type, *Visnyk Kharkiv. Univ. Ser. Mat. Prikl. Mat. i Mekh.*, No. 444, pp. 111–118. (in Russian).

[526] Kabalyants, P. S. and Rutkas, A. G. (2000). Two existence theorems for solutions of a quasilinear singular operator-differential equation, *Neliniyni Kolyvannya (Nonlinear Oscillations)* **3**, no. 3, pp. 358–364, Inst. Matem. Nats. Akad. Nauk Ukrainy. (in Russian).

[527] Kac, M. and van Moerbeke, P. (1975). On an explicitly soluble system of nonlinear differential equations related to certain Toda lattices, *Advances in Math.* **16**, pp. 160–169.

[528] Kalf, H., Okaji, T. and Yamada, O. (2003). Absence of eigenvalues of Dirac operators with potentials diverging at infinity, *Math. Nachr.* **259**, pp. 19–41.

[529] Kalyabin, G. A. (1972). On the number of $L_2(0,\infty)$ solutions of a second order self-adjoint system of differential equations, *Funktsional. Anal. i Prilozhen.* **6**, no. 3, pp. 74–76. (in Russian). (English Translation: (1972). *Funct. Anal. Appl.* **6**, pp. 237–239.)

[530] Kappeler, T. and Mityagin, B. (2001). Estimates for periodic and Dirichlet eigenvalues of the Schrödinger operator, *SIAM J. Math. Anal.* **33**, no. 1, pp. 113–152 (electronic).

[531] Kappeler, T. and Möhr, C. (2001). Estimates for periodic and Dirichlet eigenvalues of the Schrödinger operator with singular potentials, *J. Funct. Anal.* **186**, no. 1, pp. 62–91.

[532] Karataeva, T. V. and Koshmanenko, V. D. (1999). Generalized sum of operators, *Mat. Zametki* **66**, no. 5, pp. 671–681. (in Russian). (English Translation: (1999). *Math. Notes* **66**, no. 5-6, pp. 556–564.)

[533] Karpeshina, Yu. E. and Pavlov, B. S. (1986). Interaction of the zero radius for the biharmonic and the polyharmonic equation, *Mat. Zametki* **40**, no. 1, pp. 49–59. (in Russian). (English Translation: (1987). *Math. Notes* **40**, no. 1-2, pp. 528–533.)

[534] Kashaev, R. M. (2005). On self-adjoint extensions of a difference operator, *Algebra i Analiz* **17**, no. 1, pp. 209–223. (in Russian).

[535] Kats, I. S. (1983). Linear relations generated by canonical differential equations, *Funktsional. Anal. i Prilozhen* **17**, no. 4, pp. 86–87. (in Russian). (English Translation: (1983). *Funct. Anal. Appl.* **17**, no. 4, pp. 315–317.)

[536] Kats, I. S. (1992). Reconstruction of a string from its principal spectral matrix function, *Dokl. Akad. Nauk Ukrainy*, no. 1, pp. 5–10. (in Russian).

[537] Kaufman, R. M., Read T. T. and Zettl, A. (1977). The deficiency index problem for powers of ordinary differential expressions, *Lecture Notes in Math.*, Vol. 621, Springer-Verlag, Berlin-New York.

[538] Kaykina, E. I., Naumkin, P. I. and Shishmarev, I. A. (2005). Cauchy problem for an equation of Sobolev type with power nonlinearity, *Izv. Akad. Nauk.*

Ser. Mat. **69**, no. 1, pp. 61–114. (in Russian).

[539] Khalanai, A. and Shandor, Sh. (1957). Sturm type theorems for self-conjugate systems of higher order differential equations, *Dokl. Akad. Nauk SSSR* **114**, no. 3, pp. 506–507. (in Russian).

[540] Khol'kin, A. M. (1979). Oscillation theorems for the Sturm-Liouville system and the Dirac system on the line, In: *Research in the operator theory and its applications. Collection of research papers*, Naukova Dumka, Kiev, pp. 147–161. (in Russian).

[541] Khol'kin, A. M. (1980). Self-adjoint conditions at infinity for absolutely indefinite system of differential equations of arbitrary order, Manuscript deposited at the All-union Institute for Scientific and Technical Information, 11/14/1980, No. 4801-80, pp. 23. (in Russian).

[542] Khol'kin, A. M. (1981). Selfadjoint boundary conditions at infinity for a quasiregular system of even-order differential equations, In: *Theory of operators in function spaces and its applications. Collection of research papers*, Naukova Dumka, Kiev, pp. 174–183. (in Russian).

[543] Khol'kin, A. M. (1985). Description of selfadjoint extensions of differential operators of arbitrary order on an infinite interval in the absolutely indeterminate case, *Teor. Funktsii Funktsional. Anal. i Prilozhen.*, No. 44, pp. 112–122. (in Russian). (English Translation: (1990). *J. Soviet Math.* **48**, no. 3, pp. 337–345.)

[544] Khol'kin, A. M. (1995). Fundamental system of solutions of differential equations with operator coefficients and singular boundary condition, *Vestnik Priazov. Gos. Tekh. Univ.*, No. 1, pp. 262–267. (in Russian).

[545] Khrabustovskiy, V. I. (1973). The perturbation of the spectrum of selfadjoint differential operators with periodic matrix-valued coefficients, *Matematicheskaya Fizika i Funktsional'nyi Analiz*, No. 4, Fiz.-Tehn. Inst. Nizkih Temperatur, Akad. Nauk Ukrain. SSR, Kharkov, pp. 117–138. (in Russian).

[546] Khrabustovskiy, V. I. (1974). The perturbation of the spectrum of selfadjoint differential operators of arbitrary order with periodic matrix coefficients, *Matematicheskaya Fizika i Funktsional'nyi Analiz*, No. 5, Fiz.-Tehn. Inst. Nizkih Temperatur, Akad. Nauk Ukrain. SSR, Kharkov, pp. 123–140. (in Russian).

[547] Khrabustovskiy, V. I. (1977). The discrete spectrum of perturbed differential operators of arbitrary order with periodic matrix coefficients, *Mat. Zametki* **21**, no. 6, pp. 829–838. (in Russian). (English Translation: (1977). *Math. Notes* **21**, pp. 467–472.)

[548] Khrabustovskiy, V. I. (1981). The spectral matrix of a periodic symmetric system with a degenerate weight on the axis, *Teor. Funktsii Funktsional. Anal. i Prilozhen.*, No. 35, p. 111–119. (in Russian).

[549] Khrabustovskiy, V. I. (1984). Eigenfunction expansions of periodic systems with weight, *Dokl. Akad. Nauk Ukrain. SSR Ser. A*, no. 5, pp. 26–29.

[550] Khrabustovskiy, V. I. (1985). Spectral analysis of periodic systems with degenerate weight on the axis and half-axis, *Teor. Funktsii Funktsional. Anal. i Prilozhen.*, No. 44, pp. 122–133 (in Russian). (English translation: (1990). *J. Soviet Math.* **48**, no. 3, pp. 345–355.)

[551] Khrabustovskiy, V. I. (1986). Spectral analysis of periodic systems with degenerate weight (expansions in Bloch solutions), *Teor. Funktsii Funktsional. Anal. i Prilozhen.*, No. 46, pp. 122–132 (in Russian). (English Translation: (1990). *J. Soviet Math.* **48**, no. 5, pp. 598–607.)

[552] Khrabustovskiy, V. I. (2003). On the characteristic matrix of Weyl-Titchmarsh type for differential-operator equations with the spectral parameter entering linearly or by Nevanlinna type, *Mat. Fiz. Anal. Geom.* **10**, no. 2, pp. 205–227. (in Russian).

[553] Khruslov, E. Ya. and Kotlyarov, V. P. (1994). Soliton asymptotics of nondecreasing solutions of nonlinear completely integrable evolution equations, *Spectral operator theory and related topics, Adv. Soviet Math.* **19**, Amer. Math. Soc., Providence, RI, pp. 129–180.

[554] Khryashchev, S. V. (1990). Discrete spectrum for a periodic Schrödinger operator perturbed by a decreasing potential, *Oper. Theory Adv. Appl.* **46**, pp. 109–114.

[555] Khryashchev, S. V. (1997). The density of states of the spectrum of the Schrödinger operator with a radial potential, *Algebra i Analiz* **9**, no. 1, pp. 201–212 (in Russian). (English Translation: (1998). *St. Petersburg Math. J.* **9**, no. 1, pp. 147–155.)

[556] Kochubei, A. N. (1975). On extension of symmetric operators and binary relations, *Mat. Zametki* **17**, no. 1, pp. 41–48. (in Russian). (English Translation: (1975). *Math. Notes* **17**, pp. 25–28.)

[557] Kochubei, A. N. (1977). On extensions of a non-densely defined symmetric operator, *Sibirsk. Mat. Zh.* **18**, no. 2, pp. 314–320. (in Russian). (English Translation: (1977). *Siberian Math. J.* **18**, pp. 225–229.)

[558] Kochubei, A. N. (1979). Extensions of a positive definite symmetric operator, *Dokl. Akad. Nauk Ukrain. SSR Ser. A*, No. 3, pp. 168–171. (in Russian).

[559] Kochubei, A. N. (1991). Selfadjoint extensions of the Schrödinger operator with a singular potential, *Sibirsk. Mat. Zh.* **32**, no. 3, pp. 60–69. (in Russian). (English Translation: (1991). *Siberian Math. J.* **32**, no. 3, pp. 401–409.)

[560] Kogan, V. I. (1970). Operators that are generated by J_p-matrices in the case of maximal defect indices. Teor. Funktsii Funktsional. Anal. i Prilozhen., No. 11, pp. 103–107. (in Russian).

[561] Kogan, V. I. and Rofe-Beketov, F. S. (1971). On the question of the deficiency indices of differential operators with complex coefficients, *Matematicheskaya Fizika i Funktsional'nyi Analiz*, Physics and Techical Institute of Low Temperatures of the Academy of Sciences of the Ukrainian SSR, pp. 45–60. (English Translation: (1974). *Proc. Roy. Soc. Edinburgh Sect. A* **72**, pp. 281–298.)

[562] Kogan, V. I. and Rofe-Beketov, F. S. (1973). On square integrable solutions of systems of differential equations of arbitrary order, Physics and Technical Institute of Low Temperatures of the Academy of Sciences of the Ukrainian SSR, preprint, 60pp. (in Russian). (English Translation: (1975). *Proc. Roy. Soc. Edinburgh Sect. A* **74**, pp. 5–40.)

[563] Kohn, A. T. (1983). Un résultat de Plemelj, *Mathematique et physique,*

Semin. Ec. Norm. Super., Paris 1979-1982, Prog. Math. **37**, Birkhäuser, Boston, MA, pp. 307–312. (in French).

[564] Kolmogorov, A. N. (2004). Mathematical models of the turbulent motion of an incompressible viscous fluid, *Uspekhi Mat. Nauk* **59**, no. 1(355), pp. 5–10. (in Russian).

[565] Kolmogorov, A. N. (2004). Almost horizontal turbulence, *Uspekhi Mat. Nauk* **59**, no. 2, pp. 3–8. (in Russian). (English Translation: (2004). *Russian Math. Surveys* **59**, no. 2, pp. 197–202.)

[566] Kolmogorov, A. N., Petrovskiy, I. G. and Piskunov, N. S. (1937). Investigation of the equation of diffusion combined with the increase in the quantity of the substance and its application to one biological problem, *Bulletin of Moscow University Ser. A*, No. 1(6), pp. 1–26. (in Russian). (see also the book [90, p. 221–246].)

[567] Komatsu, H. (1973). An introduction to the theory of hyperfunctions, *Hyperfunctions and pseudo-differential equations. Including the Proceedings of a Conference at Katata, 12–14 October 1971 and part of the Proceedings of a Conference on the Theory of Hyperfunctions and Analytic Functions and Applications, Research Institute for Mathematical Sciences, Kyoto University, Kyoto, 27–30 September 1971. Dedicated to the memory of André Martineau. Edited by Hikasaburo Komatsu, Lecture Notes in Math.* **287**, Springer-Verlag, Berlin-New York, 1973, pp. 3–41.

[568] Komatsu, H. (1973). Ultradistributions and hyperfunctions, *Hyperfunctions and pseudo-differential equations. Including the Proceedings of a Conference at Katata, 12–14 October 1971 and part of the Proceedings of a Conference on the Theory of Hyperfunctions and Analytic Functions and Applications, Research Institute for Mathematical Sciences, Kyoto University, Kyoto, 27–30 September 1971. Dedicated to the memory of André Martineau. Edited by Hikasaburo Komatsu, Lecture Notes in Math.* **287**, Springer-Verlag, Berlin-New York, 1973, pp. 164–179.

[569] Kondrat'ev, V. A. (1961). On the solvability of the first boundary-value problem for elliptic equations, *Dokl. Akad. Nauk SSSR* **136**, no. 4, pp. 771–774. (in Russian). (English Translation: (1961). *Soviet Math. Dokl.* **2**, pp. 127–130.)

[570] Kondrat'ev, V. A. (1967). The solvability of the first boundary value problem for strongly elliptic equations, *Tr. Mosk. Mat. Obs.* **16**, pp. 293–318. (in Russian). (English Translation: (1967). *Trans. Moscow Math. Soc.* **1967**, pp. 315–341.)

[571] Korotyaev, E. L. (1997). Uniform estimates for the Hill operator, *Dokl. Akad. Nauk* **356**, no. 6, pp. 740–743. (in Russian). (English Translation: (1997). *Russian Acad. Sci. Dokl. Math.* **56**, no. 2, pp. 770–773.)

[572] Korotyaev, E. L. (1999). Inverse problems for the Hill and Dirac operators, *Dokl. Akad. Nauk* **365**, no. 6, pp. 730–733. (in Russian). (English Translation: (1999). *Russian Acad. Sci. Dokl. Math.* **59**, no. 2, pp. 292–295.)

[573] Koshmanenko, V. D. (1989). Perturbations of selfadjoint operators by singular bilinear forms, *Ukraïn. Mat. Zh.* **41**, no. 1, pp. 3–19. (in Russian). (English Translation: (1989). *Ukrainian Math. J.* **41**, no. 1, pp. 1–14.)

[574] Kostov, V. P. (1992). Fuchsian linear systems on CP^1 and the Riemann-Hilbert problem, *C. R. Acad. Sci. Paris Sér. I Math.* **315**, no. 2, pp. 143–148.

[575] Kostyuchenko, A. G. and Mirzoev, K. A. (2001). Conditions for the complete indefiniteness of Jacobi matrices with matrix elements, *Funktsional. Anal. i Prilozhen.* **35**, no. 4, pp. 32–37. (in Russian). (English Translation: (2001). *Funct. Anal. Appl.* **35**, no. 4, pp. 265–269.)

[576] Kotani, S. (1984). Ljapunov indices determine absolutely continuous spectra of stationary random one-dimensional Schrödinger operators, *Stochastic analysis (Katata/Kyoto, 1982)*, North-Holland Math. Library, 32, North-Holland, Amsterdam, pp. 225–247.

[577] Kotani, S. (1985). On an inverse problem for random Schrödinger operators, *Particle systems, random media and large deviations (Brunswick, Maine, 1984)*, *Contemp. Math.* **41**, Amer. Math. Soc., Providence, RI, pp. 267–280.

[578] Kotani, S. (1987). One-dimensional random Schrödinger operators and Herglotz functions, *Probabilistic methods in mathematical physics (Katata/Kyoto, 1985)*, Academic Press, Boston, MA, pp. 219–250.

[579] Kotani, S. (1997). Generalized Floquet theory for stationary Schrödinger operators in one dimension, *Chaos, Solitons and Fractals* **8**, no. 11, pp. 1817–1854.

[580] Kotlyarov, V. P. (1989). Inversion of the Miura transformation, *Mat. Zametki* **46**, no. 4, pp. 14–24. (in Russian). (English Translation: (1989). *Math. Notes* **46**, no. 3-4, pp. 762–769.)

[581] Kotlyarov, V. P. (1995). Asymptotic analysis of the Marchenko integral equation and soliton asymptotics of a solution of the nonlinear Schrödinger equation (Chernogolovka, 1994), *Phys. D* **87**, no. 1-4, pp. 176–185.

[582] Kovalishina, I. V. (1983). Analytic theory of a class of interpolation problems, *Izv. Akad. Nauk SSSR Ser. Mat.* **47**, no. 3, pp. 455–497. (in Russian). (English Translation: (1984). *Math. USSR-Izv.* **22**, pp. 419–463 (1984).)

[583] Kovalishina, I. V. and Potapov, V. P. (1981). Radii of the Weyl circle in the matrix Nevanlinna–Pick problem. *Theory of operators in function spaces and its applications*, Vol. 188, Naukova Dumka, Kiev, pp. 25–49. (in Russian).

[584] Kozel, V. A., Lundina, D. Sh. and Marchenko, V. A. (1976). A generalized spectral matrix for a three-dimensional non-selfadjoint Schrödinger operator, *Teor. Funktsii Funktsional. Anal. i Prilozhen.*, No. 27, pp. 71–90. (in Russian).

[585] Krein, M. G. (1937). On self-adjoint differential operators and their symmetric Green's functions, *Mat. Sb.* **2**, no. 6, pp. 1023–1072. (in Russian).

[586] Krein, M. G. (1947). Theory of self-adjoint extensions of semi-bounded Hermitian operators and its applications. I, II, *Mat. Sb.* **20**, no. 3, pp. 431–495; **21**, no. 3, pp. 365–404. (in Russian).

[587] Krein, M. G. (1950). On a one dimensional singular boundary problem of even order in the interval $(0, \infty)$, *Dokl. Akad. Nauk SSSR* **74**, no. 1, pp. 9–

12. (in Russian).

[588] Krein, M. G. (1952). On the indeterminate case of the Sturm-Liouville boundary problem in the interval $(0, \infty)$, *Izv. Akad. Nauk SSSR Ser. Mat.* **16**, pp. 293–321. (in Russian).

[589] Krein, M. G. (1953). On the transition function of the one-dimensional boundary problem of the second order, *Dokl. Akad. Nauk SSSR* **88**, no. 3, pp. 405–408. (in Russian).

[590] Krein, M. G. (1955). On criteria of stable boundedness of solutions of periodic canonical systems, *Prikl. Mat. Meh.* **19**, no. 6, pp. 641–680. (in Russian).

[591] Krein, M. G. (1965). Introduction to the geometry of indefinite J-spaces and to the theory of operators in those spaces, *The Second Summer Mathematical School, Katsiveli (June-July 1964)*, Naukova Dumka, Kiev, pp. 15–92. (in Russian).

[592] Krein, M. G. and Langer, H. (1971). The defect subspaces and generalized resolvents of a Hermitian operator in the space Π_κ, *Funktsional Anal. i Prilozhen.* **5** no. 2, pp. 59–71; **5**, no. 3, pp. 54–69 (in Russian). (English Translation: (1971). *Funct. Anal. Appl.* **5**, pp. 136–146; **5**, pp. 217–228.)

[593] Krein, M. G. and Langer, H. (1973). Über die Q-Funktion eines π-hermiteschen Operators im Raume Π_κ, *Acta Sci. Math. (Szeged)* **34**, pp. 191–230. (in German).

[594] Krein, M. G. and Ovcharenko, I. E. (1976). On the theory of generalized resolvents of nondensely defined Hermitian contractions : *Dokl. Akad. Nauk Ukrain. SSR Ser. A*, no. 10, pp. 881–885. (in Russian).

[595] Krein, M. G. and Ovcharenko, I. E. (1976). On generalized resolvents and resolvent matrices of positive Hermitian operators *Dokl. Akad. Nauk SSSR* **231**, no. 5, pp. 1063–1066. (in Russian). (English Translation: (1976). *Soviet Math. Dokl.* **17**, pp. 1705–1709.)

[596] Krein, M. G. and Ovcharenko, I. E. (1977). Q-functions and sc-resolvents of nondensely defined Hermitian contractions, *Sibirsk. Mat. Zh.* **18**, no. 5, pp. 1032–1056. (in Russian). (English Translation: (1977). *Siberian Math. J.* **18**, pp. 728–746.)

[597] Krein, M. G. and Ovcharenko, I. E. (1978). Inverse problems for Q-functions and resolvent matrices of positive Hermitian operators, *Dokl. Akad. Nauk SSSR* **242**, no. 3, pp. 521–524. (in Russian). (English Translation: (1978). *Soviet Math. Dokl.* **19**, pp. 1131–1134.)

[598] Krein, M. G. and Ovcharenko, I. E. (1982). On the theory of inverse problems for the canonical differential equation, *Dokl. Akad. Nauk Ukrain. SSR Ser. A*, no. 2, pp. 14–18. (in Russian).

[599] Krein, S. G. and Laptev, G. I. (1962). Boundary value problems for an equation in Hilbert space, *Dokl. Akad. Nauk SSSR* **146**, no. 3, pp. 535–538. (in Russian). (English Translation: (1962). *Soviet Math. Dokl.* **3**, pp. 1350–1354.)

[600] Krylov, B. L. (1956). Solution in the finite form of the Riemann problem for the Gaussian system, *Proc. Kazan. Aviat. Inst.* **31**, pp. 203–445. (in Russian).

[601] Krys'ko, V. A., Awrejcewicz, J. and Bruk, V. M. (2002). On the solution of a coupled thermo-mechanical problem for non-homogeneous Timoshenko-type shells, *J. Math. Anal. Appl.* **273**, no. 2, pp. 409–416.

[602] Krys'ko, V. A., Awrejcewicz, J. and Bruk, V. M. (2002). The existence and uniqueness of solution of one coupled plate thermomechanics problem, *J. Appl. Anal.* **8**, no. 1, pp. 129–139.

[603] Kuchment, P. A. (1982). Floquet theory for partial diferential equations, *Uspekhi Mat. Nauk* **37**, no. 4, pp. 3–52. (in Russian). (English Translation: *Russian Math. Surveys* **37**, no. 4, pp. 1–60.)

[604] Kuchment, P. (2001). The mathematics of photonic crystals, *Mathematical modeling in optical science*, Frontiers Appl. Math., 22, SIAM, Philadelphia, PA, pp. 207–272.

[605] Kuchment, P. (2002). Graph models for waves in thin structures, *Waves Random Media* **12**, no. 4, pp. 1–24.

[606] Kudryavtsev, M. A. (1996). Direct and inverse problems of spectral analysis for five-diagonal symmetric matrices, *Dopov. Nats. Akad. Nauk Ukraïni*, no. 11, pp. 55–59. (in Ukrainian).

[607] Kudryavtsev, M. A. (1998/99). The direct and the inverse problem of spectral analysis for five-diagonal symmetric matrices. I., *Mat. Fiz. Anal. Geom.* **5**, no. 3-4, pp. 182–202; II., *Mat. Fiz. Anal. Geom.* **6**, no. 1-2, pp. 55–80. (in Russian).

[608] Kudryavtsev, M. (2002). The Cauchy problem for the Toda lattice with a class of non-stabilized initial data, *Mathematical results in quantum mechanics (Taxco, 2001), Contemp. Math.* **307**, Amer. Math. Soc., Providence, RI, pp. 209–214.

[609] Kuks, L. M. (1974). A Sturm-type comparison theorems for systems of fourth order ordinary differential equations, *Differ. Uravn.* **10**, no. 4, pp. 751–754. (in Russian). (English Translation: (1974). *Differ. Equ.* **10**, pp. 583–585.)

[610] Kulenović, M. R. S. (1982). On a result of Etgen and Lewis, *Czechoslovak Math. J.* **32**, no. 3, pp. 373–376.

[611] Ladyzhenskaya, O. A. (1956). On solution of nonstationary operator-equations, *Mat. Sb.* **39**, no. 2, pp. 491–524. (in Russian). (English Translation: (1967). *Amer. Math. Soc. Transl. Ser. II* **65**, pp. 200–236.)

[612] Ladyzhenskaya, O. A. (1958). On nonstationary operator-equations and their applications to linear problems of mathematical physics, *Mat. Sb.* **45**, no. 4, pp. 123–158. (in Russian).

[613] Landis, E. M. (1956). On some properties of solutions of elliptic equations, *Dokl. Akad. Nauk SSSR* **107**, pp. 640–643. (in Russian).

[614] Landis, E. M. (1974). The behavior of the solutions of higher order elliptic equations in unbounded domains, *Tr. Mosk. Mat. Obs.* **31**, pp. 35–58. (in Russian).

[615] Langer, H. and Textorius B. (1977). On generalized resolvents and Q-functions of symmetric linear relations (subspaces) in Hilbert space, *Pacific J. Math.* **72**, no. 1, pp. 135–165.

[616] Last, Y. and Simon, B. (1999). Eigenfunctions, transfer matrices, and absolutely continuous spectrum of one-dimensional Schrödinger operators,

Invent. Math. **135**, no. 2, pp. 329–367.

[617] Lavrent'ev, M. M. (1957). On the Cauchy problem for linear elliptic equations of the second order, *Dokl. Akad. Nauk SSSR* **112**, pp. 195–197. (in Russian).

[618] Lee, S. J. (1976). Formally self-adjoint systems of differential operators, *Math. Anal. Appl.* **55**, no. 1, pp. 90–101.

[619] Lesch, M. and Malamud, M. (2000). On the number of square integrable solutions and self-adjointness of symmeteric first order systems of differential equations, Mathem. Inst. der Univ. zu Köln, Jan. 2001, Preprint, arXiv: math.FA/0012080 11 Dec. 2000, pp. 1–50.

[620] Lesch, M. and Malamud, M. M. (2001). On the deficiency indices of Hamiltonian systems, *Dokl. Akad. Nauk* **381**, 5, pp. 599–603. (in Russian). (English Translation: (2001). *Russian Acad. Sci. Dokl. Math.* **64**, no. 3, pp. 393–397.)

[621] Lesch, M. and Malamud, M. M. (2003). On the deficiency indices and self-adjointness of symmetric Hamiltonian systems, *J. Differential Equations* **189**, no. 2, pp. 556–615.

[622] Levin, B. Ya. and Mirochnik, L. Ya. (1994). On the estimates of solutions of the Sturm-Liouville equations, *Ukraïn. Mat. Zh.* **46**, no. 3, pp. 244–278. (in Russian). (English Translation: (1994). *Ukrainian Math. J.* **46**, no. 3, pp. 251–289.)

[623] Levitan, B. M. (1981). Almost periodicity of infinite gap, *Izv. Akad. Nauk SSSR Ser. Mat.* **45**, no. 2, pp. 291–320. (in Russian). (English Translation: (1981). *Math. USSR-Izv.* **18**, pp. 249–273.)

[624] Levitan, B. M. (1982). Approximation of infinte-zone by finite-zone potentials, *Izv. Akad. Nauk SSSR Ser. Mat.* **46**, no. 1, pp. 56–87. (in Russian). (English Translation: (1982). *Math. USSR-Izv.* **20**, pp. 55–87.)

[625] Levitan, B. M. (1982). The inverse Sturm-Liouville problem in the case of finite-range and infinite-range potentials, *Tr. Mosk. Mat. Obs.* **45**, pp. 3–36. (in Russian). (English Translation: (1984). *Trans. Moscow Math. Soc.* **1984**, pp. 1–35.)

[626] Levitan, B. M. (1984). The closure of a set of finite-gap potentials, *Mat. Sb.* **123**, no. 1, pp. 69–91. (in Russian). (English Translation: (1985). *Sb. Math.* **51**, pp. 67–89.)

[627] Levitan, B. M. (1994). To the theory of finite gap and perturbed finite gap potentials, *Differential equations and mathematical physics (Birmingham, AL, 1994)*, Int. Press, Boston, MA, pp. 143–156.

[628] Levitan, B. M. (1995). Some special Tauberian theorems for Fourier type integrals and eigenfunction expansions of Sturm-Liouville equations on the whole line, *Math. Nachr.* **176**, pp. 179–198.

[629] Levitan, B. M. and Khasanov, A. B. (1992). An estimate of Cauchy function in the case of finite-zone non-periodic potentials, *Funktsional. Anal. i Prilozhen.* **26**, no. 2, pp. 18–28. (in Russian). (English Translation: (1992). *Funct. Anal. Appl.* **26**, no. 2, pp. 91–98.)

[630] Levitan, B. M. and Savin, A. V. (1984). Examples of Schrödinger operators with almost periodic potential and nowhere dense absolutely continuous

spectrum, *Dokl. Akad. Nauk SSSR* **276**, no. 3, pp. 539–542. (in Russian). (English Translation: (1984). *Soviet Math. Dokl.* **29**, pp. 541–544.)

[631] Lidskiy, V. B. (1954). On the number of square integrable solutions of the system of differential equations $-y'' + p(t)y = \lambda y$, *Dokl. Akad. Nauk SSSR* **95**, no. 2, pp. 217–220. (in Russian).

[632] Lidskiy, V. B. (1955). Oscillatory theorems for a canonical system of differential equations, *Dokl. Akad. Nauk SSSR* **102**, no. 5, pp. 877–880. (in Russian).

[633] Lidskiy, V. B. (1960). A non-selfadjoint operator of Sturm-Liouville type with a discrete spectrum, *Tr. Mosk. Mat. Obs.* **9**, pp. 45–79. (in Russian).

[634] Limić, N. (1962). Theory of scattering on highly singular potential, *Nuovo Cimento (10)* **26**, pp. 581–596.

[635] Liouville, J. and Sturm, J. C. F. (1837). Extrait d'une mémoire sur le développement des fonctions en serie, *J. Math. Pures et Appl. de Liouville*, vol. II, pp. 220–223. (in French).

[636] Lomonosova, O. M. (1982). The inverse scattering problem for the Schrödinger equation on the real line and a new class of solutions of Korteweg-de Vries equation, *Dokl. Akad. Nauk SSSR* **264**, no. 4, pp. 823–826. (in Russian). (English Translation: (1982). *Soviet Math. Dokl.* **25**, pp. 751–754.)

[637] Lundina, D. Sh. (1985). Compactness of sets of reflectionless potentials, *Teor. Funktsii Funktsional. Anal. i Prilozhen.*, No. 44, pp. 57–66. (in Russian). (English Translation: (1990). *J. Soviet Math.* **48**, no. 3, pp. 290–297.)

[638] Luzin, N. N. (1951). On the method of approximate integration by academician S. A. Chaplygin, *Uspekhi Mat. Nauk* **6**, no. 6, pp. 3–27. (in Russian).

[639] Lyubarskiy, M. G. (1973). On a generalization of Floquet-Lyapunov Theorem, *Dokl. Akad. Nauk SSSR* **213**, no. 4, pp. 780–782. (in Russian). (English Translation: (1973). *Soviet Math. Dokl.* **14**, pp. 1788–1791.)

[640] Lyubarskiy, M. G. (1977). On solutions of homogeneous sytems of differential equations with almost periodic coefficients, *Funktsional. Anal. i Prilozhen.* **11**, no. 4, pp. 90–91. (in Russian). (English Translation: (1977). *Funct. Anal. Appl.* **11**, pp. 321–323.)

[641] Lyubarskiy, M. G. (2003). Calculation of the Lyapunov exponent of linear ordinary equation with rapidly oscillating coefficients, *Visnyk Kharkiv. Univ. Ser. Mat. Prikl. Mat. Mekh.*, no. 582, vyp. 52, pp. 26–48.

[642] Lyubich, Yu. I. (1956). On fundamental solutions for linear partial differential equations of the elliptic type, *Mat. Sb.* **39**, no. 1, pp. 23–36. (in Russian).

[643] Lyubich, Yu. I. (1960). On inequalities between powers of a linear operator, *Izv. Akad. Nauk SSSR Ser. Mat.* **24**, pp. 825–864. (in Russian). (English Transaltion: (1964). *Amer. Math. Soc. Transl. Ser. II* **40**, pp. 39–84.)

[644] Makin, A. S. (2004). An eigenvalue problem for nonlinear Sturm–Liouville equations with eigenparameter dependent boundary conditions, *Dokl. Akad. Nauk* **399**, no. 2, pp. 161–164. (in Russian).

[645] Maksudov, F. G. and Allakhverdiev, B. P. (1993). On the extension of Schrödinger operator with matrix potentials, *Dokl. Akad. Nauk* **332**, no. 1,

pp. 18–20. (in Russian). (English Translation: (1994). *Russian Acad. Sci. Dokl. Math.* **48**, no. 2, pp. 240–243.)

[646] Malamud, M. M. (1992). On a formula for the generalized resolvents of a non-densely defined Hermitian operator, *Ukraïn. Mat. Zh.* **44**, no. 12, pp. 1658–1688. (in Russian). (English Translation: (1992). *Ukrainian Math. J.* **44**, no. 12, pp. 1522–1547.)

[647] Malamud, M. M. (1992). Some classes of extensions of a Hermitian operator with lacunae, *Ukraïn. Mat. Zh.* **44**, no. 2, pp. 215–233. (in Russian). (English Translation: (1992). *Ukrainian Math. J.* **44** (1992), no. 2, pp. 190–204.)

[648] Malamud, M. M. (1999). Questions of uniqueness in inverse problems for systems of differential equations on a finite interval, *Tr. Mosk. Mat. Obs.* **60**, pp. 199–258. (in Russian). (English Translation: (1999). *Trans. Moscow Math. Soc.* **1999**, pp. 173–224.)

[649] Malamud M. M. and Malamud S. M. (2003). Spectral theory of operator measures in a Hilbert space, *Algebra i Analiz* **15**, no. 3, pp. 1–77. (in Russian). (English Translation: (2004). *St. Petersburg Math. J.* **15**, no. 3, pp. 323–373.)

[650] Malamud, M. M. and Mogilevskiy, V. I. (1997). On extensions of dual pairs of operators, *Dopov. Nats. Akad. Nauk Ukr. (Mat. Prirodozn. Tekh. Nauki)*, no. 1, pp. 30–37.

[651] Malamud, M. M. and Mogilevskiy, V. I. (2002). Krein type formula for canonical resolvents of dual pairs of linear relations, *Methods Funct. Anal. Topology* **8**, no. 4, pp. 72–100.

[652] Mantlik, F. and Schneider, A. (1990). Note on the absolutely continuous spectrum of Sturm-Liouville operators, *Math. Z.* **205**, no. 3, pp. 491–498.

[653] Marchenko, V. A. (1950). Some questions of the theory of second order differential operators, *Dokl. Akad. Nauk SSSR* **72**, no. 3, pp. 457–460. (in Russian).

[654] Marchenko, V. A. (1955). Theorems of Tauberian type in spectral analysis of differential operators, *Izv. Akad. Nauk SSSR Ser. Mat.* **19**, pp. 381–422. (in Russian).

[655] Marchenko, V. A. (1990). The Cauchy problem for the Korteweg-de Vries equation with nondecreasing initial data, In: *Bar'yakhtar, V. G., Zakharov, V. E. and Chernousenko, V.M. (Editors), Integrability and kinetic equations for solitons*, Naukova Dumka, Kiev, pp. 168–212. (in Russian). (English Translation: (1990). Springer series in Nonlinear Dynamics, What is integrability? Ed. by Zakharov V. E., pp. 273–318.)

[656] Marchenko, V. A. (1992). Characteristic properties of Weyl solutions, *Dokl. Akad. Nauk* **324**, no. 2, pp. 261–264. (in Russian). (English Translation: (1992). *Russian Acad. Sci. Dokl. Math.* **45**, no. 3, pp. 563–566.)

[657] Marchenko, V. A. (1996). Operator algebras, non-linear equations and Darboux-like transforms, *Algebraic and geometric methods in mathematical physics (Kaciveli, 1993), Math. Phys. Stud.* **19**, Kluwer Acad. Publ., Dordrecht, pp. 323–342.

[658] Marchenko, V. A. and Ostrovskiy, I. V. (1975). A characterization of the

spectrum of Hill's operator, *Mat. Sb.* **97**, no. 4, pp. 540–606. (in Russian). (English Translation: (1975). *Sb. Math.* **26**, no. 4, pp. 493–554.)

[659] Marchenko, V. A. and Ostrovskiy, I. V. (1980). Approximation of periodic potentials by finite-zone potentials, *Vestink Khar'kov Gos. Univ.* **205**, pp. 4–40. (in Russian). (English Translation: (1987). *Selecta Math. Sovietica* **6**, pp. 101–136; (1988). *Selecta Math. Sovietica* **7**, pp. 99–100.)

[660] Marchenko, V. A. and Rofe-Beketov, F. S. (1958). Expansion in eigenfunctions of non-self-adjoint singular differential operators, *Dokl. Akad. Nauk SSSR* **120**, no. 5, pp. 963–966. (in Russian).

[661] Markus A. S. and Matsaev V. I. (1975). Spectral theory of holomorphic operator-valued functions in the Hilbert space, *Funktsional. Anal. i Prilozhen.* **9**, no. 1, pp. 76–77 (in Russian). (English Translation: (1975). *Funct. Anal. Appl.* **9**, no. 1, pp. 73–74.)

[662] Martynov, V. V. (1965). Conditions for the discreteness and continuity of spectrum in the case of a self-conjugate system of even order differential equations, *Differ. Uravn.* **1**, no. 12, pp. 1578–1591. (in Russian). (English Translation: (1965). *Differ. Equ.* **1**, pp. 1241–1251.)

[663] Martynov, V. V. (1965). Conditions of discreteness and continuity of the spectrum in the case of a self-conjugate system of first order differential equations, *Dokl. Akad. Nauk SSSR* **165**, pp. 996–999. (in Russian). (English Translation: (1965). *Soviet Math. Dokl.* **6**, pp. 1546–1550.)

[664] März, R. (1995). On linear differential-algebraic equations and linearizations, *Seventh Conference on the Numerical Treatment of Differential Equations (Halle, 1994)*, *Appl. Numer. Math.* **18**, no. 1-3, pp. 267–292.

[665] Maslov, V. P. (1968). The existence of a solution which decreases as $t \to \infty$ of Sobolev's equation for small vibrations of a rotating fluid in a cylindrical region, *Sibirsk. Mat. Zh.* **9**, no. 6, pp. 1351–1359. (in Russian). (English Translation: (1968). *Siberian Math. J.* **9**, pp. 1013–1020.)

[666] Matkovskiy, A. P. and Mukhachev, V. A. (1970). Fundamental systems for differential equations with operator coefficients, *Mat. Zametki* **8**, no. 6, pp. 777–782. (in Russian). (English Translation: (1970). *Math. Notes* **8**, pp. 910–913.)

[667] McGarvey, D. (1962). Operators commuting with translation by one. I. Representation theorems, *J. Math. Anal. Appl.* **4**, pp. 366–410.

[668] McGarvey, D. (1965). Operators commuting with translation by one. II. Differential operators with periodic coefficients in $L_p(-\infty, \infty)$, *J. Math. Anal. Appl.* **11**, pp. 564–596.

[669] McGarvey, D. (1965). Operators commuting with translation by one. II. Perturbation results for periodic differential operators, *J. Math. Anal. Appl.* **12**, pp. 187–234.

[670] McKean, H. and Van Moerbeke, P. (1975). The spectrum of Hill's equations, *Invent. Math.* **30**, pp. 217–274.

[671] McLeod, J. B. (1966). The number of integrable-square solutions of ordinary differential equations, *Quart. J. Math. Oxford Ser. (2)* **17**, pp. 285–290.

[672] Melik-Adamyan, F. E. (1977). Canonical differential operators in a Hilbert space, *Izv. Akad. Nauk Armyan. SSR Ser. Mat.* **12**, no. 1, pp. 10–31. (in

Russian).
[673] Mennicken, R. and Shkalikov, A. A. (1996). Spectral decomposition of symmetric operator matrices, *Math. Nachr.* **179**, pp. 259–273.
[674] Mikhailets, V. A. (1974). Solvable and sectorial boundary value problems for the operator Sturm-Liouville equation, *Ukraïn. Mat. Zh.* **26**, no. 4, pp. 450–459. (in Russian). (English Translation: (1974). *Ukrainian Math. J.* **26**, pp. 370–376.)
[675] Mikhailets, V. A. (1980). Spectra of operators and boundary value problems, *Spectral analysis of differential operators*, Akad. Nauk Ukrain. SSR, Inst. Mat., Kiev, pp. 106–131. (in Russian).
[676] Mikhailets, V. and Molyboga, V. (2004). Singular eigenvalue problems on the circle, *Methods Funct. Anal. Topology* **10**, no. 3, pp. 44–53.
[677] Mikhailets, V. and Molyboga, V. (2004). Uniform estimates for the semi-periodic eigenvalues of the singular differential operators, *Methods Funct. Anal. Topology* **10**, no. 4, pp. 30–57.
[678] Mikhailets, V. A. and Molyboga, V. M. (2004). Spectral problems on classes of periodic generalized functions, *Institute of Mathematics of National Academy Sciences of Ukraine*, Kiev, preprint 2004.10, 46 pp. (in Ukrainian).
[679] Mikhailets, V. A. and Sobolev, A. V. (1999). Common eigenvalue problem and periodic Schrödinger operators, *J. Funct. Anal.* **165**, no. 1, pp. 150–172.
[680] Mirzoev, G. A. (1980). Fourth-order quasi-regular differential operator, *Dokl. Akad. Nauk SSSR* **251**, no. 3, pp. 550–553. (in Russian). (English Translation: (1980). *Soviet Math. Dokl.* **21**, pp. 480–483.)
[681] Mirzoev, G. A. (1981). The description of self-adjoint extensions of quasi-regular operators generated by binomial differential expressions, *Mat. Zametki* **29**, no. 2, pp. 225–233. (in Russian). (English Translation: (1981). *Math. Notes* **29**, pp. 116–121.)
[682] Mishchenko, V. O. (1977). Generalized spectral matrix and inverse problem for a non-selfadjoint abstract hyperbolic differential operator, *Dokl. Akad. Nauk Ukrain. SSR Ser. A*, No. 11, pp. 981–984. (in Russian).
[683] Mishchenko, V. O. (1979). A generalized spectral matrix for a nonselfadjoint abstract hyperbolic differential operator and an inverse problem, *Teor. Funktsii Funktsional. Anal. i Prilozhen.*, No. 31, pp. 110–119. (in Russian).
[684] Mishchenko, V. O. (1980). Expansion in eigenfunctions of a symmetric abstract elliptic differential operator, *Differ. Uravn.* **16**, no. 8, pp. 1474–1483. (in Russuan). (English Translation: (1980). *Differ. Equ.* **16**, no. 8, pp. 949–955.)
[685] Mishchenko, V. O. (1983). Reconstruction of the operator potential from the generalized spectral matrix of an abstract elliptic differential operator, *Teor. Funktsii Funktsional. Anal. i Prilozhen.*, No. 40, pp. 115–118. (in Russian).
[686] Misyura, T. V. (1978). Characteristics of the spectra of the periodic and antiperiodic boundary value problems that are generated by the Dirac operator, I, II, *Teor. Funktsii Funktsional. Anal. i Prilozhen.*, No. 30, pp. 90-101;

(1979). *Teor. Funktsii Funktsional. Anal. i Prilozhen.*, No. 31, pp. 102–109. (in Russian).

[687] Misyura, T. V. (1980). Finite-zone Dirac operators, *Teor. Funktsii Funktsional. Anal. i Prilozhen.*, No. 33, pp. 107–111. (in Russian).

[688] Misyura, T. V. (1991). Asymptotic formula for Weyl solutions of Dirac equation, *Dokl. Akad. Nauk Ukrain. SSR Ser. A*, No. 5, pp. 26–28. (in Russian).

[689] Mityagin, B. (2003). Convergence of expansions in eigenfunctions of the Dirac operator, *Dokl. Akad. Nauk* **393**, no. 4, pp. 456–459. (in Russian).

[690] Mityagin, B. (2004). Spectral expansions of one-dimensional periodic Dirac operators, *Dynamics of PDE* **1**, no. 2, pp. 125–191.

[691] Mogilevskiy, V. I. (1994). On proper exentsions of a singular differential operator in a space of vector functions, *Dopov./Dokl. Akad. Nauk Ukraïni*, no. 9, pp. 29–33. (in Russian).

[692] Mogilevskiy, V. I. (1995). On proper exentsions of a singular differential operator in a space of vector functions, *Ukraïn. Mat. Zh.* **47**, no. 5, pp. 671–685. (in Russian). (English Translation: (1995). *Ukrainian Math. J.* **47**, no. 5, pp. 773–788.)

[693] Mogilevskiy, V. I. (2000). On not self-adjoint differential operators in a vector function space, *Dopov. Nats. Akad. Nauk Ukr. (Mat. Prirodozn. Tekh. Nauki)*, no. 7, pp. 35–40.

[694] Möhr, C. (2001). Schrödinger operators with singular potentials on the circle: spectral analysis and applications, Thesis, University of Zürich, 134 p.

[695] Molchanov, S. A. and Chulaevskiy, V. A. (1984). The structure of a spectrum of the lacunary-limit-periodic Schrödinger operator, *Funktsional. Anal. i Prilozhen.* **8**, no. 4, pp. 90–91. (in Russian). (English Translation: (1984). *Funct. Anal. Appl.* **18**, pp. 343–344.)

[696] Molchanov, S. A. and Novitskiy, M. V. (1986). Spectral invariants of the Schrödinger operator on a Euclidean torus, *Dokl. Akad. Nauk SSSR* **286**, no. 2, pp. 287–291. (in Russian). (English Translation: (1986). *Soviet Math. Dokl.* **33**, no. 1, pp. 82–85.)

[697] Molyboga, V. (2003). Estimates for periodic eigenvalues of the differential operator $(-1)^m d^{2m}/dx^{2m} + V$ with V-distribution, *Methods Funct. Anal. Topology* **9**, no. 2, pp. 163–178.

[698] Morse, M. (1930). A generalization of the Sturm Theorems in n-space, *Math. Ann.* **103**, pp. 52–69.

[699] Moser, J. (1981). An example of a Schrödinger equation with almost periodic potential and nowhere dense spectrum, *Comment. Math. Helv.* **56**, pp. 198–224.

[700] Muckenhoupt, B. (1972). Hardy's inequality with weights, Collection of articles honoring the completion by Antoni Zygmund of 50 years of scientific activity, I, *Studia Math.* **44**, pp. 31–38.

[701] Myshkis, A. D. (1948). Uniqueness of solution of a Cauchy problem, *Uspekhi Mat. Nauk* **3**, no. 2, pp. 3–46. (in Russian).

[702] Myshkis, A. D. (1979). Differential inequality, Mathematical Encyclopaedia, vol. 2., Soviet Encyclopaedia, pp. 279–280. (in Russian).

[703] Nayman, P. B. (1962). On the theory of periodic and limit-periodic Jacobian

matrices, *Dokl. Akad. Nauk SSSR* **143**, pp. 277–279. (in Russian). (English Translation: (1962). *Soviet Math. Dokl.* **2**, pp. 383–385.)

[704] Neyman-zade, M. I. and Shkalikov, A. A. (1999). Schrödinger operators with singular potentials from spaces of multipliers, *Mat. Zametki* **66**, no. 5, pp. 723–733. (in Russian). (English Translation: (1999). *Math. Notes* **66**, no. 5-6, pp. 599–607.)

[705] Novikov, S. P. (1970). Algebraic construction and properties of Hermitian analogues of K-theory over rings with invloution from the viepoint of Hamiltonian formalism. Applications to differential topology and characteristic classes, I, II, *Izv. Akad. Nauk. SSSR Ser. Mat.* **34**, no. 2, pp. 253–288; no. 3, pp. 475–500. (in Russian). (English Translation: (1970). *Math. USSR-Izv.* **4** pp. 257–292; pp. 479–505.)

[706] Novikov, S. P. (1974). Periodic Korteweg-de Vries problem I, *Funktsional. Anal. i Prilozhen.* **8**, no. 3, pp. 54–66. (in Russian). (English Translation: (1974). *Funct. Anal. Appl.* **8**, pp. 236–246.)

[707] Novikov, S. P. (1986). Bloch homology. Critical points of functions and closed 1-forms, *Dokl. Akad. Nauk SSSR* **287**, no. 6, pp. 1321–1324. (in Russian). (English Translation: (1986). *Soviet Math. Dokl.* **33**, no. 2, pp. 551–555.)

[708] Novikov, S. P. (1997). Schrödinger operator on graphs and topology, *Uspekhi Mat. Nauk* **52**, no. 6, pp. 177–178. (in Russian). (English Translation: (1997). *Russian Math. Surveys* **52**, no. 6, pp. 1320–1321.)

[709] Novikov, S. P. (1999). The discrete Schrödinger operator, *Tr. Mat. Inst. Steklova* **224**, pp. 275–290. (in Russian). (English Translation: (1999). *Proc. Steklov Inst. Math.* **224**, no. 1, pp. 250–265.)

[710] Novikov, S. P. (2004). Topology in the 20th century: a view from the inside, *Uspekhi Mat. Nauk* **59**, no. 5, pp. 3–28. (in Russian). (English Translation: (2004). *Russian Math. Surveys* **59**, no. 5, pp. 803–829.)

[711] Novikov, S. P. and Shubin, M. A. (1986). Morse inequalities and von Neumann II_1-factors, *Dokl. Akad. Nauk SSSR* **289**, no. 2, pp. 289–292. (in Russian). (English Translation: (1987). *Soviet Math. Dokl.* **34**, no. 1, pp. 79–82.)

[712] Novitskiy, M. V. (1985). Reconstructing the function of the number of rotations for the Schrödinger operator with almost periodic potential from a countable collection of polynomial conservation laws, *Funktsional. Anal. i Prilozhen.* **19**, no. 3, pp. 90–91. (in Russian). (English Translation: (1985). *Funct. Anal. Appl.* **19**, no. 3, pp. 243–245.)

[713] Novitskiy, M. V. (1989). On an inverse problem for smooth random Schrödinger operators, *Schrödinger operators, standard and nonstandard, Papers from the conference held in Dubna, September 6–10, 1988, Edited by Pavel Exner and Petr Sheba*, World Scientific Publishing Co., Inc., Teaneck, NJ, pp. 397–399.

[714] Novitskiy, M. V. (1990). On a complete description of the principal discrete series of spectral invariants of the Hill operator, *Order, disorder and chaos in quantum systems (Dubna, 1989)*, *Oper. Theory Adv. Appl.* **46**, Birkhäuser, Basel, pp. 115–117.

[715] Novitskiy, M. V. (1994). Quasianalytic classes and isospectral Hill's oper-

ators, *Spectral operator theory and related topics*, *Adv. Soviet Math.* **19**, Amer. Math. Soc., Providence, RI, pp. 27–39.

[716] Novitskiy, M. V. (1996). The motion integrals for the KdV equation and related inverse problems, *Inverse Problems* **12**, no. 1, pp. 55–75.

[717] Ohtsuki, M. (1982). On the number of apparent singularities of a linear differential equation, *Tokyo J. Math.* **5**, no. 1, pp. 23–29.

[718] Orlov, S. A. (1953). On the deficiency index of linear differential operators, *Dokl. Akad. Nauk SSSR* **92**, no. 3, pp. 483–486. (in Russian).

[719] Orlov, S. A. (1976). Nested matrix discs that depend analytically on a parameter, and theorems on the invariance of the ranks of the radii of the limit matrix discs, *Izv. Akad. Nauk. SSSR, Ser. Mat.* **40**, no. 3, pp. 593–644. (in Russian). (English Translation: (1976). *Math. USSR-Izv.* **10**, pp. 565–613.)

[720] Orlov, S. A. (1989). Description of the Green functions of canonical differential systems I, II, *Teor. Funktsii Funktsional. Anal. i Prilozhen.*, No. 51, pp. 78–88; No. 52, pp. 33–39. (in Russian). (English Translation: (1990). *J. Soviet Math.* **52**, no. 6, pp. 3500–3508; *J. Soviet Math.* **52**, no. 5, pp. 3372–3377.)

[721] Orochko, Yu. B. (1988). The hyperbolic equation method in the theory of operators of Schrödinger type with locally integrable potential, *Uspekhi Mat. Nauk* **43**, no. 2(260), pp. 43–86. (in Russian). (English Translation: (1988). *Russian Math. Surveys* **43**, no. 2, pp. 51–102.)

[722] Orudzhev, E. G. (1999). Boundary value problems for even-order differential equations with multiple characteristics, *Dokl. Akad. Nauk* **368**, no. 1, pp. 14–17. (in Russian). (English Translation: (1999). *Russian Acad. Sci. Dokl. Math.* **60**, no. 2, pp. 157–160.)

[723] Orudzhev, E. G. (2000). The resolvent and the spectrum of a class of differential operators with periodic coefficients, *Funktsional. Anal. i Prilozhen.* **34**, no. 3, pp. 87–90. (in Russian). (English Translation: (2000). *Funct. Anal. Appl.* **34**, no. 3, pp. 232–234.)

[724] Orudzhev, E. G. (2003). On a boundary value problem for an even-order equation with multiple characteristic roots, *Dokl. Akad. Nauk* **390**, no. 2, pp. 158–161. (in Russian).

[725] Orudzhev, E. G. (2005). On spectral decomposition by main functions of a quadratic pencil on the whole axis, *Mat. Fiz. Anal. Geom.* **12**, no. 1, pp. 107–113. (in Russian).

[726] Palais, R. S. and Smale, S. (1964). A generalized Morse theory, *Bull. Amer. Math. Soc.* **70**, pp. 165–172.

[727] Pastur, L. A. (1973). Spectra of random self-adjoint operators, *Uspekhi Mat. Nauk* **28**, no. 1, pp. 3–64. (in Russian). (English Translation: (1973). *Russian Math. Surveys* **28**, no. 1, pp. 1–67.)

[728] Pastur, L. A. (1980). Spectral properties of disordered systems in the one-body approximation, *Comm. Math. Phys.* **75**, no. 2, pp. 179–196.

[729] Pastur, L. A. and Tkachenko, V. A. (1984). On the spectral theory of the one-dimensional Schrödinger operator with limit-periodic potential, *Dokl. Akad. Nauk SSSR* **279**, no. 5, pp. 1050–1053. (in Russian). (English Translation: (1984). *Soviet Math. Dokl.* **30**, pp. 773–776.)

[730] Pastur, L. A. and Tkachenko, V. A. (1988). Spectral theory of a class of one-dimensional Schrödinger operator with limit-periodic potential, *Tr. Mosk. Mat. Obs.* **51**, pp. 114–168. (in Russian). (English Translation: (1989). *Trans. Moscow Math. Soc.* **1989**, pp. 115–166.)

[731] Pastur, L. A. and Tkachenko, V. A. (1990). The inverse problem for a class of one-dimensional Schrödinger operator with periodic complex potential, *Izv. Akad. Nauk SSSR Ser. Mat.* **54**, no. 6, pp. 1252–1269. (in Russian). (English Translation: (1990). *Math. USSR-Izv.* **37**, no. 3, pp. 611–629.)

[732] Pastur, L. A. and Tkachenko, V. A. (1991). On the geometry of spectrum of one-dimensional Schrödinger operator with complex periodic potential, *Mat. Zametki* **50**, no. 4, pp. 88–95. (in Russian). (English Translation: (1991). *Math. Notes* **50**, no. 3-4, pp. 1045–1050.)

[733] Pavinskiy, P. P. (1961). Analytic expression for the effective mass of a particle moving in one-dimensional periodic field, *Vestnik Leningradskogo Universiteta, Ser. Fiz. Khim.* **16**, No. 2, pp. 8–12. (in Russian).

[734] Pavlov, B. S. (1961). On the non-selfadjoint operator $-y" + q(x)y$ on a semiaxis, *Dokl. Akad. Nauk SSSR* **141**, no. 4, pp. 807–810. (in Russian). (English Translation: (1962). *Soviet Math. Dokl.* **2**, pp. 1565–1568.)

[735] Pavlov, B. S. (1962). On the spectral theory of non-self-adjoint differential operators, *Dokl. Akad. Nauk SSSR* **146**, no. 6, pp. 1267–1270. (in Russian). (English Translation: (1962). *Soviet Math. Dokl.* **3**, pp. 1483–1486.)

[736] Pavlov, B. S. (1966). On a non-selfadjoint Schrödinger operator I, *Probl. Math. Phys.*, No. 1, Spectral Theory and Wave Processes, Izdat. Leningrad. Univ., Leningrad, pp. 102–132; (1967). On a non-selfadjoint Schrödinger operator II, *Probl. Math. Phys.*, No. 2, Spectral Theory, Diffraction Problems, Izdat. Leningrad. Univ., Leningrad, pp. 133–157; (in Russian). (English Translation: (1967). *Topics in Mathematical Physics* **1**, Consultants Bureau, New York, pp. 87–113; (1968). *Topics in Mathematical Physics* **2**, Consultants Bureau, New York, pp. 111–134.)

[737] Pavlov, B. S. (1977). Selfadjoint dilation of a dissipative Schrödinger operator, and expansion in its eigenfunction, *Mat. Sb.* **102**, no. 4, pp. 511–536. (in Russian). (English Translation: (1977). *Sb. Math.* **31**, no. 4, pp. 457–478.)

[738] Pavlov, B. S. (1987). The theory of extensions and explicitly solvable models, *Uspekhi Mat. Nauk* **42**, no. 6, pp. 99–131. (English Translation: (1987). *Russian Math. Surveys* **42**, no. 6, pp. 127–168.)

[739] Perkolab, L. V. (1984). An inverse problem for a periodic Jacobi matrix, *Teor. Funktsii Funktsional. Anal. i Prilozhen.*, No. 42, (1984), pp. 107–121. (in Russian).

[740] Pesin, Ya. B. and Yurchenko, A. A. (2004). Some physical models of the reaction-diffusion equation, and coupled map lattices, *Uspekhi Mat. Nauk* **59**, no. 3, pp. 81–114. (in Russian). (English Translation: (2004). *Russian Math. Surveys* **59**, no. 3, pp. 481–513.)

[741] Petrina, D. Ya. and Mishchenko, A. V. (1988). Exact solutions of a class of Boltzmann equations, *Dokl. Akad. Nauk SSSR* **298**, no. 2, 338–342. (in Russian). (English Translation: (1988). *Soviet Phys. Dokl.* **33**, no. 1,

pp. 32–33.)

[742] Petrina, D. Ya. and Petrina, K. D. (1998). Stochastic dynamics and Boltzmann hierarchy. I, II, III. *Ukraïn. Mat. Zh.* **50**, no. 2, pp. 195–210; no. 3, pp. 372–387; no. 4, pp. 552–569. (in Russian). (English Translation: (1998). *Ukrainian Math. J.* **50**, no. 2, pp. 224–240; no. 3, pp. 425–441; no. 4, pp. 626–645.)

[743] Petrovisch, M. (1901). Sur une manier d'étendre le théoreme de la moyenne aux équations differentielles du premier ordre, *Math. Ann.* **54**, no. 3, pp. 417–436. (in French).

[744] Phillips, R. S. (1959). Dissipative operators and hyperbolic systems of partial differential equations, *Trans. Amer. Math. Soc.* **90**, pp. 193–254. (Russian Translation: (1962). *Matematika* **6**, no. 4, pp. 11–70.)

[745] Piven', A. L. (1999) Solvability of Cauchy problem and estimates of initial manifold for an implicit operator-differential equation *Visnyk Kharkiv. Univ. Ser. Mat. Prikl. Mat. i Mekh.*, No. 458, pp. 101–108. (in Russian).

[746] Piven', A. L. (2005). Completness conditions for elementary solutions of higher-order operator-differential equations, *Mat. Fiz. Anal. Geom.* **12**, no. 1, pp. 86–102. (in Russian).

[747] Piven', A. L. and Rutkas, A. G. (2003). On the well-posedness of the Cauchy problem for higher-order operator-differential equations, *Dopov. Nats. Akad. Nauk Ukr. (Mat. Prirodozn. Tekh. Nauki)*, no. 5, pp. 32–37. (in Russian).

[748] Pleijel, A. (1975). A survey of spectral theory for pairs of ordinary differential operators, *Spectral theory and differential equations (Proc. Sympos., Dundee, 1974; dedicated to Konrad Jörgens)*, *Lecture Notes in Math.* **448**, Springer, Berlin, pp. 256–272.

[749] Plemelj, J. (1908). Riemannsche Funktionenscharen mit gegebener Monodromiegruppe, *Monatsch. für Math. u. Phys.* **19**, pp. 211–246. (in German).

[750] Pokornyi, Yu. V. and Pryadiev, V. L. (2004). Some problems of the qualitative Sturm–Liouville theory on a spatial network, *Uspekhi Mat. Nauk* **59**, no. 3, pp. 115–150. (in Russian). (English Translation: (2004). *Russian Math. Surveys* **59**, no. 3, pp. 515–552.)

[751] Posilicano, A. (2004). Boundary triples and Weyl functions for singular perturbations of self-adjoint operators, *Methods Funct. Anal. Topology* **10**, no. 2, pp. 57–63.

[752] Povzner, A. Ya. (1953). On the expansion of arbitrary functions in characteristic functions of the operator $-\Delta u + cu$, *Mat. Sb.* **32**, no. 1, pp. 109–156. (in Russian). (English Translation: (1967). *Amer. Math. Soc. Transl. Ser. II* **60**, pp. 109–156.)

[753] Povzner, A. Ya. (1962). On the Boltzmann equation in the kinetic theory of gases, *Mat. Sb.* **58**, no. 1, pp. 65–86. (in Russian). (English Translation: (1962). *Amer. Math. Soc. Transl. Ser. II* **47**, pp. 193–216.)

[754] Pozharskiy, A. A. (2000). A crystal with a singular potential in a homogeneous electric field, *Teoret. Mat. Fiz.* **123**, no. 1, pp. 132–149. (in Russian). (English Translation: (2000). *Theoret. and Math. Phys.* **123**, no. 1,

pp. 524–538.)

[755] Pozharskiy, A. A. (2002). On operators of Wannier-Stark type with singular potentials, *Algebra i Analiz* **14**, no. 1, pp. 158–193. (in Russian). (English Translation: (2003). *St. Petersburg Math. J.* **14**, no. 1, pp. 119–145.)

[756] Pozharskiy, A. A. (2004). On the nature of the spectrum of Stark-Wannier operator *Algebra i Analiz* **16**, no. 3, pp. 171–200. (in Russian).

[757] Reid, W. T. (1938). A system of ordinary linear differential equations with two point boundary conditions, *Trans. Amer. Math. Soc.* **44**, no. 3, pp. 508–521.

[758] Rekalo, A. M. (2002). Stabilization of solutions of nonlinear parabolic equations on thin two-layer domains, *Mat. Fiz. Anal. Geom.* **9**, no. 3, pp. 446–454.

[759] Rekalo, A. M. (2003). Asymptotic behavior of solutions of nonlinear parabolic equations on two-layer thin domains, *Nonlinear Anal.* **52**, no. 5, pp. 1393–1410.

[760] Rekalo, A. M. and Cheshov, I. D. (2004). The global attractor of a contact parabolic problem with a thin two-layer domain, *Mat. Sb.* **195**, no. 1, pp. 103–128. (in Russian).

[761] Rellich, F. (1954). Halbbeschränkte Differentialoperatoren höherer Ordnung, *Proceedings of the International Congress of Mathematicians*, bd. III, Amsterdam, pp. 243–250. (in German).

[762] Riemann, B. (1857). Zwei allgemeine Lehrsätze über lineare Differentialgleichungen mit algebraischen Koefficienten, *Gesammelte mathematische Werke, wissenschaftlicher Nachlaß und Nachträge. Collected papers. Nach der Ausgabe von Heinrich Weber und Richard Dedekind neu hrsgg. von Raghavan Narasimhan. (Collected mathematical works, scientific posthumous works and supplements. Collected papers. According to the edition by H. Weber and R. Dedekind newly edited by R. Narasimhan), 1990*, Springer-Verlag, Berlin; BSB B. G. Teubner Verlagsgesellschaft, Leipzig. (in German). (Russian Translation: (1948). Works, Moscow, Gostekhizdat, pp. 176–186.)

[763] Riemann, B. (1857). Beitrage zur Theorie der durch die Gauβsche Reine $F(\alpha, \beta, \gamma, x)$ darstellbaren Funktionen, *Gesammelte mathematische Werke, wissenschaftlicher Nachlaß und Nachträge. Collected papers. Nach der Ausgabe von Heinrich Weber und Richard Dedekind neu hrsgg. von Raghavan Narasimhan. (Collected mathematical works, scientific posthumous works and supplements. Collected papers. According to the edition by H. Weber and R. Dedekind newly edited by R. Narasimhan), 1990*, Springer-Verlag, Berlin; BSB B. G. Teubner Verlagsgesellschaft, Leipzig. (in German). (Russian Translation: (1948). Works, Moscow, Gostekhizdat, pp. 159–175.)

[764] Rofe-Beketov, F. S. (1957). On the boundary-value problem for a non-linear differential equation, *Mat. Prosveschenie*, No. 1, pp. 149–154. (in Russian).

[765] Rofe-Beketov, F. S. (1960). On the limit distribution of characteristic frequencies of an unordered chain (the mathematical ground of Dyson's method), *Zap. Mat. Otd. Fiz-Mat. Fak. Univ. i Mat. Obs.* **26**, ser. 4, pp. 143–153. (in Russian).

[766] Rofe-Beketov, F. S. (1960). Expansion in eigenfunctions of infinite systems of differential equations in the non-self-adjoint and self-adjoint cases, *Mat. Sb.* **51**, 3, pp. 293–342. (in Russian).

[767] Rofe-Beketov, F. S. (1963). On the spectrum of non-self-adjoint operators with periodic coefficients, *Dokl. Akad. Nauk SSSR* **152**, no. 6, pp. 1312–1315. (in Russian). (English Translation: (1963). *Soviet Math. Dokl.* **4**, pp. 1563–1566.)

[768] Rofe-Beketov, F. S. (1964). Expansion in eigenfunctions of systems with summable potential, *Dokl. Akad. Nauk SSSR* **156**, no. 5, pp. 1029–1032. (in Russian). (English Translation: (1964). *Soviet Math. Dokl.* **5**, pp. 772–775.)

[769] Rofe-Beketov, F. S. (1964). A finiteness test for the number of discrete levels which can be introduced into the gaps of the continuous spectrum by perturbations of a periodic potential, *Dokl. Akad. Nauk SSSR* **156**, no. 3, pp. 515–518. (in Russian). (English Translation: (1964). *Soviet Math. Dokl.* **5**, pp. 689–692.)

[770] Rofe-Beketov, F. S. (1966). Non-semibounded differential operators, *Teor. Funktsii Funktsional. Anal. i Prilozhen.*, No. 2, pp. 178–184. (in Russian).

[771] Rofe-Beketov, F. S. (1967). The spectral matrix and the inverse Strum-Liouville problem on the axis $(-\infty, \infty)$, *Teor. Funktsii Funktsional. Anal. i Prilozhen.*, No. 4, pp. 189–197. (in Russian).

[772] Rofe-Beketov, F. S. (1969). Self-adjoint extensions of differential operators in a space of vector-valued functions, *Teor. Funktsii Funktsional. Anal. i Prilozhen.*, No. 8, pp. 3–24. (in Russian).

[773] Rofe-Beketov, F. S. (1969). Self-adjoint extensions of differential operators in a space of vector-valued functions, *Dokl. Akad. Nauk SSSR* **184**, no. 5, 1034–1037. (in Russian). (English Translation: (1969). *Soviet Math. Dokl.* **10**, no. 1, pp. 188–192.)

[774] Rofe-Beketov, F. S. (1970). Conditions for the self-adjointness of the Schrödinger operator, *Mat. Zametki* **8**, no. 6, pp. 741–751. (in Russian). (English Translation: (1970). *Math. Notes* **8**, pp. 888–894.)

[775] Rofe-Beketov, F. S. (1973). A perturbation of a Hill's operator, that has a first moment and a non-zero integral, introduces a single discrete level into each of the distant spectral lacunae, *Matematicheskaya Fizika i Funktsional'nyi Analiz*, No. 4, Fiz.-Tehn. Inst. Nizkih Temperatur, Akad. Nauk Ukrain. SSR, Kharkov, pp. 158–159. (in Russian).

[776] Rofe-Beketov, F. S. (1975). Deficiency indices and properties of spectrum of some classes of differential operators, *Lecture Notes in Math.*, vol. 448, Springer Verlag, Berlin, pp. 273–293.

[777] Rofe-Beketov, F. S. (1975). Self-adjoint boundary conditions for differential operators in a space of vector-valued functions, *Proceedings of the Summer School on Spectral Theory of Operators and Representation Theory of Groups (15 May–3 June 1968)*, Èlm, Baku, pp. 180–185. (in Russian).

[778] Rofe-Beketov, F. S. (1975). The inverse Sturm-Liouville problem on the axis $(-\infty, \infty)$, *Proceedings of the Summer School on Spectral Theory of Operators and Representation Theory of Groups (15 May–3 June 1968)*,

Èlm, Baku, pp. 186–194. (in Russian).

[779] Rofe-Beketov, F. S. (1977). Square-integrable solutions, self-adjoint extensions and spectrum of differential systems, *Differential equations (Proc. Internat. Conf., Uppsala, 1977), Sympos. Univ. Upsaliensis Ann. Quingentesimum Celebrantis*, No. 7, Almqvist and Wiksell, Stockholm, pp. 169–178.

[780] Rofe-Beketov, F. S. (1977). Spectral analysis of the Hill operator and its perturbations, *Functional Analysis*, No. 9, Ul'yanovsk, pp. 144–155. (in Russian).

[781] Rofe-Beketov, F. S. (1979). Generalization of the Prüfer transformation and the discrete spectrum in gaps of the continuous spectrum, *Spectral theory of operators (Proc. Second All-Union Summer Math. School, Baku, 1975)*, Èlm, Baku, pp. 146–153. (in Russian).

[782] Rofe-Beketov, F. S. (1980). Perturbations and Friedrichs extensions of semibounded operators on variable domains, *Dokl. Akad. Nauk SSSR* **255**, no. 5, pp. 1054–1058. (in Russian). (English Translation: (1980). *Soviet Math. Dokl.* **22**, no. 3, pp. 795–799.)

[783] Rofe-Beketov, F. S. (1983). Oscillatory properties of solutions of systems of ordinary and partial differential equations, *Proceedings of the Conference: Constructive Theory of Functions, 1981, Sofia*, pp. 144–149. (in Russian).

[784] Rofe-Beketov, F. S. (1984). On positive differential operators (deficiency indices, factorization, perturbations), *Proc. Roy. Soc. Edinburgh Sect. A* **97**, pp. 247–257. (see also, Preprint 23-83, Khar'kov: Fiz. Tekh. Inst. Niz. Temp. Akad. Nauk Ukrain. SSR, 1983, 23pp.)

[785] Rofe-Beketov, F. S. (1984). Spectrum perturbations, the Kneser-type constants and the effective masses of zone-type potentials, *Proceedings of the Conference: Constructive Theory of Functions, 1984, Sofia*, pp. 757–766. (in Russian).

[786] Rofe-Beketov, F. S. (1984). Kneser-type constants and effective masses for zone-type potentials, *Dokl. Akad. Nauk SSSR* **276**, no. 2, pp. 356–359. (in Russian). (English Translation: (1984). *Soviet Phys. Dokl.* **29**, pp. 391–393.)

[787] Rofe-Beketov, F. S. (1985). Necessary and sufficient conditions for a finite rate of propagation for elliptic operators, *Ukraïn Mat. Zh.* **37**, no. 5, pp. 668–670. (in Russian).

[788] Rofe-Beketov, F. S. (1985). The numerical range of a linear relation and maximum relations, *Teor. Funktsii Funktsional. Anal. i Prilozhen.*, No. 44, pp. 103–112. (in Russian). (English Translation: (1990). *J. Soviet Math.* **48**, no. 3, pp. 329–336.)

[789] Rofe-Beketov, F. S. (1991). The inverse Sturm-Liouville problem for the spectral matrix on the whole axis and associated problems. *Pitman Res. Notes Math. Ser.*, No. 235, *Petkov V. and Lazarov R. (eds.), Integral equations and inverse problems (Varna, 1989)*, Longman Sci. Tech., Harlow, pp. 234–238.

[790] Rofe-Beketov, F. S. (1994). On an estimate for the growth of solutions of canonical almost periodic systems, *Mat. Fiz. Anal. Geom.* **1**, no. 1, pp. 139–

148. (in Russian).

[791] Rofe-Beketov, F. S. (1994). A factored phase matrix of a perturbed system and the discrete spectrum, *Functional Analysis*, No. 35, Ul'yanovsk, pp. 71–81. (in Russian).

[792] Rofe-Beketov, F. S. (2005). Operator-theoretical proof of the Arnold alternation theorem and its generalization, *Mat. Fiz. Anal. Geom.* **12**, no. 1, pp. 119–125. (in Russian).

[793] Rofe-Beketov, F. S. and Grinshpun, E. Z. (1990). On self-adjointness of differential operators, Appendix to the book [151], pp. 163–189. (in Russian).

[794] Rofe-Beketov, F. S. and Khol'kin, A. M. (1977). The connection between the spectral and oscillation properties of the matrix Sturm-Liouville problem, *Mat. Sb.* **102**, no. 3, pp. 410–424. (in Russian). (English Translation: (1977). *Sb. Math.* **31**, no. 3, pp. 365–378.)

[795] Rofe-Beketov, F. S. and Khol'kin, A. M. (1981). The connection between spectral and oscillation properties of differential systems of arbitrary order, *Dokl. Akad. Nauk SSSR* **261**, no. 3, pp. 551–555. (in Russian). (English Translation: (1981). *Soviet Math. Dokl.* **24**, no. 3, pp. 555–559.)

[796] Rofe-Beketov, F. S. and Khol'kin, A. M. (1984). The fundamental system of solutions of an operator differential equation with a boundary condition at infinity, *Mat. Zametki* **36**, no. 5, pp. 697–709. (in Russian). (English Translation: (1984). *Math. Notes* **36**, no. 5–6, pp. 846–853.)

[797] Rofe-Beketov, F. S. and Khol'kin, A. M. (1985). The connection between spectral and oscillatory properties of arbitrary-order systems, Preprint Fiz. Tekh. Inst. Niz. Temp. Akad. Nauk Ukrain. SSR, Khar'kov, 8-85, 31pp. (in Russian).

[798] Rofe-Beketov, F. S. and Khol'kin, A. M. (1985). Dependence of the spectrum of an operator boundary value problem on the change in the interval, *Teor. Funktsii Funktsional. Anal. i Prilozhen.*, No. 43, pp. 107–119. (in Russian). (English Translation: (1990). *J. Soviet Math.* **48**, no. 2, pp. 210–220).

[799] Rofe-Beketov, F. S. and Khol'kin, A. M. (1987). The connection between the spectral and oscillation properties of operator-differential equations of arbitrary order, I. Theorems of Sturm type, *Teor. Funktsii Funktsional. Anal. i Prilozhen.*, No. 48, pp. 101–111. (in Russian). (English Translation: (1990). *J. Soviet Math.* **49**, no. 2, pp. 919–926.)

[800] Rofe-Beketov, F. S. and Khol'kin, A. M. (1988). The connection between spectral and oscillation properties of operator-differential equations of arbitrary order, II. Application of factorization and squaring, *Teor. Funktsii Funktsional. Anal. i Prilozhen.*, No. 49, pp. 101–111. (in Russian). (English Translation: (1990). *J. Soviet Math.* **49**, no. 4, pp. 1106–1113.)

[801] Rofe-Beketov, F. S. and Khol'kin, A. M. (2002). Oscillation theory of differential-operator equations (Plenary Lecture), Ukrainian Math. Congress, 2001, *Funct. Anal. Proc.*, Kyiv, pp. 254–269.

[802] Rofe-Beketov, F. S. and Khol'kin, A. M. (2004/2005). Theorems of Sturm type for differential equations with operator coefficients, *Visnyk Pryazovs'kogo Derzhavnogo Tekhnichnogo Universytetu*, submitted. (in Russian).

[803] Rofe-Beketov, F. S. and Khrabustovskiy, V. I. (1971). The stability of the solutions of Hill's equation with an operator coefficient, *Teor. Funktsii Funktsional. Anal. i Prilozhen.*, No. 13, pp. 140–147. (in Russian).
[804] Rofe-Beketov, F. S. and Khrabustovskiy, V. I. (1971). The stability of the solutions of Hill's equation with an operator coefficient that has a nonnegative mean value, *Teor. Funktsii Funktsional. Anal. i Prilozhen.*, No. 14, pp. 101–105. (in Russian).
[805] Rofe-Beketov, F. S. and Khrabustovskiy, V. I. (1971). A letter to the Editorial Board, *Teor. Funktsii Funktsional. Anal. i Prilozhen.*, No. 14, pp. 195. (in Russian).
[806] Rofe-Beketov, F. S. and Khristov, E. Kh. (1966). Transformation operators and the scattering function in the case of a highly singular potential, *Dokl. Akad. Nauk SSSR* **168**, no. 6, pp. 1265–1268. (in Russian). (English Translation: (1966). *Soviet Math. Dokl.* **7**, pp. 834–837.)
[807] Rofe-Beketov, F. S. and Khristov, E. Kh. (1969). Certain analytical questions and the inverse Sturm-Liouville problem for an equation with highly singular potential, *Dokl. Akad. Nauk SSSR* **185**, no. 4, pp. 768–771. (in Russian). (English Translation: (1969). *Soviet Math. Dokl.* **10**, no. 2, pp. 432–435.)
[808] Rofe-Beketov, F. S. and Khristov, E. Kh. (1971). Asymptotic and analytic questions connected with scattering by a highly singular potential, *Collection of Proceedings of Fiz. Tekh. Inst. Niz. Temp. Akad. Nauk Ukrain. SSR. Mathematics, Physics and Functional Analysis Series, Khar'kov*, No. 2, pp. 122–168. (in Russian).
[809] Rofe-Beketov, F. S. and Kogan, V. I. (1979). Deficiency indices for systems with a Nevanlinna weight, Spectral theory of operators, *Proc. Second All-Union Summer Math. School, Baku, 1975*, Èlm, Baku, pp. 153–161. (in Russian).
[810] Röhrl, H. (1957). Das Riemann-Hilbertsche Problem der Theorie der linearen Differentialgleichungen, *Math. Ann.* **133**, pp. 1–25. (in German).
[811] Roitberg, I. and Sakhnovich, A. (2005). D'Alembert-Liouville-Ostrogradskiy formula and related results, *Mat. Fiz. Anal. Geom.* **12**, no. 1, pp. 114–118.
[812] Rudykh, G. A. and Sinitsyn, A. V. (2005). On solvability of a nonlinear boundary value problem arising when plasma diffusion across magnetic field and its equilibria configurations are modeled, *Mat. Zametki* **77**, no. 2, pp. 219–234. (in Russian).
[813] Russakovskiy, E. M. (1992). The matrix Sturm-Liouville problem with a spectral parameter in the boundary conditions. The theory of V-Bezoutians, *Dokl. Akad. Nauk* **325**, no. 6, pp. 1124–1128. (in Russian). (English Translation: (1993). *Russian Acad. Sci. Dokl. Math.* **46**, no. 1, pp. 82–186.)
[814] Russakovskiy, E. M. (1993). Matrix Sturm-Liouville problem with a spectral parameter in boundary conditions, *Funtksional. Anal. i Prilozhen.* **27**, no. 1, pp. 86–88. (in Russian). (English Translation: (1993). *Funct. Anal. Appl.* **27**, no. 1, pp. 73–74.)
[815] Russakovskiy, E. M. (1994). The theory of V-Bezotians and its applications,

Linear Algebra Appl. **212/213**, pp. 437–460.

[816] Russakovskiy, E. M. (1996). Matrix Sturm-Liouville problem with a spectral parameter in boundary conditions. Algebraic and operator aspects, *Tr. Mosk. Mat. Obs.* **57**, pp. 171–198. (in Russian). (English Translation: (1996). *Trans. Moscow Math. Soc.* **1996**, pp. 159–184.)

[817] Russakovskiy, E. M. (1997). Matrix boundary-value problems with eigenvalue dependent boundary conditions (the linear case), *Oper. Theory Adv. Appl.* **95**, pp. 453–462.

[818] Rutkas, A. G. (1975). The Cauchy problem for the equation $Ax'(t)+Bx(t) = f(t)$, *Differ. Uravn.* **11**, no. 11, pp. 1996–2010. (in Russian). (English Translation: (1975). *Differ. Equ.* **11**, no. 11, pp. 1486–1497.)

[819] Rutkas, A. G. (1989). Classification and properties of solutions of the equation $Ax'+Bx = f(t)$, *Differ. Uravn.* **25**, no. 7, pp. 1150–1155. (in Russian). (English Translation: (1989). *Differ. Equ.* **25**, no. 7, pp. 809–813.)

[820] Rutkas, A. G. (1989). Perturbation of skew-Hermitian pencils and a degenerate Cauchy problem, *Ukrain. Mat. Zh.* **41**, no. 8, pp. 1082–1088. (in Russian). (English Translation: (1989). *Ukrainian Math. J.* **41**, no. 8, pp. 931–935.)

[821] Rutkas, A. and Vlasenko, L. (2000). Implicit operator differential equations and applications to electrodynamics, *Math. Methods Appl. Sci.* **23**, no. 1, pp. 1–15.

[822] Rutkas, A. G. and Vlasenko, L. A. (2001). Existence of solutions of degenerate nonlinear differential operator equations, *Nonlinear Oscil.* **4**, no. 2, pp. 252-263.

[823] Rutkas, A. G. and Vlasenko, L. A. (2003). Existence, uniqueness and continuous dependence for implicit semilinear functional differential equations, *Nonlinear Anal.* **55**, no. 1-2, pp. 125–139.

[824] Sadovnichiy, V. A., Sultanaev, Ya. T. and Akhtyamov, A. M. (2004). On the correctness of the inverse Sturm–Liouville problem with non-separated boundary conditions, *Dokl. Akad. Nauk* **395**, no. 5, pp. 592–595. (in Russian).

[825] Safronov, O. L. (1996). The discrete spectrum in gaps of the continuous spectrum for indefinite-sign perturbations with a large coupling constant, *Algebra i Analiz* **8**, no. 2, pp. 162–194. (in Russian). (English Translation: (1996). *St. Petersburg Math. J.* **8**, no. 2, pp. 307–331.)

[826] Sakhnovich, A. L. (1997). Iterated Bäcklund-Darboux transform for canonical systems, *J. Funct. Anal.* **144**, no. 2, pp. 359–370.

[827] Sakhnovich, A. L. (2001). Generalized Bäcklund-Darboux transformation: spectral properties and nonlinear equations, *J. Math. Anal. Appl.* **262**, no. 1, pp. 274–306.

[828] Sakhnovich, A. L. (2002). Dirac type and canonical systems: spectral and Weyl-Titchmarsh matrix functions, direct and inverse problems, *Inverse Problems* **18**, no. 2, pp. 331–348.

[829] Sakhnovich, A. L. (2003). Non-Hermitian matrix Schrödinger equation: Bäcklund-Darboux transformation, Weyl functions and *PT* symmetry, *J. Phys. A* **36**, no. 28, pp. 7789–7802.

[830] Sakhnovich, A. L. and Zubelli, J. P. (2001). Bundle bispectrality for matrix differential equations, *Integral Equations Operator Theory* **41**, no. 4, pp. 472–496.

[831] Sansuc, J.-J. and Tkachenko, V. (1996). Spectral parametrization of non-self-adjoint Hill's operators, *J. Differential Equations* **125**, no. 2, pp. 366–384.

[832] Savchuk, A. M. (2001). On the eigenvalues and eigenfunctions of the Sturm-Liouville operator with a singular potential, *Mat. Zametki* **69**, no. 2, pp. 277–285. (in Russian). (English Translation: (2001). *Math. Notes* **69**, no. 1-2, pp. 245–252.)

[833] Savchuk, A. M. and Shkalikov, A. A. (1999). Sturm-Liouville operators with singular potentials, *Mat. Zametki* **66**, no. 6, pp. 897–912. (in Russian). (English Translation: (2001). *Math. Notes* **66**, no. 5-6, pp. 741–753.)

[834] Savchuk, A. M. and Shkalikov, A. A. (2003). Sturm-Liouville operators with distribution potentials, *Tr. Mosk. Mat. Obs.* **64**, pp. 159–212. (in Russian).

[835] Scharf, G. (1965). Fastperiodische Potentiale, Inauguraldissertation zur Erlangung der Würde eines Doktors der Philosophie vorgelegt der Philosophischen Fakultät II der Universität Zürich Buchdruckerei Birkhäuser AG, Basel, 34pp. (in German). (*Helv. Phys. Acta* **38**, 6, pp. 573–606).

[836] Schmidt, K. M. (1997). Absolutely continuous spectrum of Dirac systems with potentials infinite at infinity, *Math. Proc. Cambridge Philos. Soc.* **122**, no. 2, pp. 377–384.

[837] Schmidt, K. M. (1999). Oscillation of the perturbed Hill equation and lower spectrum of radially periodic Schrödinger operators in the plane, *Proc. Amer. Math. Soc.* **127**, no. 8, pp. 2367–2374.

[838] Schmidt, K. M. (2000). Critical coupling constants and eigenvalue asymptotics of perturbed periodic Sturm-Liouville operators, *Comm. Math. Phys.* **211**, no. 2, pp. 465–485.

[839] Schmidt, K. M. (2000). Relative oscillation-non-oscillation criteria for perturbed periodic Dirac systems, *J. Math. Anal. Appl.* **246**, no. 2, pp. 591–607.

[840] Schmidt, K. M. (2001). An application of Gesztesy–Simon–Teschl oscillation theory to a problem in differential geometry, *J. Math. Anal. Appl.* **261**, no. 1, pp. 61–71.

[841] Schmidt, K. M. (2003). Eigenvalue asymptotics of perturbed periodic Dirac systems in the slow-decay limit, *Proc. Amer. Math. Soc.* **131**, no. 4, pp. 1205–1214.

[842] Schmidt, K. M. and Yamada, O. (1998). Spherically symmetric Dirac operators with variable mass and potentials infinite at infinity, *Publ. Res. Inst. Math. Sci.* **34**, no. 3, pp. 211–227.

[843] Serov, M. I. (1960). Certain properties of the spectrum of a non-selfadjoint differential operator of the second order, *Dokl. Akad. Nauk SSSR* **131**, no. 1, pp. 27–29. (in Russian). (English Translation: (1960). *Soviet Math. Dokl.* **1**, pp. 190–192.)

[844] Shapiro, M., Vinnikov, V. and Yuditskiy, P. (2004). Finite difference op-

erators with a finite band spectrum, *Mat. Fiz. Anal. Geom.* **11**, no. 3, pp. 331–340.

[845] Shin, D. Yu. (1940). On solutions to linear quasi-differential equation of the n-th order, *Mat. Sb. (N.S.)* **7** (59), no. 1, pp. 479–532. (in Russian).

[846] Shin, D. Yu. (1940). On solutions of the system of quasi-differential equations, *Dokl. Akad. Nauk SSSR* **28**, no. 5, pp. 391–395.

[847] Shkalikov, A. A. (1983). Boundary value problems for ordinary differential equations with a parameter in the boundary conditions, *Tr. Semin. im. I. G. Petrovskogo*, No. 9, pp. 190–229. (in Russian). (English Translation: (1987). *J. Soviet Math.* **33**, no. 6, pp. 1311–1342.)

[848] Shkalikov, A. A. (1995). On the essential spectrum of matrix operators, *Mat. Zametki* **58**, no. 6, pp. 945–949. (in Russian). (English Translation: (1995). *Math. Notes* **58**, no. 6, pp. 1359–1362.)

[849] Shnol', E. (1954). The eigenfunction behavior and the spectrum of Sturm–Liouville operators, *Uspekhi Mat. Nauk* **9**, No. 4, pp. 113–132. (in Russian).

[850] Shtraus (=Strauss), A. V. (1957). On the generalized resolvents and spectral functions of even-order differential operators, *Izv. Akad. Nauk SSSR Ser. Mat.* **21**, no. 6, pp. 785–808. (in Russian). (English Translation: (1960). *Amer. Math. Soc. Tr .nsl. Ser. II* **16**, pp. 462–464.)

[851] Shtraus, A. V. (1957). On spectral functions of an even order differential operator, *Dokl. Akad. Nauk SSSR* **115**, no. 1, pp. 67–70. (in Russian).

[852] Shtraus, A. V. (1958). On the spectral functions of the operator of differentiation, *Uspekhi Mat. Nauk* **13**, no. 6, pp. 185–191. (in Russian).

[853] Shtraus, A. V. (1961). On some families of extensions of symmetric operators, *Dokl. Akad. Nauk SSSR* **139**, no. 2, pp. 316–319. (in Russian). (English Translation: (1961). *Soviet Math. Dokl.* **2**, pp. 943–946.)

[854] Shtraus, A. V. (1962). Some questions on the expansion theory for symmetric non-self-adjoint operators, *Proceedings of the Second Scientific Conference of Math. Chairs of Pedagogical Institutes of the Povolzh'e(=Volga reigon), Kuybyshev Pedagogical Institute*, issue no. 1, pp. 121–124. (in Russian).

[855] Shtraus, A. V. (1973). On extensions of semi-bounded operators, *Dokl. Akad. Nauk SSSR* **211**, no. 3, pp. 543–546. (in Russian). (English Translation: (1973). *Soviet Math. Dokl.* **14**, pp. 1075–1079.)

[856] Shtraus, A. V. (1976). On self-adjoint extensions of semi-bounded operators, *Functional Analysis*, No. 3, Ul'yanovsk, pp. 161–171. (in Russian).

[857] Shtraus, A. V. (1998). Functional models and generalized spectral functions of symmetric operators, *Algebra i Analiz* **10**, no. 5, pp. 1–76 (in Russian). (English Translation: (1999). *St. Petersburg Math. J.* **10**, no. 5, pp. 733–784.)

[858] Shubin, M. A. (1979) Spectral theory and the index of elliptic operators with almost-periodic coefficients, *Uspekhi Mat. Nauk* **34**, no. 2, pp. 95–135. (in Russian). (English Translation: (1979). *Russian Math. Surveys* **34**, no. 2, pp. 109–157.)

[859] Sidel'nikova, N. A. (2003). On deficiency indices of a singular differential operator generated by the expression $y^{\mathrm{IV}} - 2a(x^{\alpha}y')' + bx^{\beta}y$, *Mat. Zametki* **73**, no. 1, pp. 153–156. (in Russian). (English Translation: (2003). *Math.*

Notes **73**, 1-2, pp. 148–151.)

[860] Simon, B. (1978). A canonical decomposition for quadratic forms with application to monotone convergence theorems, *J. Funct. Anal.* **28**, no. 3, pp. 377–385.

[861] Simon, B. (1977/78). Lower semicontinuity of positive quadratic forms, *Proc. Roy. Soc. Edinburgh Sect. A* **79**, no. 3-4, pp. 267–273.

[862] Simon, B. (2005). Sturm oscillation and comparison theorems, in: *Sturm–Liouville theory: past and present (Invited papers by participants of Sturm-200 Colloquium in Geneva, 2003; Amrein, W. O., Hinz, A. M. and Pearson, D. B., Editors)*, Birkhäuser, Basel, pp. 29–43.

[863] Skachek, B. Ya. (1965). A property of the spectrum of Dirac operators, *Dopov. Akad. Nauk Ukrain. SSR*, No. 10, pp. 1272–1275. (in Ukrainian).

[864] Skriganov, M. M. and Sobolev, A. V. (2005). Asymptotic estimates for spectral zones of periodic Schrödinger operators, *Algebra i Analiz* **17**, no. 1, pp. 276–288. (in Russian).

[865] Slyusarchuk, V. E. (1996). Strengthening of Kneser's theorem on the zeros of the solutions of the equation $y'' + p(x)y = 0$, *Ukraïn. Mat. Zh.* **48**, no. 4, pp. 520–524. (English Translation: (1996). *Ukrainian Math. J.* **48**, no. 4, pp. 576–581.)

[866] Smale, S. (1965). On the Morse index theorem, *J. Math. Mech.* **14**, no. 6, pp. 1049–1055.

[867] Sobolev, A. V. (1991). Weyl asymptotic for the discrete spectrum of the perturbed Hill operator, *Estimates and asymptotics for discrete spectra of integral and differential equations (Papers from the Seminar on Math. Phys. held in Leningrad 1989-1990. Edited by M. Sh. Birman), Adv. Soviet Math.* **7**, Amer. Math. Soc., Providence, RI, pp. 159–178.

[868] Sobolev, A. V. (1991). On asymptotics of the discrete spectrum in gaps of the continuous spectrum of the perturbed Hill operator, *Funktsional. Anal. i Prilozhen.* **25**, no. 2, pp. 93–95. (in Russian). (English Translation: (1991). *Funct. Anal. Appl.* **25**, no. 2, pp. 162–164.)

[869] Sobolev, S. L. (1954). On a new problem of mathematical physics, *Izv. Akad. Nauk SSSR. Ser. Mat.* **18**, pp. 3–50. (in Russian).

[870] Sodin, M. L. and Yuditskiy, P. M. (1994). Infinite-gap Jacobi matrices with pseudocontinuable Weyl functions and homogeneous spectrum, *Dokl. Akad. Nauk* **335**, no. 4, pp. 433–436. (in Russian). (English Translation: (1994). *Russian Acad. Sci. Dokl. Math.* **49**, no. 2, pp. 364–368.)

[871] Sodin, M. L. and Yuditskiy, P. M. (1994). Almost periodic Sturm-Liouville operators with a Cantor homogeneous spectrum and pseudo-continuable Weyl functions, *Dokl. Akad. Nauk* **339**, no. 6, pp. 736–738. (in Russian). (English Translation: (1995). *Russian Acad. Sci. Dokl. Math.* **50**, no. 3, pp. 512–515.)

[872] Sodin, M. L. and Yuditskiy, P. M. (1995). Almost periodic Sturm-Liouville operators with Cantor homogeneous spectrum, *Comment. Math. Helv.* **70**, no. 4, pp. 639–658.

[873] Storozh, O. G. (1984). Extensions of symmetric operators with unequal deficiency numbers, *Mat. Zametki* **36**, no. 5, pp. 791–796. (in Russian). (En-

glish Translation: (1984). *Math. Notes* **36**, no. 5-6, pp. 897–900.)

[874] Storozh, O. G. (1987). Description of some classes of extensions of a nonnegative operator, *Dopov. Akad. Nauk Ukrain. SSR Ser. A*, No. 10, pp. 15–17. (in Russian).

[875] Storozh, O. G. (1990). Extremal extensions of a non-negative operator and accretive boundary value problems, *Ukraïn. Mat. Zh.* **42**, no. 6, pp. 858–860. (in Russian). (English Translation: (1990). *Ukrainian Math. J.* **42**, no. 6, pp. 758–760.)

[876] Storozh, O. G. and Shuvar, O. B. (1993). The closure, dense definiteness and conditions for selfadjointness of some differential-boundary operators in the space of vector functions, *Dopov./Dokl. Akad. Nauk Ukraïni*, no. 8, pp. 20–24. (in Ukrainian).

[877] Storozh, O. G. and Shuvar, O. B. (1994). Conditions for maximum dissipativity of some differential-boundary operators in a space of vector functions, *Dopov./Dokl. Akad. Nauk Ukraïni*, no. 2, pp. 20–24. (in Ukrainian).

[878] Storozh, O. G. and Shuvar, O. B. (2004). Conditions for maximal dissipativity of smooth perturbations of the operators adjoint to symmetric ones, *Ukraïn Mat. Zh.* **56**, no. 7, pp. 966–976. (in Ukrainian).

[879] Sturm, J. C. F. (1829). Extrait d'une mémoire sur integration d'un systeme d'équations linéaires, *Bull. Sci. Math. Ferussac*, XII, pp. 313–322. (in French).

[880] Sturm, J. C. F. (1836). Mémoire sur les équations différentielles linéaires du second ordre, *J. Math. Pures et Appl. de Liouville*, vol. I, pp. 106–186. (=*Liouville Journal*). (in French).

[881] Sturm, J. C. F. (1837). Mémoire sur une classe d'équations différentielles partielles, *J. Math. Pures et Appl. de Liouville*, vol. I, pp. 373–444. (in French).

[882] Suslina, T. A. (2004). Homogenization of the stationary periodic Maxwell system, *Algebra i Analiz* **16**, no. 5, pp. 162–244. (in Russian).

[883] Sviridov, V . V. (1984). On the completeness of quasienergetic states, *Dokl. Akad. Nauk SSSR* **274**, 6, pp. 1366–1367. (in Russian). (English Translation: (1984). *Soviet Phys. Dokl.* **29**, no. 2, pp. 139–140.)

[884] Swanson, R. C. (1978). Fredholm intersection theory and elliptic boundary deformation problems. I, *J. Differential Equations* **28**, no. 2, pp. 189–201.

[885] Teller, E., Ishikawa, M., Wood, L., Hyde, R. and Nuckolls, J. (1996). Completely automated nuclear reactors for long-term operation, *1996 International Conference on Emerging Nuclear Energy Systems, Obninsk, Russia, 24-28 June 1996.*

[886] Thomas, L. E. (1973). Time dependent approach to scattering from impurities in a crystal, *Comm. Math. Phys.* **33**, pp. 335–343.

[887] Tikhomirov, M. and Filonov, N. (2004). Absolute continuity of the "even" periodic Schrödinger operator with non-smooth coefficients, *Algebra i Analiz* **16**, no. 3, pp. 201–210. (in Russian).

[888] Tikhonov, I. V. and Eidel'man, Yu. S. (2005). Uniqueness criterion in the inverse problem for an abstract differential equation with a non-stationary non-homogeneous term, *Mat. Zametki* **77**, no. 2, pp. 273–290. (in Russian).

[889] Tkachenko, V. A. (1964). On the spectral analysis of a non-self-adjoint Schrödinger operator with periodic complex-valued potential, *Dokl. Akad. Nauk SSSR* **155**, no. 2, pp. 289–291. (in Russian). (English Translation: (1964). *Soviet Math. Dokl.* **5**, pp. 413–415.)

[890] Tkachenko, V. A. (1992). Spectral analysis of the non-selfadjoint Hill operator, *Dokl. Akad. Nauk* **322**, no. 2, pp. 248–252. (in Russian). (English Translation: (1992). *Soviet Math. Dokl.* **45**, no. 1, pp. 78–82.)

[891] Tkachenko, V. A. (1994). Discriminants and generic spectra of nonselfadjoint Hill's operators, *Spectral operator theory and related topics*, *Adv. Soviet Math.* **19**, Amer. Math. Soc., Providence, RI, pp. 41–71.

[892] Tkachenko, V. A. (1996). Spectra of non-self-adjoint Hill operators and a class of Riemannian surfaces, *Ann. of Math.* **143**, no. 2, pp. 181–231.

[893] Uhlenbeck, K. (1973). The Morse index theorem in Hilbert space, *J. Differential Geometry* **8**, no. 4, pp. 555–564.

[894] Vainerman, L. I. (1974). Selfadjoint boundary value problems for strongly elliptic and hyperbolic second order equations in Hilbert space, *Dokl. Akad Nauk SSSR* **218**, pp. 745–748. (in Russian). (English Translation: (1974). *Soviet Math. Dokl.* **15**, pp. 1391–1395.)

[895] Vainerman, L. I. (1980). Extensions of closed operators in Hilbert space, *Mat. Zametki* **28**, no. 6, pp. 833–842. (in Russian). (English Translation: (1980). *Math. Notes* **28**, no. 5–6, pp. 871–875.)

[896] Vainerman, L. I. and Gorbachuk, M. L. (1970). On the self-adjointness of semi-bounded abstract differential operators, *Ukraïn. Mat. Zh.* **22**, no. 6, pp. 806–808. (in Russian).

[897] Vainerman, L. I. and Gorbachuk, M. L. (1975). Boundary value problems for a second order differential equation of hyperbolic type in Hilbert space, *Dokl. Akad. Nauk SSSR* **221**, no. 4, pp. 763–766. (in Russian). (English Translation: (1975). *Soviet Math. Dokl.* **16**, no. 2, pp. 401–405.)

[898] Vereshchagin, S. D. and Yurov, A. V. (2004). The Darboux transform and exactly solvable cosmological models, *Teoret. Mat. Fiz.* **139**, no. 3, pp. 405–422. (in Russian).

[899] Vernik, A. N. and Etkin, A. E. (1979). The eigenvalue problem with a spectral parameter that enters rationally into the boundary conditions, *Functional Analysis*, No. 12 (Spectral theory), Ul'yanovsk, pp. 167–173. (in Russian).

[900] Vishik, M. I. (1951). On strongly elliptic systems of differential equations *Mat. Sb.* **29**, no. 3, pp. 615–676. (in Russian).

[901] Vishik, M. I. (1952). On general boundary value problems for elliptic differential equations, *Tr. Mosk. Mat. Obs.* **1**, pp. 187–246. (in Russian). (English Translation: (1963). *Amer. Math. Soc. Transl. Ser. II* **24**, 2, pp. 107–172.)

[902] Vishik, M. I. and Lyusternik, L. A. (1960). A solution of some perturbation problems in the case of matrices and self-adjoint and non-self-adjoint operators, *Uspekhi Mat. Nauk* **15**, no. 3, pp. 3–80. (in Russian). (English Translation: (1960). *Russian Math. Surveys* **15**, no. 3, pp. 1–73.)

[903] Vladimirov, A. A. (2004). On the convergence of sequences of ordinary differential equations *Mat. Zametki* **75**, no. 6, pp. 941–943. (in Russian).

(English Translation: (2004). *Math. Notes* **75**, no. 5–6, pp. 877–880.)

[904] Vlasenko, L. A., Piven', A. L. and Rutkas, A. G. (2004). Correctness conditions of a Cauchy problem for differential operator equations of arbitrary order, *Ukraïn. Mat. Zh.* **56**, no. 11, pp. 1484–1500. (in Russian).

[905] Vlasenko, L. A. and Rutkas, A. G. (1993). Solvability and completeness for an electrodynamic system that is not of Kovalevskaya type, *Mat. Zametki* **53**, no. 1, pp. 138–140. (in Russian). (English Translation: (1993). *Math. Notes* **53**, no. 1-2, pp. 96–97.)

[906] Vlasenko, L. A. and Rutkas, A. G. (1996). Uniqueness and approximation theorems for a degenerate operator-differential equation, *Mat. Zametki* **60**, no. 4, pp. 597–601. (in Russian). (English Translation: (1996). *Math. Notes* **60**, no. 3-4, pp. 445–449.)

[907] Vlasenko, L. A. and Rutkas, A. G. (2003). On the unique solvability of a degenerate functional-differential equation, *Dopov. Nats. Akad. Nauk Ukr. (Mat. Prirodozn. Tekh. Nauki)*, no. 3, pp. 11–16. (in Russian).

[908] Walter, J. (1972). Absolute continuity of the essential spectrum of $-d^2/dt^2+q(t)$ without monotony of q, *Math. Z.* **129**, pp. 83–94.

[909] Walter, J. (1973). Regular eigenvalue problems with eigenvalue parameter in the boundary condition, *Math. Z.* **133**, pp. 301–312.

[910] Walters, T. (1980). A characterization of positive linear functionals and oscillation criteria for matrix differential equations, *Proc. Amer. Math. Soc.* **78**, no. 2, pp. 198–202.

[911] Walther, H.-O. and Skubachevskiy, A. L. (2003). On Floquet multipliers for slowly oscillating periodic solutions of nonlinear functional-differential equations, *Tr. Mosk. Mat. Obs.* **64**, pp. 3–53. (in Russian). (English Translation: (2003). *Trans. Moscow Math. Soc.* **2003**, pp. 1–44.)

[912] Walther, H.-O. and Skubachevskiy, A. L. (2005). On the hyperbolicity of fast oscillating periodic solutions to functional-differential equations, *Funktsional Anal. i Prilozhen.* **39**, no. 1, pp. 82–85. (in Russian).

[913] Weidmann, J. (1987). Note to the paper by R. Hempel, A. M. Hinz and H. Kalf: "On the essential spectrum of Schrödinger operators with spherically symmetric potentials", *Math. Ann.* **277**, pp. 209–211.

[914] Weikard, R. (1998). Picard operators, *Math. Nachr.* **195**, pp. 251–266.

[915] Weikard, R. (1998). On Hill's equation with a singular complex-valued potential, *Proc. London Math. Soc. (3)* **76**, no. 3, pp. 603–633.

[916] Weyl, H. (1909). Über gewöhnliche lineare Differentialgleichungen mit singulären Stellen und ihre Eigenfunktionen, *Göttinger Nachrichten* **1909**, pp. 37–64.

[917] Weyl, H. (1935). Über das Pick-Nevanlinna'sche Interpolationsproblem und sein infinitesimales Analogon, *Ann. of Math.(2)* **36**, pp. 230–254. (in German).

[918] Wienholtz, E. (1958). Halbbeschränkte partielle Differentialoperatoren zweiter Ordnung vom elliptischen Typus, *Math. Ann.* **135**, pp. 50–80. (in German).

[919] Windau, W. (1921). Über lineare Differentialgleichungen vierter Ordnung mit Singularitäten und die dazugehörigen Darstellungen willkürlicher

Funktionen, *Math. Ann.* **83**, pp. 256–279. (in German).

[920] Yafaev, D. R. (1970). On negative spectrum of an operator Schrödinger equation, *Mat. Zametki* **7**, no. 6, pp. 753–763. (in Russian). (English Translation: (1970). *Math. Notes* **7**, pp. 451–457.)

[921] Yakovets, V. P. (1998). On the structure of the generalized solution of a degenerate linear system of second-order differential equations, *Ukraïn. Mat. Zh.* **50**, no. 2, pp. 292–298. (in Ukrainian). (English Translation: (1998). *Ukrainian Math. J.* **50**, no. 2, pp. 334–341.)

[922] Yakubovich, V. A. (1962). Oscillatory properties solutions of canonical equations, *Mat. Sb.* **56**, pp. 3–42. (in Russian). (English Translation: (1964). *Amer. Math. Soc. Transl. Ser. II* **42**, pp. 247–288.)

[923] Yurko, V. A. (1990). Reconstruction of differential operators from the Weyl matrix, *Dokl. Akad. Nauk SSSR* **313**, no. 6, pp. 1368–1372. (in Russian). (English Translation: (1991). *Soviet Math. Dokl.* **42**, no. 1, pp. 229–233.)

[924] Yurko, V. A. (1991). Reconstruction of nonselfadjoint differential operators on the semi-axis from the Weyl matrix, *Mat. Sb.* **182**, no. 3, pp. 431–456. (in Russian). (English Translation: (1992). *Sb. Math.* **72**, no. 2, pp. 413–438.)

[925] Yurko, V. A. (2004). Inverse spectral problem for singular non-self-adjoint differential systems, *Mat. Sb.* **195**, no. 12, pp. 123–156. (in Russian). (English Translation: (2004). *Sb. Math.* **195**, no. 12, pp. 1823–1854.)

[926] Zakharov, V. E. and Faddeev, L. D. (1971). Korteweg-de Vries equation—completely integrable Hamiltonian system, *Funktsional. Anal. i Prilozhen.* **5**, no. 4, pp. 18–27. (in Russian). (English Translation: (1971). *Funct. Anal. Appl.* **5**, pp. 280–287.)

[927] Zelenko, L. B. (1967). Discreteness conditions for the spectrum of the Sturm-Liouville operator in a space of vector functions, *Mat. Zametki* **1**, no. 3, pp. 583–588. (in Russian).

[928] Zelenko, L. B. (1976). Asymptotic distribution of eigenvalues in a gap of the coninuous spectrum of a perturbed Hill operator, *Mat. Zametki* **20**, no. 3, pp. 341–350. (in Russian). (English Translation: (1976). *Math. Notes* **20**, pp. 750–755.)

[929] Zelenko, L. (2004). On a generic topological structure of the spectrum of one-dimensional Schrödinger operators with complex limit-periodic potentials, *Integral Equations Operator Theory* **50**, pp. 393–430.

[930] Zelenko, L. B. and Rofe-Beketov, F. S. (1971). The limit spectrum of systems of first order differential equations with slowly changing coefficients, *Differ. Uravn.* **7**, no. 11, pp. 1982–1991. (in Russian). (English Translation: (1971). *Differ. Equ.* **7**, pp. 1498–1505.)

[931] Zettl, A. (1973). Explicit conditions for the factorization of n-th order linear deifferential operators, *Proc. Amer. Math. Soc.* **41**, no. 1, pp. 137–145.

[932] Zettl, A. (1974). General theory of the factorization of ordinary linear differential operators, *Trans. Amer. Math. Soc.* **197**, pp. 341–353.

[933] Zettl, A. (1975). Deficiency indices of polynomials in symmetric differential expressions II, *Proc. Roy. Soc. Edinburgh Sect. A* **73**, pp. 301–306.

[934] Zheludev, V. A. (1967). On eigenvalues of a perturbed Schrödinger opera-

tor with a peridic potential, *Problems of mathematical physics*, monograph, No. 2, Izdat. Leningrad. Univ., pp. 108–123. (in Russian). (English Translation: (1968). *Topics in Mathematical Physics* **2**, Consultants Bureau, New York, pp. 87–101.)

[935] Zheludev, V. A. (1968). On the perturbation of spectrum of a Schrödinger operator with complex-valued periodic potential, *Problems of mathematical physics*, monograph, No. 3, Izdat. Leningrad. Univ., pp. 31–48. (in Russian). (English Translation: (1969). *Topics in Mathematical Physics* **3**, Consultants Bureau, New York, pp. 25–40.)

[936] Zheludev, V. A. (1970). On the perturbation of spectrum of one-dimensional self-adjoint Schrödinger operator with periodic potential, *Problems of mathematical physics*, monograph, No. 4, Izdat. Leningrad. Univ., pp. 61–82. (in Russian). (English Translation: (1971). *Topics in Mathematical Physics* **4**, Consultants Bureau, New York, pp. 55–76.)

[937] Zhernakov, N. V. (1986). Direct and inverse problems for a periodic Jacobian matrix, *Ukraïn. Mat. Zh.* **38**, no. 6, pp. 785–788. (in Russian). (English Translation: (1986). *Ukrainian Math. J.* **38**, no. 6, pp. 665–668.)

[938] Zhernakov, N. V. (1987). Integration of Toda lattices in the class of Hilbert-Schmidt operators, *Ukraïn. Mat. Zh.* **39**, no. 5, pp. 645–648. (in Russian). (English Translation: (1987). *Ukrainian Math. J.* **39**, no. 5, pp. 527–530.)

[939] Zhikov, V. V. (2004). On lacunae in the spectrum of some divergent elliptic operators with periodic coefficients, *Algebra i Analiz* **16**, no. 5, pp. 34–58. (in Russian).

[940] Zhislin, G. M. (1960). Investigation of the spectrum of the Schrödinger operator for the multiparticle system, *Tr. Moskov. Mat. Obs.* **9**, pp. 81–120. (in Russian).

[941] Zhizhenkova, E. F. and Zhislin, G. M. (1960). On the existence of the minimum of some square functionals in the unbounded domain, *Tr. Moskov. Mat. Obs.* **9**, pp. 121–128. (in Russian).

List of Symbols

Index